Agriculture Issues and Policies

nova
science publishers
www.novapublishers.com

Agriculture Issues and Policies

Pomegranate: For Horticulture Students and Farmers
Arkalgud Nanjundaiah Ganeshamurthy, PhD (Editor)
2023. ISBN: 979-8-88697-812-4 (eBook)

Research Advancements in Organic Farming
Jatindra Nath Bhakta, PhD (Editor)
Sukanta Rana, PhD (Editor)
2023. ISBN: 979-8-88697-519-2 (Hardcover)
2023. ISBN: 979-8-88697-580-2 (eBook)

Legumes: Nutritional Value, Health Benefits and Management
Phetole Mangena, PhD (Editor)
2023. ISBN: 979-8-88697-558-1 (Softcover)
2023. ISBN: 979-8-88697-583-3 (eBook)

Jute: Cultivation, Properties and Uses
Matthieu Issa (Editor)
2022. ISBN: 979-8-88697-490-4 (Softcover)
2022. ISBN: 979-8-88697-505-5 (eBook)

Pistachios: Cultivation, Production and Consumption
Shaziya Haseeb Siddiqui, PhD (Editor)
Shoaib Alam Siddiqui, PhD (Editor)
2022. ISBN: 978-1-68507-949-9 (Hardcover)
2022. ISBN: 979-8-88697-063-0 (eBook)

More information about this series can be found at https://novapublishers.com/product-category/series/agriculture-issues-and-policies/

Rahul Kumar
Ashutosh Sharma
Yogesh Kumar
and Indu Sharma
Editors

Genome Editing for Crop Improvement

nova
science publishers
www.novapublishers.com

DOI: https://doi.org/10.52305/RZCJ7386.

NOTICE TO THE READER

Library of Congress Cataloging-in-Publication Data

Names: Kumar, Rahul (Plant geneticist), editor. | Sharma, Ashutosh (Botanist), editor. | Kumar, Yogesh (Biotechnologist), editor. | Sharma, Indu (Botanist), editor.
Title: Genome editing for crop improvement / Rahul Kumar, Ashutosh Sharma, Yogesh Kumar, Indu Sharma.
Other titles: Agriculture issues and policies series.
Description: New York : Nova Science Publishers, [2023] | Series: Agriculture issues and policies | Includes bibliographical references and index.
Identifiers: LCCN 2023021930 (print) | LCCN 2023021931 (ebook) | ISBN 9798886977400 (hardcover) | ISBN 9798886978926 (adobe pdf)
Subjects: LCSH: Plant breeding. | Plant biotechnology. | Plant genomes. | Gene editing. | Plants, Cultivated--Genetics.
Classification: LCC SB123 .G426 2023 (print) | LCC SB123 (ebook) | DDC 631.5/3--dc23/eng/20230615
LC record available at https://lccn.loc.gov/2023021930
LC ebook record available at https://lccn.loc.gov/2023021931

Published by Nova Science Publishers, Inc. † New York

Contents

Foreword

I am delighted to introduce the book, *Genome Editing for Crop Improvement*, published by Nova Science Publishers in New York. The editors, well accomplished in the subject, have attempted to update the researchers in the field or those who are interested in it about the novel and recent technological advancements in editing plant genomes for breaking yield barriers in agriculture.

Agricultural science is intricately related to human life and the living ecosystem as the food chain starts with autotrophs that include crop plants for food and fodder for humans and other farm animals respectively. According to Atharva Veda, one of the ancient sacred Hindu literature works dedicated to the art of prolonging life and healing illnesses or seeking cures from herbs, food is the basis of all creatures as life cannot survive without it (जीवन्ति स्वधयाऽन्नेन मर्त्या: | Atharva Veda, 12.1.22.)

With the evolution of human beings and their ever-increasing population, domestication of agriculture and farm animals started as the necessity of food production and its constraints became evident. In the Vedas, limitations in food production have been described as one of the major challenges for mankind. To overcome this, novel agricultural practices and varieties have been introduced in the last hundreds of years of agriculture by human beings. Also in Vedic literature, agricultural work is considered an act of pride and divinity. Along with farmers, gods, kings, scholars, and researchers also used to do agriculture (अन्ने समस्य यदसन मनीषा: | Atharva Veda, 20.76.4.)

To meet the food requirements of the ever-increasing global population, there is an urgent need to improve crop productivity and accelerate crop improvement programs using the latest technological advances e.g., various genome editing approaches. The contributing authors of this book have provided an excellent platform for scientists in reviewing and comprehending this vast and complex subject in a simple and understandable manner. I hope this book will take its readers a few steps ahead in understanding the complex technology and genome editing tools in an unpretentious, eloquent, and lucid way for the readers.

I take this opportunity to congratulate the editors, Dr. Rahul Kumar, Dr. Ashutosh Sharma, Dr. Yogesh Kumar, and Dr. Indu Sharma, for teaming up in the conceptualization and compilation of this book on relevant and futuristic approaches for crop improvement. I express my best wishes to all the contributing authors for this significant edited work on genome editing.

Professor R.K. Kohli, PhD
Vice Chancellor
Amity University Punjab
July 12, 2023

Preface

Agriculture has been one of the oldest and most fundamental human activities. We, the humans have been trying to enhance the agriculture yield and quality for thousands of years by inventing new techniques, development of selection based traditional breeding and updating our agricultural practices. However, in recent times, there has been an unprecedented expansion in the requirements of agricultural land, water for irrigation, fertilizers and pesticides. This has put huge stress on our environment, natural landscapes and resources. In the year 2009, UNFAO estimated the world food production demand to increase by 70 percent by 2050, to nourish a likely human population of 9.1 billion. Natural resources have their own limitations in quantity, exploration and exploitation. Selection-based breeding has been exploited to create new varieties which has its own limitations in terms of time and duration and other technical aspects (e.g., Linkage drag). Biotechnological advances have been applied and explored in the last three decades to pinpoint and quicken the trait improvement programmes. The efforts of researchers paved the way as the development of recent genome editing technologies has given a new hope to zoom in and speed up crop improvement programmes in addition of breaking the yield barriers as well.

Although genome editing had been pioneered in the 1990s, before the advent of the common current nuclease-based gene editing platforms, yet its use was limited by low efficiencies of editing. Just after the estimate and forecast by UNFAO in 2009, genome editing with engineered nucleases, i.e., all three major classes of these enzymes—zinc finger nucleases (ZFNs), transcription activator-like effector nucleases (TALENs) and engineered meganucleases were selected by Nature Methods as the method of the Year of 2011. The CRISPR-Cas system was selected by *Science* as "Breakthrough of the Year" in 2015.

Genome editing comprises of the technologies to precisely and efficiently modify DNA sequence of different organisms including plants. Previous generation transgenic technologies relied on random integration of DNA, less efficient gene targeting methods using random mutagenesis and homologous recombination. However, recent genome editing technologies especially the CRISPR/Cas platform has emerged as highly precise, efficacious, genome, convenient and versatile tool. The CRISPR/Cas platform is highly customizable in terms of controlling the specificity of the components to target preferred sites, and the activity of the Cas nuclease. These technologies may be used to edit genomes of any species, explant or cell type as long as we can deliver and/or express the components in those cells. This versatility of the CRISPR/Cas platform has resulted in emergence of huge number of applications, particularly in agriculture and health sectors.

This edited book reviews the methods, techniques, applications, achievements and other progress like regulations made and adopted globally in targeted genome editing.

The book is classified in three sections as follows:

1) "Genome Editing: Techniques and Regulations,"
2) "Genome Editing for Stress Tolerance,"
3) "Genome Editing in Selected Plants/Crops."

Section one contains six chapters dealing with the tools, techniques and regulatory affairs of genome editing worldwide. Section two comprises of four chapters dealing with advances made in crop improvement targeting different biotic and abiotic stresses. Section three emphasises the achievements of genome editing of made in specific crops and involves six chapters.

More than fifty authors have contributed to sixteen chapters for this book. We congratulate all the authors for spending precious time to review the literature in their respective domains and chapters. We hope that the hard work of these quality researchers will help other scholars worldwide to understand genome editing in a lucid way. We express our sincere thanks to the Nova Science Publishers for agreeing and publishing this book.

Rahul Kumar, PhD
Ashutosh Sharma, PhD
Yogesh Kumar, PhD
Indu Sharma, PhD
Editors

Section 1. Genome Editing: Techniques and Regulation

Chapter 1

Genome Editing: Opportunities and Challenges

Harsharan Singh*
PG Department of Biotechnology, Lyallpur Khalsa College, Jalandhar, Punjab, India

Abstract

Genome editing is a technique for the generation of alteration (insertions, deletions or substitutions) in the genome of any living organism. Gene modification and gene editing are synonymously used in place of genome editing. The ultimate objective of this technique is, the creation of more variation in plants, with the ultimate aim of improvement of desired traits. Genome editing can be accomplished either *in vitro* or *in vivo* by delivering the nucleotide modifying machinery *in situ*, which precisely adds, cuts, replaces or corrects the genome to perform highly targeted genomic editing.

Genome editing encompasses a wide variety of tools using either a site-specific recombinase (SSR) or a site-specific nuclease (SSN) system. Both systems require recognition of a known sequence. Nuclease-induced double-stranded breaks (DSBs) are the first step in targeted genome editing. DSBs in DNA lead to the elicitation of highly efficient recombination mechanisms of cellular DNA repair. Nuclease-induced DNA DSBs can be repaired by endogenous repair mechanisms, which occur in almost all cell types. These repair mechanisms are homology-directed repair (HDR) and non-homologous end-joining (NHEJ), resulting in targeted integration or gene disruptions, respectively. Different types of genome editing systems have been developed which include the utilization of mega-nucleases (homing endonuclease), zinc finger nucleases (ZFNs), transcription activator-like effector nucleases (TALENs), and clustered regularly interspaced short palindromic repeats/Cas9 (CRISPR/Cas9). The most predominantly used approach to genome editing nowadays is CRISPR-Cas9 based system. CRISPR-Cas9 has been modified to create a versatile genome editing technology that has a wide diversity of applications in medicine, agriculture, and basic studies of gene functions.

Genome editing provides new opportunities both in monocot and dicot plant species to enhance yield, quality, and nutritional value, to introduce or enhance tolerance to biotic and abiotic stresses, among other applications. The risks involved in altering genomes through the use of genome-editing technology are significantly lower than those associated with GM crops, because most edits alter only a few nucleotides, producing changes that are not unlike those found throughout naturally occurring populations. Each of these mentioned methods poses various challenges to researchers, likewise designing ZFN is complicated and challenging, TALE repeats remain a challenge and the efficiency of gene

* Corresponding Author's Email: harsharan@lkc.ac.in.

In: Genome Editing for Crop Improvement
Editors: Rahul Kumar, Ashutosh Sharma, Yogesh Kumar et al.
ISBN: 979-8-88697-740-0

targeting with TALENs is variable. CRISPR-Cas9 system is also associated with off target effects, which may result in unwanted mutations at the non-desired site also.

Modification of the genome at multiple sites is perhaps the biggest challenge to be addressed in genome editing. Safety of newly edited plant species, legal, intellectual property (IP), and regulatory issues need to be addressed to the satisfaction of scientists, farmers, and consumers to fully exploit the potentials of different platforms of genome-editing technology in crop improvement and other areas such as animal improvement, microbial engineering, and gene therapy, etc. Although biosafety concerns remain, genome editing is a promising technology with the gigantic potential to enhance food production for the benefit of the growing human population. Here, we review the principles, current advances, and applications of genome editing technology along with the potential opportunities and challenges involved.

Keywords: site-specific recombinase, zinc finger nucleases, transcription activator-like effector nucleases, meganucleases, off-targeting

Introduction

Genome editing (GE) is a technique for the generation of the desired alternation in the genetic makeup of a live individual. Gene modification and gene editing are synonymously used for genome editing. The ultimate objective of this technique is the creation of variation in plants to improve the desired traits. Genome editing can be performed in the laboratory on isolated cells or in an organism by employing the genome editing component, which is capable of precise addition, deletion, substitution and correction of genes as well as perform highly specific genomic alternation [1, 2]. Most of the time, these genetic components make use of either a site-specific recombinase (SSR) or a site-specific nuclease (SSN) system.

The prerequisites for these systems are the recognition of known nucleotide sequences. Targeted DNA editing commences with the nuclease-induced double-stranded breaks (DSBs), which activate recombination mechanisms of cellular DNA in respective cells [3, 4]. These DSBs can be repaired either by homology-directed repair (HDR) or non-homologous end-joining (NHEJ), hence resulting in site-specific genome editing. The NHEJ based repair resulted in the introduction of mutations, like insertions and deletions (Indel), whereas HDR results in a targeted repair of DSBs [5].

These technologies have been successfully applied in various organisms, including plants [1, 2]. The pioneer or oldest system used for GE is mega-nucleases also known as homing endonuclease. The other two systems which deserve to be mentioned here are zinc finger nucleases (ZFNs) and transcription activator-like effector nucleases (TALENs). All these three systems require protein elements for recognition of the target sequence. The relatively newer system is clustered regularly interspaced short palindromic repeats/Cas9 popularly known as CRISPR/Cas9 which performed RNA-based recognitions.

The fast-evolving and immensely used system to GE nowadays is CRISPR-Cas9 [6]. This system has been improved heavily to make it fit for GE applications in medicine, agriculture and genomic studies related to forward and reverse genetics. GE has been tested in a wide range of plants including monocot and dicot plant species for high nutritional value, high-quality metabolites, and increasing overall yield. The CRISPR/Cas9 system is also successfully tested for development of tolerance towards biotic and abiotic stresses. The GE technology significantly minimize the risks associated with Genetically Modified Organisms (GMO) as often lesser number of nucleotides are modified, and genomic alternations are made at a rate which are comparable to natural mutations occurring spontaneously in target plant species.

Every genetic element discussed here has some pitfall, due to which researchers face various challenges, like finding target sequence for designing tailor-made ZFN is tedious and cumbersome and target variability of TALENs. Apart from this perhaps the most confusing challenge is off-target effect. The other includes regulations concerned with Safety, toxicity and guidelines for the varietal release of GE species. The regulatory issue related to intellectual property (IP), patent and uses related to GE technology still require attention. These regulations must be resolved in such a manner that there should be a win-win situation for scientists, entrepreneurs and consumers. The appropriate management of these issues will unfold the superb potential of GE in areas such as cultivated plant species, domestic animal of importance and microbial modification and genetic manipulation. Although these issues remain, Still GE technology has unmatched potential to address the needs and problems of global population related to food, health and medicine. This chapter outlined the basic introduction of this technology along with possible opportunities and challenges of GE technology.

Classical Genome Editing System

Classical genome editing techniques, by definition may not strictly relate with evolving genome-editing techniques. These techniques are dependent on physiological processes like double-stranded repair system. Special modifying enzymes called restriction endonucleases and ligase extracted from bacterial and viral genomes are used for cleaving and rejoining of DNA molecules in small genomes. Manipulations of complex genomes of higher organisms like plant genomes are extremely cumbersome and highly laborious with these systems. The major problem is that target of these restriction endonucleases is the short DNA sequences. Although, such specificity is works well in short DNA of viruses and bacteria, it is not sufficient for large plant genomes. The first attempt to develop methods for the editing of complex genomes were associated with the designing of 'artificial enzymes' as oligonucleotides (short nucleotide sequences) with the capability of recognizing the specific sequences in the structure of the target DNA and having chemical groups capable of cleaving DNA. However, these approaches are now replaced by recent techniques of genome editing which are discussed in detail.

Modern Approaches for Genome Editing (Nucleases)

Meganuclease or Homing Nuclease

Meganucleases are endonucleases from different organisms such as bacteria, fungi and algae. They have a long DNA recognition site with variable size differing between 12 and 30 bp. (Table 1). Naturally occurring meganucleases can be modified by introducing changes to the amino acids found in the recognition site of the protein to adapt for the restriction of a specifically chosen sequence [7].

Table 1. Comparative account of different systems of site-specific nuclease used for genome editing. Different features of each nuclease are grouped under a common heading and each row depicts a separate parameter

Feature	MegaN	ZFN	TALEN	Cas9
Nuclease Structure	Monomer	Dimer	Dimer	Monomer
Recognition site	14 and 40 bp	9–18 bp per ZFN monomer, 18–36 bp per ZFN pair	14–20 bp per TALEN monomer, 28–40 bp per TALEN pair	22 bp (20-bp guide sequence + 2-bp protospacer adjacent motif (PAM) for Cas9; up to 44 bp for double nicking
Modification pattern	Nuclease	FokI nuclease	FokI nuclease	Cas9 nuclease
Accuracy	Positional mismatches tolerated in small number	Positional mismatches tolerated in small number	Positional mismatches tolerated in small number	Positional and multiple consecutive mismatches tolerated
Target Condition	Specific sequences required often result in low efficiency	G-rich sequences needed	T nucleotide at 5' end needed for each TALEN monomer	Targeted sequence must precede a PAM
Double standard breakage efficiency	Lower	Lower	Higher in comparison to MegaN and ZFN	Highly efficient
Off-target effects	Present	Present	Less off-target activities reported, not well studied in plants	Absent in plants, but high off-target levels reported in other systems.
Mechanism of action	DSBs in target DNA	DSBs in target DNA	DSBs in target DNA	DSBs in target DNA by Cas9 or single strand nicks by Cas9 nickase
Multiplex genome editing	Limited	Limited	Limited	Enormous multiplexing (no need for obtaining embryonic stem cells)
Affordability	Limited	Limited	Resource intensive	Highly affordable
Designing	Most difficult as every target should have matching sequence	Most difficult as ZFNs need customized protein for every DNA sequence)	Slightly difficult as more than one identical repeat is present, poses difficulty in engineering and delivery into cells	Simpler as available versions for crRNA can be easily designed

The Meganuclease Family

Meganucleases, or homing endonucleases, could be distinguished into five unique families on basis of their sequence and structure motifs: LAGLIDADG, GIY-YIG, HNH, His-Cys box and PD-(D/E)XK [8]. Among these LAGLIDADG proteins based meganuclease are abundantly used for genome editing. Their presence has been found universally in all the living creatures. These had been generally encoded within introns or the exons, but some free-standing members have been also discovered. Routinely, LAGLIDADG involved in RNA maturation perform either the splicing of their own intron or cleaving of the exon-exon junction sequence wherein the intron resides. Due to these intron-clearing activities, they are known as homing endonucleases [HE].

LAGLIDADG homing endonucleases (LHEs) are an attractive choice for genome editing due to having 20–22 bp recognition sites, and cleavage activity which is closely associated to DNA target site recognition [9, 10, 11]. Attempts have been applied to develop LHE variants with new cleavage specificities. These attempts involve the generation of 'local' variant library either through random mutation or structural-based modification of the LHE protein interface that attaches the DNA target site. This is followed by selection based on DNA cleavage or recombination activities [8, 11, 12, 13]. As a result of these efforts, various methods have been developed, which have ability to generate variants with different cleavage specificity in a 'local' region of the LHE DNA/protein interface targeting a relatively small number of adjacent base pairs. To target multiple base pair mismatches by variants, the engineering of LHE variants with high specificity and activity requires combinations of local changes that often include conflicting sets of amino acid (AA) changes in the interface.

Zinc-Finger Nucleases (ZFN)

ZFNs contain tailor-designed Cys_2-His_2 zinc-finger protein and the cleavage domain of the FokI restriction endonuclease [14]. They were the first targeted nuclease used heavily to achieve targeted GE [15, 16]. Two ZFNs domain act as dimers, each monomer recognizes a specific sequence, typically nine to 18 base pairs (bps) of DNA via the zinc-finger DNA-binding domain (Table 1). Dimerization of the ZFN proteins is dictated by the FokI cleavage domain. This action cuts DNA within a five- to seven-bp spacer sequence that separates two flanking zinc-finger binding sites [17]. Each ZFN consists of three or four zinc-finger domains, each individual ZFN domain had~30 amino acid residues that are organized in a ββα motif [18]. The residues involved in DNA recognition are located within the α-helical domain and interact with three bps of DNA, with occasional overlap from an adjacent domain [19]. The phage display [20, 21, 22], had helped to discover a large number of zinc-finger domains recognizing distinct DNA triplets [23, 24, 25, 26]. The fusion of these domains using a linker peptide [27], generate polydactyl zinc-finger proteins capable to target a wide range of possible DNA sequences [28, 29, 30]. Moreover, the "modular assembly" approach and selection-based methods have been utilized, for constructing zinc-finger proteins [31, 32, 33], including oligomerized pool engineering (OPEN) [34]. Apart from this, specialized sets of validated two-finger, zinc-finger modules have been used to assemble zinc-finger arrays [30, 35], including those that take context-dependent effects into account [36, 37].

The off-target mutations associated with the use of ZFNs for genome poses the major challenge [38, 39]. To combat this, attempts have been made to improve their specificity. As result of these attempts, obligate heterodimeric ZFN architectures has been developed that rely on charge–charge repulsion and prevent unwanted homodimerization of the FokI cleavage domain [40]. Additionally, protein-engineering methods have been applied for enhanced cleavage efficiency of the FokI cleavage domain [41]. Modified ZFN proteins endowed with improved cell-penetrating activity have been reported [42]. ZFNickases are capable of gene correction in the absence of a DSB [43, 44, 45]. These enzymes consist of two ZFN monomers one catalytically inactivated ZFN monomer and a second native ZFN monomer, which can stimulate HDR by nicking or cleaving one strand of DNA [46].

TALENs

TALENs are recombinant DNA-cutting enzymes, they have a combination of a highly modular DNA binding domain and the FokI nuclease domain [47, 48]. TALE moves to the nucleus, where it recognizes and binds to a specific DNA sequence in the promoter region of a specific gene in the host genome [49]. The DNA binding domain of a TALEN contains transcription activator like (TAL) effectors from the bacterial plant pathogen *Xanthomonas andhas* about13-28 repeat monomers of 34 amino acids [50]. Each TALE repeat recognizes only a single bp, with little to no target site overlap from adjacent domains [51, 52]. These amino acids of repeat monomers are highly conserved otherwise, but highly variable at positions 12 and 13 (hypervariable), known as repeat-variable di-residues (RVD). These two RVD repeats are specifically involved in the recognition a particular base on a DNA sequence [53]. The detailed studies on the crystal structure of TALE revealed that the first RVD residue involved in stabilizing contact with the backbone of the RVD loop, whereas the second involved in making a base-specific contact with DNA [51]. The routinely used RVDs for assembling synthetic TALE arrays are: NI for adenine, HD for cytosine, NG for thymine, and NN or HN for guanine or adenine [49, 52, 54, 55].

A pair of TALENs that recognizes adjacent genomic targets led to dimerization of the FokI nuclease domain and hence creates a DSB [50]. Dimerization of TALEN proteins is mediated by the FokI cleavage domain, which cuts within a 12- to 19-bp spacer sequence that separates each TALE binding site [47]. TALENs can be customized to target any given DNA sequence with relative simplicity [49]. The construction of TALE DNA-binding domains can be performed by different methods, but most favorite approach is Golden Gate assembly [56]. High-throughput TALE assembly can be achieved via FLASH assembly [57] iterative capped assembly [58], and ligation independent cloning [59]. Attempts have been focused on the technical advances that can enhance their performance, including specificity profiling to uncover non-conventional RVDs that improve TALEN activity [60, 61, 62], directed evolution as a means to refine TALE specificity [63] and even fusing TALE domains to homing endonuclease variants to generate chimeric nucleases with extended targeting specificity [64].

CRISPR/Cas9

The unique genome editing system, CRISPR (Clustered Regularly Interspaced Short Palindromic Repeats) was developed in 2013. This system was originally developed from the bacterium *Streptococcus pyogenes* (SpCas9). Following this its orthologs were discovered in different bacterial genomes with diverse properties [65]. CRISPRs are genetic elements for providing immunity to bacteria against viruses. The protection mechanism involves 3 main stages: adaptation, transcription and interference. At the adaptation stage, during DNA entry into bacterial cells, a short stretch of DNA is inserted into the CRISPR locus of the host genome in specialized repeat structures separated from each other by short palindromic repeats and therefore called as CRISPR. The Cas genes (CRISP associated), which are located near to the CRISPR cassette express protein and have helicase and nuclease activity [66]. In the transcription stage, the complete CRISPR locus is transcribed into a long pre-crRNA (poly-spacer precursor crRNA) and produces short crRNAs (CRISPR RNA) of 39–45 nucleotides containing one spacer sequence (Table 1). During the interference stage ternary Cas9-crRNAtracr RNAcomplex assembled at the 3' side of the crRNAtracrRNA duplex structure and binds to Cas proteins. The crRNA performs recognition of complement sequences with the protospacer sequence, whereas tracrRNA performs Cas-mediated DNA interference. The Cas-proteins cut matching foreign DNA sequences causing DNA degradation similar to RNAi. Therefore, CRISPR/Cas9 system, is an RNA-directed DNA endonuclease developed from the bacterial immune system [67, 68]. It has a CRISPR-RNA (crRNA) molecule which recognizes target sequence, a trans-activating crRNA (tracrRNA) which is involved in crRNA maturation and a Cas9 protein dedicated for DNA cleavage [69]. The Cas genes are located next to CRISPR repeat spacers arrays. Based upon sequence similarity of Cas encoded proteins, there are 93 Cas genes, which are grouped into 35 families; and out of these 35 genes, 11 form the Cas core. CRISPR Cas system can be divided into two classes based on the functioning of Cas protein. Class 1 CRISPR Cas system consists of multiple Cas proteins to degrade nucleic acids, whereas class 2 system uses only a single Cas protein for the same activity. Class 1 is grouped into types I, types III and types IV; similarly, class 2 is divided into types II, types V and types VI. These 6 system types are further divided into 33 subtypes [66-69].

In order to adapt CRISPR as RNA directed genome editing, different modifications had been performed, such as codon optimization for the Cas9 nuclease to facilitate transcription in higher eukaryotic cells fused to nuclear localization signals (NLS). Moreover, a single chimeric guide sgRNA (crRNAtracrRNAduplex) is constructed under the control of promoter type III instead of expressing non-coding RNAs (tracrRNA and pre-crRNA) [69] and the easiness of CRISPR/Cas9 system, researchers frequently use this system in eukaryotic organisms, like plants [70]. Any DNA sequence having first 20 nucleotide sequence complementary to the sgRNA can act as a target for CRISPR/Cas9 system. However, DNA cleavage occurs only when a G-rich (NGG) protospacer adjacent motif (PAM) is present at the 3' end of the DNA targets (protospacer) [4]. Different orthologs CRISPR/Cas9 systems have distinct efficiencies. For example, a smaller Cas9 derived from *Staphylococcus aureus* (SaCas9) and Cas9 derived from *Streptococcus thermophilus* (StCas9) were found to work efficiently in plants [71]. Moreover, type V CRISPR/Cas12a (also known as Cpf1) systems have also been modified for plant gene editing. A type V Cas12a systems are distinct from the Cas9 systems, in three aspects. First, they recognize T-rich PAM sequences (TTTN or TTN), which are located just upstream of the non-complementary strand of the target. Second, Cas12a proteins produce DSBs with 5 nucleotide overhangs instead of the blunt ends produced by type II CRISPR/Cas9 systems. Third, Cas12a can process its own crRNAs from primary transcripts of CRISPR arrays and no tracrRNAs are required for crRNA maturation [72, 73].

Opportunities

The revolutionary technology popularly famous as Genome Editing using engineered nucleases containing sequence-specific DNA-binding domains fused to a nonspecific DNA cleavage module has drastically improved the genome manipulation [71, 74]. This technology enables investigators to manipulate specific genes in a diverse range of cell types and organisms. These engineered nucleases enable efficient and precise genome editing with relatively simple as compared to classical genome editing tools like modifying restriction enzyme, transposons element and mutations. These technologies have lightened up novel opportunities for genome variation. Here, we have tried to discuss the major advances in site-specific genome editing and its applications in targeted genome engineering and genomic analysis.

Opportunities in Functional Genomics

A multitude of possibilities is available for genome modifications by utilization HE, ZFN, TALEN and CRISPR/Cas genome editing systems (Table 2). Researchers around the globe are now capable to mutate the existing genome by inserting of new genes at specific loci or perform modular deletion in the genome and correct or substitute the individual genetic loci and gene fragments using these editing systems [75, 76, 77, 78, 79]. These properties of genome editing systems empowered the researchers to modify the genome and investigate the role of individual genes in individual cells as well as in the whole organism. For example, the peculiar ability of CRISPR/Cas9 system to target specific DNA sites has helped to regulate gene activity [80, 81, 82, 83]. To achieve this purpose, activator or repressor proteins of gene promoters controlling the gene function are required to be attached to the catalytically inactive mutant Cas9 protein. It has also been shown that the complex binding to the target DNA can inhibit or stimulate the function of the target gene [82]. Moreover, using CRISPR/Cas9 system, it is possible to target more than one gene loci simultaneously in a respective cell [80, 83, 84]. This provides a favorable environment for investigating intergenic interaction and naturally accumulated variation in genome as more than one gene are simultaneously altered by the CRISPR/Cas9 system [82]. Genes involved in crop domestication process was successfully identified using this approach [13].

Opportunities in Crop Improvement

Genome editing technologies provide ample opportunity in vital field of modern biotechnology for the creation of new varieties of crops, with high-yield, resistance to abiotic and biotic stresses and having high nutritional value (Table 2) [85, 86, 87, 88, 89, 90, 91, 92]. To achieve this goal, genome editing system has been employed in plant breeding (1) for insertion of point mutations similar to natural SNPs [93, 94], (2) for modifications to gene function [95], (3) for integration of foreign genes, (4) for gene pyramiding and knockout, and (5) for the repression or activation of gene expression, as well as (6) epigenetic editing [77, 92].

Table 2. Genome modification performed in different plant species using genome editing technology. The type of editing is mentioned in the first column followed by modification types and references

S. No	Genome Editing Type	Plant species	Genome Element	References	Year
1	Gene modification	*Arabidopsis thaliana*	ZFN	Osakabe et al. [96]	2010
2	Gene editing	*Nicotiana tabacum*	ZFN	Townsend et al. [97]	2009
3	Targeted integration into endogenous loci	*Zea mays*	ZFN	Shukla [98]	2009
4	Targeted mutagenesis	*Oryza sativa*	TALEN	Zhang et al. [78]	2016
5	Gene editing	*Oryza sativa*	TALEN	Li et al. [104]	2012
6	Targeted transcriptional gene repression	*Arabidopsis thaliana*	TALEN	Mahfouz et al. [129]	2012
7	Herbicide resistance	*Arabidopsis thaliana*	ODM	Sauer et al. [138]	2016
8	Herbicide resistance	*Nicotiana tabacum*	ODM	Sauer et al. [138]	2016
9	Herbicide resistance	*Oryza sativa*	ODM	Sauer et al. [138]	2016
10	Herbicide resistance	*Zea mays*	ODM	Sauer et al. [138]	2016
11	Herbicide resistance	*Brassica napus*	ODM	Sauer et al.[138]	2016
12	Gene knockout or editing	*Arabidopsis thaliana*	CRISPR/Cas	Li et al. [85]	2013
				Mao et al. [95]	2013
13	Gene knockout or editing	*Nicotiana benthamiana*	CRISPR/Cas	Li et al. [85]	2013
14	Gene knockout or editing	*Nicotiana tabacum*	CRISPR/Cas	Gaj et al. [6]	2015
15	Gene knockout or editing	*Oryza sativa*	CRISPR/Cas	Mao et al. [95]	2013
16	Multiplex genome editing	*Arabidopsis thaliana*	CRISPR/Cas	Li et al. [85]	2013
				Mao et al. [95]	2013
17	Multiplex genome editing	*Nicotiana tabacum*	CRISPR/Cas	Cong et al. [80]	2013
				Gaj et al. [6]	2015
18	Gene insertion or replacement	*Arabidopsis thaliana*	CRISPR/Cas	Mao et al.[95]	2013
19	Gene insertion or replacement	*Nicotiana benthamiana*	CRISPR/Cas	Li et al. [85]	2013
20	Targeted mutagenesis	*Glycine max*	CRISPR/Cas	Jacobs et al. [93]	2015
21	Targeted mutagenesis	*Oryza sativa*	CRISPR/Cas	Xu et al. [94]	2016
22	Improvement of nutritional value	*Camelina sativa*	CRISPR/Cas	Jiang et al. [100]	2017
23	Yield improvement	*Zea mays*	CRISPR/Cas	Peng et al. [101]	1999
24	Salt resistance	*Oryza sativa*	CRISPR/Cas	Osakabe et al. [86]	2016
25	Resistance to drought	*Oryza sativa*	CRISPR/Cas	Shan et al. [111]	2013
26	Herbicide resistance	*Zea mays*	CRISPR/Cas	Svitashev et al. [103]	2015
27	Biotic stress	*Oryza sativa*	CRISPR/Cas	Li et al. [85]	2013
28	Biotic stress	*Arabidopsis thaliana*	CRISPR/Cas	Ji et al. [105]	2015
		Nicotiana benthamiana		Li et al. [85]	2013
35	Biotic stress	*Oryza sativa*	CRISPR/Cas	Liu et al. [106]	2012
36	Biotic stress	*Oryza sativa*	CRISPR/Cas	Wang et al. [107]	2016
37	Biotic stress	*Cucumis sativus*	CRISPR/Cas	Chandrasekaran et al. [81]	2016
38	Targeted Mutagenesis	*Zea mays*	Meganuclease	Gao et al. [141]	2010
39	Targeted DNA excision	*Arabidopsis thaliana*	Meganuclease (PB1 endonuclease)	Antunes et al. [142]	2012

For example, ZFN based genome editing has led to the successful development of herbicide tolerant genotypes in *Arabidopsis thaliana* [96, 97], other example is *Zea mays* [98] wherein herbicide resistance genes were inserted into targeted sites in the genome [99]. ZFN was also employed for the modification of endogenous malate dehydrogenase (*MDH*) gene in plants for increased yield [99]. The CRISPR/Cas9 technology was used to obtain "a biotech" oil from *Camelina sativa* seeds which has an improved fatty acid composition that renders this oil more beneficial to human health, more resistant to oxidation, and more appropriate for the production of certain commercial chemicals including biofuel [100]. In another attempt, CRISPR-induced mutagenesis of *OST2* gene in *Arabidopsis*, resulted in new alleles with salt stress resistance [96]. The Alternation and Modulation of metabolic pathways is another area where Genome editing had successfully used for value addition and crop improvement. The alternation in gibberellin biosynthesis by genome editing methods resulted in the creation of dwarf fruit trees [101]. These dwarf varieties are of immense importance as they led to increase productivity through higher density plantings and reduced labor costs. These dwarf varieties require less input resources like land, water, pesticide, and fertilizer [102]. Genome editing targeting ethylene biosynthesis or its signaling pathways, which has an important role in fruit ripening [102] result in the creation of new varieties with extended shelf life [87].

The genome editing approaches has been applied in plant breeding, for developing insect and pest-resistant varieties. These methods have been successfully applied for targeting the vital plant immunity stages in different crops. The favorite target for this purpose are: (1) susceptibility genes (S-genes), (2) resistance genes (R-genes), (3) gene/s concerned with regulation of effector and target, and (4) the genes related to regulation plant hormone [92]. TALEN and CRISPR/Cas9-mediated genome editing on mildew resistance locus O (MLO) had used for developing resistance against powdery mildew in wheat genotypes [103]. Similarly, the TALEN approach has been used developing variety resistant to bacterial leaf blight, caused by *Xanthomonas oryzae* pv. *Oryzae* 104]. The CRISPR/Cas9 system has been used against geminiviruses by single guide RNA/Cas9 (sgRNA/Cas9) resulted in suppression of viral genome [105].

Metabolic pathways concerning to hormonal balance can also be modified using the genome editing technologies to enhance the immunomodulatory component of the plants immune system. Deactivation of ethylene-responsive factor (ERF) by CRISPR/Cas9-mediated targeting OsERF922 gene mutations, resulted in increased resistance to *Magnaporthe oryzae* [106, 107], knock out of eIF4E gene by CRISPR/Cas9 has been used to confers resistance to viruses such as cucumber vein yellowing virus (CVYV), zucchini yellow mosaic virus (ZYMV), and papaya ring spot mosaic virus-W (PRSV-W) [81]. CRISPR/Cas9 mediated genome editing in *Phytophthora sojae* (an oomycete pathogen of Soybean), by modifying the pathogenicity gene (Avr4/6), result in pathogen control. Similarly, CRISPR/Cas9 based approaches have been successfully used to obtain plants resistant to herbicides [103]. The alternation in *ALS2* gene of maize, which code for a regulatory enzyme acetolactate synthase has an important role in the biosynthesis of amino acids in plants and allowed the development of corn plant resistant to chlorsulfuron [103]. The other breakthrough of CRISPR/Cas9 system is the development of plants capable of synthesizing human proteins like insulin and albumin required for the treatment of patients with diabetic and hemorrhagic shock, burns, hypoproteinemia, and cirrhosis, respectively [108, 109]. These proteins expressed in the edited plant can be isolated from plant and animal tissues, and after purification, can be used for medical purposes. From the above-cited studies, we may clearly inference that how these novel genome editing techniques are being widely used for the purpose of crop improvement including new bio-energy crop developments [110].

Challenges

The genome editing technologies in plants involve the specifically designed sequence-specific nucleases, delivery of editing reagents, generation and selection of editing events, and characterization and utilization of intact plants, which have further implications for public acceptance and regulatory approval of the genome edited plants. Genome editing platforms have been established and applied in more than 45 genera of plants [111]. However, the need for better understanding of the molecular and genetic mechanism of genome editing, further improvements and novel applications of genome editing technologies in plants still require major research efforts. Although these developments are promising to realize the dream of designing dream plants for the future, many challenges still remain to overcome. Specifically, plant genome editing faces major challenges are outlined below.

The Off-Target/Non-Target Effects

The native genomic structure of transformed plant via GE remains unchanged and therefore, GE considered as a safe technology for crop improvement. However, there are some concerns related to the biosafety of crops created using these techniques. The major challenge is the possibility of non-specific target effects by these nucleases during GE. The efficiency and specificity of the engineered nucleases are the two most important functional requirements and are closely related to the choice of a target site. The efficiency of DNA cleavage (both target and non-target) depends on the nuclease activity (*Fok* I domains and *Ruv* domains of the Cas9 proteins), and on the target site availability and affinity of the DNA-binding domain (e.g., TAL effector domains and guide RNA, gRNA) to the target sequence. The specificity of engineered nucleases is solely dependent on the binding affinity of nuclease-DNA, including the binding of zinc finger to DNA (ZFNs), TAL effector to DNA (TALENs) and hybridization of gRNA with DNA (CRISPR). Apart from this dimerization of *Fok*I domain (ZFNs and TALENs) and Cas9 interaction with the motif contiguous to protospacer adjacent motif (PAM) may also play critical role [112]. In case of ZFNs, the binding efficiency of canonical C2H2 binding domain containing ZFNs, and those containing C3H1 binding domain significantly differ in levels of binding efficiency [113].

To minimize the non-target effects of genome editing systems, target sites for the introduction of the double-stranded breaks should be selected with high precision. For this purpose, a prior bioinformatics analysis should be performed [114]. During selection of target sites, repetitive sequences and sites with high homology with other regions of the genome should be avoided. To facilitate the selection of the target sites for nucleases and experimental verification of the presence of non-target effects, several software packages are readily available that may prove useful in nuclease design and validation [115, 116, 112].

Overcoming Species and Genotype-Dependent Transformation

Genome editing based transformation in plants involves delivering of the editing reagents at the target site into plant cells, selection of transformed cells, and regeneration of transformed plants with desired traits. These methods and processes still need optimization for most crops; transformation and regeneration still remain tedious and remote task despite technological advances. These are, therefore, challenges in fully materializing the great potential of GE in agriculture [117]. Studies have documented the use of development regulator (DR) genes (e.g., Baby Boom and Wuschel) along with optimal composition of phytohormones to improve transformation efficiency in plants. It was found that expression of specific combinations of the DR genes in plant somatic cells leads to the formation of meristematic tissue capable of shoot regeneration [118, 119]. However, technological improvement and widespread adaption for crop plants beyond the few tested remain challenging. Research efforts are being required to make new progresses in the various types of genome editing, including site-specific mutagenesis, gene insertions, and base transition and transversion.

Broadening Targets for Crop Improvement

Genome editing in agriculture can only be successful through micro-level understanding of the gene or genetic element of interest in crop plants [120, 121]. The number of available targets for crop improvement via genome editing are in scarcity due to the absence of understanding of biological processes involving genes, pathways, networks, and their interactions with environmental factors. Although, CRISPR-based GE technologies are providing new opportunities to perform genome-wide and high-throughput functional genomics, expediting gene/trait discovery in both model or non-model crops [122, 123]. The challenge is to connect multidisciplinary approaches to identify targets that have the potential for beneficial agronomic traits suitable for genome editing [124].

Achieving High-Efficiency Base Replacement

Targeted mutagenesis is central to genome editing development and application. There are fewer numbers of the methods and applications available for the base replacement, very low efficiency in some plants and no efficiency in the majority of crop plants. In crop plants, improved or novel traits are generally governed by genetic variations in SNPs (single nucleotide polymorphisms); the challenge is in their identification and integration in elite varieties in an expedited manner. This opens up a great opportunity to addition of the superior alleles in the target genomes through base replacement. The HDR (homology directed repair) based base replacement or gene insertion can be applied to achieve this. The 'prime editing' technology that utilizes Cas9 and guides RNA variants offers opportunity to precise and specific alternation in target DNA sequence, its application can be extended to base replacement if adapted and validated in plants [125]. Prime editing is better than base editors which convert C to T and A to G within a window of a guide RNA-specified DNA site, as it alters the small genomic region [126]. Research efforts are necessary to adapt and improve these technologies for a use in diverse range of plant species.

Regulations for Genome-Edited Plants

The genome editing systems are widely used for introduction of stably inherited modifications into the plant genome and transgenic region can be easily removed after editing a target gene. This allows the creation of non-transgenic plants and improved crop varieties [127, 128, 129, 130]. These technologies are faster than breeding methods and help to obtain the null segregant lines that have lost the transgene insertion [131, 132, 133, 134]. Edited plants with targeted mutations using genome editing technology are almost similar to plants obtained by classical breeding. The safety of these plants must be assessed taking into account the resulting product rather than the process used to create them [135, 136, 137]. In this context, edited products are in many cases indistinguishable from conventionally bred or traditional mutagenesis products; therefore, such products should not be regulated in the same way as the products generated by genetic engineering methods [138]. The plant modified by use of CRISPR-Cas9 system, is generally marker-free, because they lack any marker genes of antibiotic resistance [136]. Therefore, the existing operating rules for the regulation of genetically modified plants could not be applied [131, 136, 137] on the new varieties with targeted mutations, developed using genome editing systems. Currently, genome editing technologics are being discussed by various advisory and regulatory authorities in the context of GMO legislation. Cultures and plants obtained using genome editing techniques are not considered genetically modified [131, 137, 139]. The European Commission is expected to publish a report on the regulatory uncertainty of genome editing methods [70, 136, 140].

Towards a Better Public Perception of Genome Editing in Agriculture

The transformed products via GE do not strictly fit to be called genetically modified organisms (GMOs) within most regulatory regimes [140]. The usage of GE can reduce time for crop breeding and can make desired changes with more accuracy, increase crop yields, develop resistant to pests, diseases, abiotic stresses, and enhance shelf life of food products to avoid food waste. Therefore, the challenge might be to find wider public and regulatory acceptance of GE approaches as compared to classical transgenic plants. It is important to enhance public trust and adapt regulatory frameworks, which ultimately will affect the usage of GE approaches in agriculture. This will lead to a better transparency in sharing information regarding genome editing techniques.

Future Prospects and Conclusion

GE has gigantic potential, to play dominating role to meet the huge demands of crop production for the exponentially growing human population. In comparison to conventional breeding techniques, the molecular breeding strategies by GE tools make it possible to specifically target and alter the desired traits in lesser time. GE can be used to improve crop productivity, increase nutritional value, and develop resistance against biotic as well as abiotic stressors by modifying the crop genome. These technologies have been in great demand to edit crops for a specific agricultural trait and breeding platforms for the creation of an elite local variety. The advent of meganucleases, ZFNs, and TALENs has revolutionary changed way of plant modifications.

Among the available genome editing techniques, CRISPR/Cas9 is widely intensively used for crop modification. It is used for developing resistance against abiotic and biotic stress and to improve other plant traits. CRISPR/Cas9 has higher target specificity, simpler to execute and lower input cost, which made it superior to conventional approaches. Nevertheless, off-target effect represents major hurdle in its advancement. Fortunately, off-target mutations are mostly tolerable in plants and can be detected and eliminated through segregation over successive crosses. The selection of Cas9 with longer PAMs, the designing of sgRNA having high affinity for target sequence, may minimize the off-target effects in the future time. Therefore, further research and scientific experimentation are required to cross these hurdles in order to improve the experimental versatility and application skills of the CRISPR/Cas9 system.

These approaches have been greatly acknowledged as an innovation to manipulate genomes and poses new challenges regarding the assessment of current regulatory policies. The safety of GE techniques is the matter of debate around the world. There is demand of development of regulatory policies for novel crop innovations. These policies should be multidimensional, transparent, and be capable to distinguish between GMO and GE Plants. The coordinated ideas of system biology, synthetic biology, next-generation sequencing, and the latest developments in functional genomic approaches integrated with the advanced tools of genome editing will permit the development of smart crops with higher yields and improved qualities. In the near future, genome editing technology can be integrated with speed breeding programs to revolutionize agriculture and promise global food security.

References

[1] Shrivastav M., De Haro L. P., Nickoloff J. A. (2008). Regulation of DNA double-strand break repair pathway choice. *Cell Res.* Jan;18(1):134-47.

[2] Ceccaldi R., Rondinelli B. and D'Andrea A. D. (2016). Repair pathway choices and consequences at the double-strand break. *Trends Cell Biol* 26: 52–64.

[3] Jiang W. Z., Henry I. M., Lynagh P. G., Comai L., Cahoon E. B., and Weeks D. P., (2017) "Significant enhancement of fatty acid composition in seeds of the allohexaploid, *Camelina sativa*, using CRISPR/Cas9 gene editing," *Plant Biotechnology Journal*, vol. 15, no. 5, pp. 648–657.

[4] Jinek M., Chylinski K. and Fonfara I. Hauer M., Doudna J. A., & Charpentier E. (2012). A programmable dual-RNA-guided DNA endonuclease in adaptive bacterial immunity. *Science* 2012; 337: 816– 21.

[5] Lieber, M. R. (2010). The mechanism of double-strand DNA break repair by the nonhomologous DNA end-joining pathway. *Annual Review of Biochemistry*, *79*, pp. 181-211.

[6] Gaj, T., Sirk, S. J., Shui, S. L., & Liu, J. (2016). Genome-Editing Technologies: Principles and Applications. *Cold Spring Harbor Perspectives in Biology*, *8*(12), a023754. https://doi.org/10.1101/cshperspect.a023754.

[7] Silva, G., Poirot, L., Galetto, R., Smith, J., Montoya, G., Duchateau, P., &Pâques, F. (2011). Meganucleases and other tools for targeted genome engineering: perspectives and challenges for gene therapy. *Current Gene Therapy*, *11*(1), 11–27. https://doi.org/10.2174/156652311794520111.

[8] Baxter, S., Lambert, A. R., Kuhar, R., Jarjour, J., Kulshina, N., Parmeggiani, F., Danaher, P., Gano, J., Baker, D., Stoddard, B. L. et al. (2012) Engineering domain fusion chimeras from I-OnuI family LAGLIDADG homing endonucleases. *Nucleic Acids Res.*, 40, 7985–8000.

[9] Stoddard, B. L. (2011) Homing endonucleases: from microbial genetic invaders to reagents for targeted DNA modification. *Structure*, 19, 7–15.

[10] Belfort, M. and Roberts, R. J. (1997) Homing endonucleases: keeping the house in order. *Nucleic Acids Res.*, 25, 3379–3388.

[11] Hafez, M. and Hausner, G. (2012) Homing endonucleases: DNA scissors on a mission. *Genome*, 55, 553–569.

[12] Grizot, S., Smith, J., Daboussi, F., Prieto, J., Redondo, P., Merino, N., Villate, M., Thomas, S., Lemaire, L., Montoya, G., Blanco, F. J., Pâques, F., & Duchateau, P. (2009) Efficient targeting of a SCID gene by an engineered single-chain homing endonuclease. *Nucleic Acids Res.*, 37, 5405–5419.

[13] Chen Z., Wen, F., Sun, N. and Zhao, H. (2009) Directed evolution of homing endonuclease I-SceI with altered sequence specificity. *Protein Eng. Des. Sel.*, 22, 249–256.

[14] Kim Y. G., Cha J., Chandrasegaran S. 1996. Hybrid restriction enzymes: Zinc finger fusions to Fok I cleavage domain. *Proc Natl Acad Sci* 93: 1156–1160.

[15] Porteus M. H., Carroll D. 2005. Gene targeting using zinc finger nucleases. *Nat Biotechnol* 23: 967–973.

[16] Urnov, F., Rebar, E., Holmes, M. Zhang, H. S., & Gregory, P. D. (2010). Genome editing with engineered zinc finger nucleases. *Nat Rev Genet 11*, 636–646. https://doi.org/10.1038/nrg2842.

[17] Smith J., Bibikova M., Whitby F. G., Reddy A. R., Chandrasegaran S., Carroll D. (2000). Requirements for double-strand cleavage by chimeric restriction enzymes with zinc finger DNA-recognition domains. *Nucleic Acids Res* 28: 3361–3369.

[18] Pavletich N. P., Pabo C. O. (1991). Zinc finger-DNA recognition: Crystal structure of a Zif268-DNA complex at 2.1 A. *Science* 252: 809–817.

[19] Wolfe S. A., Nekludova L., Pabo C. O. (2000). DNA recognition by Cys2His2 zinc finger proteins. *Annu Rev Biophys Biomol Struct* 29: 183–212.

[20] Choo Y., Klug A. (1994). Toward a code for the interactions of zinc fingers with DNA: Selection of randomized fingers displayed on phage. *Proc Natl Acad Sci* 91: 11163–11167.

[21] Jamieson A. C., Kim S. H., Wells J. A. (1994). *In vitro* selection of zinc fingers with altered DNA-binding specificity. *Biochemistry* 33: 5689–5695.

[22] Wu H., Yang W. P., Barbas C. F. III. (1995). Building zinc fingers by selection: Toward a therapeutic application. *Proc Natl Acad Sci* 92: 344–348.

[23] Segal D. J., Dreier B., Beerli R. R., Barbas C. F. III. (1999). Toward controlling gene expression at will: Selection and design of zinc finger domains recognizing each of the 5′-GNN-3′ DNA target sequences. *Proc Natl Acad Sci* 96: 2758–2763.

[24] Dreier B., Beerli R. R., Segal D. J., Flippin J. D., Barbas C. F. III. (2001). Development of zinc finger domains for recognition of the 5′-ANN-3′ family of DNA sequences and their use in the construction of artificial transcription factors. *J Biol Chem* 276: 29466–29478.

[25] Dreier B., Fuller R. P., Segal D. J., Lund C. V., Blancafort P., Huber A., Koksch B., Barbas C. F. III. (2005). Development of zinc finger domains for recognition of the 5′-CNN-3′ family DNA sequences and their use in the construction of artificial transcription factors. *J Biol Chem* 280: 35588–35597.

[26] Bae K. H., Kwon Y. D., Shin H. C., Hwang M. S., Ryu E. H., Park K. S., Yang H. Y., Lee D. K., Lee Y., Park J., Sun Kwon, H., Kim, H. W., Yeh, B. I., Lee, H. W., Hyung Sohn, S., Yoon, J., Seol, W., & Kim, J. S. (2003). Human zinc fingers as building blocks in the construction of artificial transcription factors. *Nat Biotechnol* 21: 275–280.

[27] Liu Q., Segal D. J., Ghiara J. B., Barbas C. F. III. (1997). Design of polydactyl zinc-finger proteins for unique addressing within complex genomes. *Proc Natl Acad Sci* 94: 5525–5530.

[28] Beerli R. R., Segal D. J., Dreier B., Barbas C. F. III. (1998). Toward controlling gene expression at will: Specific regulation of the erbB-2/HER-2 promoter by using polydactyl zinc finger proteins constructed from modular building blocks. *Proc Natl Acad Sci* 95: 14628–14633.

[29] Beerli R. R., Schopfer U., Dreier B., Barbas C. F. III. (2000). Chemically regulated zinc finger transcription factors. *J Biol Chem* 275: 32617–32627.

[30] Kim H. J., Lee H. J., Kim H., Cho S. W., Kim J. S. (2009). Targeted genome editing in human cells with zinc finger nucleases constructed via modular assembly. *Genome Res* 19: 1279–1288.

[31] Greisman H. A., Pabo C. O. (1997). A general strategy for selecting high-affinity zinc finger proteins for diverse DNA target sites. *Science* 275: 657–661.

[32] Isalan M., Klug A., Choo Y. (2001). A rapid, generally applicable method to engineer zinc fingers illustrated by targeting the HIV-1 promoter. *Nat Biotechnol* 19: 656–660.

[33] Magnenat L., Blancafort P., Barbas C. F. III. (2004). *In vivo* selection of combinatorial libraries and designed affinity maturation of polydactyl zinc finger transcription factors for ICAM-1 provides new insights into gene regulation. *J Mol Biol* 341: 635–649.

[34] Maeder M. L., Thibodeau-Beganny S., Osiak A., Wright D. A., Anthony R. M., Eichtinger M., Jiang T., Foley J. E., Winfrey R. J., Townsend J. A., Unger-Wallace, E., Sander, J. D., Müller-Lerch, F., Fu, F., Pearlberg, J., Göbel, C., Dassie, J. P., Pruett-Miller, S. M., Porteus, M. H., Joung, J. K. (2008). Rapid "open-source" engineering of customized zinc-finger nucleases for highly efficient gene modification. *Mol Cell* 31: 294–301.

[35] Bhakta M. S., Henry I. M., Ousterout D. G., Das K. T., Lockwood S. H., Meckler J. F., Wallen M. C., Zykovich A., Yu Y., Leo H., Xu, L., Gersbach, C. A., & Segal, D. J. 2013. Highly active zinc-finger nucleases by extended modular assembly. *Genome Res* 23: 530–538.

[36] Gupta A, Christensen RG, Rayla AL, Lakshmanan A, Stormo GD, Wolfe SA. (2012). An optimized two-finger archive for ZFN-mediated gene targeting. *Nat Methods* 9: 588–590.

[37] Sander J. D., Dahlborg E. J., Goodwin M. J., Cade L., Zhang F., Cifuentes D., Curtin S. J., Blackburn J. S., Thibodeau-Beganny S., Qi Y., Pierick, C. J., Hoffman, E., Maeder, M. L., Khayter, C., Reyon, D., Dobbs, D., Langenau, D. M., Stupar, R. M., Giraldez, A. J., Joung, J. K. (2011). Selection-free zinc-finger-nuclease engineering by context-dependent assembly (CoDA). *Nat Methods* 8: 67–69.

[38] Gabriel R., Lombardo A., Arens A, Miller J. C., Genovese P., Kaeppel C., Nowrouzi A., Bartholomae C. C., Wang J., Friedman G, Holmes, M. C., Gregory, P. D., Glimm, H., Schmidt, M., Naldini, L., & von Kalle, C. (2011). An unbiased genome-wide analysis of zinc-finger nuclease specificity. *Nat Biotechnol* 29: 816–823.

[39] Pattanayak V., Ramirez C. L., Joung J. K., Liu D. R. (2011). Revealing off-target cleavage specificities of zinc-finger nucleases by *in vitro* selection. *Nat Methods* 8: 765–770.

[40] Doyon Y., Vo T. D., Mendel M. C., Greenberg S. G., Wang J., Xia D. F., Miller J. C., Urnov F. D., Gregory P. D., Holmes M. C. (2011). Enhancing zinc-finger-nuclease activity with improved obligate heterodimeric architectures. *Nat Methods* 8: 74–79.

[41] Guo J., Gaj T., Barbas C. F. III. (2010). Directed evolution of an enhanced and highly efficient FokI cleavage domain for zinc finger nucleases. *J Mol Biol* 400: 96–107.

[42] Liu J., Gaj T., Wallen M. C., Barbas C. F. III. (2015). Improved cell-penetrating zinc-finger nuclease proteins for precision genome engineering. *Mol Ther Nucleic Acids* 4: e232.

[43] Kim E., Kim S., Kim D. H., Choi B. S., Choi I. Y., Kim J. S. (2012). Precision genome engineering with programmable DNA-nicking enzymes. *Genome Res* 22: 1327–1333.

[44] Ramirez C. L., Certo M. T., Mussolino C., Goodwin M. J., Cradick T. J., McCaffrey A. P., Cathomen T., Scharenberg A. M., Joung J. K. (2012). Engineered zinc finger nickases induce homology-directed repair with reduced mutagenic effects. *Nucleic Acids Res* 40: 5560–5568.

[45] Wang J., Friedman G., Doyon Y., Wang N. S., Li C. J., Miller J. C., Hua K. L., Yan J. J., Babiarz J. E., Gregory P. D., & Holmes, M. C. (2012). Targeted gene addition to a predetermined site in the human genome using a ZFN-based nicking enzyme. *Genome Res* 22: 1316–1326.

[46] McConnell Smith A., Takeuchi R., Pellenz S., Davis L., Maizels N., Monnat R. J. Jr, Stoddard B. L. (2009). Generation of a nicking enzyme that stimulates site-specific gene conversion from the I-AniI LAGLIDADG homing endonuclease. *Proc Natl Acad Sci* 106: 5099–5104.

[47] Miller J. C., Tan S., Qiao G., Barlow K. A., Wang J., Xia D. F., Meng X., Paschon D. E., Leung E., Hinkley S. J., Dulay, G. P., Hua, K. L., Ankoudinova, I., Cost, G. J., Urnov, F. D., Zhang, H. S., Holmes, M. C., Zhang, L., Gregory, P. D., & Rebar, E. J. (2011). A TALE nuclease architecture for efficient genome editing. *Nat. Biotechnol.* 29: 143–148.

[48] Zhang F., Cong L., Lodato S., Kosuri S., Church G. M., Arlotta P. (2011). Efficient construction of sequence-specific TAL effectors for modulating mammalian transcription. *Nat. Biotechnol.* 29: 149–153.

[49] Boch J., Scholze H., Schornack S., Landgraf A., Hahn S., Kay S., Lahaye T., Nickstadt A., Bonas U. (2009). Breaking the code of DNA binding specificity of TAL-type III effectors. *Science* 326: 1509–1512.

[50] Boch J., Bonas U. (2010). *Xanthomonas* AvrBs3 family-type III effectors: Discovery and function. *Annu. Rev. Phytopathol.* 48, 419–436.

[51] Deng, D., Yan, C., Wu, J., Pan, X., & Yan, N. (2014). Revisiting the TALE repeat. *Protein Cell* 5(4):297–306.

[52] Mak AN, Bradley P, Cernadas RA, Bogdanove AJ, Stoddard BL. (2012). The crystal structure of TAL effector PthXo1 bound to its DNA target. *Science* 335: 716–719.

[53] Moscou M. J., Bogdanove A. J., (2009) A simple cipher governs DNA recognition by TAL effectors. *Science* 326, 1501.

[54] Cong L., Zhou R., Kuo Y. C., Cunniff M., Zhang F. (2012). Comprehensive interrogation of natural TALE DNA-binding modules and transcriptional repressor domains. *Nat Commun* 3: 968.

[55] Streubel J., Blucher C., Landgraf A., Boch J. (2012). TAL effector RVD specificities and efficiencies. *Nat Biotechnol* 30: 593–595.

[56] Cermak T., Doyle E. L., Christian M., Wang L., Zhang Y., Schmidt C., Baller J. A., Somia N. V., Bogdanove A. J., Voytas D. F. (2011). Efficient design and assembly of custom TALEN and other TAL effector-based constructs for DNA targeting. *Nucleic Acids Res.* Jul;39(12):e82. doi: 10.1093/nar/gkr218. Epub 2011 Apr 14. Erratum in: *Nucleic Acids Res.* 2011 Sep 1;39(17):7879. PMID: 21493687; PMCID: PMC3130291.

[57] Reyon D., Tsai S. Q., Khayter C., Foden J. A., Sander J. D., Joung J. K. (2012). FLASH assembly of TALENs for high-throughput genome editing. *Nat. Biotechnol.* 30: 460–465.

[58] Briggs A. W., Rios X., Chari R., Yang L., Zhang F., Mali P., Church G. M. (2012). Iterative capped assembly: rapid and scalable synthesis of repeat-module DNA such as TAL effectors from individual monomers. *Nucleic Acids Res* 40: e117.

[59] Schmid-Burgk J. L., Schmidt T., Kaiser V., Honing K., Hornung V. (2013). A ligation-independent cloning technique for high-throughput assembly of transcription activator-like effector genes. *Nat. Biotechnol.* 31: 76–81.

[60] Yang J., Zhang Y., Yuan P., Zhou Y., Cai C., Ren Q., Wen D., Chu C., Qi H., Wei W. (2014). Complete decoding of TAL effectors for DNA recognition. *Cell Res* 24: 628–631.

[61] Miller J. C., Zhang L., Xia D. F., Campo J. J., Ankoudinova I. V, Guschin D. Y., Babiarz J. E., Meng X., Hinkley S. J., Lam S. C., Paschon, D. E., Vincent, A. I., Dulay, G. P., Barlow, K. A., Shivak, D. A., Leung, E., Kim, J. D., Amora, R., Urnov, F. D., Rebar, E. J. (2015). Improved specificity of TALE-based genome editing using an expanded RVD repertoire. *Nat Methods* 12: 465–471.

[62] Juillerat A., Pessereau C., Dubois G., Guyot V., Marechal A., Valton J., Daboussi F., Poirot L., Duclert A., Duchateau P. (2015). Optimized tuning of TALEN specificity using non-conventional RVDs. *Sci Rep* 5: 8150.

[63] Hubbard B. P., Badran A. H., Zuris J. A., Guilinger J. P., Davis K. M., Chen L., Tsai S. Q., Sander J. D., Joung J. K., Liu D. R. (2015). Continuous directed evolution of DNA-binding proteins to improve TALEN specificity. *Nat Methods* 12: 939–942.

[64] Boissel S., Jarjour J., Astrakhan A., Adey A., Gouble A., Duchateau P., Shendure J., Stoddard B. L., Certo M. T., Baker D., & Scharenberg, A. M. (2014). megaTALs: A rare-cleaving nuclease architecture for therapeutic genome engineering. *Nucleic Acids Res* 42: 2591–2601.

[65] Marraffini, L., Sontheimer, E. (2010) CRISPR interference: RNA-directed adaptive immunity in bacteria and archaea. *Nat Rev Genet* 11, 181-190.

[66] Haft D. H., Selengut J., Mongodin E. F., Nelson K. E. (2005) A Guild of 45 CRISPR-Associated (Cas) Protein Families and Multiple CRISPR/Cas Subtypes Exist in Prokaryotic Genomes. PLoS Comput. Biol. 1(6): e60. https://doi.org/10.1371/journal.pcbi.0010060.

[67] Wiedenheft B., Sternberg S. H., Doudna J. A. (2012). RNA-guided genetic silencing systems in bacteria and archaea. *Nature* 482: 331–338.

[68] Sorek R., Lawrence C. M., Wiedenheft B. 2013. CRISPR-mediated adaptive immune systems in bacteria a Marraffini L. A., Sontheimer E. J. 2010. CRISPR interference: RNA-directed adaptive immunity in bacteria and archaea. *Nat Rev Genet* 11: 181–190.

[69] Hsu, Patrick D., Eric S. Lander, and Feng Zhang.(2014) "Development and applications of CRISPR-Cas9 for genome engineering." *Cell* 157, no. 6: 1262-1278.

[70] Belhaj K., Chaparro-Garcia A., Kamoun S., Patron N. J., and Nekrasov V. (2015) "Editing plant genomes with CRISPR/Cas9," *Current Opinion in Biotechnology*, vol. 32, pp. 76–84.

[71] Steinert J., Schiml S. and Fauser F. & Puchta, H. (2015) Highly efficient heritable plant genome engineering using Cas9 orthologues from Streptococcus thermophilus and Staphylococcus aureus. *Plant J;* 84: 1295–305.

[72] Zetsche B., Gootenberg J. S. and Abudayyeh O. O., Slaymaker, I. M., Makarova, K. S., Essletzbichler, P., Volz, S. E., Joung, J., van der Oost, J., Regev, A., Koonin, E. V., & Zhang, F. (2015) Cpf1 is a single RNA guided endonuclease of a class 2 CRISPR-Cas system. *Cell*; 163: 759– 71.

[73] Zetsche B., Heidenreich M. and Mohanraju P. Fedorova, I., Kneppers, J., DeGennaro, E. M., Winblad, N., Choudhury, S. R., Abudayyeh, O. O., Gootenberg, J. S., Wu, W. Y., Scott, D. A., Severinov, K., van der Oost, J., & Zhang, F. (2017) Multiplex gene editing by CRISPR-Cpf1 using a single crRNA array. *Nat Biotechnol.*; 35: 31–4.

[74] Weeks DP, Spalding MH, and Yang B, (2016) "Use of designer nucleases for targeted gene and genome editing in plants," *Plant Biotechnology Journal*, vol. 14, no. 2, pp. 483–495.

[75] GajT, Gersbach CA, and. Barbas III C. F, (2013) "ZFN, TALEN, and CRISPR/Cas-based methods for genome engineering," *Trends in Biotechnology*; vol. 31, no. 7, pp. 397–405.

[76] Chen B, Hu J., Almeida R., Liu, H., Balakrishnan, S., Covill-Cooke, C., Lim, W. A., & Huang, B. (2016) "Expanding the CRISPR imaging toolset with Staphylococcus aureus Cas9 for simultaneous imaging of multiple genomic loci," *Nucleic Acids Research*, vol. 44, no. 8, pp 75.

[77] Kumar V. and Jain M. (2015) "The CRISPR-Cas system for plant genome editing: Advances and opportunities," *Journal of Experimental Botany*; vol. 66, no. 1, pp. 47–57.

[78] Zhang H., Gou F., Zhang J., Liu, W., Li, Q., Mao, Y., Botella, J. R., & Zhu, J. K. (2016) "TALEN-mediated targeted mutagenesis produces a large variety of heritable mutations in rice," *Plant Biotechnology Journal*; vol. 14, no. 1, pp. 186–194.

[79] Gao J., Wang G., Ma S., Xie, X., Wu, X., Zhang, X., Wu, Y., Zhao, P., & Xia, Q. (2015) "CRISPR/Cas9-mediated targeted mutagenesis in *Nicotiana tabacum*," *Plant Molecular Biology*; vol. 87, no. 1-2, pp. 99–110.

[80] Cong L., Ran F. A., Cox D., Lin, S., Barretto, R., Habib, N., Hsu, P. D., Wu, X., Jiang, W., Marraffini, L. A., & Zhang, F. (2013) "Multiplex genome engineering using CRISPR/Cas systems," *Science*, vol. 339, no. 6121, pp. 819– 823.

[81] Chandrasekaran, J., Brumin, M., Wolf, D., Leibman, D., Klap, C., Pearlsman, M., Sherman, A., Arazi, T., & Gal-On, A. (2016) "Development of broad virus resistance in non-transgenic cucumber using CRISPR/Cas9 technology," *Molecular Plant Pathology*; vol. 17, no. 7, pp. 1140–1153.

[82] Lowder L., Malzahn A., and. Qi Y. (2016) "Rapid evolution of manifold CRISPR systems for plant genome editing," *Frontiers in Plant Science*; vol. 7, no. 2016, article no. 1683.

[83] Zhang F., Wen Y., and Guo X., (2014) "CRISPR/Cas9 for genome editing: Progress, implications and challenges," *Human Molecular Genetics*; vol. 23, no. 1, pp. R40–R46.

[84] Wang H., Yang H., Shivalila C. S. Dawlaty, M. M., Cheng, A. W., Zhang, F., & Jaenisch, R. (2013) "One-step generation of mice carrying mutations in multiple genes by CRISPR/Cas mediated genome engineering," *Cell*; vol. 153, no. 4, pp. 910–918.

[85] Li J. F., Norville J. E., and Aach J. (2013) "Multiplex and homologous recombination-mediated genome editing in Arabidopsis and *Nicotiana benthamiana* using guide RNA and Cas9," *Nature Biotechnology*, vol. 31, no. 8, pp. 688–691.

[86] Osakabe Y., Watanabe T., Sugano S. S., Ueta, R., Ishihara, R., Shinozaki, K., & Osakabe, K. (2016) "Optimization of CRISPR/Cas9 genome editing to modify abiotic stress responses in plants," *Scientific Reports*, vol. 6, Article ID 26685.

[87] Xiong J, Ding J, and Li Y, (2015) "Genome-editing technologies and their potential application in horticultural crop breeding," *Horticulture Research*, vol. 2, article 15019.

[88] Kissoudis C., van de Wiel C., Visser R. G. F., and van der Linden G. (2014) "Enhancing crop resilience to combined abiotic and biotic stress through the dissection of physiological and molecular crosstalk," *Frontiers in Plant Science*, vol. 5, no 207.

[89] Jain M, (2015) "Function genomics of abiotic stress tolerance in plants: A CRISPR approach," Frontiers in *Plant Science*, vol. 6, pp. 1–4.

[90] Khatodia S, Bhatotia K, Passricha N, Khurana SMP, and N. Tuteja N, (2016) "The CRISPR/Cas genome-editing tool: Application in improvement of crops," *Frontiers in Plant Science*, vol. 7, no. 2016, article no. 506.

[91] Nongpiur RC, Singla-Pareek SL, and Pareek A, (2016) "Genomics Approaches for Improving Salinity Stress Tolerance in Crop Plants," *Current Genomics*; vol. 17, no. 4, pp. 343–357.

[92] AndolfoG, Iovieno P, Frusciante L, and. Ercolano MR, (2016) "Genome-editing technologies for enhancing plant disease resistance," *Frontiers in Plant Science*, vol. 7, no. 2016, article no. 1813.

[93] Jacobs TB, LaFayette PR, Schmitz RJ, and W. A. Parrott WA, (2015) "Targeted genome modifications in soybean with CRISPR/Cas9," *BMC Biotechnology*; pp. 1–10.

[94] Xu, R., Qin, R., Li, H., Li, D., Li, L., Wei, P., & Yang, J. (2016) "Generation of targeted mutant rice using a CRISPR-Cpf1 system," *Plant Biotechnology Journal*; vol. 14, pp. 1–5, 2016.

[95] Mao Y, Zhang H, Xu N, Zhang B, Gou F, and Zhu JK, (2013) "Application of the CRISPR-Cas system for efficient genome engineering in plants," *Molecular Plant*, vol. 6, no. 6, pp. 2008– 2011.

[96] Osakabe K, Osakabe Y, and Toki S,(2010) "Site-directed mutagenesis in Arabidopsis using custom-designed zinc finger nucleases," *Proceedings of the National Academy of Sciences of the United States of America*, vol. 107, no. 26, pp. 12034–12039.

[97] Townsend JA, Wright DA, Winfrey RJ, Fu F, Maeder ML, Joung JK, Voytas DF(2009) "High frequency modification of plant genes using engineered zinc finger nucleases," *Nature*, vol. 459, no. 7245, pp. 442–445.

[98] Shukla, V. K., Doyon, Y., Miller, J. C., DeKelver, R. C., Moehle, E. A., Worden, S. E., Mitchell, J. C., Arnold, N. L., Gopalan, S., Meng, X., Choi, V. M., Rock, J. M., Wu, Y. Y., Katibah, G. E., Zhifang, G., McCaskill, D., Simpson, M. A., Blakeslee, B., Greenwalt, S. A., Urnov, F. D. (2009) "Precise genome modification in the crop species Zea mays using zinc-finger nucleases," *Nature*, vol. 459, pp. 437–441.

[99] Shukla V., Gupta M., Urnov F., Guschin D., Jan M., and Bundock P., (2013) "Targeted modifcation of malate dehydrogenase," WO Patent Publication Number: WO 2013166315 A1.

[100] Jiang W. and Marraffini L. A. (2015) CRISPR-Cas: new tools for genetic manipulations from bacterial immunity systems. *Annu. Rev. Microbiol.* 2015; 69: 209–28.

[101] Peng J., Richards D. E., and Hartley N. M., (1999) "Green revolution genes encode mutant gibberellin response modulators," *Nature,* vol. 400, no. 6741, pp. 256–261.

[102] Hollender C. A. and Dardick C., (2015) "Molecular basis of angiosperm tree architecture," *New Phytologist*, vol. 206, no. 2, pp. 541–556.

[103] Svitashev S., Young J. K., Schwartz C., Gao H., Falco S. C., and Cigan A. M. (2015) "Targeted mutagenesis, precise gene editing, and site specific gene insertion in maize using Cas9 and guide RNA," *Plant Physiology*, vol. 169, no. 2, pp. 931–945.

[104] Li T., Liu B., Spalding M. H., Weeks D. P., and Yang B. (2012). "High-efficiency TALEN-based gene editing produces disease resistant rice," Nature Biotechnology, vol. 30, no. 5, pp. 390–392.

[105] Ji X., Zhang H., Zhang Y., Wang Y., and Gao C. (2015) "Establishing a CRISPR-Cas-like immune system conferring DNA virus resistance in plants," *Nature Plants*, vol. 1, article 15144, Article ID 15144.

[106] Liu D., Chen X., Liu J,. Ye J., and Guo Z. (2012) "The rice ERF transcription factor OsERF922 negatively regulates resistance to *Magnaporthe oryzae* and salt tolerance," *Journal of Experimental Botany*, vol. 63, no. 10, pp. 3899–3912.

[107] Wang, F., Wang, C., Liu, P., Lei, C., Hao, W., Gao, Y., Liu, Y.-G., & Zhao, K. (2016) "Enhanced rice blast resistance by CRISPR/ Cas9-Targeted mutagenesis of the ERF transcription factor gene OsERF922," *PLoS ONE*, vol. 11, no. 4, Article ID e0154027.

[108] Fang Y. and Tyler B, M. (2016) "Efficient disruption and replacement of an effector gene in the oomycete *Phytophthora sojae* using CRISPR/Cas9," *Molecular Plant Pathology*, vol. 17, no. 1, pp. 127– 139.

[109] Hastings G. E. and Wolf P. G. (1992) "The Therapeutic Use of Albumin," *Archives of Family Medicine*, vol. 1, no. 2, pp. 281–287.

[110] Bosch M and Hazen SP, (2013) "Lignocellulosic feedstocks: Research progress and challenges in optimizing biomass quality and yield," *Frontiers in Plant Science*, vol. 4, article no. 474.

[111] Shan S., Soltis P. S., Soltis D. E., Yang B. (2020). Considerations in adapting CRISPR/Cas9 in nongenetic model plant systems. *Appl. Plant Sci.* 8:e11314. 10.1002/aps3.11314

[112] Lee CM, Cradick TJ, Fine EJ, and Bao G, (2016) "Nuclease target site selection for maximizing on-target activity and minimizing off-target effects in genome editing," *Molecular Therapy*, vol. 24,no. 3, pp. 475–487.

[113] Cai Q. B., Miller J., Urnov F. Vipula K. Shukla, Joseph F. Petolino, Lisa W. Baker, Robbi J. Garrison, Ryan C. Blue, Nicole L. Arnold, Sarah E. Worden, Jon C. Mitchell. (2015) "Optimized non-canonicalzinc finger proteins," US Patent Number: 9,187,758. Publication date: Nov 17.

[114] Lombardo A., Cesana D., Genovese P., Di Stefano, B., Provasi, E., Colombo, D. F., Neri, M., Magnani, Z., Cantore, A., Lo Riso, P., Damo, M., Pello, O. M., Holmes, M. C., Gregory, P. D., Gritti, A., Broccoli, V., Bonini, C., & Naldini, L. (2011) "Site-specific integration and tailoring of cassette design for sustainable gene transfer," *Nature Methods*, vol. 8, no. 10, pp. 861–869, 2011.

[115] Khatodia S., Bhatotia K., Passricha N., Khurana S. M. P., and Tuteja N. (2016) "The CRISPR/Cas genome-editing tool: Application in improvement of crops," *Frontiers in Plant Science*, vol. 7, no. 2016, article no. 506, 2016.

[116] Koo T., Lee J., and Kim J. (2015) "Measuring and reducing off-target activities of programmable nucleases including CRISPR-Cas9," *Molecules and Cells*, vol. 38, no. 6, pp. 475–481.

[117] Altpeter F., Springer N. M., Bartley L. E., Blechl A., Brutnell T. P., Citovsky V., Conrad L., Gelvin S. B., Jackson D., Kausch A. P., Lemaux P. G., Medford J. I., Orozo-Cardenas M., Tricoli D., VanEck J., Voytas D. F., Walbot V., Wang K., Zhang Z. J., Stewart, C. Neal. (2016) Advancing Crop Transformation in the Era of Genome Editing. *The Plant Cell*. doi:10.1105/tpc.16.0019.

[118] Lowe K., Wu E., Wang N., Hoerster G., Hastings C., Cho M. J., Scelonge C., Lenderts B., Chamberlin M., Cushatt J., Wang L., Ryan L., Khan T., Chow-Yiu J., Hua W., Yu M., Banh J., Bao Z., Brink K., Igo E., Rudrappa B., Shamseer P. M., Bruce W., Newman L., Shen B., Zheng P., Bidney D., Falco S. C., RegisterIII J. C., Zhao Z. Y., Xu D., Jones T. J., Gordon-Kamm W. J. (2016) Morphogenic Regulators Baby boom and Wuschel Improve Monocot Transformation. *The Plant Cell* 28 (9):1998-2015. doi:10.1105/tpc.16.0012.

[119] Maher M. F., Nasti R. A., Vollbrecht M., Starker C. G., Clark M. D., Voytas D. F. (2020) Plant gene editing through *de novo* induction of meristems. *Nat Biotechnol.* 38:84–89.

[120] Kwon C. T., Heo J., Lemmon Z. H., Capua Y., Hutton S. F., Van Eck J., Park S. J., Lippman Z. B. (2020), Rapid customization of solanaceae fruit crops for urban agriculture *Nat Biotechnol*, 38, pp. 182-188.

[121] Oliva R., Ji C., Atienza-Grande G., Huguet-Tapia J. C., Perez-Quintero A., Li T., Eom J. S., Li C., Nguyen H., Liu B., Auguy F., Sciallano C., Luu V. T., Dossa G. S., Cunnac S., Schmidt S. M., Slamet-Loedin I. H., Vera Cruz C., Szurek B., Frommer W. B., White F. F., Yang B. (2019). Broad-spectrum resistance to bacterial blight in rice using genome editing. *Nat Biotechnol.*;37(11):1344-1350.

[122] Lu Y., Ye X., Guo R., Huang J., Wang W., Tang J., Tan L., Zhu J., Chu C., Qian Y. (2017) Genome-wide targeted mutagenesis in rice using the CRISPR/Cas9 system. *Mol. Plant.* 2017;10:1242–1245.

[123] Meng X., Yu H., Zhang Y., Zhuang F., Song X., Gao S., Gao C., Li J. Construction of a genome-wide mutant library in rice using CRISPR/Cas9. *Mol. Plant.* 10:1238–1241.

[124] Araus, J. L., Kefauver, S. C., Zaman-Allah, M., Olsen, M. S. and Cairns, J. E. (2018) Translating high-throughput phenotyping into genetic gain. *Trends Plant Sci.*23, 451–466.

[125] Anzalone A. V., Koblan L. W., Liu D. R. (2020) Genome editing with CRISPR-Cas nucleases, base editors, transposases and prime editors. *Nat Biotechnol.* Jul;38(7):824-844. doi: 10.1038/s41587-020-0561-9. Epub 2020 Jun 22. PMID: 32572269.

[126] Rees H. A., Liu D. R. (2018) Base editing: precision chemistry on the genome and transcriptome of living cells. *Nat Rev Genet.* Dec;19(12):770-788. doi: 10.1038/s41576-018-0059-1. Erratum in: *Nat Rev Genet*. 2018 Oct 19;: PMID: 30323312; PMCID: PMC6535181.

[127] Gao Y. and Zhao Y. (2014) "Specific and heritable gene editing in Arabidopsis," *Proceedings of the National Academy of Sciences of the United States of America*, vol. 111, no. 12, pp. 4357-4358.

[128] Nagamangala Kanchiswamy C., Sargent D. J., Velasco R., Maffei M. E., and Malnoy M. (2015) "Looking forward to genetically edited fruit crops," *Trends in Biotechnology*, vol. 33, no. 2, pp. 62–64.

[129] Mahfouz, M. M., Li, L., Piatek, M., Fang, X., Mansour, H., Bangarusamy, D. K., & Zhu, J. K. (2012) "Targeted transcriptionalrepression using a chimeric TALE-SRDX repressor protein," *Plant Molecular Biology*, vol. 78, no. 3, pp. 311–321.

[130] Xu, R. F., Li, H., Qin, R. Y., Li, J., Qiu, C. H., Yang, Y.C., Ma, H., Li, L., Wei, P. C., & Yang, J. B. (2015) "Generation of inheritable and "transgene clean" targeted genome-modified rice in later generations using the CRISPR/Cas9 system," *Scientific Reports*, vol. 5, Article ID 11491

[131] Araki M. and Ishii T. (2015) "Towards social acceptance of plant breeding by genome editing," Trends in *Plant Science*, vol. 20, no. 3, pp. 145–149.

[132] Podevin N., Devos Y., Davies H. V., and Nielsen K. M. (2012) "Transgenic or not? No simple answer! New biotechnology based plant breeding techniques and the regulatory landscape," *EMBO Reports*, vol. 13, no. 12, pp. 1057–1061.

[133] Schaart J. G., Van de Wiel C. C. M., Lotz L. A. P., and Smulders M. J. M. (2016), "Opportunities for Products of New Plant Breeding Techniques," *Trends in Plant Science*, vol. 21, no. 5, pp. 438–449.

[134] Woo J. W., Kim J., Kwon S. I., Corvalan C., Cho S. W., Kim H., Kim S. G., Kim S. T., Choe S., Kim J. S. (2015). DNA-free genome editing in plants with preassembled CRISPR-Cas9 ribonucleoproteins. *Nat Biotechnol.* 33: 1162–1164.

[135] Hartung F. and Schiemann J. (2014) "Precise plant breeding using new genome editing techniques: Opportunities, safety and International Journal of Agronomy 15regulation in the EU," *Plant Journal*, vol. 78, no. 5, pp. 742–752.

[136] Jones H. D. (2011) "Regulatory uncertainty over genome editing," *Nature Plants*, vol. 1, Article ID 14011.

[137] Voytas D. F. and Gao C. (2014) "Precision genome engineering and agriculture: opportunities and regulatory challenges," *PLoS biology*, vol. 12, no. 6, p. e1001877.

[138] Sauer N. J., Mozoruk J., Miller R. B., Warburg, Z. J., Walker, K. A., Beetham, P. R., Schöpke, C. R., & Gocal, G. F. W. (2016) "Oligonucleotide directed mutagenesis for precision gene editing," *Plant Biotechnology Journal*, vol. 14, no. 2, pp. 496–502.

[139] Lusser M., Parisi C., Plan D., and Rodriguez-Cerezo E. (2012) "Deployment of new biotechnologies in plant breeding," *Nature Biotechnology*, vol. 30, no. 3, pp. 231–239.

[140] Wolt J. D., Wang K., and Yang B. (2016) "The regulatory status of genome-edited crops," *Plant Biotechnology Journal*, vol. 14, no.2, pp. 510–518.

[141] Gao, H., Smith, J., Yang, M., Jones, S., Djukanovic, V., Nicholson, M. G., West, A., Bidney, D., Falco, S. C., Jantz, D., & Lyznik, L. A. (2010). Heritable targeted mutagenesis in maize using a designed endonuclease. *Plant J.* 61: 176– 187. doi: 10.1111/j.1365-313X.2009.04041.x.

[142] Antunes, M. S., Smith, J. J., Jantz, D., & Medford, J. I. (2012). Targeted DNA excision in Arabidopsis by a re-engineered homing endonuclease. *BMC Biotechnol.* 12**,** 86. https://doi.org/10.1186/1472-6750-12-86.

Chapter 2

Plant Genome Editing Using the CRISPR System

Satyananda Patel[1,*]
and Rahul Kumar[2]

[1] MITS School of Biotechnology, Bhubaneswar, Odisha, India
[2] Faculty of Agricultural Sciences, DAV University, Jalandhar, Punjab, India

Abstract

There is a need of the great leap in agricultural production by means of innovative breeding methodologies to feed the ever-growing population of this planet. The CRISPR/Cas genome editing tool is very efficient in targeted crop modifications and is accelerating the works on the crop improvement. This book chapter discusses the CRISPR/Cas9 and its variants in detail as well as it provides the applications in plant genome editing and the gene related manipulations. We emphasize base-editing useful for targeted nucleotide substitutions. We also deliberate upon various delivery systems, like DNA-free methods, which is very important in genome editing led crop breeding. The list of trait improvement related genome editing applications is also reviewed in this chapter. The fine-tuning gene regulation and the breeding for virus resistance strategies is also touched upon. The use of high-throughput mutant libraries is also reviewed. This chapter outlines the future perspectives of genome editing in plant synthetic biology, the delivery systems, gene editing specificity, homology-directed repair, and gene drives. In the end, the challenges and the opportunities of precision plant breeding and its role in the future of agriculture have been accommodated.

Keywords: genome editing, CRISPR/Cas, delivery of CRISPR/Cas, stress resistance

Introduction

Plants are the primary source of food and are essential for human life. Anticipated world population will be 9.6 billion by 2050. This will be result in increased global demand for crops [1]. To fulfill the demand of growing population, there is a need of innovative crop breeding technology to increase agricultural productivity. Cross breeding, mutation breeding and

---

* Corresponding Author's Email: satyanandapatel@gmail.com.

In: Genome Editing for Crop Improvement
Editors: Rahul Kumar, Ashutosh Sharma, Yogesh Kumar et al.
ISBN: 979-8-88697-740-0

transgenic breeding are current major methodologies for crop improvement (Figure 1). Out of these, cross breeding methodology is time consuming [2]. It is in practice since millennium, thus large parts of the genomes of major crops are fixed, limiting the potential of crop improvement. Chemical mutagens or physical irradiation based mutation breeding has enhanced the genetic variations of plants [3] however, its stochastic nature is a limitation. Simultaneously, screening of large numbers of mutants is also a challenging task. Transgenic breeding has the potential that it can break the bottleneck of reproductive isolation. The major concern in this approach is the regulation and public fear, which prevented its commercial materialization [4].

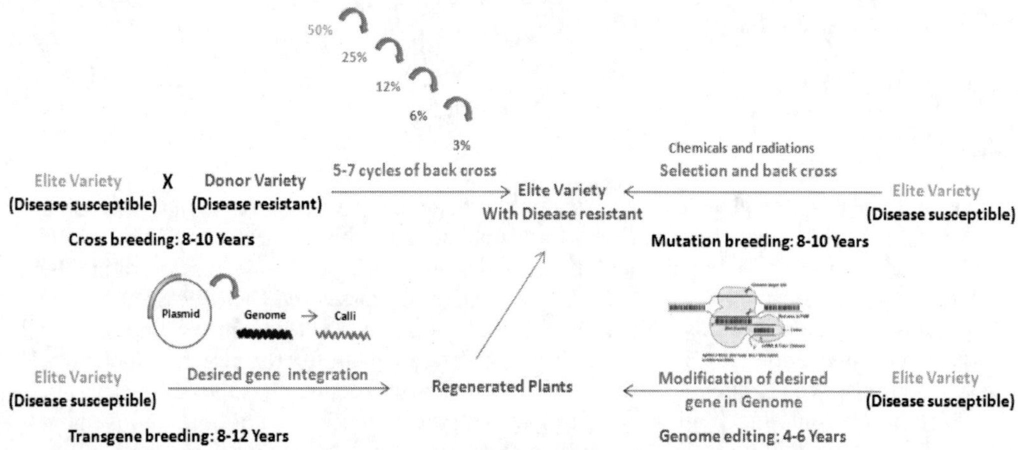

Figure 1. Breeding methods comparison in agriculture.

In 1988, the first gene-targeting experiment in tobacco (*Nicotiana tabacum*) protoplasts was performed [5]. It was discovered that DNA double-strand breaks (DSBs) enhance gene-targeting efficiency [6]. Based on these foundations scientists have developed tools to precisely manipulate plant genomes. Zinc finger nucleases (ZFN) were used for trait improvement in few tobacco plants [7]. By 2010, transcription activator-like effector nucleases (TALENs) were used for plant genome manipulation [8]. Despite their advantages these gene editing tools have their own limitations and are thus far from achieving widespread success. By 2013, clustered regularly interspaced short palindromic repeat (CRISPR) system was used for Genome editing in rice (*Oryza sativa*), wheat (*Triticum aestivum*), *Nicotiana benthamiana*, and *Arabidopsis thaliana* [9]. With this development, plant breeders got the ability to precisely edit the targeted sequence. It opens the portal for rapid plant genome manipulation. Since then, CRISPR/Cas systems have been in continuous evolution with the development of system like CRISPR/Cpf1 [10] and nucleotide substitution tools for base editing [11-12]. It is less expensive and user friendly so widely used for genome manipulation (Figure 1). CRISPR/Cas has the potential and has given us the hope for global food security and sustainable agriculture through rapid improvement of targeted traits like yield, quality, and biotic and abiotic stress tolerance, etc.

CRISPR/Cas Systems for Plant Genome Editing

CRISPR/Cas Systems

The CRISPR/Cas system, that contains CRISPR repeat-spacer arrays and Cas nuclease, is an RNA-mediated bacteria/archaea adaptive immune system, that provides protection against plasmid, phages and other invasive genetic elements by digesting the invader's genome. Considering the Cas genes and the character of the interference complex, CRISPR/Cas systems have been divided into two classes that have been further subdivided into six types. Class 1 CRISPR/Cas systems (types I, III, and IV) recruits multi-Cas protein complexes for interference. Class 2 systems (types II, V, and VI) execute interference with single effector protein in a complex with CRISPR RNAs (crRNAs). The CRISPR system is based on single RNA-guided interference with DNA [14]. Type II CRISPR/Cas9 from *Streptococcus pyogenes* was shown to specifically cut DNA *in vitro* and in eukaryotic cells [15-18]. After improvement, the CRISPR/Cas9 system has two components: the Cas9 nuclease and a single guide RNA (sgRNA) consisting of an artificial fusion of a crRNA and a fixed transactivating crRNA (Figure 2a). The sgRNA and Cas9 protein form a Cas9/sgRNA complex, and 20 nucleotides guide at the 5' end of the sgRNA identifies the target DNA using Watson-Crick base pairing. Following Cas9 introduces the DSB adjacent to 5'-NGG-3' protospacer-adjacent motifs (PAMs). There are reports that claims SpCas9 also cleaved target sites with the non-canonical NAG PAMs in mammalian cells [19] and rice [20]. Cas9 has nuclease domains, RuvC and HNH, which cleave DNA strands that are complementary and non-complementary, respectively [15] (Figure 2a). Catalytic residues surgical removal in either domain produces a DNA nickase (nCas9). The domains inactivation produces a deactivated Cas9 protein (dCas9) that retains the ability to target specific genomic loci and thus act as a scaffold for recruiting effector proteins [21]. Cas9 derived from *Staphylococcus aureus* (SaCas9), *Streptococcus thermophiles* (StCas9), and *Neisseria meningitides* (NmCas9), are also used for genome editing [22]. Cas9 has been re-engineered to recognize different PAMs, such as VQR-Cas9 (NGA PAM), EQR-Cas9 (NGAG PAM), VRERCas9 (NGCG PAM), SaKKH-Cas9 (NNNRRT PAM) [23], xCas9 (NG, GAA, and GTA PAM) [24], and SpCas9-NG (NG PAM) [25] to expand the targeting scope.

Class 2 type V-A Cas enzyme Cpf1 (Cas13), including variants from *Francisella novicida* (FnCpf1), *Acidaminococcus* sp. (AsCpf1), and *Lachnospiraceae bacterium* (LbCpf1), are also used as gene-editing tools [10] (Figure 2a). Cpf1 uses a T-rich PAM sequence and thus expands the editing sites beyond NGG. The guide RNA of Cpf1 is shorter than that for Cas9 sgRNA (~43 versus ~80 nucleotides), and the Cpf1 cleavage site is located distal and downstream of the PAM sequence (Figure 2a). Cpf1 cleaves DNA in a staggered fashion, creating a 5-nucleotide-5' overhang starting at 18 nucleotides 3' of the PAM [10] (Figure 2a). Cpf1 has the limitations that it recognizes only TTTV PAM. To address this Cpf1 has been engineered to recognize AsCpf1-RR (TYCV PAM), AsCpf1-RVR (TATV PAM), LbCpf1-RR (CCCC and TYCV PAM), and LbCpf1-RVR (TATG PAM) [26-28]. Cms1 (CRISPR of Microgenomates and Smithella), is very efficient in generating rice indel mutations. Cms1 nucleases do not require a transactivating crRNA, and have an AT-rich PAM site requirement [29]. Recently, a CRISPR-Cas12b system from *Alicyclobacillus acidiphilus* (AaCas12b), a distinct type V-B system, has been characterized and repurposed to engineer mammalian genomes, and

AaCas12b maintains optimal nuclease activity over a wide temperature range (31–59°C) [30]. LwaCas13a [31] and PspCas13b [32], have been engineered to target RNA or to perform RNA editing (Figure 2d). C2c2 is guided by a single crRNA and can cleave single-strand RNA targets carrying complementary protospacers *via* its HEPN domains [32].

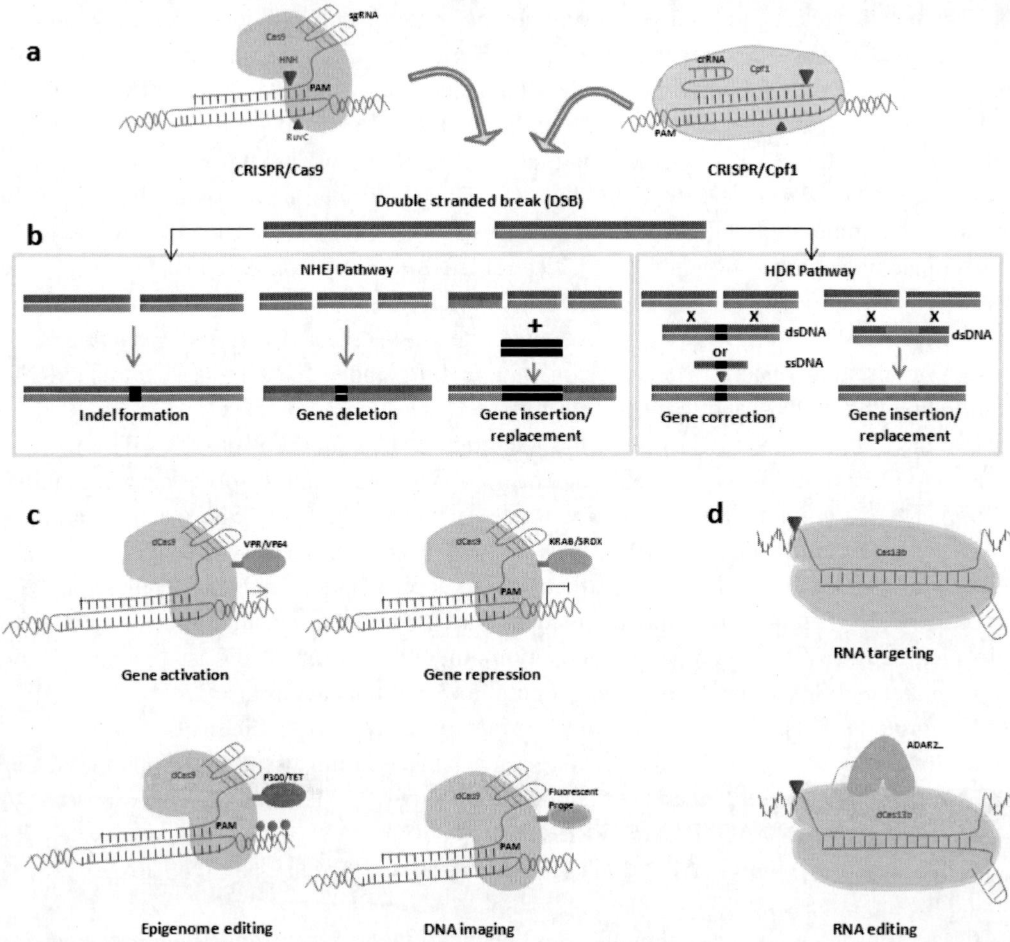

Figure 2. CRISPR/Cas systems for genome editing. (*a*) Cas9 and Cpf1 systems. (*b*) DSB following repair results in several outcomes as depicted. NHEJ repair pathway outcomes: I, II, and III; HDR pathway where DNA donor template is used: IV & V. (*c*) dCas9 fusion-based gene editing. dCas9 fused with proteins like transcriptional activator/repressors, epigenetic modulators, and fluorescent proteins for manipulating gene regulation, epigenome engineering, and genomic labeling. (*d*) Cas13b usually used for RNA targeting, and dCas13b-ADAR designed for RNA editing. *Abbreviations*: ADAR, adenosine deaminase acting on RNA; crRNA, CRISPR RNA; dCas, deactivated Cas; DSB, double-strand break; dsDNA, double-strand DNA; HDR, homology-directed repair; NHEJ, nonhomologous end joining; PAM, protospacer-adjacent motif; sgRNA, single guide RNA; ssODN, single-strand oligodeoxynucleotide; TET, ten-eleven translocation.

Genome Editing via CRISPR-Induced DNA Double-Strand Breaks

CRISPR/Cas generated DSB is repaired by two main pathways: nonhomologous end joining (NHEJ) and homology-directed repair (HDR) [33] (Figure 2b).

Genome Editing by Nonhomologous End Joining

The NHEJ repair pathway does not require a homologous template. So, gene disruption is performed by creating target specific small insertions or deletions (Figure 2b). In CRISPR systems multiple sgRNAs can be used to simultaneously target multiple sites. Multiplexing is possible by creating multigene knockouts, chromosomal deletions and translocations, and gene knock-ins [34]. One promoter can be used to achieve uniform expression of gRNA. This is possible in a polycistronic gene construct in which the gRNA is interspersed with ribozyme sites [35], Csy4 recognition sites [36], or transfer RNA sequences [37], are processed in the plant cell followed by gRNA release for editing. Cpf1 has its own crRNA and thus is used for multiplex genome editing in plants [38]. Since it facilitates simultaneous modification of multiple traits, the CRISPR system represents a highly efficient method for pyramid breeding.

Homology-Directed Repair Pathway and Precision Genome Editing

Precision is missing in NHEJ let genome engineering. In HDR, it is possible to precisely introduce point mutations and to insert or replace desired sequences at the target DNA (Figure 2b). HDR initiates in the S- and G2-phase of the cell cycle. HDR Repair requires a homologous template to the break site. The template can be the sister chromatid or an exogenous donor template, such as an exogenous DNA or single-strand DNA. The donor template sequence can replace the target DNA sequence [34] (Figure 2b). CRISPR/Cas9 based gene targeting has been achieved in rice [39]. The donor template quantity is directly proportional to HDR efficiency. Geminivirus replicons based on bean yellow dwarf virus and wheat dwarf virus increases the number of copies of the donor and thus, increases the frequency of HDR-mediated gene insertion [40-41]. In addition, the use of chimeric sgRNA molecules, also increases the HDR efficiency in rice [42].

dCas9 can provide a unique platform for recruiting proteins for sequence-specific gene regulation, epigenome editing, and genome imaging. Transcriptional repressor like KRAB domain and SRDX domain fused to dCas9 or transcriptional activators (e.g., VP64, p65AD, and VPR) can be used for gene regulation [43] (Figure 2c). Repressor system (pCo-dCas9-3X-SRDX) was used on the Arabidopsis cleavage stimulating factor 64 gene and on nonprotein-coding genes (redundant microRNAs: miR159A and miR159B) [44]. dCas9-TALEs (a potent dCas9-TV) fusion protein provides much stronger transcriptional activation [45]. dCas9 recruits epigenetic effectors such as the histone demethylase LSD1, histone acetyltransferase p300, and ten-eleven translocation proteins to modify epigenetic marks at their DNA or histone targets (Figure 2c); this brings chromatin modification and thus gene expression and cell differentiation [46]. Human demethylase TET1 catalytic domain fused with dCas9 and target the FWA promoter, which results in demethylation in Arabidopsis [47]. Repetitive regions like centromeres, pericentric regions, and telomeres of DNA can be visualized by dCas9-GFP [48] (Figure 2c). *N. benthamiana* leaf cells telomere repeats were visualized by dCas9 -eGFP/mRuby [49].

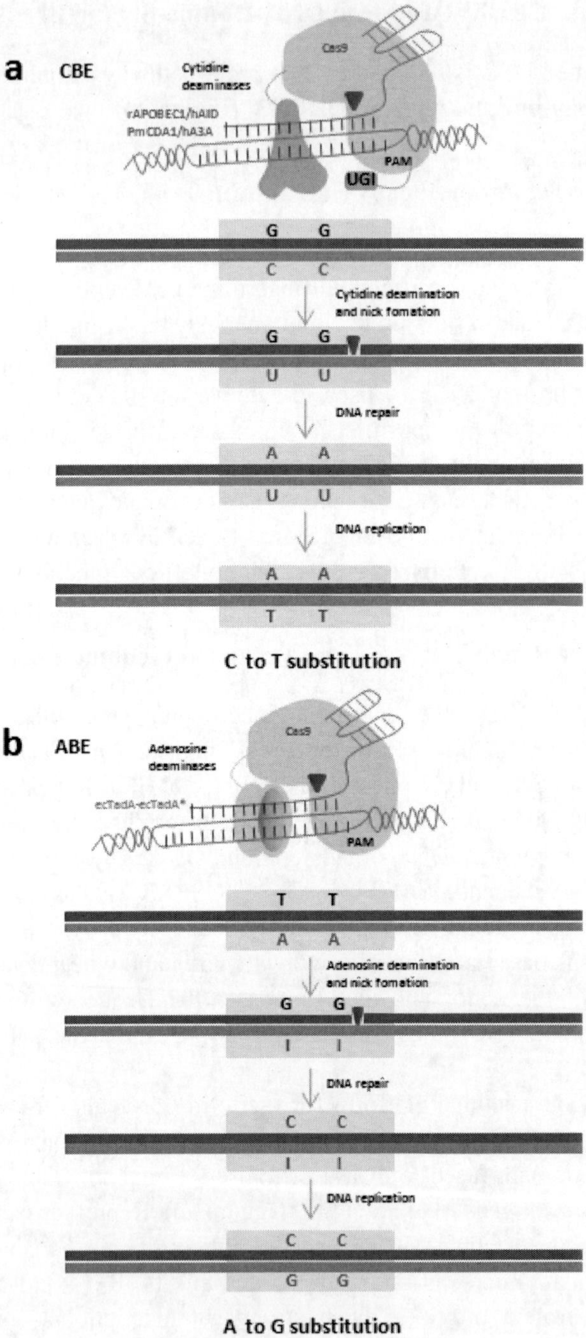

Figure 3. Base editing mechanisms. (*a*) CBE-based C-to-T base-editing. The deaminases: rAPOBEC1, hAID, PmCDA1, and hA3A. (*b*) ABE-based A-to-G base-editing. The deaminase is the fusion protein ecTadA-ecTadA∗. *Abbreviations*: ABE, adenine base editor; CBE, cytosine base editor; hAID, human activation-induced cytidine deaminase; nCas9, a DNA nickase; PAM, protospacer-adjacent motif; UGI, uracil glycosylase inhibitor. Base-editing systems is advantageous over non-DSB-mediated genome editing in plants: (a) BES are more efficient and generate fewer undesired products compared to DSB mediated systems; (b) There is no chromosomal rearrangements, such as large deletions and inversions; and (c) BES can be used to create nonsense mutations to avoid DSB-induced in-frame indels.

Base Editing in Plants

Base editing systems can generate specific base changes and do not depend on HDR or donor DNA (Figure 3). The cytosine base-editor (CBE) system, which consists of a cytidine deaminase-nCas9 (D10A) and a uracil glycosylase inhibitor, modifies cytosine to uracil in genomic DNA (50) (Figure 3a). Cytidine deaminases converts cytosine to uracil. The uracil is replaced by thymine during DNA repair. In this procedure, fused uracil glycosylase inhibitor inhibits uracil DNA glycosylase, thus blocks uridine excision, and ensures base excision repair pathway to increase base-editing efficiency. Base-editor 3 (BE3) systems involving rat cytidine deaminase APOBEC1 has been widely used for gene editing [51]. The human APOBEC3A-based plant cytidine base editor has been used to convert Cs to Ts in wheat, rice, and potato (*Solanum tuberosum*) [52-53]. Adenine base editors (ABEs) mediate A to G conversion in genomic DNA [54] (Figure 3b). Wheat and rice ABE systems have also been optimized. Enhanced sgRNAs [sgRNA(F+E)] combined with three copies of nuclear localization sequences at the C terminus of nCas9 achieved A-to-G conversion efficiencies of up to 60% in rice and wheat [55].

Delivery of CRISPR/Cas Reagents to Plants

The delivery of Genome editing tools into plant cells to edit genes is an important step. Delivery of CRISPR-related editing reagents can be achieved into plant cells by protoplast transfection, *Agrobacterium tumifaciens*-mediated transformation, or biolistic approach.

Plant Genome Editing via CRISPR/Cas DNA

Stable Expressed CRISPR/Cas DNA-Mediated Genome Editing

CRISPR/Cas is a widely used plant editing tool. A marker gene is selected to verify the DNA integration into the plant genome and its expression to execute genome editing (Figure 4a). The off-target changes could limit commercial applications. To avoid off-targeting, transgene-free derivatives can be obtained through genetic segregation by selfing and crossing. Gao et al. [26] considered a fluorescent cassette marker for validating the presence of the CRISPR/Cas9 construct. The suicide genes CMS2 and BARNASE can kill transgene-containing pollen and embryos produced by the T0 plant [56]. Despite the genetic segregation efficiency for obtaining transgene-free mutants, segregation cannot be used for asexually propagated crops such as potato, cassava (*Manihot esculenta*), and banana (*Musa* spp.).

Genome Editing by Transient Expression of CRISPR/Cas

CRISPR mediated transient gene expression is an alternative delivery method to achieve transgene free editing (Figure 4b). This method is free from any canonical selection, e.g., use of herbicides or antibiotics. So regenerated plants are the edited ones without any integration of foreign DNA into the genome. The first report was on wheat [57] where the transformation was achieved by biolistic approach. The indiscernibility of transgenes was up to 86.8% of the T0 mutants. The transient method delivery of plant cytidine base editor and plant adenine base editor led to transgene-free nucleotide substitutions [12, 27, 53]. Protoplast transformations in tetraploid potato and tobacco protoplasts [58-59] and *Agrobacterium*-mediated transient CRISPR/Cas expression [60] in tobacco resulted in transgene-free editing.

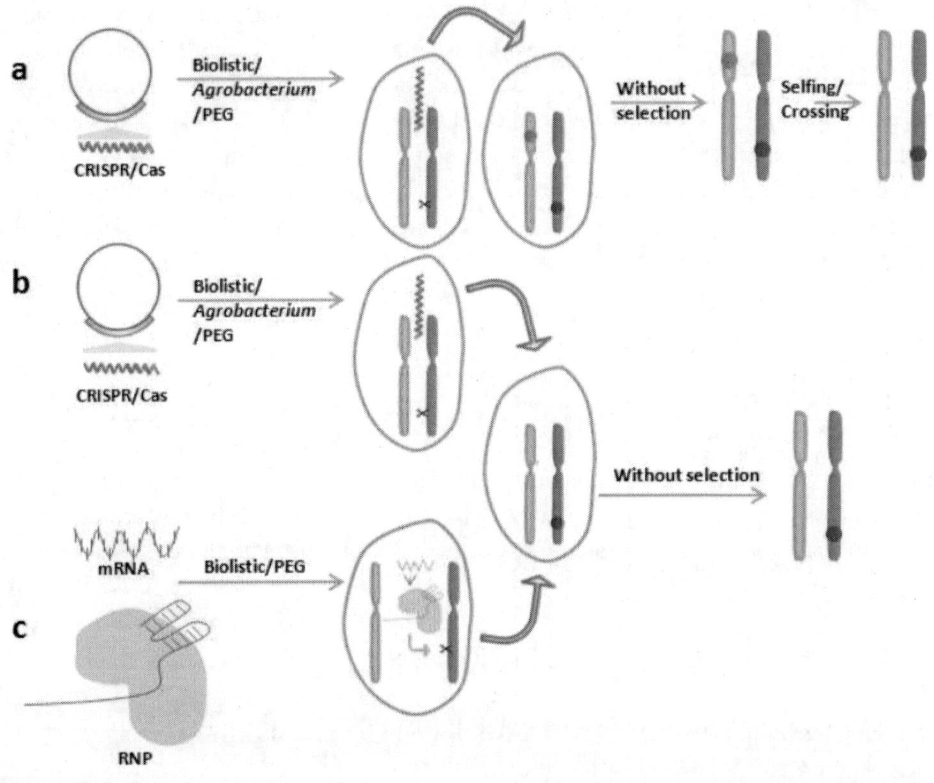

Figure 4. CRISPR/Cas systems delivery strategies to plants. (*a*) Traditionally CRISPR/Cas DNA combined with herbicide or antibiotic selection. Transgene-free plants traditionally obtained through genetic segregation using selfing and crossing. (*b* and *c*) Transgene-free and DNA-free genome editing was performed using transient delivery systems. CRISPR reagents: DNA, RNA, and RNPs. Following transient expression, CRISPR/Cas DNA, RNA, or RNP degraded, in contrast the edited plants can be regenerated without selection pressure. *Abbreviations*: mRNA, messenger RNA; PEG, polyethylene glycol; RNP, ribonucleoprotein.

DNA-Free Genome Editing via CRISPR/Cas *In Vitro* Transcripts or Ribonucleoproteins

While the transient expression of CRISPR/Cas DNA succeeds in reducing transgene integration, it does not completely eradicate it; moreover, degraded DNA fragments may still be integrated into the plant genome. However, Zhang et al. [57] transferred *in vitro* transcripts of Cas9 and sgRNAs into immature wheat embryos by biolistic approach and obtained DNA-free edited wheat (Figure 4c). There was reduced off-target effects. There is an efficient DNA-free genome-editing system developed using Cas9/sgRNA RNPs in plants to avoid the disadvantages of plasmid and messenger RNA (mRNA)-based expression of Cas9/sgRNA [61-64] (Figure 4c). RNP can digest the target immediately upon delivery to reduce off-target without requiring the cellular transcription and translational machinery and is further degraded quickly. It has been used for genome editing in rice, *Arabidopsis*, tobacco, and lettuce (*Lactuca* sativa). Protoplasts fusion was done using polyethylene glycol–mediated RNP transfection. CRISPR/Cas9 RNP has also been efficiently delivered into grape (*Vitis vinifera*), apple, *Petunia hybrida*, and potato protoplasts [58, 65-66]. Kim et al. [67] added Cpf1 to the RNP editing toolbox in plants, delivering LbCpf1/crRNA and AsCpf1/crRNA RNPs into soybean and wild tobacco protoplasts. Still regeneration from protoplasts is challenging for most cereal crops, particularly monocots.

CRISPR/Cas9 RNP delivery by biolistic approach has been documented in maize and wheat [61-63]. Svitashev et al. [63] transferred preassembled Cas9/gRNA RNPs into maize embryos and regenerated maize plants. There were no mutations at the off-target sites. Liang et al. [61] delivered RNPs by biolistic approach into immature wheat embryos. It has been successfully shown BE3 and high-fidelity BE3 in animals [68-69] that shows promising prospects in plants also. A3A plant cytidine base editor without uracil glycosylase inhibitor and assembled with *in vitro*–transcribed sgRNA for successful base editing in wheat [53].

Applications for Precision Plant Breeding

Knockout-Mediated Crop Trait Improvement

Elimination of negative elements by knocking out undesirable traits is the common application of CRISPR/Cas9 (Figure 5a). Yield, quality, and biotic- and abiotic-stress resistance, and hybrid-breeding techniques have been enhanced using CRISPR/Cas. The complex trait "Yield" is dependent on many factors. Removal of negative regulators that affect yield-determining factors like grain number (OsGn1a), grain size (OsGS3), grain weight (TaGW2, OsGW5, OsGLW2, or TaGASR7), panicle size (OsDEP1, TaDEP1), and tiller number (OsAAP3) generated the expected phenotypes in plants with loss-of-function mutations in these genes. CRISPR/Cas9 is a useful tool for increasing yield-related traits [57, 70-73]. Simultaneous removal of three grain weight–related genes (GW2, GW5, and TGW6) in rice resulted in trait pyramiding which consequently increased grain weight [74].

Improving Quality

Genome editing has modified starch content, fragrance, nutritional value, and storage quality in crops. CRISPR/Cas9 mediated knockout of waxy generated low amylose content and improved rice eating and cooking quality [75-76]. Similarly, CRISPR/Cas9 was used to knockout waxy corn line [77] and to produce high-amylose and resistant starch rice by mutating the starch branching enzyme gene SBEIIb [78].

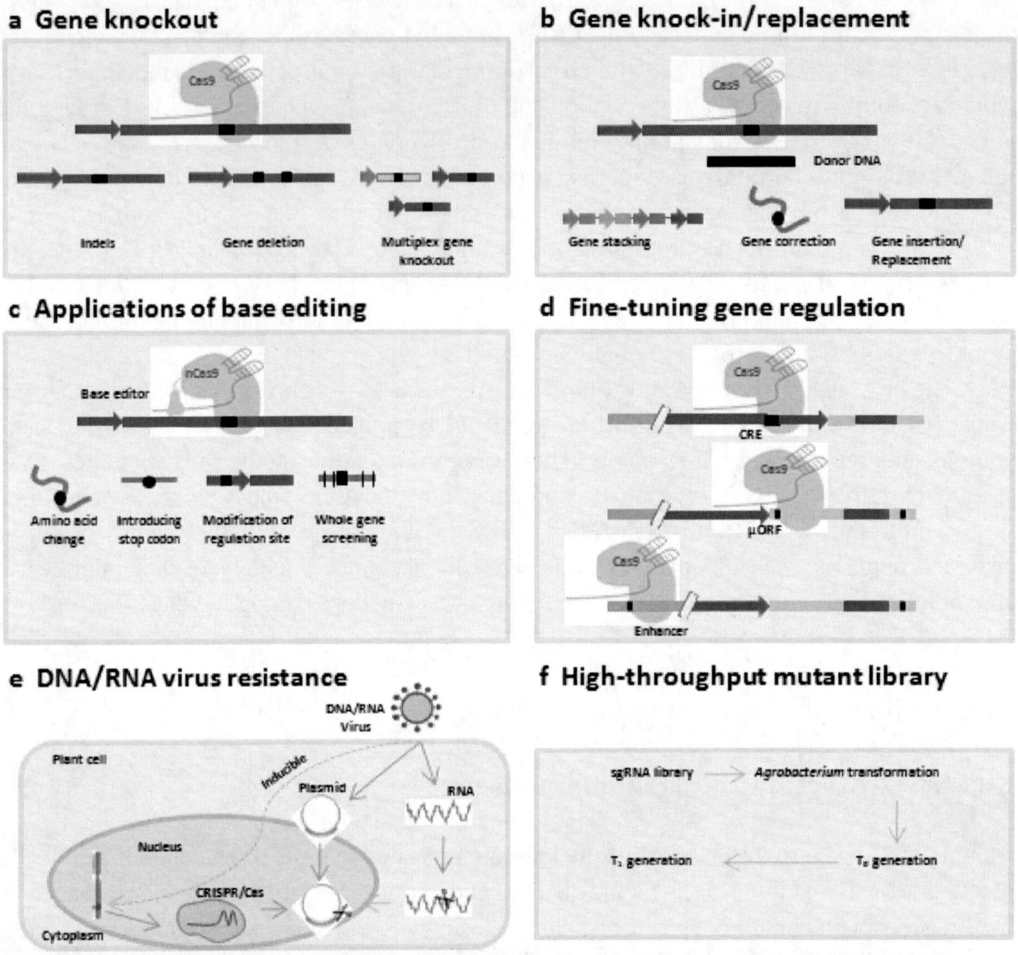

Figure 5. Summary: CRISPR/Cas-mediated plant breeding applications. CRISPR/Cas-based crop improvement broadly focuses on yield, quality, and biotic and abiotic resistance. (a) CRISPR/Cas-based mutation results in indels, gene deletions, and multiplex gene knockout. (b) Gene insertion and replacement achieved by either homology-directed repair or nonhomologous end joining results in gene stacking for multiple traits, gene correction for gain-of-function, and gene insertion or replacement to produce new traits in breeding. (c) Base editing applications for crop trait improvement, like precise amino acid substitution, gene disruption by introducing a stop codon, gene regulation, and whole-gene screening. (d) CRISPR/Cas system-mediated gene regulation by editing the regulatory site in the untranslated region, promoter, or enhancer region. (e) CRISPR/Cas-mediated antiviral breeding approaches. The CRISPR system targeting DNA or RNA viruses is integrated into the plant genome, provides resistance to invading viruses. (f) CRISPR/Cas-mediated genome-wide screening, is a important technique for functional genomics and genetic improvement. *Abbreviations*: CRE, cis-regulatory element; sgRNA, single guide RNA; uORF, upstream open reading frame.

Fragrance is an important quality trait in rice and it increases commercial value. Defective betaine aldehyde dehydrogenase 2 (BADH2) gene results in the biosynthesis of 2-acetyl-1-pyrroline, which is a major fragrance compound. Simultaneously knocking out most conserved domains of α-gliadin family members, researchers have created low-gluten wheat [79]. CRISPR/Cas9 editing was used to produce seeds with high oleic acid oil in *Camelina sativa*

[21], [80] and *Brassica napus* [81], tomatoes with a long shelf life [82-83], high-value tomato with enhanced lycopene [84] or γ-aminobutyric acid content [85], and potato (hairy roots) with reduced levels of toxic steroidal glycoalkaloids [86].

Biotic and Abiotic Stress Resistance

Stresses are the main factors affecting crop yield and quality. CRISPR/Cas9 knockout was used to increase biotic resistance to fungal, bacterial, and viral diseases and insects. Using TALEN and CRISPR/Cas9, all six TaMLO alleles were knocked out in wheat and powdery mildew resistant plants were obtained [87]. Similarly, knockingout of MLO confers resistance to powdery mildew in tomato [88]. Blast-resistant rice was produced via knockout of OsERF922, an ethylene responsive factor transcription factor gene [89]. Deletion of the OsSWEET13 promoter led to the production of plants resistant to *Xanthomonas oryzae* pv. *oryzae* [90], tungro disease–resistant eif4g rice [91], broad potyvirus–resistant eif4e cucumber [92], and cotton leaf curl disease–resistant CLCuD cotton [93] are produced by using CRISPR/Cas. Disrupting OsCYP71A1 blocked serotonin biosynthesis and greatly increased salicylic acid levels, thereby conferring resistance to plant hoppers and stem borers, pests of rice [94]. Abiotic stresses, like accumulation of toxic heavy metals in crops are prevented by knocking out OsARM1, OsNramp5, and OsHAK1. Breeders have developed rice strains with low levels of cadmium, radioactive cesium, and arsenic, respectively [38, 95-96].

Speeding Hybrid Breeding

High-quality hybrid variety is produced by creating a male-sterile maternal line. Thermo sensitive male-sterile tm5 lines in rice [97] and maize [98], photosensitive genic male-sterile csa (*Carbon Starved Anther)* rice [99], and ms45 wheat [100] have been reported. Reproductive barriers were overcome in japonica-indica hybrids, SaF/SaM at the sterility locus Sa [101] and OgTPR1 at the S1 locus [102] were disrupted. Knockout of one or two copies of the Sc gene in the indica allele Sc-I also rescued male fertility in japonica-indica hybrids [103]. Two independent groups developed asexual propagation lines either by simultaneous activation of BBM1 in the egg cell [104] or by knocking out MTL [105], enabling a fix of heterozygosity of hybrids through seed propagation. Genome manipulation in improving haploid breeding [106-107], shortening growth times [108], increasing silique shatter resistance [109], and overcoming self-incompatibility in diploid potato [110], to fulfill breeders' desire.

Crop Trait Improvement *via* Knock-In and Replacement

Agronomic traits alteration achieved by single-nucleotide substitutions, gene expression changes. Knock-ins and replacements introduces new alleles without linkage drag or generating allelic variants [111] (Figure 2b). Knock-in is used to alter traits by stacking genes in a single variety. Thus, knock-ins and replacements have great value for crop trait improvement (Figure 5b).

For HDR, a geminivirus-based DNA replicon has been used as repair templates, which increased gene-targeting efficiency in potato [112], tomato [40, 113], rice [38], wheat [41], and cassava [114] by using geminivirus replicons [40], demonstrated a ten-fold increase in the frequency of insertion of the cauliflower mosaic virus 35S promoter upstream of ANT1 in tomato; constitutive expression of ANT1 led to the generation of a purple tomato with increased anthocyanin content. Sulfonylurea based herbicide-resistant soybean [115], maize [63], and rice [42, 116-117] plants were produced through HDR-introduced nucleotide changes in ALS. Similarly, T102I/P106S (TIPS) and T102I/P106A double amino acid substitutions of EPSPS were introduced into flax (*Linum usitatissimum*) [118] and cassava [114] by selecting for HDR-generated glyphosate resistance.

Applications of Base Editors in Plants

Several agriculturally important traits are produced by single-nucleotide polymorphisms in either coding or noncoding regions, base editing is quite useful for plant breeding and crop improvement (Figure 5c). An application of coding regions base editing is to introduce herbicide resistance. Sulfonylurea- or imidazolinone-resistant rice [11], wheat [53], *Arabidopsis*, and watermelon [119] have been generated by targeting ALS with a plant cytidine base editor, and haloxyfop-R-methyl resistant rice has been created by targeting acetyl-coenzyme-A carboxylase ACCase gene with a plant adenine base editor [55]. RNA splicing pathways is regulated also by "base editing". Mostly eukaryotic mRNA splicing processes follow the canonical GU/AG rule. The introns of these genes have a splice donor site (GU) at the 5' end and a splice acceptor site (AG) at the 3' end. Base editing may generate point mutations at these conserved nucleotides, leading to the loss of particular splice forms or mis-splicing. Kang et al. [120] disrupted the splicing acceptor site on an intron by converting A to G, which led to altered splicing of AtPDS mRNA.

Fine-Tuning Gene Regulation in Plants

Besides creating mutations in coding sequences, modulating gene expression is a beneficial approach for studying gene function and can greatly help plant breeding. Gene expression can be affected at many levels e.g., transcription, mRNA processing, and mRNA translation. These processes are under the control of a series of cis-regulatory elements, which can be modified by genome editing (Figure 5d). To date, alteration of gene expression has focused mainly on promoters, e.g., replacing promoters and deleting cis-regulatory elements [121-122]. Rodríguez-Leal et al. [123] edited the promoter regions of SlCLV3, SlS, and SlSP, creating a continuum of variation and leading to the selection of mutated alleles with improved yields. Upstream open reading frames (uORFs) are cis-elements and often have negative effects on translation and nonsense-mediated mRNA decay [124]. Enhancers contain important cis regulatory elements. Locating DNAase1-hypersensitive sites is an efficient way to identify putative enhancer elements [125]. Messenger RNA 5' leader sequences, like RNA structures and internal ribosome entry sites, can affect translation [61]. Messenger RNA 3' untranslated regions also play very important roles in gene regulation [126].

Antiviral Plant Breeding Strategies

Viruses have led to massive losses in agricultural production worldwide [127]. CRISPR/Cas system confers virus resistance in plants (Figure 5e). Geminiviruses are single strand DNA viruses with a double-stranded intermediate necessary for rolling-circle replication. Cas9 and sgRNAs targeting the geminivirus genome to inhibit its replication is an antiviral strategy in plants [128]. However, indels caused by the NHEJ help virus variants that can escape Cas9/gRNA cleavage. CRISPR/Cas limitation is the off-target mutations. But by using a virus promoter for Cas9 expression can decrease off-target effects to minimal. RNA viruses are the cause of more agricultural losses compared to DNA viruses. FnCas9 binds to RNA in a PAM-independent manner and inhibits translation and replication of hepatitis C virus in mammalian cells [129]. FnCas9 also efficiently represses replication of cucumber mosaic virus and tobacco mosaic virus in plants [130]. Unlike most Cas proteins, C2c2 can cleave single-strand RNA [31], and it interferes with turnip mosaic virus replication in plants [131].

High-Throughput Plant Mutant Libraries

Whole-genome mutant libraries are essential tools for functional genomics and genetic improvement (Figure 5f). Traditional mutant libraries are based on random mutations induced by agents such as irradiation, T-DNA insertions, ethyl methane sulfonate (EMS) mutagenesis, and transposons. However, these methods require several generations to stabilize loss-of-function mutations and determining the relationship between phenotype and genotype among mutants is a time-consuming tedious process. Large-scale CRISPR/Cas9-generated knockout rice mutant libraries have been constructed [20, 94]. Mutant libraries of immunity-associated leucine-rich repeat genes of subfamily XII, comprising 54 members in tomato have been produced [132].

Future Prospects

Exploiting the Potential of Plant Synthetic Biology

Plant synthetic biology has been used to engineer new devices that can predict behaviors. This field will have great impact on crop improvement [133]. Plants are the source of food as well as the secondary metabolites. The CRISPR/Cas system has promising potential to improve plant design and synthetic biology. Considering the importance of nitrogen nitrogen fixation (nif) genes responsible for it have been characterized [134]. It is helping in reducing the dependence on inorganic fertilizers. dCas9-mediated multiplex-based gene activation, repression, and epigenome editing offers unprecedented opportunities for designing synthetic transcription factors. CRISPR/Cas mediated multiplex gene regulation could serve as a tool for this synthetic biology project.

Given the increasing number of plant species where whole genome sequencing has been achieved, genome editing is helping us in plant domestication and thus in expanding crop diversity and increasing agricultural sustainability. Unlike other oilseed plants, pennycress has a short growing season, extreme cold tolerance, high seed oil productivity, and distinct cover crop attributes and is like advanced mustard family members [135]. Modification of the genes that control seed dormancy (DOG1), oil quality (FAE1 and FAE2), glucosinolate accumulation (HAG1 and GTR2), and oil content (DGAT genes) could greatly facilitate the development of special domesticated varieties of pennycress. There are reports on domestication of wild tomato in terms of growth habit, flower and fruit production, and nutritional traits without losing the stress tolerance of the original wild germplasm [136]. Genome editing of a wild tomato variety, the ground cherry (*Physalis pruinosa*) has shown higher yield and bigger fruit [137]. New domesticated crops with augmented tolerance to a wide range of challenging environments, like deserts, maritime areas, low-nutrient soil, and cold climates, will promote agricultural diversity and help in solving many problems associated with sustainable agriculture.

Improved Delivery Systems

Plant cell wall makes it difficult for the efficient delivery of genome-editing reagents into the cell. Improving delivery systems and developing new systems will be pivotal in reducing barriers to inexpensive application of gene editing in plants (Figure 6).

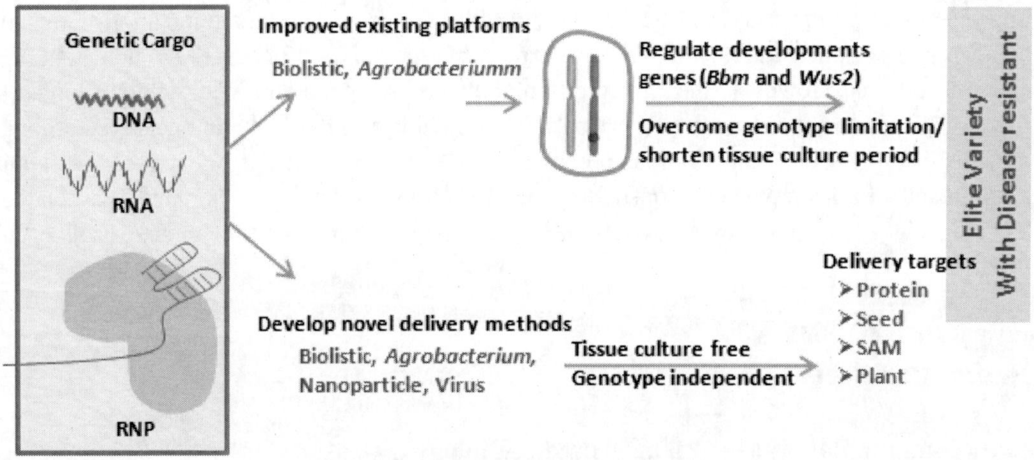

Figure 6. Optimized delivery strategies. (Upper panels) Optimization of existing delivery systems and developmental gene regulations to overcome species limitations and speed up tissue culture process. (Lower panels) Tissue culture–free delivery systems, via plant germline or meristematic cells and nanotechnology-mediated delivery systems. *Abbreviations*: Bbm, Baby boom; RNP, ribonucleoprotein; SAM, shoot apical meristem; Wus2, Wuschel2.

To expand the range of delivery systems, both Agrobacterium and plant genes could be manipulated to improve Agrobacterium-mediated transformation [138]. A further innovation will be achieving genotype-independent, tissue culture–free delivery via the plant germline or meristematic cells [139]. Sperm cells, egg cells, and zygotes are the emerging delivery targets. For example, species specificity will be enhanced and limitations will be ruled out by the use of pollen-mediated transformation. In addition, shoot apical meristems could be used for delivery, as stem cells are destined to differentiate into gametes [140]. Nanotechnology and virus particles (VLP) are the potential future delivery strategies for crop improvement. For instance, carbon nanotubes have been used to deliver DNA into mature plant leaves, leading to successful protein expression. Other nanomaterials such as layered double hydroxides [141], mesoporous silica nanoparticles [142], and polyethylenimine [142] also have great potential for expanding the availability of delivery vehicles, as they may cause little cellular damage, have low toxicity, and yield high transformation efficiencies.

Table 1. Limitations of current CRISPR/Cas9 system

1.	Limited PAM sites
2.	Results in multiple and random mutations in the Genome and may cause more off-target mutations
3.	Low HDR efficiency
4.	Difficult to market CRISPR/Cas expressed crop in various countries
5.	It needs *Agrobacterium*-mediated transformation system for creating mutant plants, so expensive, resource and time consuming

Improved Specificity of CRISPR/Cas Systems

There are concerns CRISPR/Cas-mediated genome editing led off target effects. There are reports on whole genome sequencing that detect cleavage of off-target sites by Cas9 or Cpf1 nucleases in stably transgenic *Arabidopsis* [143] and rice [144]. It was observed that both Cas9 and Cpf1 activities are highly specific. It was concluded that off-targeting could be avoided by designing highly specific sgRNAs. Several strategies have been designed to improve the specificity of Cas9-linked base editors, e.g., extending sgRNA guide sequences, linking APOBEC1 with Cas9-HF1, and delivering base editors *via* RNP [69].

Increased Efficiency of Precise Gene Editing Mediated by Homology-Directed Repair

In HDR-mediated gene editing there is simultaneous induction of DSBs and delivery of a repair template to one location within the genome. There are many potential ways to increase the frequency of HDR in plant cells e.g., manipulation of DNA repair pathways. PolQ is essential for T-DNA integration in plants [145]; and hence, this approach may eliminate the integration of donor DNA into the genome and enhances the efficiency of HDR-mediated genome editing. Along with it, heterologous expression of many critical proteins in the HDR pathway could increase HDR efficiency; like the homologous pairing and DNA strand exchange proteins RAD52, RAD54, and RPA; resection protein RecQL4; and Exo1 and Spo11 [146]. Novel plant delivery methods that enhance donor delivery could increase the efficiency of HDR-mediated genome editing. *Agrobacterium* delivers virulence effector proteins using a type IV secretion system to plant cells [147], and VirD2 protein covalently linked to single-strand T-DNA, allow the T-DNA transfer by protein transfer apparatus; this approach could co-deliver DNA or RNP along with donor templates for HDR-mediated genome editing. This process could also be combined with the non-integrating geminivirus replicon system to enhance the number of donor templates in cells.

CRISPR/Cas-Based Gene Drive in Regulating Agricultural Invasive Species

CRISPR/Cas gene-editing system led gene drive is very efficient method in spreading genetic elements through populations via sexual reproduction [148]. This technique has promising future for public health and humanitarian purposes, like reducing the burden of vector borne diseases [149]. In agricultural science, CRISPR/Cas-led gene drive could be used to suppress or eliminate invasive species (e.g., pests and weed), and it could be used to alter pathogens and introduce new traits into existing populations, e.g., pigweed (*Amaranthus*) could be made susceptible to herbicide glyphosate. For genome editing, careful explanation is required for general public acceptance, we should extend our effort for social responsibility and request researchers to be united and careful on not allowing gene drive until sufficient regulatory and social frameworks guideline are prepared and accepted for current genome-editing methodologies.

Conclusion

The remarkable ability to generate precise sequence-defined, genome-wide genetic diversity in plants through genome manipulation has led to fast and unprecedented advances in basic plant research and crop breeding. The simple, versatile, and robust CRISPR/Cas systems make genome manipulation a powerful tool for desired crop improvement via gene knockout, knock-in, replacement, point mutations, fine-tuning of gene regulation, and other modifications at any gene locus in crops. It has great potential in antiviral breeding and library construction of high-throughput mutant. There is effective delivery of CRISPR/Cas reagents to plant cells. The amalgamation of synthetic biology concept and systems biology and functional genomics, in association with the development of genome-editing technology, next-generation sequencing, and relevant techniques, can be used for engineering advanced and improved qualities of crops. However, the transfer of technologies from the lab to the field needs to be under watchful observations and regulations.

Key Points

1. CRISPR/Cas technology is simple, efficient, and a versatile tool that gives us the opportunities to study plant biology and perform precision plant breeding.
2. Genome editing via CRISPR/Cas RNPs provides a DNA-free and low off-target editing method in plant.
3. CRISPR/Cas editing of cis-regulatory elements, such as promoters and upstream open reading frames, enables the fine-tuning of quantitative trait variation and posttranscriptional upregulation of translation in eukaryotic genes.
4. CRISPR/Cas-mediated gene knockout, insertion, and replacement are being rapidly applied in attempts to enhance yield, quality, disease resistance, and stress tolerance traits in crops and to improve hybrid breeding and crop domestication.

References

[1] Tilman, D., Balzer, C., Hill, J., and Befort, B. L. (2011). Global food demand and the sustainable intensification of agriculture. *Proceedings of the National Academy of Sciences of the United States of America*, 108, 20260-20264.

[2] Scheben, A., Wolter, F., Batley, J., Puchta, H., and Edwards, D. (2017). Towards CRISPR/Cas crops - bringing together genomics and genome editing. *The New Phytologist*, 216, 682-698.

[3] Pacher, M., and Puchta, H. (2017). From classical mutagenesis to nuclease-based breeding - directing natural DNA repair for a natural end-product. *The Plant Journal: For Cell and Molecular Biology*, 90, 819-833.

[4] Prado, J. R., Segers, G., Voelker, T., Carson, D., Dobert, R., et al. (2014). Genetically engineered crops: from idea to product. *Annual Review of Plant Biology*, 65, 769-790.

[5] Paszkowski, J., Baur, M., Bogucki, A., and Potrykus, I. (1988). Gene targeting in plants. *The EMBO Journal*, 7, 4021-4026.

[6] Puchta, H., Dujon, B., and Hohn, B. (1993). Homologous recombination in plant cells is enhanced by *in vivo* induction of double strand breaks into DNA by a site-specific endonuclease. *Nucleic Acids Research*, 21, 5034-5040.

[7] Wright, D. A., Townsend, J. A., Winfrey, R. J., Jr., Irwin, P. A., Rajagopal, J., et al., (2005). High-frequency homologous recombination in plants mediated by zinc-finger nucleases. *The Plant Journal: for Cell and Molecular Biology*, 44, 693-705.

[8] Christian, M., Cermak, T., Doyle, E. L., Schmidt, C., Zhang, F., et al., (2010). Targeting DNA double-strand breaks with TAL effector nucleases. *Genetics*, 186, 757-761.

[9] Li, J. F., Norville, J. E., Aach, J., McCormack, M., Zhang, D., et al., (2013). Multiplex and homologous recombination-mediated genome editing in Arabidopsis and Nicotiana benthamiana using guide RNA and Cas9. *Nature Biotechnology*, 31, 688-691.

[10] Zetsche, B., Gootenberg, J. S., Abudayyeh, O. O., Slaymaker, I. M., Makarova, K. S., et al. (2015). Cpf1 is a single RNA-guided endonuclease of a class 2 CRISPR-Cas system. *Cell*, 163, 759-771.

[11] Shimatani, Z., Kashojiya, S., Takayama, M., Terada, R., Arazoe, T., et al. (2017). Targeted base editing in rice and tomato using a CRISPR-Cas9 cytidine deaminase fusion. *Nature Biotechnology*, 35, 441-443.

[12] Zong, Y., Wang, Y., Li, C., Zhang, R., Chen, K., et al., (2017). Precise base editing in rice, wheat and maize with a Cas9-cytidine deaminase fusion. *Nature Biotechnology*, 35, 438-440.

[13] Chen, K., Wang, Y., Zhang, R., Zhang, H., and Gao, C. (2019). CRISPR/Cas Genome Editing and Precision Plant Breeding in Agriculture. *Annual Review of Plant Biology*, 70, 667-697.

[14] Koonin, E. V., Makarova, K. S., and Zhang, F. (2017). Diversity, classification and evolution of CRISPR-Cas systems. *Current Opinion in Microbiology*, 37, 67-78.

[15] Gasiunas, G., Barrangou, R., Horvath, P., and Siksnys, V. (2012). Cas9-crRNA ribonucleoprotein complex mediates specific DNA cleavage for adaptive immunity in bacteria. *Proceedings of the National Academy of Sciences of the United States of America*, 109, E2579-2586.

[16] Jinek, M., Chylinski, K., Fonfara, I., Hauer, M., Doudna, J. A., et al., (2012). A programmable dual-RNA-guided DNA endonuclease in adaptive bacterial immunity. *Science*, 337, 816-821.

[17] Mali, P., Yang, L., Esvelt, K. M., Aach, J., Guell, M., et al., (2013). RNA-guided human genome engineering via Cas9. *Science*, 339, 823-826.

[18] Cong, L., Ran, F. A., Cox, D., Lin, S., Barretto, R., et al. (2013). Multiplex genome engineering using CRISPR/Cas systems. *Science*, 339, 819-823.

[19] Leenay, R. T., Maksimchuk, K. R., Slotkowski, R. A., Agrawal, R. N., Gomaa, A. A., et al., (2016). Identifying and Visualizing Functional PAM Diversity across CRISPR-Cas Systems. *Molecular cell*, 62, 137-147.

[20] Meng, X., Hu, X., Liu, Q., Song, X., Gao, C., et al., (2018). Robust genome editing of CRISPR-Cas9 at NAG PAMs in rice. *Science China. Life Sciences*, 61, 122-125.

[21] Jiang, F., and Doudna, J. A. (2017). CRISPR-Cas9 Structures and Mechanisms. *Annual review of biophysics*, 46, 505-529.

[22] Cebrian-Serrano, A., and Davies, B. (2017). CRISPR-Cas orthologues and variants: optimizing the repertoire, specificity and delivery of genome engineering tools. *Mammalian Genome: Official Journal of the International Mammalian Genome Society*, 28, 247-261.

[23] Kleinstiver, B. P., Prew, M. S., Tsai, S. Q., Topkar, V. V., Nguyen, N. T., et al. (2015). Engineered CRISPR-Cas9 nucleases with altered PAM specificities. *Nature*, 523, 481-485.

[24] Hu, J. H., Miller, S. M., Geurts, M. H., Tang, W., Chen, L., et al. (2018). Evolved Cas9 variants with broad PAM compatibility and high DNA specificity. *Nature*, 556, 57-63.

[25] Nishimasu, H., Shi, X., Ishiguro, S., Gao, L., Hirano, S., et al. (2018). Engineered CRISPR-Cas9 nuclease with expanded targeting space. *Science*, 361, 1259-1262.

[26] Gao, L., Cox, D. B. T., Yan, W. X., Manteiga, J. C., Schneider, M. W., et al., (2017). Engineered Cpf1 variants with altered PAM specificities. *Nature Biotechnology*, 35, 789-792.

[27] Li, S., Zhang, X., Wang, W., Guo, X., Wu, Z., et al., (2018). Expanding the Scope of CRISPR/Cpf1-Mediated Genome Editing in Rice. *Molecular Plant*, 11, 995-998.

[28] Zhong, Z., Zhang, Y., You, Q., Tang, X., Ren, Q., et al. (2018). Plant Genome Editing Using FnCpf1 and LbCpf1 Nucleases at Redefined and Altered PAM Sites. *Molecular Plant*, 11, 999-1002.

[29] Begemann, M. B., Gray, B. N., January, E., Gordon, G. C., He, Y., et al., (2017). Precise insertion and guided editing of higher plant genomes using Cpf1 CRISPR nucleases. *Scientific Reports*, 7, 11606.

[30] Teng, F., Cui, T., Feng, G., Guo, L., Xu, K., et al., (2018). Repurposing CRISPR-Cas12b for mammalian genome engineering. *Cell Discovery*, 4, 63.

[31] Abudayyeh, O. O., Gootenberg, J. S., Konermann, S., Joung, J., Slaymaker, I. M., et al. (2016). C2c2 is a single-component programmable RNA-guided RNA-targeting CRISPR effector. *Science*, 353, aaf5573.

[32] Cox, D. B. T., Gootenberg, J. S., Abudayyeh, O. O., Franklin, B., Kellner, M. J., et al., (2017). RNA editing with CRISPR-Cas13. *Science*, 358, 1019-1027.

[33] Symington, L. S., and Gautier, J. (2011). Double-strand break end resection and repair pathway choice. *Annual Review of Genetics*, 45, 247-271.

[34] Salsman, J., and Dellaire, G. (2017). Precision genome editing in the CRISPR era. *Biochemistry and Cell Biology,* 95, 187-201.

[35] Gao, Y., and Zhao, Y. (2014). Self-processing of ribozyme-flanked RNAs into guide RNAs *in vitro* and *in vivo* for CRISPR-mediated genome editing. *Journal of Integrative Plant Biology*, 56, 343-349.

[36] Cermak, T., Curtin, S. J., Gil-Humanes, J., Cegan, R., Kono, T. J. Y., et al. (2017). A Multipurpose Toolkit to Enable Advanced Genome Engineering in Plants. *The Plant Cell*, 29, 1196-1217.

[37] Xie, K., Minkenberg, B., and Yang, Y. (2015). Boosting CRISPR/Cas9 multiplex editing capability with the endogenous tRNA-processing system. *Proceedings of the National Academy of Sciences of the United States of America*, 112, 3570-3575.

[38] Wang, F. Z., Chen, M. X., Yu, L. J., Xie, L. J., Yuan, L. B., et al. (2017). OsARM1, an R2R3 MYB Transcription Factor, Is Involved in Regulation of the Response to Arsenic Stress in Rice. *Frontiers in Plant Science*, 8, 1868.

[39] Nishizawa-Yokoi, A., Endo, M., Ohtsuki, N., Saika, H., and Toki, S. (2015). Precision genome editing in plants via gene targeting and piggyBac-mediated marker excision. *The Plant Journal: For Cell and Molecular Biology*, 81, 160-168.

[40] Cermak, T., Baltes, N. J., Cegan, R., Zhang, Y., and Voytas, D. F. (2015). High-frequency, precise modification of the tomato genome. *Genome Biology*, 16, 232.

[41] Gil-Humanes, J., Wang, Y., Liang, Z., Shan, Q., Ozuna, C. V., et al. (2017). High-efficiency gene targeting in hexaploid wheat using DNA replicons and CRISPR/Cas9. *The Plant Journal : For Cell and Molecular Biology*, 89, 1251-1262.

[42] Butt, H., Eid, A., Ali, Z., Atia, M. A. M., Mokhtar, M. M., et al., (2017). Efficient CRISPR/Cas9-Mediated Genome Editing Using a Chimeric Single-Guide RNA Molecule. *Frontiers in Plant Science*, 8, 1441.

[43] Dominguez, A. A., Lim, W. A., and Qi, L. S. (2016). Beyond editing: repurposing CRISPR-Cas9 for precision genome regulation and interrogation. *Nature Reviews. Molecular Cell Biology*, 17, 5-15.

[44] Lowder, L. G., Zhang, D., Baltes, N. J., Paul, J. W., 3rd, Tang, X., et al., (2015). A CRISPR/Cas9 Toolbox for Multiplexed Plant Genome Editing and Transcriptional Regulation. *Plant Physiology*, 169, 971-985.

[45] Li, Z., Zhang, D., Xiong, X., Yan, B., Xie, W., et al., (2017). A potent Cas9-derived gene activator for plant and mammalian cells. *Nature Plants*, 3, 930-936.

[46] Kungulovski, G., and Jeltsch, A. (2016). Epigenome Editing: State of the Art, Concepts, and Perspectives. *Trends in Genetics : TIG*, 32, 101-113.

[47] Gallego-Bartolome, J., Gardiner, J., Liu, W., Papikian, A., Ghoshal, B., et al., (2018). Targeted DNA demethylation of the Arabidopsis genome using the human TET1 catalytic domain. *Proceedings of the National Academy of Sciences of the United States of America*, 115, E2125-E2134.

[48] Chen, B., Gilbert, L. A., Cimini, B. A., Schnitzbauer, J., Zhang, W., et al. (2013). Dynamic imaging of genomic loci in living human cells by an optimized CRISPR/Cas system. *Cell*, 155, 1479-1491.

[49] Dreissig, S., Schiml, S., Schindele, P., Weiss, O., Rutten, T., et al., (2017). Live-cell CRISPR imaging in plants reveals dynamic telomere movements. *The Plant Journal: For Cell and Molecular Biology*, 91, 565-573.

[50] Komor, A. C., Kim, Y. B., Packer, M. S., Zuris, J. A., and Liu, D. R. (2016). Programmable editing of a target base in genomic DNA without double-stranded DNA cleavage. *Nature*, 533, 420-424.

[51] Hess, G. T., Tycko, J., Yao, D., and Bassik, M. C. (2017). Methods and Applications of CRISPR-Mediated Base Editing in Eukaryotic Genomes. *Molecular Cell*, 68, 26-43.

[52] Li, Z., Xiong, X., and Li, J. F. (2018). New cytosine base editor for plant genome editing. *Science China. Life sciences*, 61, 1602-1603.

[53] Zong, Y., Song, Q., Li, C., Jin, S., Zhang, D., et al. (2018). Efficient C-to-T base editing in plants using a fusion of nCas9 and human APOBEC3A. *Nature Biotechnology*.

[54] Gaudelli, N. M., Komor, A. C., Rees, H. A., Packer, M. S., Badran, A. H., et al. (2017). Programmable base editing of A*T to G*C in genomic DNA without DNA cleavage. *Nature*, 551, 464-471.

[55] Li, C., Zong, Y., Wang, Y., Jin, S., Zhang, D., et al. (2018). Expanded base editing in rice and wheat using a Cas9-adenosine deaminase fusion. *Genome Biology*, 19, 59.

[56] He, Y., Zhu, M., Wang, L., Wu, J., Wang, Q., et al. (2018). Programmed Self-Elimination of the CRISPR/Cas9 Construct Greatly Accelerates the Isolation of Edited and Transgene-Free Rice Plants. *Molecular Plant*, 11, 1210-1213.

[57] Zhang, Y., Liang, Z., Zong, Y., Wang, Y., Liu, J., et al. (2016). Efficient and transgene-free genome editing in wheat through transient expression of CRISPR/Cas9 DNA or RNA. *Nature Communications*, 7, 12617.

[58] Andersson, M., Turesson, H., Olsson, N., Falt, A. S., Ohlsson, P., et al. (2018). Genome editing in potato via CRISPR-Cas9 ribonucleoprotein delivery. *Physiologia Plantarum*, 164, 378-384.

[59] Lin, C. S., Hsu, C. T., Yang, L. H., Lee, L. Y., Fu, J. Y., et al. (2018). Application of protoplast technology to CRISPR/Cas9 mutagenesis: from single-cell mutation detection to mutant plant regeneration. *Plant Biotechnology Journal*, 16, 1295-1310.

[60] Chen, L., Li, W., Katin-Grazzini, L., Ding, J., Gu, X., et al. (2018). A method for the production and expedient screening of CRISPR/Cas9-mediated non-transgenic mutant plants. *Horticulture Research*, 5, 13.

[61] Liang, Z., Chen, K., Li, T., Zhang, Y., Wang, Y., et al. (2017). Efficient DNA-free genome editing of bread wheat using CRISPR/Cas9 ribonucleoprotein complexes. *Nature Communications*, 8, 14261.

[62] Liang, Z., Chen, K., Zhang, Y., Liu, J., Yin, K., et al. (2018). Genome editing of bread wheat using biolistic delivery of CRISPR/Cas9 *in vitro* transcripts or ribonucleoproteins. *Nature Protocols*, 13, 413-430.

[63] Svitashev, S., Schwartz, C., Lenderts, B., Young, J. K., and Mark Cigan, A. (2016). Genome editing in maize directed by CRISPR-Cas9 ribonucleoprotein complexes. *Nature Communications*, 7, 13274.

[64] Woo, J. W., Kim, J., Kwon, S. I., Corvalan, C., Cho, S. W., et al. (2015). DNA-free genome editing in plants with preassembled CRISPR-Cas9 ribonucleoproteins. *Nature Biotechnology*, 33, 1162-1164.

[65] Malnoy, M., Viola, R., Jung, M. H., Koo, O. J., Kim, S., et al. (2016). DNA-Free Genetically Edited Grapevine and Apple Protoplast Using CRISPR/Cas9 Ribonucleoproteins. *Frontiers in Plant Science*, 7, 1904.

[66] Subburaj, S., Chung, S. J., Lee, C., Ryu, S. M., Kim, D. H., et al. (2016). Site-directed mutagenesis in Petunia x hybrida protoplast system using direct delivery of purified recombinant Cas9 ribonucleoproteins. *Plant Cell Reports*, 35, 1535-1544.

[67] Kim, H., Kim, S. T., Ryu, J., Kang, B. C., Kim, J. S., et al. (2017). CRISPR/Cpf1-mediated DNA-free plant genome editing. *Nature Communications*, 8, 14406.

[68] Kim, K., Ryu, S. M., Kim, S. T., Baek, G., Kim, D., et al. (2017). Highly efficient RNA-guided base editing in mouse embryos. *Nature Biotechnology*, 35, 435-437.

[69] Rees, H. A., Komor, A. C., Yeh, W. H., Caetano-Lopes, J., Warman, M., et al. (2017). Improving the DNA specificity and applicability of base editing through protein engineering and protein delivery. *Nature Communications*, 8, 15790.

[70] Li, M., Li, X., Zhou, Z., Wu, P., Fang, M., et al. (2016). Reassessment of the Four Yield-related Genes Gn1a, DEP1, GS3, and IPA1 in Rice Using a CRISPR/Cas9 System. *Frontiers in Plant Science*, 7, 377.

[71] Liu, J., Chen, J., Zheng, X., Wu, F., Lin, Q., et al. (2017). GW5 acts in the brassinosteroid signalling pathway to regulate grain width and weight in rice. *Nature Plants*, 3, 17043.

[72] Lu, K., Wu, B., Wang, J., Zhu, W., Nie, H., et al. (2018). Blocking amino acid transporter OsAAP3 improves grain yield by promoting outgrowth buds and increasing tiller number in rice. *Plant Biotechnology Journal*, 16, 1710-1722.

[73] Zhang, Y., Li, D., Zhang, D., Zhao, X., Cao, X., et al. (2018). Analysis of the functions of TaGW2 homoeologs in wheat grain weight and protein content traits. *The Plant Journal: For Cell and Molecular Biology*, 94, 857-866.

[74] Xu, R., Yang, Y., Qin, R., Li, H., Qiu, C., et al. (2016). Rapid improvement of grain weight via highly efficient CRISPR/Cas9-mediated multiplex genome editing in rice. *Journal of Genetics and Genomics*, 43, 529-532.

[75] Ma, X., Zhang, Q., Zhu, Q., Liu, W., Chen, Y., et al. (2015). A Robust CRISPR/Cas9 System for Convenient, High-Efficiency Multiplex Genome Editing in Monocot and Dicot Plants. *Molecular Plant*, 8, 1274-1284.

[76] Zhang, J., Zhang, H., Botella, J. R., and Zhu, J. K. (2018). Generation of new glutinous rice by CRISPR/Cas9-targeted mutagenesis of the Waxy gene in elite rice varieties. *Journal of Integrative Plant Biology*, 60, 369-375.

[77] Waltz, E. (2016). CRISPR-edited crops free to enter market, skip regulation. *Nature Biotechnology*, 34, 582.

[78] Sun, Y., Jiao, G., Liu, Z., Zhang, X., Li, J., et al. (2017). Generation of High-Amylose Rice through CRISPR/Cas9-Mediated Targeted Mutagenesis of Starch Branching Enzymes. *Frontiers in Plant Science*, 8, 298.

[79] Sanchez-Leon, S., Gil-Humanes, J., Ozuna, C. V., Gimenez, M. J., Sousa, C., et al. (2018). Low-gluten, nontransgenic wheat engineered with CRISPR/Cas9. *Plant Biotechnology Journal*, 16, 902-910.

[80] Morineau, C., Bellec, Y., Tellier, F., Gissot, L., Kelemen, Z., et al. (2017). Selective gene dosage by CRISPR-Cas9 genome editing in hexaploid Camelina sativa. *Plant Biotechnology Journal*, 15, 729-739.

[81] Okuzaki, A., Ogawa, T., Koizuka, C., Kaneko, K., Inaba, M., et al. (2018). CRISPR/Cas9-mediated genome editing of the fatty acid desaturase 2 gene in Brassica napus. *Plant Physiology and Biochemistry: PPB*, 131, 63-69.

[82] Ito, Y., Nishizawa-Yokoi, A., Endo, M., Mikami, M., and Toki, S. (2015). CRISPR/Cas9-mediated mutagenesis of the RIN locus that regulates tomato fruit ripening. *Biochemical and Biophysical Research communications*, 467, 76-82.

[83] Li, R., Fu, D., Zhu, B., Luo, Y., and Zhu, H. (2018). CRISPR/Cas9-mediated mutagenesis of lncRNA1459 alters tomato fruit ripening. *The Plant Journal : For Cell and Molecular Biology*, 94, 513-524.

[84] Li, X., Wang, Y., Chen, S., Tian, H., Fu, D., et al. (2018). Lycopene Is Enriched in Tomato Fruit by CRISPR/Cas9-Mediated Multiplex Genome Editing. *Frontiers in Plant Science*, 9, 559.

[85] Nonaka, S., Arai, C., Takayama, M., Matsukura, C., and Ezura, H. (2017). Efficient increase of -aminobutyric acid (GABA). content in tomato fruits by targeted mutagenesis. *Scientific Reports*, 7, 7057.

[86] Nakayasu, M., Akiyama, R., Lee, H. J., Osakabe, K., Osakabe, Y., et al. (2018). Generation of alpha-solanine-free hairy roots of potato by CRISPR/Cas9 mediated genome editing of the St16DOX gene. *Plant Physiology and Biochemistry : PPB*, 131, 70-77.
[87] Wang, Y., Cheng, X., Shan, Q., Zhang, Y., Liu, J., et al. (2014). Simultaneous editing of three homoeoalleles in hexaploid bread wheat confers heritable resistance to powdery mildew. *Nature Biotechnology*, 32, 947-951.
[88] Nekrasov, V., Wang, C., Win, J., Lanz, C., Weigel, D., et al. (2017). Rapid generation of a transgene-free powdery mildew resistant tomato by genome deletion. *Scientific Reports*, 7, 482.
[89] Wang, F., Wang, C., Liu, P., Lei, C., Hao, W., et al. (2016). Enhanced Rice Blast Resistance by CRISPR/Cas9-Targeted Mutagenesis of the ERF Transcription Factor Gene OsERF922. *PloS One*, 11, e0154027.
[90] Zhou, J., Peng, Z., Long, J., Sosso, D., Liu, B., et al. (2015). Gene targeting by the TAL effector PthXo2 reveals cryptic resistance gene for bacterial blight of rice. *The Plant Journal: For Cell and Molecular Biology*, 82, 632-643.
[91] Macovei, A., Sevilla, N. R., Cantos, C., Jonson, G. B., Slamet-Loedin, I., et al. (2018). Novel alleles of rice eIF4G generated by CRISPR/Cas9-targeted mutagenesis confer resistance to Rice tungro spherical virus. *Plant Biotechnology Journal*, 16, 1918-1927.
[92] Chandrasekaran, J., Brumin, M., Wolf, D., Leibman, D., Klap, C., et al. (2016). Development of broad virus resistance in non-transgenic cucumber using CRISPR/Cas9 technology. *Molecular Plant Pathology*, 17, 1140-1153.
[93] Iqbal, Z., Sattar, M. N., and Shafiq, M. (2016). CRISPR/Cas9: A Tool to Circumscribe Cotton Leaf Curl Disease. *Frontiers in Plant Science*, 7, 475.
[94] Lu, H. P., Luo, T., Fu, H. W., Wang, L., Tan, Y. Y., et al. (2018). Resistance of rice to insect pests mediated by suppression of serotonin biosynthesis. *Nature Plants*, 4, 338-344.
[95] Nieves-Cordones, M., Mohamed, S., Tanoi, K., Kobayashi, N. I., Takagi, K., et al. (2017). Production of low-Cs(+). rice plants by inactivation of the K(+) transporter OsHAK1 with the CRISPR-Cas system. *The Plant Journal : For Cell and Molecular Biology*, 92, 43-56.
[96] Tang, L., Mao, B., Li, Y., Lv, Q., Zhang, L., et al. (2017). Knockout of OsNramp5 using the CRISPR/Cas9 system produces low Cd-accumulating indica rice without compromising yield. *Scientific Reports*, 7, 14438.
[97] Zhou, H., He, M., Li, J., Chen, L., Huang, Z., et al. (2016). Development of Commercial Thermo-sensitive Genic Male Sterile Rice Accelerates Hybrid Rice Breeding Using the CRISPR/Cas9-mediated TMS5 Editing System. *Scientific Reports*, 6, 37395.
[98] Li, J., Zhang, H., Si, X., Tian, Y., Chen, K., et al. (2017). Generation of thermosensitive male-sterile maize by targeted knockout of the ZmTMS5 gene. *Journal of Genetics and Genomics*, 44, 465-468.
[99] Li, Q., Zhang, D., Chen, M., Liang, W., Wei, J., et al. (2016). Development of japonica Photo-Sensitive Genic Male Sterile Rice Lines by Editing Carbon Starved Anther Using CRISPR/Cas9. *Journal of Genetics and Genomics*, 43, 415-419.
[100] Singh, M., Kumar, M., Albertsen, M. C., Young, J. K., and Cigan, A. M. (2018). Concurrent modifications in the three homeologs of Ms45 gene with CRISPR-Cas9 lead to rapid generation of male sterile bread wheat (Triticum aestivum L.). *Plant Molecular Biology*, 97, 371-383.
[101] Xie, Y., Niu, B., Long, Y., Li, G., Tang, J., et al. (2017). Suppression or knockout of SaF/SaM overcomes the Sa-mediated hybrid male sterility in rice. *Journal of Integrative Plant Biology*, 59, 669-679.
[102] Xie, Y., Xu, P., Huang, J., Ma, S., Xie, X., et al. (2017). Interspecific Hybrid Sterility in Rice Is Mediated by OgTPR1 at the S1 Locus Encoding a Peptidase-like Protein. *Molecular Plant*, 10, 1137-1140.
[103] Shen, R., Wang, L., Liu, X., Wu, J., Jin, W., et al. (2017). Genomic structural variation-mediated allelic suppression causes hybrid male sterility in rice. *Nature Communications*, 8, 1310.
[104] Khanday, I., Skinner, D., Yang, B., Mercier, R., and Sundaresan, V. (2019). A male-expressed rice embryogenic trigger redirected for asexual propagation through seeds. *Nature*, 565, 91-95.
[105] Wang, C., Liu, Q., Shen, Y., Hua, Y., Wang, J., et al. (2019). Clonal seeds from hybrid rice by simultaneous genome engineering of meiosis and fertilization genes. *Nature Biotechnology*, 37, 283-286.

[106] Dong, L., Li, L., Liu, C., Liu, C., Geng, S., et al. (2018). Genome Editing and Double-Fluorescence Proteins Enable Robust Maternal Haploid Induction and Identification in Maize. *Molecular Plant*, 11, 1214-1217.

[107] Yao, L., Zhang, Y., Liu, C., Liu, Y., Wang, Y., et al. (2018). OsMATL mutation induces haploid seed formation in indica rice. *Nature Plants*, 4, 530-533.

[108] Li, X., Zhou, W., Ren, Y., Tian, X., Lv, T., et al. (2017). High-efficiency breeding of early-maturing rice cultivars via CRISPR/Cas9-mediated genome editing. *Journal of Genetics and Genomics = Yi chuan xue bao*, 44, 175-178.

[109] Braatz, J., Harloff, H. J., Mascher, M., Stein, N., Himmelbach, A., et al. (2017). CRISPR-Cas9 Targeted Mutagenesis Leads to Simultaneous Modification of Different Homoeologous Gene Copies in Polyploid Oilseed Rape (Brassica napus). *Plant Physiology*, 174, 935-942.

[110] Ye, M., Peng, Z., Tang, D., Yang, Z., Li, D., et al. (2018). Generation of self-compatible diploid potato by knockout of S-RNase. *Nature Plants*, 4, 651-654.

[111] Luo, M., Gilbert, B., and Ayliffe, M. (2016). Applications of CRISPR/Cas9 technology for targeted mutagenesis, gene replacement and stacking of genes in higher plants. *Plant Cell Reports*, 35, 1439-1450.

[112] Butler, N. M., Baltes, N. J., Voytas, D. F., and Douches, D. S. (2016). Geminivirus-Mediated Genome Editing in Potato (Solanum tuberosum L.) Using Sequence-Specific Nucleases. *Frontiers in Plant Science*, 7, 1045.

[113] Dahan-Meir, T., Filler-Hayut, S., Melamed-Bessudo, C., Bocobza, S., Czosnek, H., et al. (2018). Efficient in planta gene targeting in tomato using geminiviral replicons and the CRISPR/Cas9 system. *The Plant Journal : For Cell and Molecular Biology*, 95, 5-16.

[114] Hummel, A. W., Chauhan, R. D., Cermak, T., Mutka, A. M., Vijayaraghavan, A., et al. (2018). Allele exchange at the EPSPS locus confers glyphosate tolerance in cassava. *Plant Biotechnology Journal*, 16, 1275-1282.

[115] Li, Z., Liu, Z. B., Xing, A., Moon, B. P., Koellhoffer, J. P., et al. (2015). Cas9-Guide RNA Directed Genome Editing in Soybean. *Plant Physiology*, 169, 960-970.

[116] Endo, M., Mikami, M., and Toki, S. (2016). Biallelic Gene Targeting in Rice. *Plant Physiology*, 170, 667-677.

[117] Sun, Y., Zhang, X., Wu, C., He, Y., Ma, Y., et al. (2016). Engineering Herbicide-Resistant Rice Plants through CRISPR/Cas9-Mediated Homologous Recombination of Acetolactate Synthase. *Molecular Plant*, 9, 628-631.

[118] Sauer, N. J., Narvaez-Vasquez, J., Mozoruk, J., Miller, R. B., Warburg, Z. J., et al. (2016). Oligonucleotide-Mediated Genome Editing Provides Precision and Function to Engineered Nucleases and Antibiotics in Plants. *Plant Physiology*, 170, 1917-1928.

[119] Tian, S., Jiang, L., Cui, X., Zhang, J., Guo, S., et al. (2018). Engineering herbicide-resistant watermelon variety through CRISPR/Cas9-mediated base-editing. *Plant Cell Reports*, 37, 1353-1356.

[120] Kang, B. C., Yun, J. Y., Kim, S. T., Shin, Y., Ryu, J., et al. (2018). Precision genome engineering through adenine base editing in plants. *Nature Plants*, 4, 427-431.

[121] Peng, A., Chen, S., Lei, T., Xu, L., He, Y., et al. (2017). Engineering canker-resistant plants through CRISPR/Cas9-targeted editing of the susceptibility gene CsLOB1 promoter in citrus. *Plant Biotechnology Journal*, 15, 1509-1519.

[122] Piatek, A., Ali, Z., Baazim, H., Li, L., Abulfaraj, A., et al. (2015). RNA-guided transcriptional regulation in planta via synthetic dCas9-based transcription factors. *Plant Biotechnology Journal*, 13, 578-589.

[123] Rodriguez-Leal, D., Lemmon, Z. H., Man, J., Bartlett, M. E., and Lippman, Z. B. (2017). Engineering Quantitative Trait Variation for Crop Improvement by Genome Editing. *Cell*, 171, 470-480 e478.

[124] von Arnim, A. G., Jia, Q., and Vaughn, J. N. (2014). Regulation of plant translation by upstream open reading frames. *Plant Science : An International Journal of Experimental Plant Biology*, 214, 1-12.

[125] Zhu, B., Zhang, W., Zhang, T., Liu, B., and Jiang, J. (2015). Genome-Wide Prediction and Validation of Intergenic Enhancers in Arabidopsis Using Open Chromatin Signatures. *The Plant Cell*, 27, 2415-2426.

[126] Li, W., Ma, M., Feng, Y., Li, H., Wang, Y., et al. (2015). EIN2-directed translational regulation of ethylene signaling in Arabidopsis. *Cell*, 163, 670-683.

[127] Zaidi, S. S., Tashkandi, M., Mansoor, S., and Mahfouz, M. M. (2016). Engineering Plant Immunity: Using CRISPR/Cas9 to Generate Virus Resistance. *Frontiers in Plant Science*, 7, 1673.

[128] Baltes, N. J., Hummel, A. W., Konecna, E., Cegan, R., Bruns, A. N., et al. (2015). Conferring resistance to gemini viruses with the CRISPR-Cas prokaryotic immune system. *Nature Plants*, 1.

[129] Price, A. A., Sampson, T. R., Ratner, H. K., Grakoui, A., and Weiss, D. S. (2015). Cas9-mediated targeting of viral RNA in eukaryotic cells. *Proceedings of the National Academy of Sciences of the United States of America*, 112, 6164-6169.

[130] Zhang, T., Zheng, Q., Yi, X., An, H., Zhao, Y., et al. (2018). Establishing RNA virus resistance in plants by harnessing CRISPR immune system. *Plant Biotechnology Journal*, 16, 1415-1423.

[131] Nayak, A., Tassetto, M., Kunitomi, M., and Andino, R. (2013). RNA interference-mediated intrinsic antiviral immunity in invertebrates. *Current Topics in Microbiology and Immunology*, 371, 183-200.

[132] Jacobs, T. B., Zhang, N., Patel, D., and Martin, G. B. (2017). Generation of a Collection of Mutant Tomato Lines Using Pooled CRISPR Libraries. *Plant Physiology*, 174, 2023-2037.

[133] Nemhauser, J. L., and Torii, K. U. (2016). Plant synthetic biology for molecular engineering of signalling and development. *Nature Plants*, 2, 16010.

[134] Temme, K., Zhao, D., and Voigt, C. A. (2012). Refactoring the nitrogen fixation gene cluster from Klebsiella oxytoca. *Proceedings of the National Academy of Sciences of the United States of America*, 109, 7085-7090.

[135] Sedbrook, J. C., Phippen, W. B., and Marks, M. D. (2014). New approaches to facilitate rapid domestication of a wild plant to an oilseed crop: example pennycress (Thlaspi arvense L.). *Plant Science: An International Journal of Experimental Plant Biology*, 227, 122-132.

[136] Zsogon, A., Cermak, T., Naves, E. R., Notini, M. M., Edel, K. H., et al. (2018). De novo domestication of wild tomato using genome editing. *Nature Biotechnology*.

[137] Lemmon, Z. H., Reem, N. T., Dalrymple, J., Soyk, S., Swartwood, K. E., et al. (2018). Rapid improvement of domestication traits in an orphan crop by genome editing. *Nature Plants*, 4, 766-770.

[138] Gelvin, S. B. (2010). Plant proteins involved in Agrobacterium-mediated genetic transformation. *Annual Review of Phytopathology*, 48, 45-68.

[139] Ran, Y., Liang, Z., and Gao, C. (2017). Current and future editing reagent delivery systems for plant genome editing. *Science China. Life Sciences*, 60, 490-505.

[140] Hamada, H., Linghu, Q., Nagira, Y., Miki, R., Taoka, N., et al. (2017). An in planta biolistic method for stable wheat transformation. *Scientific Reports*, 7, 11443.

[141] Mitter, N., Worrall, E. A., Robinson, K. E., Li, P., Jain, R. G., et al. (2017). Clay nanosheets for topical delivery of RNAi for sustained protection against plant viruses. *Nature Plants*, 3, 16207.

[142] Cunningham, F. J., Goh, N. S., Demirer, G. S., Matos, J. L., and Landry, M. P. (2018). Nanoparticle-Mediated Delivery towards Advancing Plant Genetic Engineering. *Trends in Biotechnology*, 36, 882-897.

[143] Feng, Z., Mao, Y., Xu, N., Zhang, B., Wei, P., et al. (2014). Multigeneration analysis reveals the inheritance, specificity, and patterns of CRISPR/Cas-induced gene modifications in Arabidopsis. *Proceedings of the National Academy of Sciences of the United States of America*, 111, 4632-4637.

[144] Tang, X., Liu, G., Zhou, J., Ren, Q., You, Q., et al. (2018). A large-scale whole-genome sequencing analysis reveals highly specific genome editing by both Cas9 and Cpf1 (Cas12a) nucleases in rice. *Genome Biology*, 19, 84.

[145] van Kregten, M., de Pater, S., Romeijn, R., van Schendel, R., Hooykaas, P. J., et al. (2016). T-DNA integration in plants results from polymerase-theta-mediated DNA repair. *Nature Plants*, 2, 16164.

[146] McVey, M., Khodaverdian, V. Y., Meyer, D., Cerqueira, P. G., and Heyer, W. D. (2016) Eukaryotic DNA Polymerases in Homologous Recombination. *Annual Review of Genetics*, 50, 393-421.

[147] Voth, D. E., Broederdorf, L. J., and Graham, J. G. (2012) Bacterial Type IV secretion systems: versatile virulence machines. *Future Microbiology*, 7, 241-257.

[148] Tian, X., Gu, T., Patel, S., Bode, A. M., Lee, M. H., et al. (2019). CRISPR/Cas9–An evolving biological tool kit for cancer biology and oncology. *npj Precision Oncology*, 3(8), 1-8

[149] Esvelt, K. M., Smidler, A. L., Catteruccia, F., and Church, G. M. (2014) Concerning RNA-guided gene drives for the alteration of wild populations. *eLife*, 3.

Chapter 3

Recent Progression in New Approaches and Technologies in Plants Using the CRISPR/Cas9 System for Genome Editing

Pooja Kumari
Preeti Sharma
Poornima Pandey
and Narendra Kumar Sharma*
Department of Bioscience and Biotechnology, Banasthali Vidyapith, Tonk Rajasthan, India

Abstract

As the population increases, so does the need for larger quantities and better quality of food. To achieve this, we intend to manipulate crops to increase production by mitigating the effects of numerous biotic and abiotic stressors. With latest improvements during editing of genome, it's now easier to manipulate the genes in many different living beings. The CRISPR system makes it easy and inexpensive for users to target any gene, unlike additional genome editing methods such as TALENs and ZFN. In addition to model plants, this genome editing technique is also applied in other crops like maize, rice, wheat. With the CRISPR system, the quality, quantity and nutritional importance of crops can be increased. Originally, CRISPR-mediated gene targeting was performed in rice, wheat, corn, tomatoes, soybeans, citrus, cotton, and potatoes. There is another genome editing tool known as the mini-CRISPR genome editing system, which acts as molecular scissors that select and cut off stretches of target DNA. Various studies have revealed the application of Cas Mini, which can be used to clear, activate and edit genetic code. The main reason for developing Mini is to make it easier to administer not only to plants but also to humans. Various researchers around the world are working on the CRISPR-Cas system to create crops that protect crops from disease. This has been observed in monocotyledonous and dicotyledonous plant species. This system has an extensive range of applications in plants, animals and medicine.

Keywords: CRISPER, plant genome editing, NHEJ, sgRNAs, tRNA, crRNA

---

* Corresponding Author's Email: snarendrakumar@banasthali.ac.in.

In: Genome Editing for Crop Improvement
Editors: Rahul Kumar, Ashutosh Sharma, Yogesh Kumar et al.
ISBN: 979-8-88697-740-0

Introduction

Food safety is one of the key problems in the current scenario and will continue to be so in the future due to land scarcity and population growth. Therefore, it is very important to manage with these pressures on plants. To manage plant stress, the CRISPR-Cas9 system is currently used, which employs a genetic modification to obtain the desired gene. Genetic modification has been practiced for several years through mutagenesis and through various conventional methods that make it possible to precisely determine the genetic composition of an organism and to change the genome in predictable ways. Due to the unexpected off-target mutations, the CRISPR/Cas9 technology for editing of genome continued more of an experimental approach than for uses like the treatment of genetic illnesses or agricultural improvement. The Cas9-induced DSBs are main causes of the mutation of off-target. The cell's DNA repair machinery repairs these DSBs, but this process usually involves small insertions and deletions that result in unintentional mutations. To achieve this, two new technologies known as base editing and prime editing were developed as an improvement on the current CRISPR/Cas9 technique [1]. Among all gene editing systems, ZFNs is the oldest, which was developed by Sangamo Bioscience in the 1990s but unfortunately was withdrawn after a while due to various limitations including the fact that it is time-consuming and expensive. Meanwhile, the CRISPR-Cas9 system has emerged, which has been widely adopted because it was applied to target an endogenous gene. It also offers the advantage of gene knockdown over the RNAi approach [2]. This technique allows for faster genetic modifications than others due to its low cost, efficiency, simplicity, and ability to target genes of interest or multiple genes. The previously neglected plants are also modified by this technique.

The goal of CRISPR-Cas9 is to modify genes by precisely cutting DNA and repairing it with natural DNA repair. This technique allows for further editing and modification in different crops, as the studies show that the ZFNs and TALENs, both techniques used for the genome mutagenizing process, require two different DNA-binding proteins, making them less acceptable to everyone. In 2012, scientists discovered this novel CRISPER-Cas9 technique for editing the plant genome, but it was first applied to different plants such as wheat and corn in 2013. This system depends on the guide RNA designed and is the more convenient and simple method among the ZFNs and TALENs [3]. Genetic engineering technologies were found as a result of advances in molecular biology. With time as the development of genetic engineering took place, various researchers proposed the application of these tools in the proofreading of genome of plants like Arabidopsis and Tobacco plants were used as models to alter specific gene(s). In addition, these genetic engineering tools help introduce the DSBs (double strand breaks) into the genome. Here below, Figure 1 shows the various genome editing tools currently available, including ZFNs, TALENs, Cas9, cpf1, and base editors. Normally, ZFNs and TALENs generate sequence-specific recognition in target DNA via protein-DNA interaction, and CRISPR/Cpf1 cleaves both part of target DNAs in a staggered pattern, while Cas9 requires multiple Pol III promoters to drive different g RNAs.

Meanwhile, in 1996, Zinc Fingers Nuclease had revolutionized the area of editing of genome, prompting researchers to open the doors to both basic and applied research. The primary information about Zinc Fingers Nuclease is that it has been used effectively in numerous organisms including plants and has also been used in trait stacking in mice. ZFN proved more effective owing to its high specificity and negligible non-target effects. Similarly, the gene altered by TALEN has been adopted effectively in many plant and animal species and was constructed by fusing the DNA-binding domain from transcription activator-like effectors (TALEs) and foci endonuclease. The comparative study between ZFNs and TALENs has shown that TALENs are considered superior and efficient to ZNF and are widely used for genome editing [4]. Among all genome editing tools, it has been found that they all have some limitations, while CRISPR-Cas9 system emerges as the recent genome editing tool by offering a simple, efficient and precise approach to genome editing.

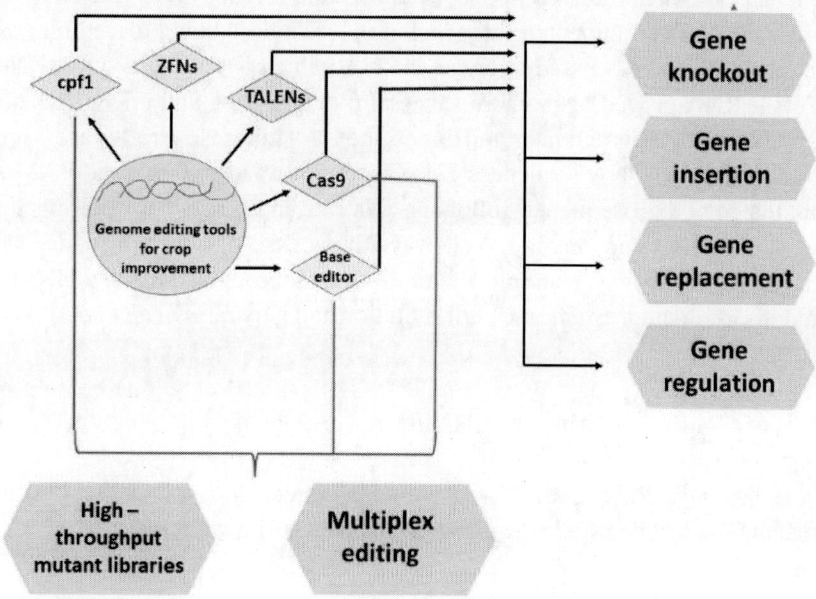

Figure 1. Depicted presently existing genome editing tool.

History

The CRISPR-Cas9 system was first proposed in 1987. It has been found that when bacteria come close to phage DNA, their 32-nucleotide spacer is inserted between CRISPR loci of 29 nucleotides of the repeat sequence. Later in 2005 it was determined that the spacer sequences were not inserted by bacteria. It was generated from the phage genome. Therefore, the year 2005 was recognized as the most significant year for the CRISPR-Cas9 system and the deletion of spacer DNA identical to phage DNA was documented. After the discovery of the Cas genes, many other factors such as PAM (Protospacer Adjacent Motif), CRISPER RNA (crRNA), and trans-activating RNA (trRNA) were uncovered, however, the exact mechanism behind them has not been well characterized later when focusing on the crRNA and trRNA processes discovered. Berkeley and his colleagues conducted this research at the University of California in August 2012 and explained the molecular mechanism.

Genome Editing

The process of editing of genome begins with the identification of target genes, and these target genes help determine the phenotype of interest. As previously mentioned, the CRISPR-Cas9 system relies entirely on the designed guide RNA, and these guide RNAs are further artificially engineered to target the gene to be edited. The CRISPER molecule has a short, regularly spaced sequence and some proteins are present that help in the unwinding and cutting of the DNA sequence. It only requires three components: Cas9, crRNA, and trRNA, and in Type II, the plasmid or viral DNA is fragmented into smaller pieces before being placed into the CRISPR locus. This locus is now processed to create crRNA, which is then joined to trRNA to create sgRNA. This sgRNA proved to be more successful than trRNA and crRNA when it came to targeting gene changes. This activates and directs the Cas9 nuclease. Protospacers are the newly created double-stranded DNA sequences [2]. These protospacers target the PAM site, which usually has the 5-NGG-3 sequence following Watson and Crick base pairing up to 17–20 nucleotides at the gRNA end, leading to cleavage. Now DSB has been introduced to stimulate the DNA repair mechanism as genomic alteration occurs through NHEJ and HR. This random deletion and insertion induces a frame shift mutation and generates gene knockouts.

Mechanism of Genome Editing by CRISPER-Cas9

In bacteria and archaea, these CRISPR-Cas systems protect against invading plasmids or viral DNA elements. The detailed CRISPR-Cas9 mechanism is shown in Figure 2.

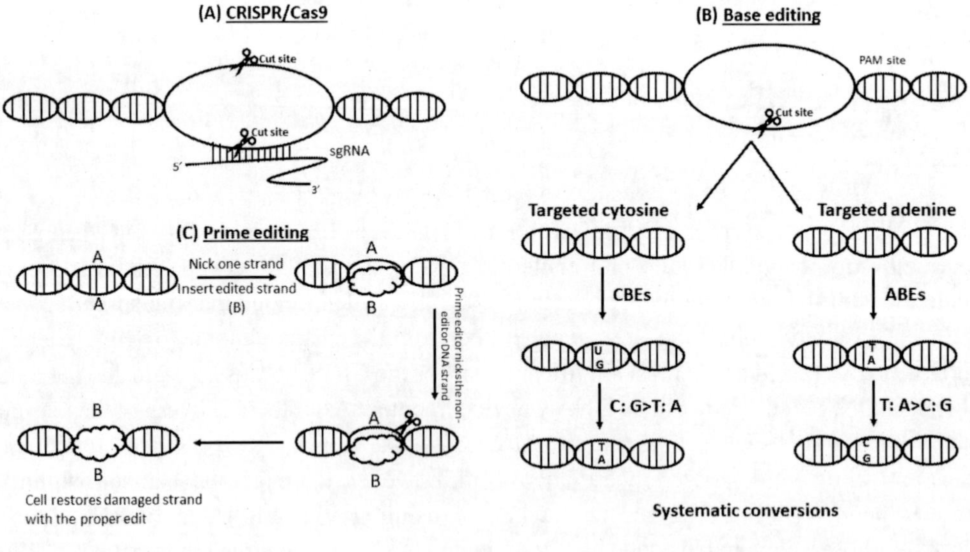

Figure 2. Overview of the mechanism involved in the CRISPR/Cas9-based precise genome editing (Adopted from Li et al., 2020).

CRISPR-Cas9

Here, Figure 2 above shows the CRISPR-Cas9 mediated editing of genome. This method is further classified into six categories depending upon the variety of Cas genes and the characteristics of the interference complex, three of which are based on mature crRNA, tracrRNA, and *Cas 9*, which are primarily in charge of cleaving the invasive element in type II CRISPR/Cas systems. In addition, tracrRNA and crRNA were combined to create a single guide RNA (sgRNA) that was used to direct the production of DSB by Cas9 *in vitro*. Sg RNA and Cas9 endonuclease are two crucial elements in this system. The second one is Cas9 endonuclease, which is an RNA guided endonuclease and is accountable for detecting the target DNA in the first instance.The existence of a protospacer-adjacent motif (PAM; NGG for SpCas9) next to the series of interest is the only requirement for deploying CRISPR-Cas9 to that location. We could only vary the sequence of guide in sgRNA.

Base Editing

Base editing which is analogous to the CRISPR-Cas system is a precise genome editing method. The base editing system makes directed point mutations without DSB, HR and foreign DNA templates.The majority of the current base editing methods include sgRNA as well as Cas9 nucleases with catalytically decreased activity or Cas9 nickases coupled to ssDNA deaminases. The modified Cas9-deaminase is directed to the target locus by the sgRNA, exposing and making the ssDNA R-loop accessible to the deaminase. According to many varieties of deaminase, there are two types of DNA base editors: one is adenine base editors (ABEs) and other one is cytidine base editors (ABEs). Within the editing window of CBE, cytidine is converted to uridine by cytidine deaminase, resulting with a mismatching base pair with guanine on the other side of strand. As we all know, uridine intermediates show mutagenesis so most organisms evolve themselves with a uracil base excision repair mechanism to excise U from genomic DNA with the help of Uracil DNA N glycosylase.In adenine base editing dCas9/nCas9 fused with adenosine deaminase and form ABE. This method was first applied on yeast and human cells before being transferred to many different plants, including maize, wheat, rice tomatoes, and arabidopsis. A desirable and effective way to hasten crop improvement is through the prudent editing of a single nucleotide in the plant. All four transition mutations (C to T, A to G, T to C, and G to A) at the targeted location are effectively mediated by the CBEs and ABEs.

Prime Editing

Here in this schematic view, the prime edit shows the main processing system, also borrowed from CRISPR-Cas. The prime editors consist mainly of Catalytically deficient Cas9 combined with modified reverse transcriptase and prime editing guide RNA make up the majority of the prime editors (pregRNA). This complex binds to the desired DNA location, nicks the strand that contains PAM, and produces a 3' end. The reverse transcription of the template with the necessary alterations is then primed after it attaches to the primer binding site of the pegRNA.In this manner, DNA gets stably edited. This mechanism does not need DSB and donor DNA templates, making it a potent and efficient tool for editing of genome [6].

Applications

While we are all aware that in the age of genetically modified crop techniques, CRISPR-Cas9 has become the most important technology among all others. There are still many effective CRISPR-Cas9 applications yet to be disclosed. Recently, as the whole world was dealing with COVID 19, CRISPR-Cas9 once again shows its value by trying to perform a successful experimental test to detect the novel coronavirus at this unfortunate moment [7]. The CRISPR-Cas9 technique has various applications in the field of biological sciences, such as the endogenous gene expression regulation, editing of genome, and the marking of living cells for chromosomal loci, high-throughput gene screening and use for chromatin immune-precipitation. It also takes over *in-vitro* mutation process used in tomatoes to edit recessive mutations. It is mainly used to enhance the yielding capacity (excellence, amount and nutritional value) in monocot and dicot species. It is able to change any gene of interest in different plant species.

This technique has wide application in animals and humans. Another application of this system is that we can simultaneously add multiple targeting genes by (non-homologous end-joining) NHEJ gene knockout. Gene targeting can also be done by the CRISPR-Cas9 system however, *via* the HRD pathway that uses homologous donor DNA to repair DNA damage (Figure 3).

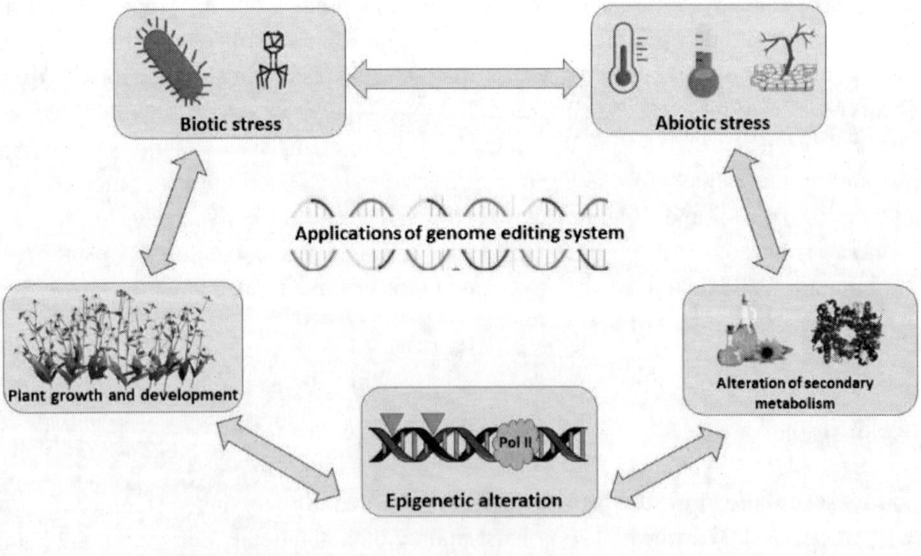

Figure 3. Application of CRISPR/Cas genome editing system.

This mechanism was explored in tomato genomic editing. The other application is their off-target effect. There are many *in-vivo* and *in-vitro* methods used to detect off-target mutation. It will also help produce stress-free plants. This technique is also used in metabolic engineering, where plant cells are selected for the production of specialized metabolites [8]. Some of the startling cases, such as thermos-sensitive genetic male sterility in corn and wheat, have attracted a great deal of interest. Endogenous gene expression was also regulated using CRISPR-Cas9 technology.

In Figure 3, different types of applications of this system that has recently been used to improve plants are depicted. Targeting an individual DNA sequence can be employed to add and/or modify expression of an individual gene that affects plant tolerance to abiotic and biotic pressures and plant growth and development. CRISPR-Cas can also alter secondary metabolism which regulates the biosynthesis of proteins, carbohydrates, oils, and functional components. As we all know, agriculture is subject to a variety of stresses such as drought, flood, heat and cold which we call climatic or abiotic pressure, instead of biological pressures including bacterial, viral and fungal infections in plants.

One reason for the emergence of these pressures is the use of pesticides on crops. This makes supplying the population a major challenge. Many traditional strategies are still being developed to address these issues, and the most extensively applied technique today is CRISPR-Cas9, which still suggests a brighter prospect for crop development, be it in terms of quality or quantity [9]. This technique has been shown to help with stress tolerance. Drought, cold, heat, flooding and other climatic variability are examples of abiotic stress, and a major concern for crops is uptake of heavy metals and salinity, all of which reduced production. Recently, the International Corn and Wheat Improvement Center (CIMMYT) also reported on drought-resistant corn (zea corn), which requires low soil fertility. This was accomplished by using a repair pathway to replace the ethylene negative response regulator with a constitutive maize promoter. Abiotic stress reduces crop development, quality and quantity of crops, making feeding such a large population somewhat difficult. The excessive accumulation of salt in the crops resulted in crop death and increased extra root osmotic pressure can be extremely detrimental to the crops. These stresses are believed to be toxic to crops [10]. So, the important task is to investigate the root cause of the problem, which CRISPR-Cas9 has successfully done.

In biotic stress, infections caused by fungi, viruses and bacteria are considered biotic stress. *Cucumis sativus* is an example of a virus-resistant plant because it inhibits the function of the eIF4E (eukaryotic translation initiation factor 4E) gene. Cas9/sgRNA was used to target the N and C termini of the eIF4E gene. Similarly, Tungro-spherical virus (RTVS)-resistant rice is also being developed, which was a recessive trait controlled by the translation initiation factor 4 (elF4G) gamma gene. A fungal-pathogenic cultivated tomato (*Oidium neolycopersici*) was also developed using the CRISPR-Cas9 system. Fungi are involved in many diseases such as powdery mildew, as mentioned above, smut and rot, which damage both the crop and its quality [7]. *Botrytis cinerea* is a plant pathogen that causes a gray mold disease that results in large crop losses. Various CRISPR-edited crops have been studied by researchers to improve crop yield. Recently, various studies have been experimentally conducted on crop improvement using the CRISPR-Cas system, some of which are listed below.

Wheat

A successful modification in wheat was performed by the CRISPR-Cas9 system. The effect of CRISPR on wheat reduces the amount of naturally occurring asparagines, amino acids and acrylamide. Following this approach, it was then used in many staple foods to increase productivity. The goal of making a modification in wheat was due to the scarcity of crops. The crop is damaged by the disease caused by insects or weeds. The huge amount of wheat stocks were taken by the government every year, indicating the need for wheat for this extreme population. This brings the development of genetically modified wheat to the fore, which can in some way lead to an increment in yield. Wheat is the most highly significant crop grown as a staple food globally. Some researchers have reported that the CRISPR Cas9 gene edition of gluten in wheat detected a reduction in gluten content upon exposure [11]. This reduced gluten content has been framed by the development of wheat lines with lower gluten genes or with inactive CD epitopes. A disease caused by ingesting gluten proteins is known as celiac disease (CD) and the only cure for this disease is a gluten-free diet. So, inactivating CD epitopes somehow prevents the gluten-producing protein from synthesizing it [12]. Apart from that, Cas9 was also found to remove editing of alpha and gamma *gliadin* genes in wheat as well. Thereafter, the edited genes produce plants with unaffected genes. Along with CD, some diseases are induced in wheat by fungi, bacteria and viruses, which led to a reduction in wheat yield and wheat quality. To overcome this, Cas9 was used for gene knockout to produce disease resistant wheat.

Apart from these, some fungal diseases such as g., powdery mildew, are significantly cured when a factor called (enhanced disease resistance1) *EDR1* is present. This *EDR1* gene turns out to be a bad factor in contradiction of powdery mildew [12]. There are also some methods to create CRISPR-edited DNA-free plants. One of them is the vector-based method; the vector is transferred into wheat callus using *Agrobacterium* or bombardment. It then enters the genome and enables targeted gene knockout. This knockout wheat is now self-crossed for separation. Similarly, some non-vector methods also involve transcribing the Cas9 and sgRNAs *in vitro* and then passing them to the immature wheat embryos. *Agrobacterium*-mediated transformation using immature embryos as explants plays a very crucial role in wheat genetic engineering [14]. In wheat, gene editing is performed for two genes associated with abiotic stress, one of the genes responsible for dehydration element-binding protein (*TaDREB2*) and the second is wheat ethylene-responsive factor 3 *(TaERF3)* [15]. After all these discoveries, scientists emphasize that GM wheat is only the direct manipulation of its genome using biotechnology and is not considered a GM crop by the Indian government.

Rice

After wheat, rice is the most important staple food in India. The oleic acid found in rice is beneficial to health and also helps prevent disease. Rice quality is improved by using the CRISPR-Cas9 system, which includes changing the oil and starch composition. The accumulation of heavy metals is also a major cause of some diseases we got from rice, CRISPR-Cas9 has proven to successfully overcome these limitations. Various studies found that the CRISPR-*Cas* system in rice increased the size and weight of the grain. Another study reported that the *FAD2* gene is a tri-functional gene present in rice as *OsFAD2-1* [16]. By disrupting these three functional *FAD2* genes, they obtained oleic acid content. In a new study, the CRISPR-Cas9 system was applied to rice. After disrupting these genes, they obtained *OsFAD-1* knockout rice plants that showed double oleic acid production. The rice that has been in the limelight recently has been fragrant rice because of its excellent market value and high quality. The main reason for the odor in rice is due to betaine aldehyde dehydrogenase *(Badh2)* [17].

As we all know, rice is a staple food in most of the states of India, so the need for large storage also plays a very important role. CRISPR-Cas9 overcomes this problem by using lipoxygenase. These lipoxygenases (*LOXs*) are responsible for regulating seed shelf life. In Asian countries, people are influenced by rice-tungro disease. The illness is caused by the combination of two viruses called *Tungro bacilliform* and *Tungro spherical*. This technique would be useful to solve this problem also where CRISPR-Cas9 worked on *RTSV*-susceptible rice varieties by targeting eIF4G [18]. At the time of genetic manipulation, some phenotypes have disappeared, in order to keep all phenotypes from generation to generation, the meiosis process is terminated by genetic editing techniques, so that it only goes through mitosis [19] and all chromosomes are now transferred to the generation and no phenotypic disappearance was observed. Rice is a major source of cadmium, and the ingestion of heavy metals could harm the human body, and the elimination of cadmium from rice was done by turning off the genes that transport the metals i.e., *OsNramp5*. These are some rice genes targeted by the CRISPR-Cas9 system, such as *OsDERF1, OsPMS3, OsEPSPS, OsMSH1, and OsMYB5* [19].

Maize

Corn is the third main crop after wheat and rice. According to the 2012 Annual Progress Report, production is expected to be 22.23 million tons. Therefore, hybrid corn seed production was undertaken, resulting in high-yielding corn varieties. Corn is also an important grain crop. In multiplex genome editing, a multiplex vector can introduce a set of spacer guide RNAs into a polycistron. It targets two sets of genes, one for simplex editing (*MADS, MYBR,* and *AP2*) and another for multiplex editing (*RPL, PPR,* and *IncRNA*) [20]. The main source in corn is phytic acid, which is present in its seeds and is not digested by ruminants. To solve this phytic acid problem, these *ZmIPK1A, ZmIPK* and *ZmMRP4* genes in *Z. mays* are targeted by the CRISPR-Cas9 system. Recently, some researchers looked at corn and applied CRISPR-Cas9 to turn normal-sized corn into a waxy version. This waxy maize produces amylopectin starch and this was retained by targeting the Wx locus in *ZC01*. If we talk about the usefulness of the CRISPR-Cas9 system in corn, it was achieved by targeting the albino marker, which is the *Zmzb7* gene present in the protoplast. Zmzb7 was targeted to its eighth exon by designed sgRNA and expression was assisted by a maize U3 promoter. The CRISPR-Cas9 system also targeted male sterility genes. Three guide RNAs were used for this knockout. The first gene targeted the first and second exons and the third the second exon. Keeping diabetics in mind, the variety and quality of corn should be increased, since corn is the only grain eaten for them due to its higher content of vitamins, minerals and fiber, and also has lower fat and sodium content. The fiber helps control the amount of sugar in the blood and can reduce the risk of further complications. Another variant of corn is BT corn, which has been genetically modified to express one or more proteins from the bacterium *Bacillus thuringiensis*.

The CRISPR-Cas9 system has also been implemented on other crops such as barley, banana, etc. In barley, the Cas9 knockout ENGase (endo-N-acetyl-b-D-glucosaminidase) and a set of five designed guide RNAs were used. Similarly, in bananas, it targets RAS-PDS (phytoene desaturase rasthali), resulting in thirteen mutant lineages targeted for carotenoid and chlorophyll content [21]. Recently, many other plants are being developed for genetic modification, and scientists have used the CRISPR-Cas9 system in cereal crops due to the growing demand for these plants. Consequently, better features and amount of crops are needed in the market. The value of such crops is sometimes higher because of the inadequacies of unedited crops. Scientists are still working on next-level CRISPR-Cas9 modification techniques such as Crispr 3.0 [22].

Cotton

Cotton is a dicot and the successful application of CRISPR-Cas9 to cotton plants demonstrated that Cas9 is present in both monocots and dicots. As we all know, cotton is a fiber crop that is used to make any type of clothing. In addition to being a fiber crop, cotton is also known for its biofuel production. This is due to the oil reserves present in the cottonseed. In 1996, genetically modified cotton was grown in Australia for the very first time. Australia has proven the worth of a developed country because it used this technique for upgrading. About 99.5% of the cotton grown in Australia is genetically modified. Genetically modified cotton was first introduced in India in 2002. It was known as BT cotton, which was produced by applying genetic engineering techniques to *Bacillus thuringiensis* (Bt) bacteria [23]. It has been grown in India for twenty years. Some of the studies found that there is no impact of BT cotton on humans and the environment as well. Another variety of BT cotton has been genetically modified, namely HT-BT cotton. It makes the plant resistant to a commonly used herbicide.

Cotton plants have a very complex genome structure (hetero-tetraploid nature); hence it becomes more difficult to modify. The CRISPR-Cas9 system was used in cotton crops to enhance sgRNA expression using a U6 promoter. After deploying the CRISPR-Cas9 system in the cotton plant, it was found to have a very high efficiency with no off-target effects. Genome editing in cotton allows targeting of two guide RNAs. One is for chloroplast (*GhCLA1*) and the other for the vacuolar gene (*GhVP*) [24]. Apart from that, the CRISPR-Cas9 system alters the homologous cotton genes and also some multiple genes. The transgene-free cotton plants can be easily obtained by self-pollination or backcrossing in the CRISPR-Cas9 system. For better cotton plant development, some methods need to be modified, including pollen tube pathway-mediated method, biolistic particle delivery system, and *Agrobacterium*-mediated genetic transformation. The bollworm is a major pest in cotton crops that can lead to the introduction of genetically modified cotton. The capsule worm arises from the capsules being eaten by their larvae from the base and also injuring the squares resulting in crop distortion. The goal of modifying cotton is not to increase yield, but to limit damage [25]. Reducing crop damage automatically increases yield. This genetically modified cotton has been shown to bring two of the benefits to farmers. One is that when the bollworm did not affect the crop or damage the seed pods, and eventually crop growth or yield was not affected, and the other, if there is no damage to the seed pods or crop, it means there is no need for insecticides at all [22].

Potato

The main problem for growers and industry is the enzymatic browning of potatoes. This problem is controlled by using chemicals that do not harm in any way. Enzymatic browning is the process in which the oxidation of phenolic compounds occurs by polyphenol oxidase enzymes. These PPOs can self-polymerize, be in the lead to the creation of a dark coloration [26]. This has also happened when PPOs react with amino acids or free radicals in proteins.

The enzymatic browning changes the taste, color and nutritional value of potatoes. The five PPO genes have been described in potatoes (StPPO). The CRISPR-Cas9 technique is used on the potato species *Solanum tuberosum* L., a heterozygous tetraploid crop. The endogenous U6 promoter plays a very important role in the editing process as it is replaced by the Arabidopsis U6 promoter and the reagents used for editing were in the form of RNPs. For most crops, the most convenient method for the gene transfer process is the *Agrobacterium*-mediated gene transfer process.

Due to the heterozygous nature of potatoes, self-crossing is not an appropriate method [27]. Therefore, some of the traits are lost in the next generation and could be preserved through multiple backcrosses. The NHEJ is a repair mechanism that performs some important deletion and insertion at the breakpoint by rendering the target gene nonfunctional, and similarly the HR works for the insertion of a homologous DNA template at the target site that does not disrupt the traits for generations [28]. A disease called light rot disease is caused by *Phytophthora infestans*, which causes a sharp drop in potato production. Therefore, the focus is on the production of late blight resistant potatoes. To overcome this problem, the genomic editing technique is editing specific amino acids in R genes and also by targeting disease susceptibility genes (S genes). There are many other approaches to potatoes such as herbicide resistance, post-harvest shelf life, starch composition modification, production of potatoes free of steroidal glycoalkaloids (SGAs), etc. [29]. CRISPR-Cas9 can be used in two different ways, either by directly attacking the viral genome or by altering the host plant genome. Both methods are suitable for genetic modification in potatoes.

Grape

The grape is one of the highly significant fruit crops universally. It is primarily applied for wine, juice, and as fresh fruit, but is more affected by stress. For the biotic stress response in grapes, the site VvWRKY52 has been targeted and also increases disease freedom from fungal infections and this site is on the 16th chromosome in grapes. Downy mildew, in turn, is a disease caused by Oomycete *P. viticola*. It is a very significant problem, or we can say fungal diseases, for the grapevine industry [6]. The appearance of albino leaves is due to the targeting of phytoene desaturase (VvPDS). Five types of target sites have been identified in most grapes grown. No off-target mutations at off-target sites were detected in genome editing in grapes. The mutations detected in grapes are 1 bp insertions or 1 to 3 nucleotide deletions. Two main strategies are used for grape transformation, organogenesis and somatic embryogenesis. Among both strategies, the most commonly used grape regeneration strategy is somatic embryogenesis [30].

Soybean

Soybeans are high in fiber and protein, low in saturated fat, cholesterol-free, lactose-free, etc. Two binary vectors are constructed for gene editing in soybeans, i.e., *sgRNAs* and Cas9 and CaMV *35s* are used in both the vectors. The researchers then examined three genes (*Glyma06g14180, Glyma08g02290 and Glyma12g37050*) in soybeans for the activity observed in both the vectors. A different sgRNA was designed for each gene, and then the vectors were transferred into soybean hair roots using *Agrobacterium rhizogenes* infection. As a result, mutations were observed in soybean hairy roots. The CRISPR-Cas9 technique in soybean modification helps to drive the genomic action especially in the roots and tubers. The U6 promoter is mostly used for the expression of various plants. These U6 promoters have an upstream sequence element (USE RTCCCACATCG) and a TATA-like box, and for *U6* gene transcription these elements are necessary for genetic engineering [31].

Conclusion

Under these drastic climatic conditions and with many plant diseases, it becomes very challenging to offer the population with better food. CRISPR-Cas9 is the most frequently used technique today. It rendered some plants stress-free and some disease-resistant, including both monocots and dicots. It has improved the productivity, quality, quantity and taste of the crops. CRISPR-Cas9 has proven to be the best among all other genome editing techniques due to its working efficiency, and the latest breeding techniques have given scientists more scope for conventional breeding. The simplest thing about CRISPR-Cas9 is that we can target specific genes of interest and edit or modify our RNA [32]. These new techniques quickly insert and delete desired features. From this it can be concluded that genome editing plays a very important role in crop improvement. Further epigenomic editing and DNA tagging have been reported but do not apply to plants. Therefore, it becomes interesting to see epigenomic marking in plants via fluorescently labelled Cas9 protein. There are some promising experiments underway that could offer plants a brighter future. As we all know, genetically modifying potatoes is not a bone breaker due to their complicated structure. This successful modification of the potato by CRISPR-Cas9 gives us future scope for improving a crop with such a complex structure. The technique has high efficiency and can improve the range of other desirable features. The off-targets affect priors as they can cause unwanted mutations [33]. Transgenic-free wheat plays a very important role in environmental issues. It will certainly give a new direction to the orthogonal gene-targeting system. This orthogonal gene targeting system is still being tested in plant systems. Beyond genome editing, the CRISPR-Cas9 system provides some additional information about gene functions.

Recent studies indicate that CRISPR-P 2.0 [34] is an improved version of the CRISPR-Cas9 genome editing tool in plants. The endogenous genes are modified to improve characteristics in genetic engineering processes. The most nutritious fruit crops such as apples, grapes, oranges, bananas, etc. have also been modified by CRISPR-Cas9 technologies. It sometimes becomes more difficult to improve the nutritional quality of fruit due to the complex genomic structure, slow growth cycle and availability of less genetic information. No exogenous DNA was found in the plant modified by CRISPR-Cas9 techniques [35]. It has certain distinctive qualities like simplicity of usage, broad applicability, and great efficacy. Recently, the genome sequence of certain medicinal plants was completed. Therefore, research on these plants is required to identify the genes responsible for plant toxicity and illness. The CRISPR-Cas9 method cleans the plants of their harmful elements. The quality of the CRISPR-Cas9 genes will be completely changed by the next research. Various plants have been modified using the CRISPR-Cas9 method, although there are still certain difficulties. Off-target rates and mechanism minimization are examples of such difficulties. Therefore, additional research is still required to enhance the experimental evaluation of the CRISPR-Cas9 system. Base editing is a brand-new editing method included in genome engineering toolkits. It is an effective technique for genome editing. It is a potent technique for genome editing that enables programmable nucleotide substitutions without the use of a donor template or DSB.

Over the past three years, efficient base editors based on cytidine and adenine deaminase have been developed and used for base editing in both plants and mammals. Reducing the catalytic window and employing Cas9 variants to improve the present CBE and ABE base editors can expand the spectrum of base editing in agricultural plants. These enhanced base editors and mutants of the cytidine deaminase can enhance DNA selectivity and lessen off-target actions [33]. With the aid of extremely accurate base editors, precision breeding may be done on model plants and crops. The field of emerging base editing technology is still in its infancy, and much work needs to be done to enhance, broaden, and raise the efficiency of editing. However, it is a revolutionary editing strategy that could one day be used to precisely alter crops and speed up agricultural progress. There is still more to be done to improve, expand, and increase the efficiency of editing in the realm of new base editing technology. However, it is a ground-breaking editing technique that one day might be applied to precisely alter crops and advance agricultural development.

References

[1] Amritha, P. P., and Jasmine M. Shah. "Can genetic engineering-based methods for gene function identification be eclipsed by genome editing in plants? A comparison of methodologies." *Molecular Genetics and Genomics* 296, no. 3 (2021): 485-500.

[2] Arora, L., and A. Narula. 2017. "Gene Editing and Crop Improvement Using CRISPR-Cas9 System." *Front Plant Sci* 8:1932. doi: 10.3389/fpls.2017.01932.

[3] Pan, C., D. Yang, X. Zhao, Y. Liu, M. Li, L. Ye, M. Ali, F. Yu, A. T. Lamin-Samu, Z. Fei, and G. Lu. 2021. "PIF4 negatively modulates cold tolerance in tomato anthers via temperature-dependent regulation of tapetal cell death." *Plant Cell* 33 (7):2320-2339. doi: 10.1093/plcell/koab120.

[4] Razzaq, A., F. Saleem, M. Kanwal, G. Mustafa, S. Yousaf, H. M. Imran Arshad, M. K. Hameed, M. S. Khan, and F. A. Joyia. 2019. "Modern Trends in Plant Genome Editing: An Inclusive Review of the CRISPR/Cas9 Toolbox." *Int J Mol Sci* 20 (16). doi: 10.3390/ijms20164045.

[5] Cong, L., F. A. Ran, D. Cox, S. Lin, R. Barretto, N. Habib, P. D. Hsu, X. Wu, W. Jiang, L. A. Marraffini, and F. Zhang. 2013. "Multiplex genome engineering using CRISPR/Cas systems." *Science* 339 (6121):819-23. doi: 10.1126/science.1231143.

[6] Li, M. Y., Y. T. Jiao, Y. T. Wang, N. Zhang, B. B. Wang, R. Q. Liu, X. Yin, Y. Xu, and G. T. Liu. 2020. "CRISPR/Cas9-mediated VvPR4b editing decreases downy mildew resistance in grapevine (*Vitis vinifera* L.)." *Hortic Res* 7:149. doi: 10.1038/s41438-020-00371-4.

[7] Bao, A., D. J. Burritt, H. Chen, X. Zhou, D. Cao, and L. P. Tran. 2019. "The CRISPR/Cas9 system and its applications in crop genome editing." *Crit Rev Biotechnol* 39 (3):321-336. doi: 10.1080/07388551.2018.1554621.

[8] Zhu, H., C. Li, and C. Gao. 2020. "Applications of CRISPR-Cas in agriculture and plant biotechnology." *Nat Rev Mol Cell Biol* 21 (11):661-677. doi: 10.1038/s41580-020-00288-9.

[9] Luo, M., B. Gilbert, and M. Ayliffe. 2016. "Applications of CRISPR/Cas9 technology for targeted mutagenesis, gene replacement and stacking of genes in higher plants." *Plant Cell Rep* 35 (7):1439-50. doi: 10.1007/s00299-016-1989-8.

[10] Liu, X., S. Wu, J. Xu, C. Sui, and J. Wei. 2017. "Application of CRISPR/Cas9 in plant biology." *Acta Pharm Sin B* 7 (3):292-302. doi: 10.1016/j.apsb.2017.01.002.

[11] Sanchez-Leon, S., J. Gil-Humanes, C. V. Ozuna, M. J. Gimenez, C. Sousa, D. F. Voytas, and F. Barro. 2018. "Low-gluten, nontransgenic wheat engineered with CRISPR/Cas9." *Plant Biotechnol J* 16 (4):902-910. doi: 10.1111/pbi.12837.

[12] Kim, D., B. Alptekin, and H. Budak. 2018. "CRISPR/Cas9 genome editing in wheat." *Funct Integr Genomics* 18 (1):31-41. doi: 10.1007/s10142-017-0572-x.

[13] Zhang, Y., Y. Bai, G. Wu, S. Zou, Y. Chen, C. Gao, and D. Tang. 2017. "Simultaneous modification of three homoeologs of TaEDR1 by genome editing enhances powdery mildew resistance in wheat." *Plant J* 91 (4):714-724. doi: 10.1111/tpj.13599.

[14] Zhang, Z., L. Hua, A. Gupta, D. Tricoli, K. J. Edwards, B. Yang, and W. Li. 2019. "Development of an Agrobacterium-delivered CRISPR/Cas9 system for wheat genome editing." *Plant Biotechnol J* 17 (8):1623-1635. doi: 10.1111/pbi.13088.

[15] Ahmad, S., X. Wei, Z. Sheng, P. Hu, and S. Tang. 2020. "CRISPR/Cas9 for development of disease resistance in plants: recent progress, limitations and future prospects." *Brief Funct Genomics* 19 (1):26-39. doi: 10.1093/bfgp/elz041.

[16] Abe, K., E. Araki, Y. Suzuki, S. Toki, and H. Saika. 2018. "Production of high oleic/low linoleic rice by genome editing." *Plant Physiol Biochem* 131:58-62. doi: 10.1016/j.plaphy.2018.04.033.

[17] Ashokkumar, S., D. Jaganathan, V. Ramanathan, H. Rahman, R. Palaniswamy, R. Kambale, and R. Muthurajan. 2020. "Creation of novel alleles of fragrance gene OsBADH2 in rice through CRISPR/Cas9 mediated gene editing." *PLoS One* 15 (8):e0237018. doi: 10.1371/journal.pone.0237018.

[18] Wang, W., S. Ma, P. Hu, Y. Ji, and F. Sun. 2021. "Genome Editing of Rice eIF4G Loci Confers Partial Resistance to Rice Black-Streaked Dwarf Virus." *Viruses* 13 (10). doi: 10.3390/v13102100.

[19] Borrelli, V. M. G., V. Brambilla, P. Rogowsky, A. Marocco, and A. Lanubile. 2018. "The Enhancement of Plant Disease Resistance Using CRISPR/Cas9 Technology." *Front Plant Sci* 9:1245. doi: 10. 3389/fpls.2018.01245.

[20] Feng, C., J. Yuan, R. Wang, Y. Liu, J. A. Birchler, and F. Han. 2016. "Efficient Targeted Genome Modification in Maize Using CRISPR/Cas9 System." *J Genet Genomics* 43 (1):37-43. doi: 10.1016/ j.jgg.2015.10.002.

[21] Wang, X., Y. Xu, J. Huang, W. Jin, Y. Yang, and Y. Wu. 2020. "CRISPR-Mediated Knockout of the ABCC2 Gene in Ostrinia furnacalis Confers High-Level Resistance to the *Bacillus thuringiensis* Cry1Fa Toxin." *Toxins (Basel)* 12 (4). doi: 10.3390/toxins12040246.

[22] Zhong, Z., S. Liu, X. Liu, B. Liu, X. Tang, Q. Ren, J. Zhou, X. Zheng, Y. Qi, and Y. Zhang. 2020. "Intron-Based Single Transcript Unit CRISPR Systems for Plant Genome Editing." *Rice (N Y)* 13 (1):8. doi: 10.1186/s12284-020-0369-8.

[23] Gao, W., L. Long, X. Tian, F. Xu, J. Liu, P. K. Singh, J. R. Botella, and C. Song. 2017. "Genome Editing in Cotton with the CRISPR/Cas9 System." *Front Plant Sci* 8:1364. doi: 10.3389/fpls.2017.01364.

[24] Chen, X., X. Lu, N. Shu, S. Wang, J. Wang, D. Wang, L. Guo, and W. Ye. 2017. "Targeted mutagenesis in cotton (Gossypium hirsutum L.) using the CRISPR/Cas9 system." *Sci Rep* 7:44304. doi: 10.1038/srep44304.

[25] Jin, M. H., Y. T. Xiao, Y. Cheng, J. Hu, C. B. Xue, and K. M. Wu. 2019. "Chromosomal deletions mediated by CRISPR/Cas9 in Helicoverpa armigera." *Insect Sci* 26 (6):1029-1036. doi: 10.1111/1744-7917.12570.

[26] Wang, S., S. Zhang, W. Wang, X. Xiong, F. Meng, and X. Cui. 2015. "Efficient targeted mutagenesis in potato by the CRISPR/Cas9 system." *Plant Cell Rep* 34 (9):1473-6. doi: 10.1007/s00299-015-1816-7.

[27] Andersson, M., H. Turesson, N. Olsson, A. S. Falt, P. Ohlsson, M. N. Gonzalez, M. Samuelsson, and P. Hofvander. 2018. "Genome editing in potato via CRISPR-Cas9 ribonucleoprotein delivery." *Physiol Plant* 164 (4):378-384. doi: 10.1111/ppl.12731.

[28] Johansen, I. E., Y. Liu, B. Jorgensen, E. P. Bennett, E. Andreasson, K. L. Nielsen, A. Blennow, and B. L. Petersen. 2019. "High efficacy full allelic CRISPR/Cas9 gene editing in tetraploid potato." *Sci Rep* 9 (1):17715. doi: 10.1038/s41598-019-54126-w.

[29] Kieu, N. P., M. Lenman, E. S. Wang, B. L. Petersen, and E. Andreasson. 2021. "Mutations introduced in susceptibility genes through CRISPR/Cas9 genome editing confer increased late blight resistance in potatoes." *Sci Rep* 11 (1):4487. doi: 10.1038/s41598-021-83972-w.

[30] Wang, X., M. Tu, D. Wang, J. Liu, Y. Li, Z. Li, Y. Wang, and X. Wang. 2018. "CRISPR/Cas9-mediated efficient targeted mutagenesis in grape in the first generation." *Plant Biotechnol J* 16 (4):844-855. doi: 10.1111/pbi.12832.
[31] Jacobs, T. B., P. R. LaFayette, R. J. Schmitz, and W. A. Parrott. 2015. "Targeted genome modifications in soybean with CRISPR/Cas9." *BMC Biotechnol* 15:16. doi: 10.1186/s12896-015-0131-2.
[32] Gupta, S., A. Kumar, R. Patel, and V. Kumar. 2021. "Genetically modified crop regulations: scope and opportunity using the CRISPR-Cas9 genome editing approach." *Mol Biol Rep* 48 (5):4851-4863. doi: 10.1007/s11033-021-06477-9.
[33] Mishra, R., R. K. Joshi, and K. Zhao. 2020. "Base editing in crops: current advances, limitations and future implications." *Plant Biotechnol J* 18 (1):20-31. doi: 10.1111/pbi.13225.
[34] Liu, H., Y. Ding, Y. Zhou, W. Jin, K. Xie, and L. L. Chen. 2017. "CRISPR-P 2.0: An Improved CRISPR-Cas9 Tool for Genome Editing in Plants." *Mol Plant* 10 (3):530-532. doi: 10.1016/j.molp.2017.01.003.
[35] Manghwar, H., K. Lindsey, X. Zhang, and S. Jin. 2019. "CRISPR/Cas System: Recent Advances and Future Prospects for Genome Editing." *Trends Plant Sci* 24 (12):1102-1125. doi: 10.1016/j.tplants.2019.09.006.

Chapter 4

A DNA-Free Genome Editing System

Anuj Sharma
and Karan Paul*
Department of Biochemistry, DAV University, Jalandhar, India

Abstract

Techniques for genome editing have been frequently used for basic research in plant sciences including investigation of gene function as well as for crop improvement via modification of desired traits, resulting in plants with increased tolerance to biotic or abiotic stress, enhanced food quality, herbicide tolerance or improved agronomic characteristics. DNA-free editing allows targeted genetic modifications without disturbing the genome. Classical methods for genome editing involve the delivery of DNA into plant cells followed by its integration into cellular genomes. DNA-free editing makes use of proteins and protein-RNA complex for targeted genetic modifications without disturbing the genome, thereby completely avoiding the integration of foreign DNA into the cellular genome. The development of engineered endonucleases like TALEN and CRISPR/Cas9 has allowed the development of genetically modified plants with recombinant DNA. The chapter elaborates on these techniques including their history, development, applications, and emerging trends.

Keywords: mutagenesis, SSN, NHEJ, HR, ZFNs, CRISPR/Cas9, TALEN, sgRNA

Introduction

Advancement in knowledge of genome and their subsequent sequencing and processing technologieshave made scientific minds more curious to use and administer this vast gain of information forthe betterment of humankind. Since most stress is given upon the food security of their populace by the nation states, the scope of using advanced techniques in improving the growth cycle, yield, and content of food crops is immense. The importance of manipulating the genes encoding these aspects of crop plants is highly stressed upon and the subsequent

* Corresponding Author's Email: karan10206@davuniversity.org.

In: Genome Editing for Crop Improvement
Editors: Rahul Kumar, Ashutosh Sharma, Yogesh Kumar et al.
ISBN: 979-8-88697-740-0

generation of mutants meeting the requirements is the goal. Induced mutagenesis, insertion mutagenesis [1] or RNA interference [2] based splicing are some of the ways to determine the gene function. But these techniques are flooded with shortcomings, are usually time consuming and cost inefficient too.

To overcome these, a need for targeted, precise and efficient method has been felt. Targeted mutagenesis based on sequence-specific nuclease (SSN) induced double strand breaks (DSBs) is one such method involving either non-homologous end joining (NHEJ) or homologous recombination (HR) [3, 4]. Zinc finger nucleases (ZFNs) [5], type II CRISPR/Cas9 system (clustered regularly interspaced short palindromic repeats/ CRISPR associated protein) [6] and transcription activator like effector nucleases (TALENs) [7] are the most widely used plant genome editing tools, at present. The CRISPR/Cas9 system is considered the simplest of them all. Construction of ZFNs and TALENs is complex and time-consuming which limits their application. CRISPR/Cas9 system on the other hand consists of Cas9 protein and a single-guide RNA (sgRNA) [8]. The targeted modification can be achieved by altering thesgRNA sequence accordingly, a process much simple than the other two techniques. The area of plant genome editing has been reinvigorated with the simplistic application of CRISPR/Cas9 system [9-11]. In this study, we discuss the mechanism of action CRISPR/Cas9 system, its simple construction and some it's most practical applications in advancing the cause of food security of all in the face of changing biotic and abiotic stresses as well as some of the future prospects this technique holds owingto its high efficiency and specificity.

Mechanism

Discovered for the first time in *Escherichia coli* in the year 1987 [12], CRISPR/Cas has been reported as an adaptive immunity system. CRISPR is a DNA fragment of non-contiguous repeats interspaced by spacers with CRISPR associated genes (Cas) genes associated with CRISPR loci [13]. A 2005 finding suggested the spacers to be derivatives of invading viruses and plasmids [14, 15]. The orientation andintegration of spacer sequences into the repeated arrays are guided by proto-spacer adjacent motifs (PAMs). Proto-spacers are the spacer precursors whereas the PAMs are short di or tri nucleotides placed at immediate locations surrounding the proto-spacers [16, 17].

A 2011 study suggested a three-stage CRISPR/Cas mediated immune process [18]. The first stage, adaption, is characterized by the integration of plasmid or virus-derived spacer sequences into the repeated arrays. The second stage, expression, is marked by the onset of the transcription and maturation process of crRNAs. The final interference stage is characterized by the destruction of plasmid and/or viral target sequences complementing the spacer sequences. This process is guided by crRNAs. Markova et al. proposed a triple-category CRISPR/Cas system based on the evolution of each type associated with a specific signature protein(s) [18]. The three types differ in their administration of Cas proteins. While type I and III are associated with a complex of Cas proteins as the signature proteins, the type II system is associatedwith a single large Cas9 multifunctional signature protein. Owing to this simplicity type II system is more malleable to function as a genome editing tool.

Construction of the CRISPR/Cas9 System

The simplicity of the type II CRISPR/Cas9 system enables it to be used for targeted gene editing in crop plants. However, the editing efficiency can be prone to certain factors which include the higher genomic GC content of target DNA, codons of Cas9, expression levels of sgRNA (single guide RNA), Cas9 and lower secondary structure content of sgRNAs and target DNA [19]. The sgRNA expression in plants is affected by RNA polymerase III promoters U3 and U6 [19] whereas the maize ubiquitin and cauliflower mosaic virus 35S promoters affect Cas9 expression. The plant usage bias codon optimization of the Cas9 gene has been performed to enable the efficient plant genome targeted editing. However, despite all the efficiency enhancing methods, the presence of off-target effects is a commonissue associated with genome editing techniques. To overcome this, highly specific sgRNAs with a 12 bpsequence have been designed to match with specific site in the genome to ward off the problem of off-targeting [20, 21]. It can be assumed that progressive improvements in the efficiency, specificity and fidelity of Cas9 will lead to a greater reduction of off-target activity. This can be evident from the efficiency of Cas9 nucleases like eSpCas9 and SpCas9-HF1 (modified from native SpCas9) which showcase lowered non-specific DNA editing [22, 23]. The role of temperature in affecting the CRISPR/Cas9 editing efficiency has been probed where the mutagenesis efficiency was observed to be higher in *Arabidopsis thaliana* and *Citrus sinensis* when grown at room temperature as compared to the lower temperature [24]. Some bioinformatics tools such as CRISPR-P 2.0 [25] and CRISPR-GE [26] have been developed to perform a range of operations including off-target analysis and designing sgRNAs for higher efficiency.

Methods for CRISPR/Cas9-Induced Mutant Screening

Using CRISPR/Cas9 technology several mutant libraries have been constructed such as human cells [27], rice [28] and tomato [29] to name a few. The onset of gene editing techniques has made the screening of mutants much more crucial. This screening process involves the identification of on-target edits, off-target modifications and elimination of transgenes in offspring plants. For this purpose, techniques such as restriction enzyme site loss assay [30], the surveyor nuclease and endonuclease [31], PAGE-based genotyping approach [32] and high-resolution melting analysis [33] have been developed. The screening of mutants is much simpler if the phenotype is distinct. The phenotype aspect of mutant screening has been deployed to identify tobacco and rice mutants with edited *phytoene desaturase* (PDS) gene which leads to an albino phenotype [34, 35]. Antibiotic and herbicide resistance genes have the potential as selectable markers for the screening of transgenic plant [36, 37]. While using phenotype distinctions to screen mutants is rather easy, it's difficult to link the target gene and its obvious phenotype for accurate results [28]. Additionally, whole gene sequencing can be used to identify transgene free plants produced by gene editing but the process demands longer timeperiod and is expensive.

Applications of CRISPR/Cas9 System in Crops

Research into Function of Genes

CRISPR/Cas9 system associated genome editing can aide to the development of crops more resilient to different biotic and abiotic stresses. The increasing global population andharsher environmental conditions are making food security as big worry [38]. Genome editing can help in development of higher yield, better quality and more stress resilient crops to overcome potential food shortages.

Table 1. Examples of use of CRISPR/Cas9 in crop improvement

Plant	Application	Gene Targeted	
Tomato	Biotic Stress	SIMLO	Gene knock-out resulted in enhanced resistance t powdery mildew
	Abiotic Stress	SIMAPK3	Gene knock-out resulted in drought tolerance
	Yield	SP5G	Gene known-out resulted in rapid flowering
	Quality	SIGAD2, SIGAD3	Gene knock-out resulted in increased level of gamma aminobutyric acid
Rice	Biotic Stress	eIF4G	Gene knock-out enhanced resistance to *Rice tungro mosaic virus*
	Abiotic Stress	SAPK2	Gene knock-out resulted in ABA-insensitive phenotype
	Yield	Gn1a, GS3	Gene knock-out resulted in larger grain size and increased grain number
	Quality	Waxy	Gene knock-out resulted in enhanced glutinosity and a decreased amylose content

Table 2. Screening methods for CRISPR/Cas9-induced mutagenesis

Method	Feature(s)
RE site loss assay	Destruction of target gene RE site by CRISPR/Cas9 induced mutagenesis, rendering the mutated sequence resistant to RE digestion. Both homozygous and heterozygous mutants can be detected.
Surveyor nuclease and T7E1 assays	Detection and subsequent digestion of mismatched DNA duplex formed by mutated and wild-type strands. Restricted to detection of heterozygous mutants.
ACT-PCR	Annealing between mutated strand – template is suppressed at optimum temperature resulting in no amplification. Restricted to detection of homozygous mutants.
HRMA	Based on the unique melting temperature of homozygous and the mutated heterozygous DNA. Better efficiency at detection of heterozygous mutants.
PAGE-based genotyping approach	Based on differences in the migration pattern of homoduplex and heteroduplex DNA. Homoduplex DNA migrates faster on native PAGE. Restricted to detection of heterozygous mutants.

Modifications of Genes for Yield Traits

By changing the flowering pattern in crops plants, the geographical cultivation range of the crop can be improved. For this CRISPR/Cas9 systems like FLOWER LOCUS T and SELF PRUNING 5G have been used to modify the flowering time in soybean [39] and tomato plants [40]. 11 news lines of thermo-sensitive genetic male sterile (TGMS) lines of rice have been developed to accelerate hybrid rice breeding [41].

Using this system rice cultivars with denser erect panicles, higher grain content and size have been developed. A similar modification system has been used to change the leaf angle in maize plants to obtain maximum light exposure. The *ZmLG*-1 targeting sgRNA/Cas9 targets desired target mutator (DTM) in T0 transgenic maize plants. The zmLG1 editing achieved lesser leaf angles [30].

The modification system has also been used to achieve high-tillering, dwarf habit rice yield. This has been achieved by using CRISPR/Cas9 system to edit out *Carotenoid cleavage dioxygenase* (*CCD7*) gene [42]. This gene is crucial in the regulation of strigo-lactones, which is a plant hormone influencing shootbranching or tillering and root development. This loss-of-function mutation method of genome editinghas been used in multiple crops.

Modifications of Genes for Improved Produce Quality

Rice varieties with low amylose content and higher glutinosity were developed using CRIPSR/Cas9 system by editing out the amylose content regulator *Waxy* gene, improving the quality of produce [43]. Similarly, a potato variety with lower amylose and higher amylopectin/amylose ratio was obtained by knocking out *granule bound starch synthase (GBSS)* gene [44].

New rice varieties with lower cadmiumcontent have been developed using CRISPR/Cas9 system by editing *OsNramp5* gene which is a metal transporter [45]. Similarly, CRISPR/Cas9 system has been used to improve the nutrition quotient of tomatoes by modifying the genes in GABA (γ- Aminobutyric acid) shunt [46, 47].

Genome Editing for Biotic and Abiotic Stress Tolerance

Powdery mildew disease is a big nuisance among cereal crops. Using CRISPR/Cas9 system a mutant wheat variety with knocked out gene for regulating the disease resistance called *enhanced disease resistance1* gene, has been developed which displayed greater resistance against powdery mildew disease [48]. Similar knocking out of *Mildew resistance locus* genes using an RNA-guided Cas9 system resulted in powdery mildew-tolerant wheat [49] and tomato plants [50].

This system has also been usedto develop canker-resistant citrus mutants where the gene responsible for canker development, *CsLOB1* was modified [51]. A similar modification to the *eukaryotic initiation factor* genes has resulted in virus-resistant cucumber [52] and *Arabidopsis* plants [53].

Knocking out of *eIF4G* gene using CRISPR/Cas9 system has led to the rice tungro spherical virus (RTSV) resistant rice varieties [54].

CRISPR/Cas9 system has been successfully demonstrated in developing abiotic stress resistant plants.*Mitogen-activated protein kinase3* (*SIMAPK3)* mutant has been shown to exhibit drought resistance in tomato plants [55]. A drought resistant maize variety with higher grain yield has been developed bymodifying the expression of an ethylene response regulator gene *ARGOS8*, using CRISPR/Cas9 system [56].

Regulation of Gene Expression

Gene expression manipulation is a crucial aspect of plant breeding [57]. Using CRISPR/Casp9 system to modify regulatory elements of gene expression, novel alleles with enhanced quantitative traits can bedeveloped. A modified lettuce plant variety with increased vitamin C content and elevated stress tolerance [58] has been developed by modifying and regulating the upstream open reading frame and primary open reading frames.

Transcription Activator-Like Effector Nucleases (TALENs)

TALEN as a genome editing system is associated with the study of genus *Xanthomonas.* It consists ofa FokI nuclease domain bound to a DNA-binding domain [59]. This bacterium secretes transcription activator-like effectors (TALE) which make the host plant more susceptible to pathogens. TALE possesses DNA binding activity as well as mimics eukaryotic transcription factors [60]. TALE has three components, a DNA binding domain, a target gene transcription activation domain and a nuclear localization signal. DNA-binding domain consists of monomers that recognize specific nucleotide. These monomers are composed of 34 amino acid residues in a tandem repeat pattern [61, 62]. The injection of TALEs into a host plant is mediated by type III secretion system and modifies it'stranscription by binding to the genomic DNA [63].

A number of repeats in the array correspond to the target site's length which results in observing the correlation between hyper variable residues and the base linking to each repeat. It has also been noted that the position at 5' end of the base that is bound to the first repeat in the array is conserved with the presence of thymine [61]. 3' end of the nucleotide is boundto the last repeat of the last array is called a half-repeat as it only contains 20 amino acid residues. To construct a chimeric TALEN, DNA-binding domain of TALE was inserted into a plasmid vector [64]. Thegenetic constructs that resulted from this conjugation expressed the DNA binding domain and FokI endonuclease domain. The application of this system in engineering artificial nucleases for different nucleotide sequences is promising which can be achieved with the different combinations of DNA- binding domain monomers with repeat variable diresidue (RVD). TALENs work in pairs with binding sites selected on both strands of the DNA.

Zinc Finger Nucleases (ZFN)

ZFN is another gene-targeting system which has separate DNA-binding and DNA-cleavage domains. These proteins were first observed by Chandrasegaran, where it he saw physically separable binding and cleavage activities of *Fok*I, a restriction enzyme [65]. The first successful demonstration of targeted mutagenesis and gene replacement was performed in some at the *yellow* locus using a pair of ZFN pair [66, 67]. Success of the gene editing depends upon efficiency of ZFNs delivery.

Conclusion

With ever changing abiotic stresses such as abnormal seasonal changes, disrupted rain patterns and aggravated global warming combined with increasing global population the prospects of genome editing to obtain a more nutritious yield from the same arable area is a necessity and something to betaken up on a war footing. In this regard, there are only so many genome editing tools as promising as the CRISPR/Cas system with its high efficiency, amenability to multiplexing, simplicity, high specificity and low off-target effects. The highly flexible multiplexing property of the CRISPR/Cas system can target a variety of genes and their loci to get a better understanding of gene-pathwayrelation and subsequently, the information can be utilized to manipulate as per needs. In the medical field, the CRISPR/Cas system has shown potential in targeted RNA-guided research in developing treatments for non-mendelian pathologies, offsetting the risk of permanent genetic modifications. The system has a great potential in developing a more sophisticated imaging platform involving the fluorescent protein. Such methods can not only help in localizing genes, and nucleic acids, but also visualize the chromosomal changes more efficiently.

References

[1] Jeon JS, Lee S, Jung K-H, Jun S-H, Jeong D-H, Lee J, Kim C, Jang S, Lee S, Yang K, Nam J, An K, Han M-J, Sung R-J, Choi H-S, Yu J-H, Choi J-H, Cho S-Y, Cha S-S, Kim S-I, An G. "T-DNA insertional mutagenesis for functional genomics in rice," *Plant J CellMol Biol*, vol. 22, no. 6, pp. 561-570, Jun. 2000, doi: 10.1046/j.1365-313x.2000.00767.x.

[2] Wang L, Zheng J, Luo Y, Xu T, Zhang Q, Zhang L, Xu M, Wan J, Wang M-B, Zhang C, Fan Y. "Construction of a genomewide RNAi mutant library in rice," *Plant Biotechnol J*,vol. 11, no. 8, pp. 997-1005, Oct. 2013, doi: 10.1111/pbi.12093.

[3] Puchta H, Dujon B and Hohn B. "Two different but related mechanisms are used in plants forthe repair of genomic double-strand breaks by homologous recombination," *Proc Natl Acad Sci USA*, vol. 93, no. 10, pp. 5055-5060, May 1996, doi: 10.1073/pnas.93.10.5055.

[4] Bleuyard J-Y, Gallego ME and White CI. "Recent advances in understanding of the DNA double-strand break repair machinery of plants," *DNA Repair*, vol. 5, no. 1, pp. 1-12, Jan. 2006,doi: 10.1016/j.dnarep. 2005.08.017.

[5] Osakabe K, Osakabe Y and Toki S. "Site-directed mutagenesis in Arabidopsis using custom- designed zinc finger nucleases," *Proc Natl Acad Sci USA*, vol. 107, no. 26, pp. 12034-12039,Jun. 2010, doi: 10.1073/pnas.1000234107.

[6] Kumar V and Jain M. "The CRISPR-Cas system for plant genome editing: Advances and opportunities," *J Exp Bot*, vol. 66, no. 1, pp. 47-57, Jan. 2015, doi: 10.1093/jxb/eru429.

[7] Nature.com. "A TALE nuclease architecture for efficient genome editing|Nature Biotechnology." https://www.nature.com/articles/nbt.1755 (accessed Mar. 11, 2022).

[8] Jinek M, Chylinski K, Fonfara I, Hauer M, Doudna JA and Charpentier E. "A programmabledual-RNA-guided DNA endonuclease in adaptive bacterial immunity," *Science*, vol. 337, no. 6096, pp. 816-821, Aug. 2012, doi: 10.1126/science.1225829.

[9] Soda N, Verma L and Giri J. "CRISPR-Cas9 based plant genome editing: Significance, opportunities and recent advances," *Plant Physiol Biochem PPB*, vol. 131, pp. 2-11, Oct. 2018. doi: 10.1016/j.plaphy.2017.10.024.

[10] Abdelrahman M, Al-Sadi AM, Pour-Aboughadareh A, Burritt DJ and Tran L-SP. "Genomeediting using CRISPR/Cas9–targeted mutagenesis: An opportunity for yield improvements of crop plants grown under environmental stresses," *Plant Physiol Biochem*, vol. 131, pp. 31-36, Oct. 2018. doi: 10.1016/j.plaphy.2018.03.012.

[11] Schindele P, Wolter F and Puchta H. "Transforming plant biology and breeding with CRISPR/Cas9, Cas12 and Cas13," *FEBS Lett*, vol. 592, no. 12, pp. 1954-1967, Jun. 2018, doi:10.1002/1873-3468.13073.

[12] Ishino Y, Shinagawa H, Makino K, Amemura M and Nakata A. "Nucleotide sequence of the *iap* gene, responsible for alkaline phosphatase isozyme conversion in *Escherichia coli*, and identification of the gene product," *J Bacteriol*, vol. 169, no. 12, pp. 5429-5433, Dec. 1987, doi:10.1128/jb.169.12.5429-5433.1987.

[13] Jansen R, van Embden JDA, Gaastra W and Schouls LM. "Identification of genes that areassociated with DNA repeats in prokaryotes," *Mol Microbiol*, vol. 43, no. 6, pp. 1565-1575, Mar. 2002. doi: 10.1046/j.1365-2958.2002.02839.x.

[14] Mojica FJM, Díez-Villaseñor C, García-Martínez J and Soria E. "Intervening sequences of regularly spaced prokaryotic repeats derive from foreign genetic elements," *J Mol Evol*, vol. 60,no. 2, pp. 174-182, Feb. 2005. doi: 10.1007/s00239-004-0046-3.

[15] Pourcel C, Salvignol G and Vergnaud G. "CRISPR elements in *Yersinia pestis* acquire new repeats by preferential uptake of bacteriophage DNA, and provide additional tools for evolutionary studies," *Microbiol Read Engl*, vol. 151, no. Pt 3, pp. 653-663, Mar. 2005. doi:10.1099/mic.0.27437-0.

[16] Bolotin A, Quinquis B, Sorokin A and Ehrlich SD. "Clustered regularly interspaced short palindrome repeats (CRISPRs) have spacers of extrachromosomal origin," *Microbiol Read Engl*, vol. 151, no. Pt 8, pp. 2551-2561, Aug. 2005. doi: 10.1099/mic.0.28048-0.

[17] KuninV, Sorek R and Hugenholtz P. "Evolutionary conservation of sequence and secondarystructures in CRISPR repeats," *Genome Biol*, vol. 8, no. 4, p. R61, Apr. 2007. doi: 10.1186/gb-2007-8-4-r61.

[18] Makarova KS, Haft DH, Barrangou R, Brouns SJJ, Charpentier E, Horvath P, Moineau S, Mojica FJM, Wolf YI, Yakunin AF, van der Oost J, Koonin EV. "Evolution and classification of the CRISPR-Cas systems," *Nat RevMicrobiol*, vol. 9, no. 6, pp. 467-477, Jun. 2011. doi: 10.1038/nrmicro2577.

[19] Ma X, Zhang Q, Zhu Q, Liu W, Chen Y, Qiu R, Wang B, Yang Z, Li H, Lin Y, Xie Y, Shen R, Chen S, Wang Z, Chen Y, Guo J, Chen L, Zhao X, Dong Z, Liu Y-G. "A Robust CRISPR/Cas9 System for Convenient, High-Efficiency Multiplex GenomeEditing in Monocot and Dicot Plants," *Mol Plant*, vol. 8, no. 8, pp. 1274-1284, Aug. 2015. doi: 10.1016/j.molp.2015.04.007.

[20] Fu Y, Foden JA, Khayter C, Maeder ML, Reyon D, Joung JK, Sander JD. "High-frequency off-target mutagenesis induced by CRISPR-Cas nucleases in humancells," *Nat Biotechnol*, vol. 31, no. 9, Art. no. 9, Sep. 2013. doi: 10.1038/nbt.2623.

[21] Cameron P, Fuller CK, Donohoue PD, Jones BN, Thompson MS, Carter MM, Gradia S, Vidal B, Garner E, Slorach EM, Lau E, Banh LM, Lied AM, Edwards LS, Settle AH, Capurso D, Llaca V, Deschamps S, Cigan M, Young JK, May AP. "Mapping the genomic landscape of CRISPR-Cas9 cleavage," *Nat Methods*,vol. 14, no. 6, pp. 600-606, Jun. 2017. doi: 10.1038/nmeth.4284.

[22] Kleinstiver BP, Pattanayak V, Prew MS, Tsai SQ, Nguyen NT, Zheng Z, Joung K. "High-fidelity CRISPR-Cas9 variants with undetectable genome-wide off-targets," *Nature*, vol. 529, no. 7587, pp. 490-495, Jan. 2016. doi: 10.1038/nature16526.

[23] Slaymaker M, Gao L, Zetsche B, Scott DA, Yan WX and Zhang F. "Rationally engineered Cas9 nucleases with improved specificity," *Science*, vol. 351, no. 6268, pp. 84-88, Jan. 2016. doi:10.1126/science.aad5227.

[24] LeBlanc C, Zhang F, Mendez J, Lozano Y, Chatpar K, Irish VF, Jacob V. "Increased efficiency of targeted mutagenesis by CRISPR/Cas9 in plants usingheat stress," *Plant J Cell Mol Biol*, vol. 93, no. 2, pp. 377-386, Jan. 2018. doi: 10.1111/tpj.13782.

[25] Liu H, Ding Y, Zhou Y, Jin W, Xie K and Chen L-L. "CRISPR-P 2.0: An Improved CRISPR-Cas9Tool for Genome Editing in Plants," *Mol Plant*, vol. 10, no. 3, pp. 530-532, Mar. 2017. doi: 10.1016/j.molp.2017.01.003.

[26] Xie X, Ma X, Zhu Q, Zeng D, Li G and Liu Y-G. "CRISPR-GE: A Convenient Software Toolkitfor CRISPR-Based Genome Editing," *Mol Plant*, vol. 10, no. 9, pp. 1246-1249, Sep. 2017. doi:10.1016/j.molp.2017.06.004.

[27] Shalem O, Sanjana NE, Hartenian E, Shi X, Scott DA, Mikkelson T, Heckl D, Ebert BL, Root DE, Doench JG, Zhang F. "Genome-Scale CRISPR-Cas9 Knockout Screening in Human Cells," *Science*, vol.343, no. 6166, pp. 84-87, Jan. 2014, doi: 10.1126/science.1247005.

[28] Lu Y, Ye X, Guo R, Huang J, Wang W, Tang J, Tan L, Zhu J-K, Chu C, Qian Y. "Genome-wide Targeted Mutagenesis in Rice Using the CRISPR/Cas9 System," *MolPlant*, vol. 10, no. 9, pp. 1242-1245, Sep. 2017. doi: 10.1016/j.molp.2017.06.007.

[29] Jacobs TB, Zhang N, Patel D and Martin GB. "Generation of a Collection of Mutant Tomato Lines Using Pooled CRISPR Libraries," *Plant Physiol*, vol. 174, no. 4, pp. 2023-2037, Aug. 2017. doi: 10.1104/pp.17.00489.

[30] Shan Q, Wang Y, Li J and Gao C. "Genome editing in rice and wheat using the CRISPR/Cassystem," *Nat Protoc*, vol. 9, no. 10, Art. no. 10, Oct. 2014. doi: 10.1038/nprot.2014.157.

[31] Vouillot L, Thélie A and Pollet N. "Comparison of T7E1 and surveyor mismatch cleavage assaysto detect mutations triggered by engineered nucleases," *G3 Bethesda Md*, vol. 5, no. 3, pp. 407-415, Jan. 2015. doi: 10.1534/g3.114.015834.

[32] Zhu X, Xu Y, Yu S, Lu L, Ding M, Cheng J, Song G, Gao X, Yao L, Fan D, Meng S, Zhang X, Hu S and Tian Y. "An Efficient Genotyping Method for Genome-modified Animals and Human Cells Generated with CRISPR/Cas9 System," *Sci Rep*, vol. 4, no. 1, Art. no. 1, Sep. 2014. doi: 10.1038/srep06420.

[33] Thomas HR, Percival SM, Yoder BK and Parant JM. "High-Throughput Genome Editing and Phenotyping Facilitated by High Resolution Melting Curve Analysis," *PLOS ONE*, vol. 9, no.12, p. e114632, Dec. 2014. doi: 10.1371/journal.pone.0114632.

[34] Chen L, Li W, Katin-Grazzini L, Ding J, Gu X, Li Y, Gu T, Wang R, Lin X, Deng Z, McAvoy RJ, Gmitter Jr FG., Deng Z, Zhao Y and Li Y. "A method for the production and expedient screening of CRISPR/Cas9-mediatednon-transgenic mutant plants," *Hortic Res*, vol. 5, no. 1, Art. no. 1, Mar. 2018. doi: 10.1038/s41438-018-0023-4.

[35] Zhang H, Zhang J, Wei P, Zhang B, Gou F, Feng Z, Mao Y, Yang L, Zhang H, Xu N, Zhu J-K. "The CRISPR/Cas9 system produces specific and homozygous targeted gene editing in rice in one generation," *Plant Biotechnol J*, vol. 12, no. 6, pp. 797-807, Aug. 2014. doi: 10.1111/pbi.12200.

[36] Xu R, Li H, Qin R, Wang L, Li L, Wei P, Yang J. "Gene targeting using the Agrobacterium tumefaciens-mediated CRISPR-Cas systemin rice," *Rice NYN*, vol. 7, no. 1, p. 5, 2014. doi: 10.1186/s12284-014-0005-6.

[37] Lu H-P, Liu S-M, Xu S-L, Chen W-Y, Zhou X, Tan Y-Y, Huang J-Z, Shu Q-W. "CRISPR-S: an active interference element for a rapid and inexpensive selection ofgenome-edited, transgene-free rice plants," *Plant Biotechnol J*, vol. 15, no. 11, pp. 1371-1373, Nov. 2017. doi: 10.1111/pbi.12788.

[38] Zaidi S S-E-A, Tashkandi M, Mansoor S and Mahfouz MM. "Engineering Plant Immunity:Using CRISPR/Cas9 to Generate Virus Resistance," *Front Plant Sci*, vol. 7, p. 1673, 2016, doi: 10.3389/fpls.2016.01673.

[39] Cai Y, Chen L, Liu X, Guo C, Sun S, Wu C, Jiang B, Han T, Hou W. "CRISPR/Cas9-mediated targeted mutagenesis of GmFT2a delays flowering time insoya bean," *Plant Biotechnol J*, vol. 16, no. 1, pp. 176-185, Jan. 2018. doi: 10.1111/pbi.12758.

[40] Soyk S, Müller NA, Park SJ, Schmalenbach I, Jiang K, Hayama R, Zhang L, Van Eck J, Jiménez-Gómez JM, Lippman ZB. "Variation in the flowering gene SELF PRUNING 5G promotes day-neutrality and early yield in tomato," *Nat Genet*, vol. 49, no. 1, pp. 162-168, Jan. 2017. doi: 10.1038/ng.3733.

[41] Zhou H, He M, Li J, Chen L, Huang Z, Zheng S, Zhu L, Ni E, Jiang D, Zhao B, Zhuang C. "Development of Commercial Thermo-sensitive Genic Male Sterile Rice Accelerates Hybrid Rice Breeding Using the CRISPR/Cas9-mediated TMS5 Editing System," *Sci Rep*, vol. 6, no. 1, Art. no. 1, Nov. 2016. doi: 10.1038/srep37395.

[42] Butt H, Jamil M, Wang JY, Al-Babili S and Mahfouz M. "Engineering plant architecture via CRISPR/Cas9-mediated alteration of strigolactone biosynthesis," *BMC Plant Biol*, vol. 18, no. 1,p. 174, Aug. 2018. doi: 10.1186/s12870-018-1387-1.

[43] Zhang H. Zhang J. Botella R and Zhu J-K. "Generation of new glutinous rice by CRISPR/Cas9-targeted mutagenesis of the Waxy gene in elite rice varieties," *J Integr Plant Biol*, vol. 60, no. 5,pp. 369-375, May 2018. doi: 10.1111/jipb.12620.

[44] Andersson M, Turesson H, Nicolia A, Fält A-S, Samuelsson M and Hofvander P. "Efficient targeted multiallelic mutagenesis in tetraploid potato (Solanum tuberosum) by transient CRISPR- Cas9 expression in protoplasts," *Plant Cell Rep*, vol. 36, no. 1, pp. 117-128, Jan. 2017, doi: 10.1007/s00299-016-2062-3.

[45] Nature.com. "Knockout of OsNramp5 using the CRISPR/Cas9 system produces low Cd-accumulating indica rice without compromising yield|Scientific Reports." https://www.nature.com/articles/s41598-017-14832-9 (accessed Mar. 11, 2022).

[46] Nonaka S, Arai C, Takayama M, Matsukura C and Ezura H. "Efficient increase of ɣ- aminobutyric acid (GABA) content in tomato fruits by targeted mutagenesis," *Sci Rep.*, vol. 7,no. 1, p. 7057, Aug. 2017. doi: 10.1038/s41598-017-06400-y.

[47] Li R, Li R, Li X, Fu D, Zhu B, Tian H, Luo Y, Zhu H. "Multiplexed CRISPR/Cas9-mediated metabolic engineering of γ-aminobutyric acid levels in Solanum lycopersicum," *Plant Biotechnol J*, vol. 16, no. 2, pp. 415-427, Feb. 2018. doi:10.1111/pbi.12781.

[48] Zhang Y, Bai Y, Wu G, Zou S, Chen Y, Gao C, Tang D. "Simultaneous modification of three homoeologs of TaEDR1 by genome editingenhances powdery mildew resistance in wheat," *Plant J Cell Mol Biol*, vol. 91, no. 4, pp. 714-724, Aug. 2017. doi: 10.1111/tpj.13599.

[49] Nature.com. "Simultaneous editing of three homoeoalleles in hexaploid bread wheat confers heritable resistance to powdery mildew|Nature Biotechnology." https://www.nature.com/articles/nbt.2969 (accessed Mar. 11, 2022).

[50] Nekrasov V, Wang C, Win J, Lanz C, Weigel D and Kamoun S. "Rapid generation of a transgene-free powdery mildew resistant tomato by genome deletion," *Sci Rep*, vol. 7, no. 1,Art. no. 1, Mar. 2017. doi: 10.1038/s41598-017-00578-x.

[51] "Genome editing of the disease susceptibility gene CsLOB1 in citrus confers resistance to citruscanker - PubMed." https://pubmed.ncbi.nlm.nih.gov/27936512/ (accessed Mar. 11, 2022).

[52] Chandrasekaran J, Brumin M, Wolf D, Leibman D, Klap C, Pearlsman M, Sherman A, Arazi T, Gal-On A. "Development of broad virus resistance in non-transgenic cucumberusing CRISPR/Cas9 technology," *Mol Plant Pathol*, vol. 17, no. 7, pp. 1140-1153, 2016. doi: 10.1111/mpp.12375.

[53] Pyott DE, Sheehan E and Molnar A. "Engineering of CRISPR/Cas9-mediated potyvirus resistance in transgene-free Arabidopsis plants," *Mol Plant Pathol*, vol. 17, no. 8, pp. 1276-1288, Oct. 2016. doi: 10.1111/mpp.12417.

[54] Macovei A, Sevilla NR, Cantos C, Jonson GB, Slamet-Loedin I, Čermák T, Voytas DF, Choi I-R, Chadha-Mohanty P. "Novel alleles of rice eIF4G generated by CRISPR/Cas9-targeted mutagenesis confer resistance to Rice tungro spherical virus," *Plant Biotechnol J*, vol. 16, no. 11, pp. 1918-1927, Nov. 2018. doi: 10.1111/pbi.12927.

[55] Wang L, Chen L, Li R, Zhao R, Yang M, Sheng J, Shen L. "Reduced Drought Tolerance by CRISPR/Cas9-Mediated SlMAPK3 Mutagenesis inTomato Plants," *J Agric Food Chem*, vol. 65, no. 39, pp. 8674-8682, Oct. 2017. doi: 10.1021/acs.jafc.7b02745.

[56] Shi J, Gao H, Wang H, Lafitte HR, Archibald RL, Yang M, Hakimi SM, Mo H, Habben JE. "ARGOS8 variants generated by CRISPR-Cas9 improve maize grain yield under fielddrought stress conditions," *Plant Biotechnol J*, vol. 15, no. 2, pp. 207-216, Feb. 2017. doi: 10.1111/pbi.12603.

[57] Mahas A, Stewart CN and Mahfouz MM. "Harnessing CRISPR/Cas systems for programmable transcriptional and post-transcriptional regulation," *Biotechnol Adv*, vol. 36, no.1, pp. 295-310, Jan. 2018. doi: 10.1016/j.biotechadv.2017.11.008.

[58] Zhang H, Si X, Ji X, Fan R, Liu J, Chen K, Wang D, Gao C. "Genome editing of upstream open reading frames enables translational controlin plants," *Nat Biotechnol*, vol. 36, no. 9, Art. no. 9, Oct. 2018. doi: 10.1038/nbt.4202.

[59] Joung JK and Sander JD. "TALENs: a widely applicable technology for targeted genome editing," *Nat Rev Mol Cell Biol*, vol. 14, no. 1, pp. 49-55, Jan. 2013. doi: 10.1038/nrm3486.

[60] Nemudryi AA, Valetdinova KR, Medvedev SP and Zakian SM. "TALEN and CRISPR/CasGenome Editing Systems: Tools of Discovery," *Acta Naturae*, vol. 6, no. 3, pp. 19-40, 2014, Accessed: Apr. 04, 2022. [Online]. Available: https://www.ncbi.nlm.nih.gov/pmc/articles/PMC4207558/.

[61] Boch J, Scholze H, Schornack S, Landgraf A, Hahn S, Kay S, Lahaye T, Nickstadt A, Bonas U. "Breaking the code of DNA binding specificity of TAL-type III effectors," *Science*,vol. 326, no. 5959, pp. 1509-1512, Dec. 2009. doi: 10.1126/science.1178811.

[62] Moscou J and Bogdanove AJ. "A simple cipher governs DNA recognition by TAL effectors," *Science*, vol. 326, no. 5959, p. 1501, Dec. 2009. doi: 10.1126/science.1178817.

[63] Boch J and Bonas U. "Xanthomonas AvrBs3 family-type III effectors: discovery and function,"*Annu Rev Phytopathol*, vol. 48, pp. 419-436, 2010, doi: 10.1146/annurev-phyto-080508- 081936.

[64] Christian M, Cermak T, Doyle EL, Schmidt C, Zhang F, Hummel A, Bogdanove AJ, Voytas DF. "Targeting DNA double-strand breaks with TAL effector nucleases," *Genetics*,vol. 186, no. 2, pp. 757-761, Oct. 2010. doi: 10.1534/genetics.110.120717.

[65] Li L, Wu LP and Chandrasegaran S. "Functional domains in Fok I restriction endonuclease,"*Proc Natl Acad Sci USA*, vol. 89, no. 10, pp. 4275-4279, May 1992. doi: 10.1073/pnas.89.10.4275.

[66] Bibikova M, Golic M, Golic KG and Carroll D. "Targeted chromosomal cleavage and mutagenesis in Drosophila using zinc-finger nucleases," *Genetics*, vol. 161, no. 3, pp. 1169-1175,Jul. 2002. doi: 10.1093/genetics/161.3.1169.

[67] Bibikova M, Beumer K, Trautman JK and Carroll D. "Enhancing gene targeting with designedzinc finger nucleases," *Science*, vol. 300, no. 5620, p. 764, May 2003. doi: 10.1126/science.1079512.

Chapter 5

Experimental and Bioinformatics Approaches to Predict Off-Targeting of Programmable Nucleases (ZFNs, TALENs and CRISPR-Cas9)

Purnima Sharma[1]
Harpreet Singh[1,*]
Rupinder Preet Kaur[2]
Kamini[1]
Rakhi[1]
Taniya[1]
and Nandni Kanwar[1]
[1] PG Department of Bioinformatics,
Hans Raj Mahila Maha Vidayala, Jalandhar, Punjab, India
[2] Department of Chemistry,
Guru Nanak Dev University College Verka, Amritsar, Punjab, India

Abstract

Genome editing is a field of study that aims to modify genes in live animals in order to better understand gene function and create treatments for hereditary and acquired disorders. Genome editing has also been applied to plants to enhance yield, nutritional value, quality, stress tolerance, etc. In different types of cells and species, genome editing may be used to fix, insert, or delete practically any DNA sequence. The revelation that a broken portion of DNA in a gene stimulates a cell's repair machinery to patch the split together led to the development of genome editing. Researchers can use genome editing to emulate this natural process.

While DNA editing techniques have been around for decades, new technologies have been evolving from zinc finger nucleases to transcription activator-like effector nucleases, LAGLIDADG homing endonucleases, oligonucleotide-directed mutagenesis and finally CRISPR/Cas in order to achieve requisite precision, simplicity, and cost effectiveness. Now-a-days, CRISPR/Cas9 has been the most extensively used genome editor for

---

* Corresponding Author's Email: harpreetsingh05@gmail.com.

In: Genome Editing for Crop Improvement
Editors: Rahul Kumar, Ashutosh Sharma, Yogesh Kumar et al.
ISBN: 979-8-88697-740-0

deciphering gene function. Because CRISPR/Cas9 is an RNA-based system, it can be tweaked more quickly and easily than protein-based systems, and it can target many locations. CRISPR/Cas9 was discovered as a result of fundamental research into how bacteria defend themselves against viruses.

CRISPR/Cas9 changes a single base pair of DNA, vast chunks of chromosomes, or gene expression levels by cutting a DNA sequence at a specified genetic region and deleting or inserting DNA sequences. For precise editing, this robust and simple method requires only two main components (sgRNA and Cas9). The Cas9 enzyme cuts target DNA (on target) under the supervision of sgRNA, but it can also cut non-target regions, resulting in off-target effects and uncontrollable mutations. The possibility of off-target effects in CRISPR technology is the key limiting issue that prevents CRISPR technology from being widely used. Therefore, it is essential to predict the off-target activity of the designed sgRNAs to avoid unexpected outcomes.

Initially, the seed region of SgRNA was a focus for improving specificity, and various rules for developing effective gRNAs were discovered. These guidelines were applicable for smaller organisms, but complex organisms, such as plants with higher ploidy levels, have substantial specificity issues. Both experimental and bioinformatics approaches have been developed extensively for detection, analysis and prediction of off-targets in Programmable nucleases. A variety of *in silico* strategies have been developed ranging from simple homology based to formula/rule based to machine learning/deep learning, and molecular dynamics simulations. This chapter therefore reviews and discusses various experimental as well as bioinformatics approaches to predict off-targets, their applications, limitations, availability, and usage.

Keywords: genome editing, ZFNs, TALENs, CRISPR-Cas9, off-target prediction, machine learning

Introduction

Genome editing, which involves making a desired modification to the sequence of genomic DNA, is causing a revolution within the biomedical sciences and has the potential to treat a good range of human disorders with a particular change in the genetic component that's liable for that disease in the future. Researchers can alter or edit DNA sequences virtually in any organism of choice, allowing them to elucidate the functional organization of the genome at the systems level and identify causal genetic variations instead of studying DNA taken out of the context of the genome [1, 2]. In recent years, the rapid advancement of genome editing has changed human genome research, allowing researchers to better understand the contribution of a single-gene product to a disease in an organism. The invention of genetic engineering (DNA or RNA modification) revolutionized the concept of genome editing [3]. Genome editing can be achieved *in vitro* or *in vivo* by delivering the editing machinery *in situ*, which powerfully adds, deletes and "corrects" genes as well as performs other highly targeted genomic modifications [4, 5].

The concept of site-specific recognition of DNA sequences by oligonucleotides, micro molecules, or self-splicing introns was utilised in initial attempts. Genome editing methods, which are supported by engineered or bacterial nucleases, have developed significantly within the last ten years and have begun to indicate remarkable applicability in a variety of domains, spanning from fundamental research to applied biotechnology and biomedical research [6]. These genome editing tools include ZFNs (zinc finger nucleases), TALENs (transcription activator-like effector nucleases), CRISPR (clustered regulatory interspaced short palindromic repeats)/Cas9 (CRISPR-associated proteins), and related CRISPR/Cas systems. Compared to other approaches, CRISPR/Cas systems are widely used among these genome editing tools because they're cost-effective, simple to use, and don't require professional skills [7]. Despite the fact that the technique has been widely adopted by academic and pharmaceutical communities, CRISPR-activity Cas9's (on-target) and specificity (off-target) are critical elements in widespread use of this technology. So many *in silico* gRNA activity and sensitivity prediction models and web tools have been developed, making CRISPR gene editing techniques much more convenient and precise [8].

The ideal genome editing tool would alter any genomic locus efficiently, with high DNA sequence specificity and few or no unwanted consequences. Given that naturally occurring types of genome-editing proteins evolved to achieve only partially related activities such as control of gene expression or protection against viral infection, such a perfect agent has yet to be found and is unlikely to exist in nature. As a result, scientists have realized that new methods are needed to expand the scope and effectiveness of genome editing, particularly in eukaryotic cells and animal models of human disease. Recent initiatives have resulted in significant progress toward this goal in a short period of time [2].

The Emergence of Genome Editing Technology

Since the discovery of the DNA double helix, technologies for manufacturing and manipulating DNA have encouraged advancements in biology. However, site-specific alterations in the genomes of cells and organisms have remained a challenge. In the 1970s, the development of recombinant DNA technology signalled the start of a new phase in biology when, in the laboratory of Paul Berg in 1972, scientists combined the *E. coli* genome with the genes of a bacteriophage and the SV40 virus [9, 10]. Molecular biologists obtained the power to modify DNA molecules for the first time, allowing them to study genes and harness them to generate medicines and innovation [1]. After that, the idea to use homologous recombination to replace a small piece of a living cell's genome with an exogenous donor DNA sequence enabled early genome editing efforts.

The exogenous DNA sequence must be homologous to the target genomic sequence to apply this technique for genome editing [11, 12]. Homologous recombination has been utilised to generate germline knockout or knocking mice in mouse embryonic stem cells. The conventional procedure takes more than a year to develop a genetically modified mouse, which is very time-consuming [13, 14]. Aside from the limited effectiveness of spontaneous homologous recombination, this method could also lead to undesirable genome editing events, in which the exogenous DNA sequence was incorporated into the genome at random places rather than at the targeted site [15]. Furthermore, similar attempts at using homologous recombination in human cells have proven to be far more challenging, and alternative approaches to knock down gene expression, such as antisense oligonucleotides and short interfering RNAs, have instead become standard. However, these approaches only transiently reduce gene expression, and the effect is usually incomplete and can often affect off-target genes [16, 17]. These shortcomings have fueled the demand for more effective methods of gene modification.

This demand has been met by a new wave of technology known as "gene editing," "genome editing," or "genome engineering," which allows researchers to precisely and efficiently introduce a variety of genetic alterations into mammalian cells, ranging from single nucleotide variant to gene insertion to chromosomal region deletion [18]. The revelation, first in yeast and then in human cells, that introducing a double-stranded break (DSB) into the genomic locus using a meganuclease, an endonuclease that detects and cleaves a lengthy DNA sequence, encouraged homology-driven DNA incorporation [19, 20].

The introduction of nuclease-induced double-stranded breaks (DSBs) leads to the stimulation of very effective cellular DNA recombination processes in mammalian cells, which leads to targeted DNA modifications [21, 22]. The "homology-directed repair" (HDR) technique increased the efficiency of the desired genomic modification event by two to three orders of magnitude, resulting in targeted insertion being significantly more efficient than random incorporation [23, 24]. Homology-directed repair (HDR) and non-homologous end-joining (NHEJ), both of which occur in practically all cell types and organisms, can repair nuclease-induced DNA DSBs, resulting in targeted insertion or gene disruption, respectively [25]. Major milestones in the evolution of various genome editing technologies have been summarized in Figure 1.

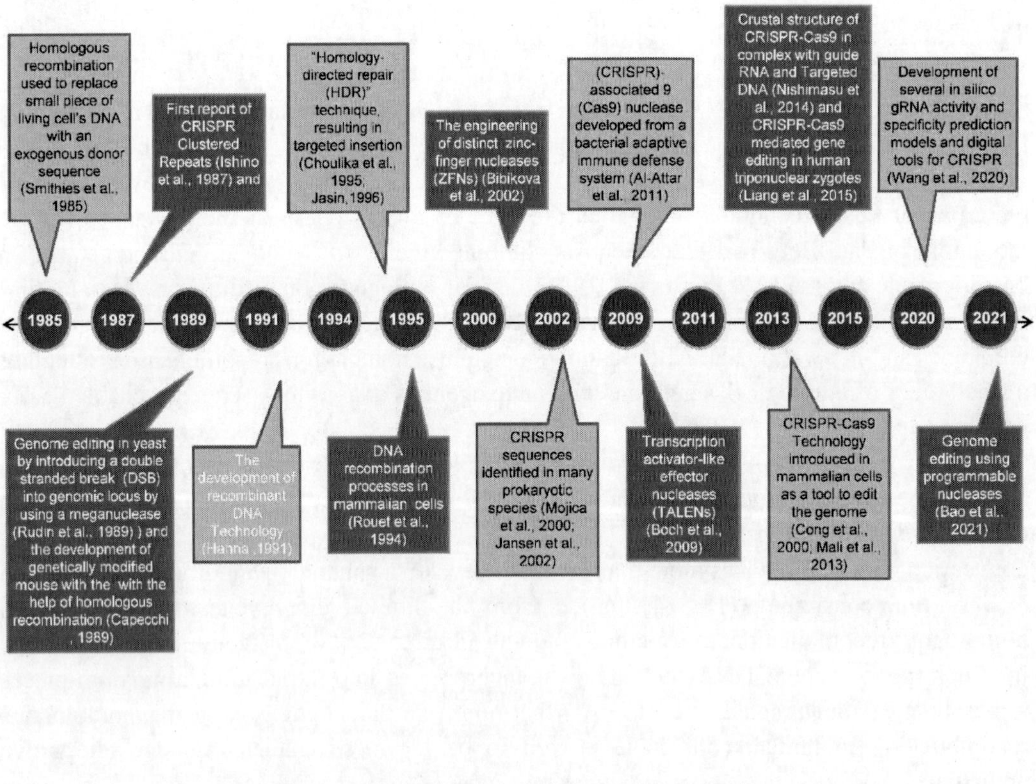

Figure 1. Timeline showing the emergence of different genome editing technologies *viz.*, rDNA technology, CRISPER-Cas9, meganucleases, zinc finger nucleases (ZFNs), transcription activator-like effector nucleases (TALENs), programmable nucleases.

In the initial stages of genome editing, to induce the appropriate DSBs at each specific DNA target site, the engineering of distinct zinc-finger nucleases (ZFNs) or meganucleases has been the research emphasis [26, 27, 28, 29, 30]. These nuclease-based methods needed specific expertise to create artificial proteins with sequence-specific DNA-binding domains coupled to a nonspecific nuclease for target cleavage, allowing researchers to undertake genetic manipulations never before possible [31].

Following that, the discovery of a new class of *Flavobacterium okeanokoites* (FokI) catalytic domains generated from bacterial proteins known as transcription activator-like effectors (TALEs) has opened new avenues for targeted genome editing [32]. TALE-based programmable nucleases have a high frequency of cleavage of any DNA sequence of interest. The construction of a sophisticated molecular clone for each novel DNA target, as well as the low effectiveness of genome screening in successfully targeted cells, are the key constraints for transcription activator-like effector nucleases (TALEN) techniques [33, 34]. The clustered regularly interspaced short palindromic repeat (CRISPR)-associated 9 (Cas9) nuclease, developed from a bacterial adaptive immune defence system, is a newly discovered, robust gene editing tool [35]. This technology has emerged as an alternative approach to ZFNs and TALENs which uses an RNA-guided DNA cleavage module to alter the genome of eukaryotic cells by inducing specific genetic mutations [36].

The CRISPR cluster repeats were reported for the first time in 1987 [37]. When an unusual region with five identical sequences of 29 nucleotides spaced by 32 nucleotides downstream of the iap gene was found in *E. coli* K12 strain, the seeds of one of the biggest breakthroughs in the history of biology were sown. Due to a paucity of high-throughput resources, the biologists were enveloped by a fog of uncertainty concerning the mechanism and applications of their discovery [38]. Emmanuelle Charpentier, in collaboration with Jennifer Doudna of the University of California Berkeley, demonstrated the use of CRISPR/Cas9 to target and cleaves five specific genomic sites of the gene encoding the green fluorescent protein (GFP) in August 2012, and they generalized their findings to potentially any target DNA sequence [39]. The multifaceted CRISPR/Cas9 technology has been rising exponentially in modifying gene expression, extending from genomic sequence repair or change to epigenetic and transcriptional modifications, since it was initially introduced in mammalian cells as a tool to edit the genome in 2013 [40, 41].

In the field of genome editing, the CRISPR-Cas9 system has emerged as the most effective and versatile technology. The discovery of CRISPR-Cas9's mechanism and its adoption to be used in eukaryotic genome modification has had a huge influence on the life sciences. Due to the simplicity with which new DNA sequences may be targeted for genome editing, scientists have been able to rapidly identify new gene functions, generate new cell and animal models of diseases, and make significant progress toward human therapies [42, 43, 44, 40, 45, 46]. Following that the discovery of programmable nucleases has significantly enhanced the process of genetic manipulation from clinical research to providing scientists with a strong tool to manipulate virtually any gene in a wide range of cell types and species for the first time. Life science and health are being transformed by genome editing using programmable nucleases. Off-target editing by these nucleases is still a major concern, particularly in clinical application [47]. CRISPR/Cas9 off-target effects are being reduced by creating a variety of CRISPR/Cas systems with high fidelity and accuracy. Off-target effects have been detected and evaluated using a variety of methods [48].

Genome Editing Approaches (Tools)

Genome editing (GE) techniques have revolutionized genetics because of their ability to efficiently manipulate genomic DNA. Several approaches of genome editing have been developed to date. We describe here, three most popular genome-editing tools (i) Zinc finger nucleases (ZFNs); (ii) Transcription activator-like effectors (TALEs); (iii) CRISPR-Cas9. This description is not meant to be a comprehensive account of the work that led to the development of the tools; rather, it is meant to give readers a deep understanding of the techniques and the capacity to choose among them for specific tasks. The major classes of programmable nucleases *viz.* ZFNs, TALENs, and CRISPR-Cas9 systems are described in Figure 2.

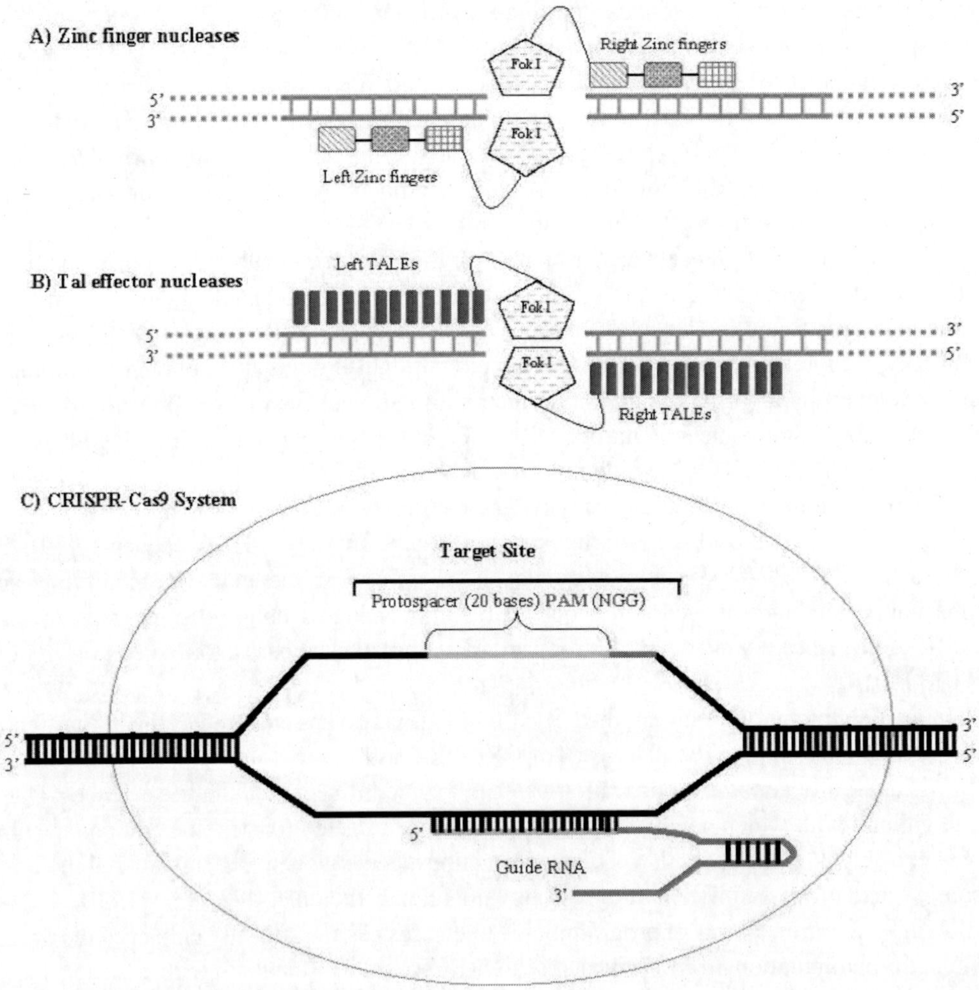

Figure 2. Schematic diagram depicting three major classes of programmable nucleases *viz.* ZFNs (A), TALENs (B) and CRISPR-Cas9 systems (C).

Zinc Finger Nucleases (ZFNs)

The engineering of unique zinc-finger nucleases (ZFNs) or meganucleases was used in the early stages of genome editing to generate the necessary DSBs at each specific DNA target site. ZFNs are fusion proteins made up of an array of site-specific DNA-binding domains modified from zinc finger–containing transcription factors and coupled to the bacterial FokI restriction enzyme's endonuclease domain. Each zinc finger domain recognises a 3- to 4-bp DNA sequence, and tandem domains can possibly attach to a longer nucleotide sequence that is unique within a cell's genome (generally a multiple of 3, commonly 9 bp to 18 bp) [29, 30]. ZFNs are designed as a pair to cleave a specific location in the genome by recognising two flanking sequences, one on the forward strand and the other on the reverse strand. The pair of FokI domains dimerize and cleave the DNA at the site when the ZFNs on either side of the site bind, resulting in a double-strand break (DSB) with 5′ overhangs [49]. Cells are repairing themselves. DSBs are repaired using one of two methods: (a) non-homologous end joining (NHEJ), which can occur at any time during the cell cycle but can occasionally result in erroneous repair; or (b) homology-directed repair (HDR), which typically occurs during the late S phase or G2 phase when a sister chromatid is available to serve as a repair template [21].

Three important aspects influence ZFNs' target sequence identification and specificity: (a) the amino acid sequence of each finger; (b) the number of fingers; and (c) the nuclease domain's interaction. Because of ZFNs' modular nature, scientists can independently adjust both the DNA-binding and catalytic domains, allowing them to construct innovative modular assemblies with adequate affinity and specificity for genome engineering. Individual ZFNs with 3-6 fingers were employed in early investigations to connect with a 9-18 nucleotide target, allowing ZFN dimmers to specify 18-36 bps of DNA at each cleavage site [50]. Despite the benefits of genome editing with ZFNs, there are a number of drawbacks. It has not been easy to put together zinc finger domains that can bind a long stretch of nucleotides with high affinity and high cost of creating protein domains for each genomic region, as well as the risk of inaccurate target DNA cleavage due to single nucleotide changes or unsuitable domain interaction. As a result, the constant search for new genome editing technologies has continued [51].

Transcription Activator-Like Effector Nucleases (TALENs)

The genome editing revolution was sparked by TALE nucleases (TALEN), which followed in the footsteps of zinc finger nucleases (ZFN). TALEs (transcription activator-like effectors) are proteins found in *Xanthomonas* bacteria that cause plant disease. They are injected into plant cells and go to the nucleus, where they attach to specific promoters and trigger gene expression. There are three functional domains in TALE proteins. Their N-terminal domain contains the bacterial secretion signal as well as a nonspecific DNA-binding activity required for the protein's general affinity to DNA [52, 53, 54]. An interacting interface for the plant transcription factor IIA, two functional nuclear localization signals, and an acidic activation domain are all found in their C-terminal domain [55]. A core repeat region, which consists of a variable number of generally 33-35 amino acid-long tandem repeats, is crucial to their specific and programmed DNA-binding. The two amino acids at positions 12 and 13 of each repeat, the repeat-variable di-residues (RVD), determine the DNA specificity of a TALE or TALEN [32].

All potential RVD combinations were decoded, revealing particularly specific ones that recognise only a single nucleotide as well as more flexible ones that tolerate two, three, or all four nucleotides. The DNA binding specificities of TALEs and TALEN can be modified as desired by rearranging the repetitions [56, 57].

TALEs have two distinct qualities that enhance their applicability. The first focuses on the employment of unusually long repeat units seen in nature. When one of these aberrant repeats is included in a repeat array, a TALE or TALEN can recognise not just normal target sequences but also those with a 1 bp deletion near the aberrant repeat [58]. The second feature is based on the ability of some RVDs to discriminate between nucleotides with different methylation states [59]. TALEs can even distinguish between two target DNA sequences that are only separated by a single methylated nucleotide, implying that TALEs and TALEN can be utilised to manipulate methylation-dependent genes [60, 61]. TALE nucleases (TALEN) were the first device that could be easily designed and manufactured to target any specific genomic location with high precision and efficiency. TALEN was quickly adopted for genome editing in crops, animals, and a variety of other model and non-model organisms. As a result, in 2015, the TALEN technique was the first genome editing tool to save a human life by curing cancer, and in 2019, the first genome-edited crop was introduced to the market [62, 63].

Clustered Regularly Interspaced Short Palindromic Repeats (CRISPR-Cas9)

The field of genomics is now experiencing a revolutionary phase with the arrival of simple genome engineering in animals and plants adopting RNA-programmable CRISPR-Cas9. The CRISPR-Cas9 technology originates from type II CRISPR-Cas system, which provides bacteria with adaptive immunity to viruses and plasmids [64]. CRISPR-Cas9, which stands for clustered regularly interspaced short palindromic repeats and CRISPR-associated protein 9, is an adaptive immune system found in bacteria and archaea that was discovered in 2012 and used for programmable and precise gene editing. CRISPR arrays are DNA segments created by bacteria that catch fragments of DNA from invading viruses. Bacteria can "remember" viruses thanks to CRISPR arrays (or closely related ones). If the viruses return, the bacteria synthesise RNA segments from the CRISPR arrays to attack the viruses' DNA. The bacteria then employ Cas9 or a similar enzyme to cleave the virus's DNA apart, making it ineffective. In the lab, the CRISPR-Cas9 system operates similarly. A little piece of RNA with a short "guide" sequence that binds to a specific target sequence of DNA in a genome is created by researchers. RNA interacts with the Cas9 enzyme as well. The modified RNA recognises the DNA sequence, just like bacteria, and the Cas9 enzyme cuts the DNA at the desired spot.

Although Cas9 is the most commonly employed enzyme, other enzymes (such as Cpf1) can also be used. Researchers employ the cell's own DNA repair mechanism to add or delete fragments of genetic material or to make modifications to the DNA by replacing an existing segment with a modified DNA sequence after the DNA has been cut [42, 43, 44, 39, 65]. Three forms of CRISPR-Cas genome editing have been found since 1987. There are three ways for Cas protein expression based on the gene sequence: type I for Cas3 and Cas6, type II for Cas9, and type III for Cas10. The functional differences between these types are due to the varied Cas cascades utilised for cleaving, which can involve single or double proteins [66]. The value of CRISPR/Cas9 in genome editing was finally recognised when Emmanuelle Charpentier and Jennifer Doudna were awarded the Nobel Prize in Chemistry in 2020 [67].

A variety of molecular tools are emerging for genome editing in a diverse range of living organisms, including humans, animals, and important plant species. These techniques are proving to be more accurate, effective, economical, and consistent. Because of their accuracy, reliability, cost-effectiveness, and accessibility, CRISPR/Cas9 and other CRISPR/Cas methods are considered revolutionary technologies that permit modifications in nuclear genomes with unprecedented precision. Genome editing (GE) tools, such as zinc finger nucleases, transcription activator-like effector nucleases, and clustered regulatory interspaced short palindromic repeats (CRISPR-associated) systems, have revolutionised life sciences by allowing for the targeted modification of genomic DNA in all organisms. In addition to animal disease modeling, material science, genetically modified plant technology, biofuel technology, gene therapy, and pharmaceutical research, it now holds significant potential in a variety of sectors. Furthermore, CRISPR/Cas9 technology has contributed in the building of robust causal linkages between genetic variants and biological phenotypes by speeding up the study of the functional structure of the genome at the systems level [1, 64, 36, 68]. Various kinds of CRISPR/Cas systems have been developed; however, off-target effects may limit their widespread application [48].

Off-Target Effects

The concept of programmable nucleases is gaining momentum due to its intended applications. However, there are still a number of obstacles to overcome, including pre-existing immunity, *in vivo* delivery efficiency, and the possibility of off-target consequences. The off-target effects have emerged as one major limiting factor for the precise use of this technology, particularly in therapeutics [47]. A programmed nuclease-induced DNA cleavage at a place in the genome other than the intended target site is known as an off-target event. An off-target event is followed by a DNA repair event via the NHEJ pathway. This DNA repair event often results in mutations or gene activation, depending upon whether the off-target event occurred alone or along with another on-target or an off-target event. In the former case, DNA repair results in an indel, while in the latter case it can result in a variety of chromosomal rearrangements including inversions [69, 70], translocations [71, 72], deletions [73] and local mutations [74]. Off-target effects can be broadly categorized into two categories. The first category of off-target effects is governed by the sequence homology at the target site, while the other type of off-target effects occurs at non-homologous genomic locations [75].

Factors Responsible for CRISPR/Cas9 Off-Target Effects

As reviewed earlier [76], a number of factors are responsible for the off-target effects are summarised below:

I. *Frequency of homologous sequences:* The frequency of homologous sequences in the genome of interest plays an important role in influencing the off-target effects. More the number of sequences similar to the intended target region the more likely it leads to off-target cleave events [77]. A variety of bioinformatics tools are available to predict multiple potential target sites in the genome [40, 78].

II. *The expression level of nuclease:* The expression level of nucleases is very important, as a higher concentration of nucleases increases the risk of off-target site binding and cleavage. People have tried to control the nuclease expression by lowering the amount of nuclease expression plasmid [77]. However, this strategy does not always work as nuclease expression can only be lowered to a certain extent without affecting the on-target cleavage efficiency [76].

III. *The amount of time exposed to nuclease:* Although stable production of designed nucleases via plasmid transfection or integrated viral vectors such as lentiviral vectors is generally harmless to cells; extended nuclease expression is expected to promote off-target breakage. The ideal technique for gene editing is nuclease expression using either transiently transfected plasmid DNA or employing the non-integrating viral vectors [76].

IV. *Accessibility of the target site:* The accessibility of the target site, which is governed by the epigenetic status of the target site, plays an important role in correct recognition and cleavage by nucleases. Improper accessibility may lead to potential off-target effects, as revealed in many studies [79, 80].

V. *Molecular interaction:* Molecular interaction between the target site and programmable nucleases as well as various components of the programmable nucleases system, play a very important role in identifying correct target sites and precise editing.

In the case of CRISPR/Cas9 cleavage, the Cas9's targeting fidelity is determined by the 20 nts long "single guide RNA" (sgRNA) as well as the PAM sites adjacent to the intended target sequence. More than three mismatches between the 20 nucleotides of gRNA and the target sequence may give rise to off-target effects [81]. Four mismatches in the PAM-distal part of the sgRNA and the target have been demonstrated to result in off-target cleavage [82]. Moreover, the number of mismatches, type of mismatch and its distance from the PAM sequence (mismatch position) also play an important role in off-target events and has been used by various researchers to develop off-target prediction strategies. Also, off- and on-target cleavage is also influenced by the structure of the gRNA [48]. In addition, GC content of the target sequence and the method of delivery may also determine off-target effects [83].

Strategies to Counter Off-Targets

As discussed in the previous section, the off-target cutting activity of a programmable nuclease at a specific site in a genome depends upon its homology with the sequence of the target site, molecular interaction, expression levels, exposure time, and accessibility. The first step dealing with off-targets is to identify the potential off-target sites [47]. Several computational as well as experimental strategies have been developed to detect off-target sites for different categories of programmable nucleases. This section highlights some of the most commonly employed methods to deal with the off-targets. Owing to the widespread use of CRISPR-Cas9 technology for gene editing, the off-target detection methods catering to CRISPR-Cas9 will be the focus of the discussion, while briefly touching on strategies related to ZFNs and TALENs.

Experimental Methods

Both *in vitro* and *in vivo* experimental techniques have been used to identify off-target sites (Table 1). These experimental techniques, also known as unbiased approaches, can be broadly classified into the following three categories [47].

Detection of Nuclease Binding

Initially developed for ZFNs, this approach relies upon characterizing the specificity with which the monomeric zinc finger proteins bind to the DNA. Various experimental techniques including ELISA [84], microarrays [85], bacteria-1-hybrid screening [86], SELEX and its variants [87, 88, 89] have been used for this purpose. The complexes of the individual zinc finger proteins and the DNA could then be used to scan the genome under consideration to locate off-target sites (homodimeric or heterodimeric) for ZFNs. The detection of Cas9 binding to the target DNA sequences has also been characterized using DNA sequencing technologies under *in vivo* [90, 91] as well as *in vitro* [92] conditions. However, this seemingly simple and straight forward approach is least informative due to the fact that nuclease binding is necessary but not sufficient to cleave DNA. In addition, these techniques tend to generate a large number of false positives, limiting their use [47].

Detection of Nuclease-Induced DSBs

This approach is based on detecting the nuclease activity, i.e., measuring double-stranded breaks (DSBs), rather than measuring their binding specificity [47]. This approach of directly discovering and quantifying off-targets is also called *in vitro* detection or *in vitro* genome-wide assay. The researchers employed DNA substrates created by rolling circle amplification of a random library to discover which sequences could be cleaved *in vitro* by a pair of ZFNs [93]. This data was then utilised to create a statistical model known as an energy compensation model that might predict off-target genomic cutting based on the observation that off-target cleavage by ZFNs leads to excess binding energy. This strategy of using synthetic DNA substrates for off-target identification has also been employed for the CRISPR-Cas9 system by many researchers [79, 94, 95]. The most commonly used methods in this category include Digenome-seq, SITE-seq, CIRCLE-seq, and DIG-seq (Figure 3).

Digenome-Seq

Digenome-seq is an *in vitro*, unbiased approach for detecting off-target cleavage [96]. This strategy involves *in vitro* digestion of genomic DNA using CRISPR-Cas9 followed by whole genome sequencing at a typical sequencing depth of 30x [97]. The difference in alignment patterns of the targeted sites as compared to other staggered reads upon nuclease cleavage is used to identify off-targets. This is due to the fact that endogenous DSBs occur at random places, but nuclease-induced targeted DSBs occur at specific sequence positions [98]. Therefore, distinct signatures of the *in vitro* cleavage sites can be employed to match on- and off-target sites using web-based software. Digenome-seq achieves target enrichment by giving nuclease cleavage targets a distinct signature, which enhances cleavage detection resolution to single-nucleotide precision. For detecting off-target locations, Digenome-seq is more sensitive and extensive than cell-based approaches. Moreover, unlike other cell-free approaches, it does not rely on PCR amplification to enrich DNA ends. This technique can also be used to understand the effect of chromatin structure on off-target nuclease activity by using histone-free genomic DNA in addition to chromatin DNA [97].

Table 1. An overview of software tools for sequence homology based off-site prediction

Name	Description	Mode	Webpage/download page link	Strengths	Limitations	Reference
Blast	The Basic Local Alignment Finder (BLAST) optimizes the measurement of local similarity; maximum fragment pair score (MSP). It can be implemented in many ways and applied in a variety of contexts, including simple searches in DNA and protein sequence databases, pattern searches, gene identification searches, and in the analysis of DNA sequences. The accumulation of many similar regions in a long DNA sequence.	Command-line, web server	https://blast.ncbi.nlm.nih.gov/Blast.cg	Time-optimized alignment of sequences using a seed- based algorithm, allows bulges	Low accuracy (may miss potential non-target locations), Mismatches allowed to a limited number, and user-defined PAM not allowed	[133]
Burrows-Wheeler Aligner	Reference Genome Alignment with predefined limit for the number of mismatches.	Command-line	https://github.com/lh3/bwa	Fast and accurate	Not straight forward to use	[150]
Bowtie	Bowtie is an ultra-fast, memory-efficient alignment program for aligning short DNA sequences with large genomes. For the human genome, Burrows Wheeler indexing allows Bowtie to perform over 25 million reads per hour on a CPU with around 1.3 GB of memory.	Command- line	https://github.com/BenLangmead/bowtie	Time-optimized alignment of sequences	Low accuracy (may miss potential non-target locations), Mismatches allowed to a limited number, and user-defined PAM not allowed; bulges not allowed	[134]
Bowtie2	Bowtie2 uses pseudo-random numbers internally. When performing multiple runs to estimate variability, up to five different starting values were used. With the right termination sequence, theI and X parameters provide the allowable separation range between the two ends.	Command- line	https://github.com/BenLangmead/bowtie2	Time-optimized alignment of sequences, allows bulges	Tendency to overlook off-targets with fewer mismatches, Mismatches allowed to a limited number, and user-defined PAM not allowed	[135]

Name	Description	Mode	Webpage/download page link	Strengths	Limitations	Reference
CasOT	CasOT is a local tool designed to find potential off-target sites in any genome or user-supplied sequence.	Command- line	http://eendb.zfgenetics.org/casot/.	Allows custom mismatch number in seed and non-seed regions, allows user-defined PAM as well as gRNA mode	Slow, bulges are not allowed	[151]
Cas-OFFinder	Cas-OFFinder is OpenCL based, ultrafast and versatile program that searches for potential off-target sites of CRISPR/Cas-derived RNA-guided endonucleases. Requires an OpenCL device to run properly.	Command- line, web server	http://www.rgenome.net/ Cas-offinder	Allows custom mismatch number, allows user-defined PAM, allows bulges, Availability of web support	Tendency to miss DNA/RNA regions with complex bulges, moderate speed	[149]
Off-Spotter	Off-Spotter identifies potential off-target sites by using a reference genome index of restriction sites for PAMs.	Command- line, web server	https://cm.jefferson.edu/ Off-Spotter/	Fast, flexible, optimize gRNA design using suitable PAMs, flexible output, allows use of pre-defined number of mismatches	Do not consider DNA/RNA bulges	[148]
dsNickFury	This program is designed to look for potential off-target effects by any type of CRISPR-Cas9 system, including effects of surrogate species with different PAM sites. This program requires python3 and a cluster system with a job queue like openSGE or similar, and an active internet connection.	Command- line	https://github.com/ michael-weinstein/dsNickFury	Sequence screening tool for CRISPR–Cas9 system, customized PAM, user-defined mismatch number.	Bulges are not allowed	[147]
FlashFry	FlashFry is a rapid and ductile Command- line tool for characterizing a large number of potential CRISPR target sequences.	Command- line	https://github.com/mckennalab/ FlashFry	A time optimized series screening device for CRISPR–Cas Systems, custom PAM, user-described mismatch number. Good for huge datasets	Bulges are not allowed	[146]

Table 1. (Continued)

Name	Description	Mode	Webpage/download page link	Strengths	Limitations	Reference
	FlashFry can be used with any genome and can be active against non-traditional model organisms or transcriptomes.					
Crisflash	Crisflash is a rapid design tool and potential off-target discovery tool for CRISPR-Cas9 guided RNA. Crisflash is designed to improve speed, gRNA matching, and scoring accuracy by providing the ability to integrate user-provided variant data.	Command- line	https://github.com/crisflash/crisflash	Custom Sequence Screening Tool for CRISPR-Cas9 System Accept PAM, custom mismatch numbers, Allow genetic haplotype level variation data	Bulges are not allowed	[152]
CRISPRitz	Software tools to aid in the design and analysis of CRISPR/CRISPR associated (Cas) experiments. By using efficient data structures combined with parallel computation, we provide a fast, reliable and complete search engine for listing a complete list of hypothetical sites off-target sites.	Command- line	https://github.com/pinellolab/CRISPRitz	A time-optimized collection screening device for CRISPR–Cas9 system, custom PAM, user-described mismatch number, bulges allowed, accepts genetic variant data	Genetic variation allowed but not at haplotype level	[145]

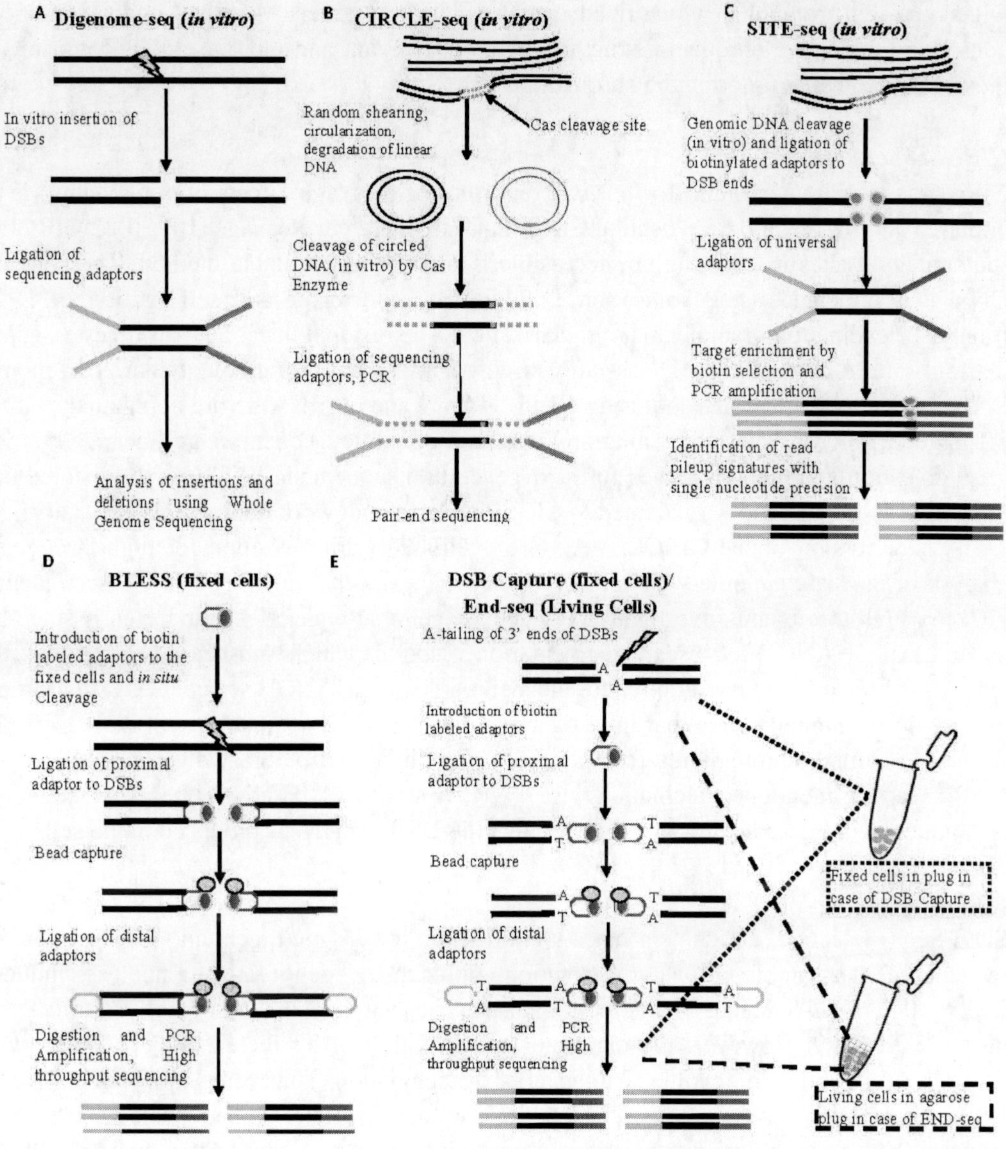

Figure 3. Schematic diagrams showing overall methodology used by various *in vitro* (A-C) and *in vivo* (D-E) methods used for detection of nuclease-induced DSBs.

Researchers have tried many modifications to further improve the efficiency of Digenome-seq. One such approach, multiplex Digenome-seq [99], has been recently introduced, which allows the simultaneous use of multiple gRNAs on the same sample. As compared to the original algorithm that produces blunt ended fragments, the multiplex version of Digenome-seq allows 1-2 overhangs, thereby enabling the identification of cleave events that produce different end moieties. The multiplex Digenome-seq strategy detects more targets as compared to the parent method and reduces false-negatives as well as false-positives, thereby improving accuracy [98]. Another modification of the characteristic Digenome-seq called DIG-seq has been introduced, which keeps the chromatin architecture intact [100]. In this modification, the native chromatin is isolated using nuclease extraction followed by cleavage using the standard

Digenome-seq protocol. It is assumed that the cleavage targets identified under conditions conserving the native chromatin structure are more relevant and more useful as compared to those obtained while ignoring the chromatin effect [98].

CIRCLE-Seq

CIRCLE-seq (Circularization for *in vitro* reporting of cleavage effects by sequencing) is an unbiased *in vitro* method for detecting CRISPR-Cas9 off-target cleavage [101, 102]. CIRCLE-seq strategy makes it a very rapid, accessible, sensitive, and scalable method. The genomic DNA is fragmented using sonication, followed by end-repair and self-ligation of DNA fragments leading to intramolecular circularization. The residual linear DNA is digested with a plasmid-safe, ATP-dependent DNase after circularization. This library of circularized genomic DNA fragments is then digested using CRISPR/Cas9 and a gRNA before being analyzed for off-target affinity. Circles containing on-target and off-target sequences are linearized during Cas9 digestion and then processed for next-generation sequencing [98]. It is to be noted that the CIRCLE-seq method was adapted to CRISPR-Cas9 from work done on ZFNs [93, 79].

A new version of the CIRCLE-seq, CHANGE-seq (circularization for high-throughput analysis of nuclease genome-wide effects by sequencing) technique has been released recently [103] for high-throughput investigation of nuclease genome-wide effects. In the early stages of its process, CHANGE-seq uses a tagmentation reaction, which substantially decreases the labour and preparation time required for the methodology. CHANGE-seq enables faster sample processing for higher-throughput investigations than CIRCLE-seq, and it will most likely be the preferred method for any future experiment including *in vitro* library digestion [98].

The above-mentioned techniques use *in vitro* strategies based on DSBs. Some other techniques as discussed below attempt to detect the DSBs *in vivo* as produced in the cells.

SITE-Seq

SITE-Seq (Selective enrichment and Identification of Tagged genomic DNA Ends by sequencing) is an unbiased *in vitro* detection approach for identifying the nuclease-induced DSBs [104]. SITE-Seq, like Digenome-seq, also carries out *in vitro* digestion of genomic DNA by CRISPR-Cas9. However, the cleavage step is followed by the ligation of DSBs with biotinylated Illumina-compatible adapters after 3′ adenylation. This results in a pool of labeled DSBs in genomic DNA, most of which are caused by gRNA-guided nuclease cleavage, allowing for selective enrichment of sequence around cleavage sites. After the strand break sites are labeled, the genomic DNA is fragmented, followed by end-repairing and 3′ adenylation, enabling another round of Illumina-compatible adaptor ligation. As a result, segments carrying sequences from one side of a DSB are bookended exclusively by the P5 and P7 binding sites required for Illumina sequencing. Following biotin selection and PCR amplification, a selectively enriched deep-sequencing library containing sequences primarily surrounding nuclease cleavage targets is generated. Similar to Digenome-seq, SITE-seq also uses the distinguishing patterns of alignments of nuclease induced DSBs of target sites as compared to the random sites. The selective enrichment of nuclease cleavage of target sequences leads to a higher signal-to-noise ratio, thereby improving accuracy and sensitivity (0.1%) in comparison to Digenome-seq [98].

BLESS

BLESS (Breaks Labeling Enrichment on Streptavidin and next generation Sequencing) was perhaps the first *in vivo* method to detect DSBs [105]. To purify intact nuclei, cells containing either endogenous or induced DSBs are cross-linked with formaldehyde, followed by lysis and incubation with proteinase K. In the following step, DSBs are blunted, 5′-phosphorylated before being *in situ* labeled by a short hairpin-like proximal biotinylated adapter and are then enriched on streptavidin beads [106]. These captured DSBs are then ligated to another hairpin-like distal adapter, followed by PCR amplification using primers bound to the proximal and distal adaptors. The amplified products are finally subjected to high-throughput sequencing.

DSBCapture

DSBCapture (Double Stranded Breaks Capture) follows the strategy of BLESS with some modifications, including the addition of a non-template nucleotide to the 3' end of the blunted DSBs (called A-tailing) and the addition of Illumina RA5 sequence directly to the BLESS adaptor [107]. These modifications lead to better ligation efficiency for the BLESS adaptors and also increased sensitivity of DSB detection in comparison to the parent BLESS technique [106].

END-Seq

END-seq (End Sequencing) method works in a similar fashion to BLESS and DSBCapture with a few modifications. In END-seq [108], low-melting agarose plugs are used to immobilize cells. The DSBs are blunted using exonucleases, which are then directly ligated with sequencing adaptors. An advanced version of END-seq enables the addition of non-templated nucleotides (A-tailing) to the 3' end of blunted DSBs leading to an increase in ligation efficiency. This advanced version also allows integration of the Illumina RA5 adapter sequence into the adapter. END-seq has been used to identify off-targets generated using RAG recombinase as well as gene-editing nucleases, including ZFNs, in addition to many other applications [109]. END-seq can also be used for detecting off-targets generated through CRISPR systems [106] and other therapeutic nucleases [108].

BLISS

BLISS, an acronym for Breaks Labeling *In Situ* and Sequencing (BLISS), is an *in vivo* method to identify genome-wide occurrences of DSBs [110]. In this procedure, the cell or tissue samples are immobilized on a solid surface allowing further steps to be carried out *in situ* as well as minimizing the artificial DNA breaks and sample loss. T4 DNA polymerase is used to blunt DSBs *in situ*, followed by ligation of a double-stranded DNA oligonucleotide linker to the blunt ends (Figure 4). This linker consists of the following [106]:

a) T7 promoter sequence
b) RA5 Illumina sequencing adapter
c) A short random stretch of about 8–12 nucleotides to act as unique molecular identifier (UMI)
d) A barcode sequence to enable multiplexing of the sample

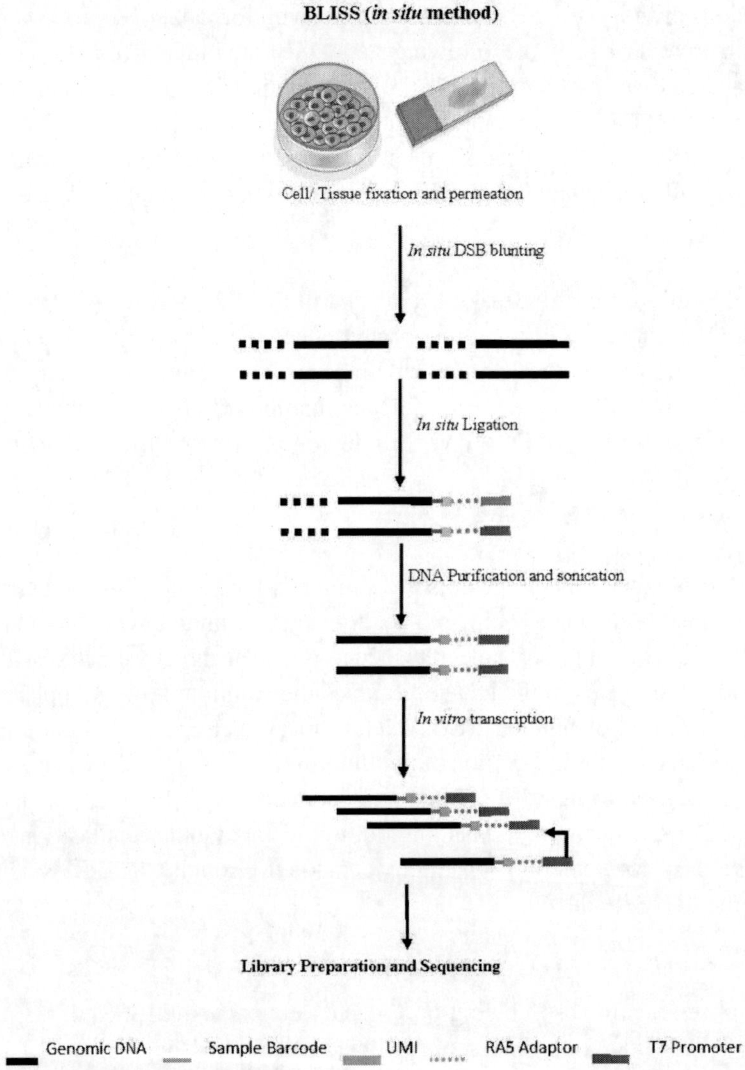

Figure 4. Schematic diagram depicting overall strategy employed for *in situ* BLISS off-target detection method.

The T7 promoter sequence, which is incorporated in the BLISS linker, is used to linearly transcribe tagged DSBs *in vitro*. The material is PCR amplified after the addition of the second sequencing adapter. Finally, high-throughput sequencing is performed on this BLISS library. The design of the BLISS linker has many advantages. Inclusion of the T7 promoter leads to linear amplification of only tagged DSB sites, thereby increasing the sensitivity of the method in addition to reducing PCR amplification biases. Integrating UMI filters out PCR duplicates and allows distinguishing and quantifying breaks occurring at the same nucleotide position in different alleles or cells. The use of barcodes helps label different samples for multiplexing as well as reducing the cost.

Detection of Repair Products Arising from DSBs

The cells have an ability to repair DSBs, and the repair products accumulate over the time. This category of off-target detection utilizes this ability to detect repair products consisting of mutated sequences [47]. Initially, this technique was tried for ZFNs by using a technique called IDLV (integrase deficient lentiviral vector) capture [111]. However, this approach proved low in sensitivity and was found to detect off-target sites showing >1% activity. Also, the performance of this technique was inferior to *in vitro* ZFNs cutting assays [112]. Recently, many advanced techniques to detect the repair products have been introduced, as discussed below (Figure 5).

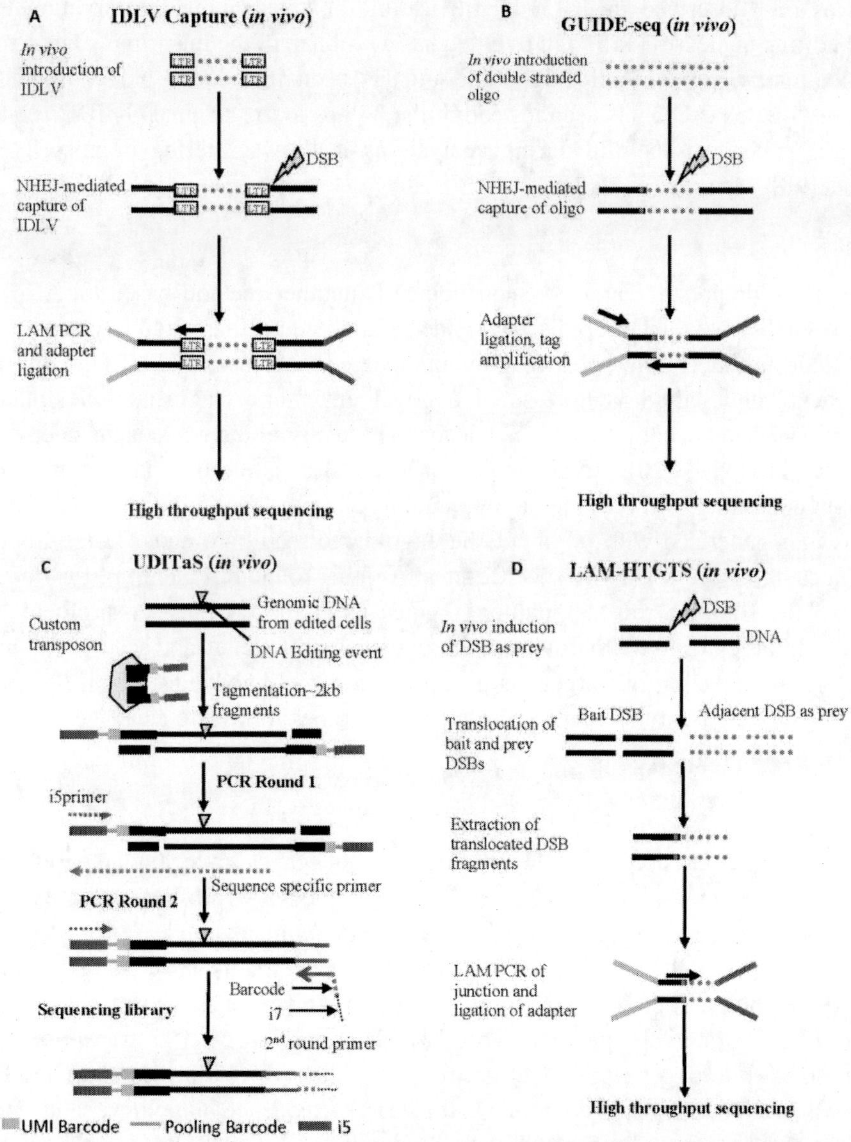

Figure 5. Schematic diagrams showing overall methodology used by various *in vivo* (A-D). Detection of repair products arising from DSBs.

GUIDE-Seq

GUIDE-seq (Genome-wide Unbiased Identification of DSBs Enabled by Sequencing) is a technique for labelling and enriching the sequence around DSBs in preparation for deep sequencing [113]. GUIDE-seq evolved from IDLV and anchored multiplex PCR (AMP) [114]. The tagging of the Cas9 generated DSBs is carried out *in vivo* using blunt, end-protected, double-stranded oligodeoxynucleotide (dsODN) through NHEJ-mediated pathway. Fragmentation of the extracted genomic DNA is then carried out by using either enzymatic cleavage or sonication. The resulting fragments are then subjected to end-repair and dA-tailing followed by addition of universal adapters at both the ends. These labeled DSBs are specifically amplified using PCR to achieve target enrichment. As a result, the amplified library is made up of strands that each carry half of the sequence around an NHEJ-repaired DSB. The technique, which was first described in 2015, is still useful for determining the targeting fidelity of genome-editing nucleases [115]. GUIDE-seq has a problem of the mispriming and to overcome this issue, a new version called iGUIDE-seq has been introduced [116]. In the iGUIDE technique, a 46 bps dsODN is used instead of the 34 bps in the original GUIDE-seq approach. This extra sequence is helpful during analysis as it allows filtering of misprimed library fragments [98].

UDiTaS

UDiTaS (Uni-Directional Targeted Sequencing) is another method based on AMP strategy. Similar to GUIDE-seq, UDiTaS also implements anchored primers and universal adaptors to detect repair products after cleavage by engineered nucleases [117]. One of the major advancements in UDiTaS is the use of a novel enzymatic fragmentation strategy called tagmentation instead of sonication, which leads to base miscalling while using deep sequencing technology [118, 119, 120]. An advancement in UDiTaS is the introduction of a novel Tn5 transposon consisting of a barcode, a UMI, and an Illumina forward adapter (i5). A genomic library is thus generated with fragments having adaptors on both ends. These fragments are then subjected to PCR to amplify specific sites. Another round of PCR amplification is carried out to add the Illumina reverse adapter (i7) just as in the GUIDE-seq protocol. Although UDiTaS technology leads to improvement in robustness, efficiency and accuracy, it has limited applicability in detecting off-targets as it requires *a priori* knowledge about the target sites. However, UDiTaS' ability to characterise repair outcomes following cleavage by nucleases is extremely useful [98].

TTISS

TTISS (Tagmentation-based Tag Integration Site Sequencing) is a technique based on GUIDe-seq, modified for multiplex detection of off-target cleavage profile nucleases [121]. The determination of the off-target cleavage profile of programmable nucleases plays an important role in genome editing experiments, and many Cas9 variants have been published with improved specificity. Therefore, it becomes helpful to analyze cleavage patterns of DSBs using different Cas9 variants in parallel. The TTISS strategy uses Tn5 transposes [122] for tagmentation, while target enrichment is carried out through two cycles of nested PCR after purifying the DNA with a spin column. On the basis of sequence homology, sequenced targets are attributed to a specific Cas9-gRNA pair. The authors applied TTISS to eight Cas9 variants covering 59 targets [121]. In general, the results show a trade-off between specificity and activity, with the exact ratio varying among the Cas9 variations [98].

HTGTS/LAM-HTGS

HTGTS (High-Throughput Genome-wide Translocation Sequencing) is a technique for detecting and sequencing translocations caused by DSBs [123]. HTGTS was originally introduced as a method for studying mechanisms of translocation and has now been adapted to analyze off-target events [124]. The technique relies on nucleases inducing DSB in a known genomic region *in vivo* and then using this DSB as "bait" that translocates to another DSB from a different genomic region as "prey." After then, the translocated junction is PCR-amplified and sequenced in high-throughput. Using HTGTS, the authors were able to demonstrate that DSBs tend to translocate with other actively transcribed regions [123]. Recently, a modified HTGTS methodology known as LAM-HTGTS (linear amplification mediated high throughput genome-wide translocation sequencing) has been introduced to enhance target enrichment [124]. Two notable modifications in the original HTGTS method are the introduction of linear amplification using LAM PCR and bridge adaptor ligation [98]. In LAM PCR, linear amplification across junction sites is carried out by targeting the known half (side of the DSB at the bait site) of each captured junction using a 5′ biotinylated primer. Streptavidin based selection is then carried out to isolate target sequences from genomic DNA. Bridge adapter ligation makes use of a double-stranded linker with a nucleotide-variable 3′ overhang to make it easier to connect adapters to a single-stranded library produced by linear PCR [125, 124, 126]. When compared to the original HTGTS technique, LAM-HTGTS is reported to generate 10–50 times more junctions [126].

ChIP-Seq

The ChIP process starts with the cross linking of a single-cell suspension using formaldehyde [127] followed by the extraction of nuclei and their fragmentation by sonication. Then, bead-bound antibodies are used to pull down fragments of interest enabling the analysis of protein-DNA interactions [128, 129]. These pulled-down DNA fragments are then processed and subjected to deep sequencing [129]. ChIP has been widely used to capture the sequence around DSBs and therefore study and characterize the genomic landscape of DSBs [98]. The immunoprecipitation based off-target detection ChIP-seq technique used for fixed cells is represented in Figure 6 (A).

DISCOVER-Seq

DISCOVER-seq is a method to detect off-targets by employing ChIP-seq technology based on chromatin immunoprecipitation. This method selectively amplifies CRISPR/Cas9 cleavage sites through recognition of signatures of endogenous DNA repair processes [129]. DISCOVER-seq is an improvement over the original ChIP-seq method and utilizes a DNA repair protein called Meiotic Recombination 11 Homolog 1 (MRE11). MRE11 is a part of the MRE11-RAD50-NBS1 (MRN) complex, which takes part in a variety of DNA damage responses (DDRs) including DSB repair [98]. The way the MRE11 subunit handles distinct DSB end-moieties, may give some indications about the pathway (HR or NHEJ) used to repair DSBs [130, 131]. Similar to other approaches employed for off-target detection, DISCOVER-seq finds DSBs with single-nucleotide resolution. DISCOVER-seq, on the other hand, is said to have a sensitivity threshold of 0.3 percent, which is slightly greater than other modern approaches [98]. Although other techniques may have higher sensitivity, DISCOVER-seq along with VIVO [132] is the only two strategies that have been reported to detect off-target events in an animal model *in vivo* [129]. The immunoprecipitation based off-target detection DISCOVER-seq technique used for fixed cells is represented in Figure 6 (B).

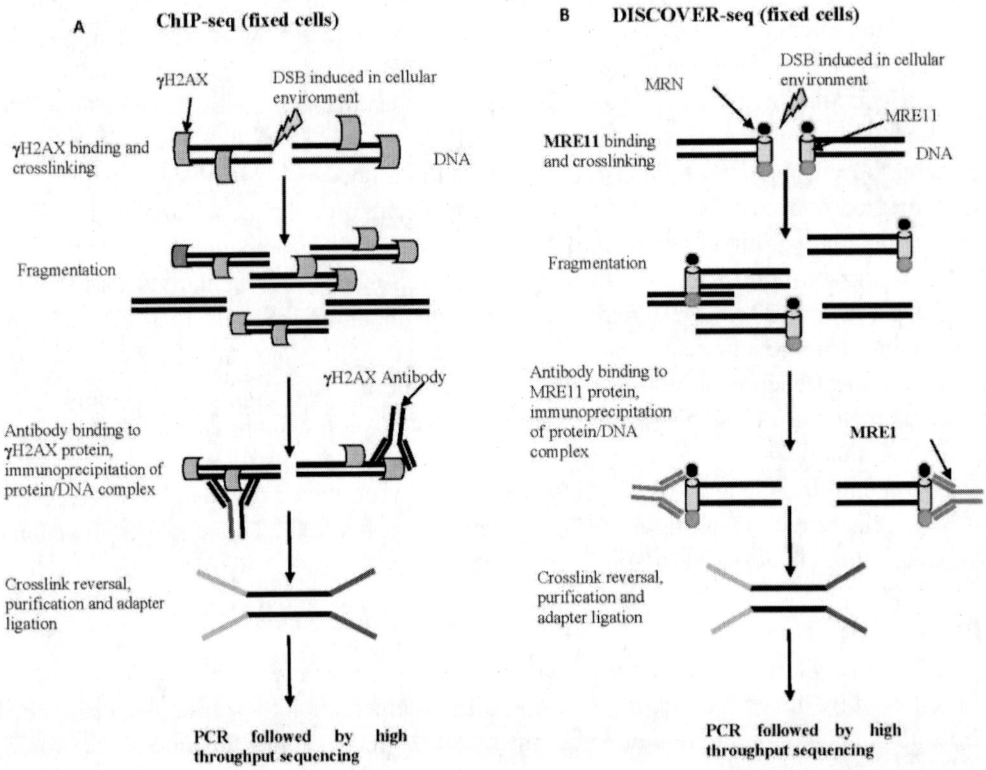

Figure 6. Schematic diagram showing the immunoprecipitation based off-target detection techniques, ChIP-seq (A) and DISCOVER-seq (B) used for fixed cells. Adapted from [106, 98].

Bioinformatics Approaches for Off-Target Detection

In addition to unbiased experimental approaches for off-target identification, a large number of *in silico* strategies have been introduced recently focused on detection as well as analysis of off-targets. Bioinformatics based approaches can be divided into two stages [47]. In the first stage, known as site detection, a homology-based search is carried out on the target genome to predict potential off-target sites. The query used for this homology search is the on-target sequence. This stage of off-target prediction has evolved from using simple homology search algorithms such as BLAST [133] to more sophisticated read mapping algorithms such as Bowtie [134] and Bowtie2 [135] which have been implemented in a number of software applications developed for this purpose (Table 2). These programmes however, have some inherent limitations as per their application to off-target detection. These programmes have been developed to locate homologous regions for relatively larger sequences instead of fragments of smaller size (12-24 bp). Moreover, they may lead to a large number of mismatches and indels, rendering them not much suitable for off-target detection. Therefore, more relevant and accurate strategies have been proposed for both ZFNs/TALENs as well as the CRISPR-Cas9 system.

The first stage of sequence homology-based search provides a number of (tens to hundreds) off-target sites which are scored/ranked in the second stage based on the degree sequence homology or cutting activity of the programmable nuclease. The second stage of analysis gives an opportunity to further explore and validate the top-ranking sites using experimental techniques such as deep sequencing. The second step of analysis can be achieved by following two strategies. One strategy used a predefined formula/feature, while the other strategy used Machine Learning (ML) and Deep Learning (DL) algorithms based on already existing off-target data. These two approaches largely depend on the availability of experimentally validated data to be used for training. Limited availability experimentally validated data sets related to true off-target sites limits the accuracy of both of these strategies [47]. In a nutshell, it can be ascertained that there are two categories of off-target prediction and evaluation methods for programmable nucleases *viz.,* alignment based and scoring based. These off-target detection tools differ in their strategy and design for various classes of programmable nucleases, *viz.,* ZFNs/TALENs and CRISPR-Cas9 systems [47], as discussed separately in the following sections.

In Silico Strategies for Evaluating Off-Targets for ZFNs and TALENs

Off-target detection strategies for ZFNs and TALENs have been inspired by the design of ZFNs and TALENs and the way they bind to their targets. Both ZFNs and TALENs consist of two binding domains separated by a spacer region made up of about 5-7 nucleotides. Off-targets detection algorithms rely upon variations in these nuclease binding sites and the range of spacer regions the off-targets exhibit. Initial strategies were based on homology-based detection using BLAST or Bowtie genera of algorithms, followed by their ranking as per the number of mismatches detected in the recognition domain [47]. Later on, more sophisticated tools were introduced for off-target prediction for ZFNs and TALENs.

Zinc Finger Tools

Zinc finger tools web server[1] uses user-provided DNA sequences to search for specific target sites involved in gene regulation or nuclease targeting [136]. ZFP binding to a particular target can be predicted using a database of experimentally characterized zinc finger domains. On the contrary, binding sites can also be predicted for a specific ZFP. This algorithm has a limitation of allowing less than 10 kb of input sequence.

ZFN-Site

ZFN-site web server[2] carries out genome-wide scanning for potential ZFNs on-target as well as off-target sites [137]. ZFN-site uses FetchGWI [138] for accurate search of large databases and allows user-defined spacing, mismatches, ambiguity, homo as well as hetero-dimeric binding sites. The ranking of off-targets is carried out based on sequence homology. However, this algorithm lacks any scoring system, thereby limiting its accuracy.

---

[1] https://www.scripps.edu/barbas/zfdesign/zfdesignhome.php.

[2] https://ccg.epfl.ch/tagger/targetsearch.html.

PROGNOS

PROGNOS (Predicted Report of Genome-wide Nuclease Off-target Sites) web server[3] has been designed to predict off-target effects for both ZFNs and TALENs [78]. In PROGNOS, a genome wide homology search is carried out to predict off-targets based on their degree of homology to the target region, followed by a ranking of hits on the basis of mismatches. The predicted off-target sites having the same score are further ranked on the basis of the class of the genomic region, following a specific order (Exon >Promoter >Intron >Intergenic). Finally, chromosomal location is used to further resolve a tiebreaker. PROGNOS implemented information about nuclease-DNA interactions in addition to using already existing off-target sites as a part of training datasets to derive refined scoring algorithms. PROGNOS also creates PCR primers for high-throughput analysis. PROGNOS is a robust method with relatively low false-positive rates and false-negative ratios comparable to experimental methods. However, limited data sets limit its applicability [47].

TALE-NT

TALE-NT 2.0 (TAL Effector-Nucleotide Targeter 2.0) web server[4] allows designing of pairs of TAL effectors for TAL effector nucleases (TALENs) to target a specific gene sequence [139, 140]. TALE-NT also includes tools for designing single TAL effectors that bind to a DNA sequence and identifying binding sites in a DNA sequence for a specific TAL effector. TALE-NT 2.0 is actually a bundle of tools, including TALEN Targeter (for designing TAL Effector Nucleases arrays), TAL Effector Targeter (for designing single TAL effector), Target Finder [for predicting best-scoring sites in a given nucleotide related to a specific repeat-variable di-residue (RVD) sequence] and Paired Target Finder (for predicting best-scoring sites related to a pair of RVD sequences). This algorithm uses information from binding specificities between naturally occurring TAL effectors (based on their RVDs) and the nucleotides present at the corresponding positions of the recognition domain to derive frequency matrices [47]. Another tool, TALENoffer/TALENgetter [141] also uses a similar strategy. However, it also includes information about changes in cutting efficiency of TALENs due to use of different RVDs. TALENoffer can be accessed using the GALAXY web server[5] or can be run in command-line mode.[6]

CHOPCHOP

The CHOPCHOP web server[7] is based on the Bowtie alignment algorithm and is applicable to both TALENs and CRISPR/Cas9 systems [142]. The efficient Bowtie algorithm makes the search faster and more accurate. The user-friendly web server accepts a variety of inputs (pasted sequences, genomic areas, gene identification) and offers a variety of target selection choices. Additionally, the primer candidates and restriction sites for each potential target site are shown to aid in the development of an efficient pipeline for the creation and validation of mutations.

---

[3] http://bao.rice.edu/Research/BioinformaticTools/prognos.html or https://www.expasy.org/resources/zfn-site.
[4] https://tale-nt.cac.cornell.edu/.
[5] http://galaxy.informatik.uni-halle.de/.
[6] http://www.jstacs.de/index.php/TALENoffer.
[7] https://chopchop.cbu.uib.no/.

CHOPCHOP is a useful tool for genome engineering since it allows users to quickly and simply identify the best target.[8]

In Silico Strategies for Identifying and Ranking Off-Targets for CRISPR-Cas9

As mentioned earlier, methods used for prediction of off-targets for CRISPR-Cas9 can be divided into two categories: i.e., Alignment-based and scoring- or ranking-based (strategies for off-target prediction and analysis for programmable nuclease mentioned in the Figure 7).

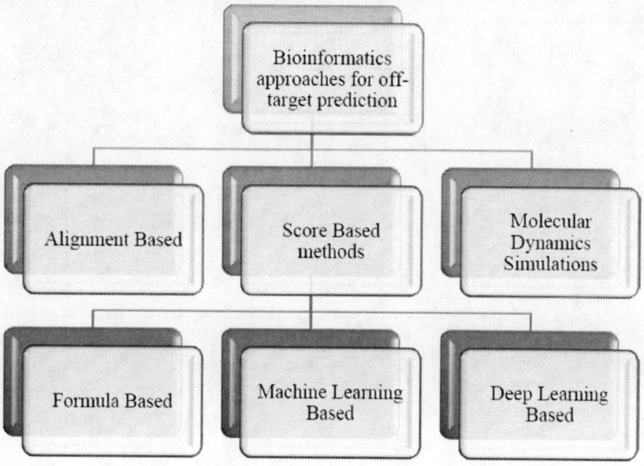

Figure 7. A layout depicting various Bioinformatics strategies for off-target prediction and analysis for programmable nucleases.

Alignment Based

In alignment-based algorithms for CRISPR-Cas9 system utilizes the sequence homology information generated by aligning gRNA with the reference genome to predict off-target effects. Earlier efforts in this direction were based on sequence alignment using BLAST [143, 144], which later evolved to use read mapping algorithms such as Bowtie/Bowie2. CCToP, CHOPCHOP and GT-Scan were based on Bowtie while CRISPRscan and E-CRISP implemented Bowtie2 in their search strategies. However, these methods suffer from some limitations discussed in the previous section. Recently, software tools such as CRISPRitz [145], FlashFry [146], dsNickFry [147], Off-Spotter [148], and Cas-OFFinder [149] have been specifically designed for off-target prediction. These new software tools implement many novel strategies, including use of sophisticated pattern recognition algorithms, taking care of proximity to the PAM site, inclusion of information about DNA/RNA bulges, and allowing genetic variation. All these advanced software tools allow custom PAM as well as mismatch numbers, but they differ in additional features. For example, Bowtie2, Cas-OFFinder, and CRISPRitz also include DNA/RNA bulges information in their design. CRISPRitz along with Crisflash allows the use of genetic variants information. FlashFry is very suitable for large data sets. Breaking-Cas is very interactive, but its applicability is confined only to eukaryotic genomes available in ENSEMBL. A description of each of the software tools is given in Table 2 along with their strengths and limitations.

[8] CHOPCHOP can also be used in command-line mode after downloading from https://bitbucket.org/valenlab/chopchop/src/master/.

Table 2. An overview of commonly used score-based software tools for off-site prediction

Name	Description	Mode	Webpage/download page link	Strengths	Limitations	Reference
E-CRISP	Online application to design gRNAs. Target sequences are identified based on their complementarities to the gRNA sequence ending in a 3' PAM [N(G or A)G] which is required to recruitCas9 nucleases to cut dsDNA. It utilizes an indexing system for searching binary interval tree as well as binding site identification to speed up annotation of putative gRNAs.	Web server	http://www.e-crisp.org/E-CRISP/index.html	Experiment-oriented design, provide options to select organism, design purpose, stringency control (relaxed, medium, and strict) to define PAM type and mismatch number	Bulges are not allowed	[172]
COSMID	CRISPR Off-target Sites with Mismatches, Insertions, and Deletions (COSMID) is a method to identify off-targets using a formula-based approach. Off-targets are predicted using information about mismatched bases and/or indels in comparison to the guide strand. The ranking of the off-targets is done based on the number and location of the mismatches. COSMID implements the FetchGWI algorithm.	Web server	https://crispr.bme.gatech.edu/	Bulges are allowed, provides Web support, option to output application specific primers	Scoring outperformed by other software tools, Supports a limited number of species (human, mouse, rat, Pig, *Rhesus macaque, C. elegans* and Zebrafish genomes)	[155]
CRISPRscan	CRISPRscan implements a sgRNA-scoring algorithm developed by Giraldez lab, based on sequence features that affects the activity of CRISPR-Cas9 *in vivo*. These features include guanine enrichment, adenine depletion, and non-canonical sgRNAs properties.	Web server	https://www.crisprscan.org/	Improved user friendly interface, allows a number of options to select targets as well as off-targets	Bulges are not allowed	[163]

Name	Description	Mode	Webpage/download page link	Strengths	Limitations	Reference
	The datsets were obtained from large scale analysis of sgRNA mutagenesis activity in zebra fish.					
CROP-IT	A widely used tool for predicting binding and cleavage of off-targets effects. The algorithm divides the photospacer sequence into three parts using specific mismatch weights. These weights were optimized using available experimental ChIP-seq data. It integrates genome wide biological information derived from existing Cas9 binding and cleavage datasets.	Web server	http://www.adlilab.org/CROP-IT/homepage.html	User-friendly interface, Improved predictions due to integration of information from ChiP seq data seta, better performance, Web support available	Lack of correlation of scores with genome-wide data related to Cas9 cleavage	[156]
Breaking-Cas	Interactive design of gRNAs for CRISPR-Cas9 specifically designed for ENSEMBL genomes.	Web server	https://bioinfogp.cnb.csic.es/tools/breakingcas/	Interactive platform, allows variable PAM sequences, customized gRNA length and scores	Specific to Eukaryotic genomes available in ENSEMBL	[173]
CCToP	Bowtie based off-target detection and scoring system.	Command-line, web server	Web server: https://cctop.cos.uni-heidelberg.de:8043/ Standalone version: https://bitbucket.org/juanlmateo/cctop_standalone/src/master/	Very interactive, allows single as well as batch mode, provides a range options (for PAM sites, mismatch number and target site lengths), Supports a large number of species, allows bulges	Allows up to only 5 mismatches as more either prevent or inhibit creation of DSBs, Scoring outperformed by other software tools	[154]
CRISPR-GE	CRISPR-GE is a collection of CRISPR gene editing applications including. Target sgRNA design off-target prediction, primer designing for sgRNA expression cassettes construction, determination of mutants and downloading specific genomic sequences from reference genomes.	Web server	http://skl.scau.edu.cn/	Allows use of custom PAMs, Allows up to 5 mismatches. Applicable to a wide range of organisms including plants	Scoring available for only CRISPR-Cas9 system. Unable to score other Cas9 variants as well as Cpf1 nucleases	[159]

Table 2. (Continued)

Name	Description	Mode	Webpage/download page link	Strengths	Limitations	Reference
MIT (Hsu-Zhang) score	MIT (Hsu-Zhang) has been one of the initial scoring systems developed for prediction of off-target effects by Hsu et al. They analyzed more than 700 gRNA variants and mutation levels (SpCas9 induced indels) for more than 1000 predicted off-target genomic loci. Improved version of Hsu score has been integrated in to the MIT web server.	Command-line (simple Hsu score), web server (MIT)		Claimed by authors to take care of more than 50% of the variance related to cutting frequency, implemented in various gRNS designing tools such as CHOPCHOP and CRISPOR	Bulges not allowed, outperformed by CFD score, The simple Hsu score lacks in correlation with the experimental data and MIT score	[77]
CFD score	CFD score is a popular scoring system for off-target prediction which uses matrices with 240 fitting parameters based on mismatch as well as sequence dependent features. It has also been used to detect off-target effects for lethal sgRNAs based on the statistically significant increase in their predicted off-targets.	Command-line		It has been validated with GUIDE-seq to perform better than MIT, implemented in many software tools (including CRISPOR, GuideScan, CRISPRscan, GPP web portal), utilizes biggest cleavage data set	Bulges not allowed	[160]
CRISPOR	CRISPOR is an online server to facilitate gRNA selection for a variety of genomes. It uses a scoring system to rank probable off-targets in the genome of interest and also predicts on-target activity.	Web server, Command-line	Web server: http://crispor.tefor.net/ Command- line: http://crispor.tefor.net/downloads/	Allows a wide range of genomes to search (759 till date), provides a wide variant of PAMs to use, option to run in batch mode	Bulges not allowed	[174]
CHOPCHOP/ CHOPCHOP V2/CHOPCHOP V3	CHOPCHOP is a Bowtie based algorithm for off-target prediction and ranking for TALENs and CRISPR-Cas9 systems.	CHOPCHOP V3 is the new available software in form of a Web server and c Command-line	Web server: https://chopchop.cbu.uib.no/ Command- line: https://bitbucket.org/valenlab/chopchop.	The user- friendly web server accepts a variety of inputs, offers a variety of target selection choices		[142, 157, 158]

Name	Description	Mode	Webpage/download page link	Strengths	Limitations	Reference
	CHOPCHOP V2 is an improved version of CHOPCHOP with additional applicability for CRISPR-Cpf1, CRISPR-Cas9/nickase systems. CHOPCHOP V3 is further improved and expanded version of CHOPCHOP toolbox to work beyond knockouts to include knockdown and repression, supports CRISPR-Cas13 systems.	(For larger datasets)		Version 2 additionally Support custom sgRNA length, include information about sequence composition for sgRNA and adjoining regions, can target promoter regions, allows truncated sgRNAs, allow off-target search up to 3 mismatches The new version 3 supports isoforms generated due to alternative transcripts, support predictions based on RNA accessibility, allows new DNA targeting modes and supports more than 200 genomes while also allowing addition of new species		
CROP	CROP (CRISPR Off-target Predictor) is software tool for off-target propensity prediction for guide RNAs and help users to select the optimum guides. CROP generates various substitution, deletion and insertion combinations which are then mapped into the reference genome. Based on these mapped variants, scoring and alignment are conducted and then reported as a table comprising the off-target propensity of all guide RNAs from a given gene sequence.	Command-line	https://github.com/vaprilyanto/crop	Users are allowed to use their own genome of interest, implements Hsu and CFD off-target scoring systems also, can run in batch mode by taking multiple sequences as input, Allows bulges	-	[164]

Table 2. (Continued)

Name	Description	Mode	Webpage/download page link	Strengths	Limitations	Reference
CRISPRroots	CRISPRroots is used to analyze on as well as off-targets based on RNAseq data. This algorithm uses a variety of features such as binding properties of Cas9 and guide RNA, changes in gene expression, sequence variants between edited and unedited cells.	Command-line	https://rth.dk/resources/crispr/crisprroots/	Outperformed previous methods as shown by results for seven public datasets	-	[165]

It is to be noted that some of the tools discussed in this section, such as CCToP, CHOPCHOP, E-CRISP, and Breaking-Cas can also be used for scoring and ranking off-targets. A timeline of the development of various software tools and algorithms for off-target detection is depicted in Figure 8.

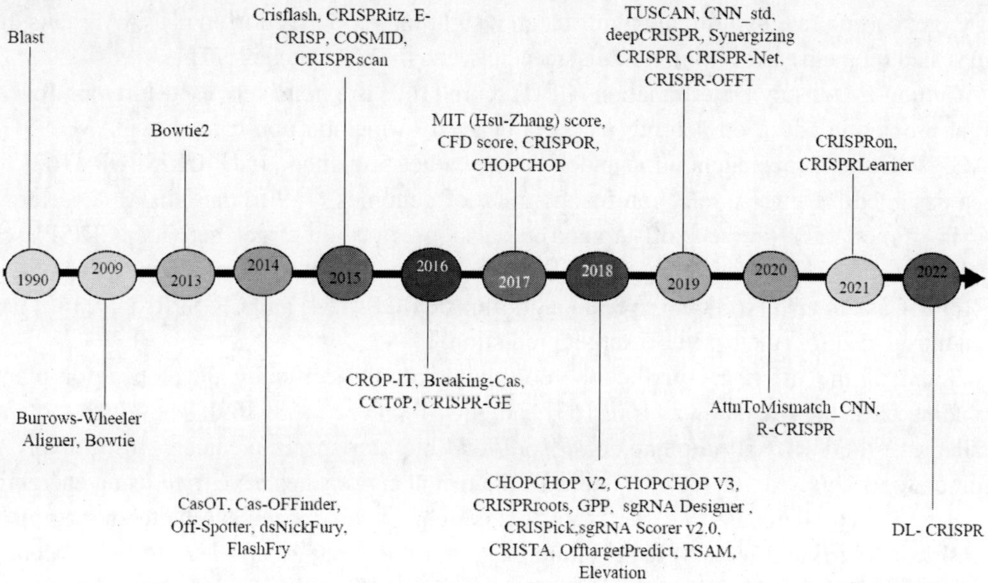

Figure 8. A timeline showing introduction various off-target detection software tools/algorithms.

Score-Based Methods

This category of *in silico* off-target prediction strategies includes algorithms that can score and rank the predicted off-targets on the basis of gRNA-DNA activity. Score-based algorithms are based on the activity of gRNA-DNA. As per their strategy, these methods can be divided into three categories, *viz.*, Formula based, Machine Learning (ML) based and Deep Learning (DL) based [153].

Formula-Based

The algorithms belonging to this category use information such as mismatch count, positions, mean distance, continuities, etc. Some of the commonly used software tools in this category include E-CRISP, CCTop (CRISPORater), COSMID, CROP-IT, CFD score, CHOPCHOP, CRISPR-GE, MIT, Hsu score, Feng Zheng Lab's Target Finder, and CRISPOR. E-CRISP is an example of one of the initial attempts to implement formula information. E-CRISP ranks off-targets based on alignment scores obtained using the Bowtie2 algorithm. In contrast, CCTop [154] and COSMID [155] rank off-targets based on the mismatch position relative to the PAM sequence. This feature is based on the fact that the closer the mismatch to the PAM, more likely is its ability to prevent Cas9 cutting [77, 156]. In addition, COSMID allows bulges (DNA/RNA) in the alignment [78]. CROP-IT integrates whole genome-level biological information derived from existing Cas9 binding and cutting data sets for more realistic predictions and accurate gRNA design [156].

CHOPCHOP was initially designed for CRISPR-Cas9 as well as TALENs [142] and has now evolved to support a wide range of CRISPR-Cas systems along with more sophisticated rules/ formulas in its new version CHOPCHOP v3 [157, 158]. CRISPR-GE is a comprehensive online tool taking care of many aspects of gene editing, including off-target identification, and analysis for a wide range of organisms including plants [159]. MIT (Hsu-Zhang) score predicts off-target cutting rate by utilizing a mismatch weight matrix derived from gRNA variants data and scaled on the basis of minimum distance between the mismatches [77].

Cutting Frequency Determination (CFD) score [160] is extensively used software for off-target prediction based on scoring matrices derived to include non-canonical PAMs (NGA, NAG, NCG etc.), mismatches and indels in the target sequence [161]. CRISPOR [162] has been designed for gRNA selection for a variety of genomes (759 to date) using a variety of PAMs. It not only predicts off-targets but also predicts on-target activity. CRISPRscan implements a sgRNA-scoring algorithm based on sequence features that affect the activity of CRISPR-Cas9 *in vivo* [163]. Two recent algorithms CROP [164] and CRISPRroots [165] have been introduced for rule based off-target prediction.

Some of the off-target prediction tools have been specifically developed for plants, including CRISPR-PLANT v2 [166, 167] and CRISPR-P v2 [168, 169]. These two tools are available at the URLs http://omap.org/crispr2/ and http://crispr.hzau.edu.cn/, respectively. In addition to using sequence features, some of the recent approaches have tried using energetics based on structural features of the sgRNA-Cas9 complexes to achieve better results. Both CRISPRoff [170] and uCRISPR [171] are based on this strategy and lead to more accurate predictions as compared other scoring methods, such as MIT and CFD. However, they need to be systematically validated through experimental procedures before used extensively [161]. CRISPRoff is implemented as a web server[1], while uCRISPR is available as a standalone application[2]. A comparative view of all the formula/rule-based software tools has been provided in Table 2.

Machine Learning-Based Methods

Due to extensive research in the CRISPR-Cas9 area, a large number of activity datasets are available. These datasets have been exploited by various researchers to develop a variety of software tools using Machine Learning (ML) and Deep Learning (DL) approaches. These models differ in the use of training data sets, features, and learning algorithms (Table 3). Some of the initial efforts were developed based on the correlation between the phenotypic changes and the gRNA activity. For example, one of the initial ideas implemented in the Rule set1 model was based on data obtained from 1841 sgRNAs belonging to humans and mice [175]. In this strategy, the support vector machine (SVM) algorithm is used to predict score, whereas a supervised learning model was used to classify data in a generalized linear model. The Rule set1 utilizes the relationship between the composition of nucleotides in the target sequences and depletion rates of gRNAs targeted to the cell surface markers in humans and mice cells.

---

[1] https://rth.dk/resources/crispr/crisproff/.
[2] http://rna.physics.missouri.edu/uCRISPR/.

Table 3. An overview of commonly used machine learning-based software tools for off-site prediction

Name	Algorithm	Mode	Webpage/download page link	Comments	Reference
Machine Learning-Based Methods					
GPP sgRNA Designer /CRISPick	Rule Set2, Azimuth and CFD score	Web server	https://portals.broadinstitute.org/gppx/crispick/public	Latest improved version is available as CRISPick	[160, 177]
sgRNA Scorer v2.0	Support vector machine	Web server, Command- line	https://sgrnascorer.cancer.gov/	Success rate varies for Cas9s (from *S. aureus*) and S. *thermophilus*) and Cpf1 CRISPR systems	[178]
CRISTA	Random forest method	Web server, Command- line		Outperforms MIT, CDF and CCTop	[181]
Off-targetPredict (ensemble learning method)	Ensemble learning method	Command- line	https://github.com/penn-hui/OfftargetPredict	Can be used to search sgRNA off-target effects across thousands of candidate genomes	[182]
TSAM	Regression, SVM, Boosting algorithm and Markov profiles	Web server, Command- line	http://www.aai-bioinfo.com/CRISPR/ https://github.com/penn-hui/TSAM	Improved performance as tested on benchmark datasets containing thousands sgRNAs from mouse, human and zebra fish.	[183]
Elevation	Two-step regression model	Web server, Command- line		Outperforms MIT, CDF and CCTop	[147]
TUSCAN	Random forest method	Command- line	https://github.com/BauerLab/TUSCAN	Highly scalable and fast method	[180]
Crispr2vec	Deep metric learning approach	Webserver	https://www.biorxiv.org/crispr2vec	Based on an offline, smart triplet sampling strategy.	[184]
Deep Learning-Based Methods					
CNN_std	Deep convolutional neural network (DCNN)	Command- line	Source code: https://github.com/MichaelLinn/off_target_prediction Datasets: https://github.com/MichaelLinn/off_target_prediction/tree/master/data	Outperformed as compared to four rule based (MIT, CFD, CCTop and CROP-IT) and three ML based algorithms	[185]
deepCRISPR	Convolutional neural network (CNN)	Web server, Command- line	Web server: http://www.deepcrispr.net/ Command- line: https://github.com/bm2-lab/DeepCRISPR https://zenodo.org/record/1246320	Implements sequence as well as epigenetic features, offers optimized sgRNA design for coding as well as non-coding regions	[186]

Table 3. (Continued)

Name	Algorithm	Mode	Webpage/download page link	Comments	Reference
DeepCpf1	A deep learning framework	Command- line	https://github.com/MyungjaeSong/Paired-Library	Specifically developed for CRISPR-Cpf1 systems but support other Cas experiments also	[187]
Synergizing CRISPR	Ensemble learning method using Adaboost classifier	Command- line	https://github.com/Alexzsx/CRISPR.	Outperformed other classifiers based on its best prediction	[188]
CRISPR-Net	Recurrent convolutional network (RCN)	Command- line	https://codeocean.com/capsule/9553651/tree/v1	Gives competitive performance on experimental datasets from CIRCLE-Seq and GUIDE-seq, outperforms the state-of-the-art off-target prediction models	[189]
CRISPR-OFFT	Attention-based CNN	Command- line	https://github.com/Peppags/CRISPRont-CRISPRofft	Competitive performance as compared to other prediction models	[190]
AttnToMismatch_CNN	Transformer-based DL method	Command- line	https://github.com/qiaoliuhub/AttnToCrispr	Useful for both CRISPR-Cas9 as well as CRISPR-Cas12a systems, Improved prediction with higher true positives rate and lower false positive rates	[191]
CRISPRLearner	CNN	Command- line	https://git-lfs.github.com/	Allows models to be trained on custom data sets	[192]
DL-CRISPR	CNN with augmented data sets	Command- line	https://github.com/yuuuuzhang/DL-CRISPR_offtarget_prediction	The original positive dataset was extended to increase competitiveness and remove data imbalance	[193]
R_CRISPR	Long-term Recurrent Convolutional Neural Network (LRCN)	Command- line	-	Effectively predicts off-target effects, surpassed six mainstream predicts as evaluated based on GUIDE-seq verified mismatch-only datasets	[194]
CRISPRon	Keras/TensorFlow2.2.0 neural network framework	Web server, Command- line	https://rth.dk/resources/crispr/ https://github.com/RTH-tools/crispron/	Based on sequence as well as thermodynamic features. Lacks consistency, limited data availability to improve models	[195, 196]
CRISPR-IP	CNN, BiLSTM (Bi-directional Long-Short Term Memory) and attention layer	Command- line	https://github.com/BioinfoVirgo/CRISPR-IP/	Very accurate and robust method	[153]

Predictions based on Rule set1 showed high correlation with the experimental data. An improved version, Rule set2 [160] combined information about changes in the expression of cell surface markers (implemented in Rule set2) with the drug resistance pathways to develop more accurate models. The sequence features, such as sgRNA location (within the gene) and position independent nucleotide compositions, were used to correlate with biological information. The optimized models thus developed were used to design gRNAs for CRISPR KO as well as CRISPRa/i experiments [176]. Moreover, a software tool by Broad Institute known as GPP sgRNA Designer was developed using these models for predicting gRNA efficiency. Recently, GPP sgRNA has been updated to CRISPick with enhanced functionalities and user-friendly web interface.[1] An algorithm implemented in GPP (Azimuth 2.0) [177] was trained on gRNAs secondary structure information and the position of the target site relative to the transcription start site (TSS) in addition to nucleotide composition data. Off-targets in GFF are predicted using the CFD score.

Some other ML algorithms used information about mutations detected by sequencing [161]. For example, sgRNA Scorer v2.0 [178], a species independent model was developed based on mutation rates of sgRNA targets in human HEK293T cells using the SVM algorithm. It has extended its application to SaCas9 and AsCpf1 Cas systems as compared to the original sgRNA model [179]. The TUSCAN model further improved this strategy by using additional information about the regions flanking the target sites and classifying the data using random forest algorithm [180]. The CRISPR Target Assessment (CRISTA) utilised off-target cutting data obtained from their widely used experimental techniques, viz., GUIDE-seq, BLESS and HTGTS to develop a random forest algorithm-based model [181]. The main features used to develop model included geometric features (RNA secondary structure), RNA thermodynamic features, DNA enthalpy epigenetic features (DNA methylation), GC contents, features related to alignments (including DNA/RNA bulges), and nucleotide identities in relation to CRISPR-Cas9 cleavage efficiency.

Off-targetPredict is an ensemble learning method which may be used to predict sgRNA off-targets from thousands of genome-wide candidates [182]. Off-targetPredict uses features such as nucleotide mismatch preferences, mismatch position specificity and GC composition to differentiate between on-targets and off-targets using an ensemble learning approach. TSAM (Two-step averaging method) is based on the regression of sgRNs cutting efficiencies and scores predicted by SVM and a boosting algorithm [183]. TSAM also implements Markov profile to capture novel features related to the global characteristics of sgRNAs. The Elevation software by Doenche laboratory utilised a two layers regression model based on sequence as well as DNA accessibility/epigenetic data [147]. Elevation can be used to predict individual gRNA off-target scores and aggregate gRNA scores. Elevation can potentially perform better than MIT, CDF, and CCTop software tools but its use is limited to predict off-target effects for human exome (GRCh38) only.

Deep Learning-Based Methods

Recently, many researchers have shifted their focus to developing off-target prediction strategies based on deep learning (DL). DL methods can learn on their own from the underlying sequence data and then use these features to predict off-targets.

---

[1] https://portals.broadinstitute.org/gppx/crispick/public.

The task of developing DL models can be divided into two stages:

1) To create a matrix/vector representation of gRNA: DNA sequence pairs and other features.
2) To use DL algorithms to learn from these matrix/vector representations and develop prediction models.

Some of the commonly used deep learning methods for off-target prediction in CRISPR-Cas systems have been listed in Table 3. CNN_std and deepCRISPR models are based on convolutional neural networks (CNN). CNN_std was trained and tested using CRISPOR off-target datasets with additional evaluation carried out with the GUIDE-seq off-target dataset. deepCRISPR implemented information related to epigenetic aspects influencing the KO/KI sgRNA efficiency in addition to the sequence features. DeepCpf1 uses deep learning methodogy based on chromatin accessibility data to predict off-targets specifically for CRISPR-Cpf1 systems [186]. DeepCpf1 also provides models for evaluating Cas9, xCas9 and Base edition experiments. Synergizing CRISPR uses a novel consensus-based strategy by combining prediction scores from five other off-target predictors (CROP-IT, CCToP, CFD, MIL website, and MIT/HSU) along with data on evolutionary conservation and chromatin state. AttnToMismatch_CNN models were developed using transformer-based DL based on cell specific responses derived from gene expression profiles and biological network data [191]. CRISPRLearner was developed using CNN and has the option to use custom datasets for training in addition to the datasets implemented in the code [192]. AttnToMismatch_CNN algorithm outperformed many state-of-the-art models. CRISPR-Net is a recurrent convolutional network to score gRNA-target pairs based on mismatches and indels [189].

Another programme from the same group, CRISPR-Net-Aggregate [189], uses a ML approach to aggregate the gRNA-target scores obtained from CRISPR-Net into a single consensus off-target score. CRISPR-OFFT combines deep CNN and attention mechanisms for the prediction of off-target activity of sgRNA using a variety of features obtained from sgRNA-DNA sequence pairs [190]. DL_CRISPR used a four-layered CNN model on augmented datasets obtained from a series of *in vitro* and cell-based assays [193]. Furthermore, 5-fold cross validation was used to check the robustness of DL_CRISPR. R_CRISPR has been modeled as Long-term Recurrent Convolutional Neural Network (LRCN) based on experimentally validated seven off-target datasets [194]. The datasets were further divided into two categories. The first one related to mismatch and indel off-target sites, while the second one contained only datasets with mismatch off-target sites. CRISPRon was modeled using TensorFlow2.2.0 neural network framework based on sequence features and thermodynamic properties [196]. However, unavailability of sufficient training data sets required to implement such a sophisticated DL methodology limits its application, extension, and consistency. CRISPR-IP is one of the most recent DL models for off-target prediction [153]. CRISPR-IP has been built using CNN, BiLSTM, and attention layer methods based on features derived from sequence pairs. This model implements the concept of function channels to differentiate between regions with different functions. CRISPR-IP was found to perform better than other predictors.

Off-Target Prediction Using Molecular Dynamics (MD) Simulations

MD simulation is one of the valuable tools for understanding the molecular basis of the CRISPR-Cas9 function. For clinical application of genome editing, CRISPR-Cas systems must achieve therapeutically relevant levels of editing at the on-target site with minimal editing at any off-target sites. Kinetic and single molecule (sm) Förster resonance energy transfer (FRET) experiments have shown that the occurrence of off-target cleavages is directly related to the conformational state adopted by the catalytic HNH domain [197, 198]. Upon DNA binding, HNH undergoes a conformational change from an inactive state to an activated state that is prone to catalysis. The inactive state of the HNH domain has been identified as a "conformational checkpoint" between DNA binding and cleavage [198]. smFRET experiments have shown that the presence of DNA mismatches at the PAM distal ends of the DNA:RNA hybrid can trap the HNH domain in the "conformational checkpoint" state, whose population increases by augmenting the number of mismatches at PAM distal sites. Early experimental characterizations also revealed that the presence of base pair mismatches at PAM distal ends leads to the formation of a stable CRISPR-Cas9 complex [82].

One major cause of the CRISPR off-target is that the minimum mismatches (up to 3 nts) between the gRNA spacer and the off-target site are tolerated. In order to understand how these mismatches, favor the inactivation of HNH site, MD simulations have been used to study the mechanism. Ricci et al., using MD simulations clarified the activation process by which the HNH domain is repositioned for cleavage and are in good agreement with experimental results. These conformational changes underlying the CRISPR-Cas9 function occur over and beyond micro-to-millisecond time scales [199]. Therefore, an accelerated molecular dynamics approach that extends the time scale of the simulation without prior knowledge of the potential energy surface can be used. MD is an enhanced sampling methodology that applies a boost potential to the simulation, thereby accelerating transitions between low-energy states. However, the use of MD for large bimolecular systems, such as CRISPR-Cas9, can suffer from high statistical noise, which hampers the characterization of the correct statistical ensemble. To overcome this limitation, a novel Gaussian accelerated MD (GaMD) method has been proposed, which uses harmonic functions to construct a boost potential that is adaptively added to the simulation.

Using GaMD Palermo et al. investigate the mechanistic basis of binding of off-target sequences at PAM distal sites. They performed all-atom GaMD simulations of the Cas9 protein in the "conformational checkpoint" state of the HNH domain (i.e., PDB 4UN3). In order to study the effects caused by base-pair mismatches, the CRISPR-Cas9 complex has been simulated in complex with the fully matched RNA:DNA hybrid (considered the reference on-target system) and in the presence of base pair mismatches at the PAM distal ends of the RNA:DNA hybrid. These simulations have been carried out for more than 1 microsecond. Palermo et al. showed that upon DNA binding, the conformation dynamics of the active domain site (HNH) of Cas9 trigger the formation of the active state, explaining how the HNH domain exerts a conformational control domain over DNA cleavage. The simulations reveal that, by introducing up to 4 base pair mismatches in the target DNA at PAM distal sites, the conserved interactions between the unwound DNA and the L2 loop of the catalytic HNH domain constitute a "lock" effectively decreasing the conformational freedom of the HNH domain and its activation for cleavage whereas up to 3 base pair mismatches at PAM distal ends are unable to stably "lock" HNH in an inactivated state, still allowing its repositioning for cleavage [200]. Similar results were observed by Zheng et al. using MD simulations on wtCas9 and its mutated form nrqqCas9 with four mismatches. After mutations, nrqqCas9 adopts open conformation and electrostatic interactions are reduced which results in lesser efficiency for target as well as off-target cleavages [201].

Conclusion

Genome editing (GE) emerged as a revolutionary concept owing to its wide range of potential applications across various living organisms. A range of molecular tools have been introduced to practically implement the concept of GE. Among programmable nucleases, CRISPR-Cas9 and other CRISPR-Cas systems have evolved as promising tools for modifying nuclear genomes due to their precision, ease of use, cost effectiveness, and efficiency in comparison to the classical ZFNs and TALENs. However, their tendency to introduce off-target effects leading to deleterious phenotypes has limited the use of GE in an extensive manner, particularly in terms of clinical implications. A significant amount of progress has been made towards the development of experimental as well as computational methods for predicting the off-target effects of programmable nucleases, including ZFNs, TALENs, and predominantly CRISPR-Cas9 systems. The success of experimental techniques to detect off-target effects varies with the type of the technique, sensitivity to detect DSBs, ability to overcome cell specific/ epigenetic effects, off-target detection limits, and the developing times. Moreover, time consumption, cost of experimentation, and noisy measurements leading to high false negative rates further limits their applicability. None of the available techniques is capable of accurately predicting low-frequency off-targets, posing a major bottleneck in therapeutic applications of genome editing technologies. This gives an opportunity to develop more accurate, sensitive and cost-effective experimental techniques to overcome these challenges. In addition, steps are being taken to explore more precise alternatives to the existing SpCas9 including new Cas variants, base editors, prime editing and anti-CRISPR proteins with fewer or negligible off-target effects.

On the other hand, a great deal of work has been carried out to develop algorithms and software tools for *in silico* prediction and analysis of off-target effects by programmable nucleases, particularly CRISPR-Cas9 systems. These algorithms have been developed based on knowledge about gRNA binding, tolerance for mismatches, and epigenetic factors. Starting from simple homology-based detection, these strategies have been expanding to develop score-based algorithms using Formulas/Rules based, Machine Learning (ML) based, and recently, Deep Learning (DL) based models. Formulas/Rule based approaches try to integrate various measurable features with pattern recognition algorithms to extract more information about the target sites and gRNAs. This information is used to derive some rules and formulas that aid in prediction of off-target effects. Rather than doing an integrated study on their interactive contributions, rule-based algorithms provide individual assessments or random combinations for various features. It becomes challenging to balance the likely conflicting outcomes of several features, thereby limiting further advancements in this category of methods. To overcome this artefact, several researchers developed prediction models by systematically integrating various features using ML approaches. ML models rely on the availability of experimental data for training and testing of models.

Advancements in NGS technologies have led to availability of sufficient datasets while in depth research to understand precise mechanisms of CRISPR-Cas9 editing have augmented potential features affecting off-target activity. Progress in both directions has provided the opportunity to develop more robust and accurate ML models. However, inconsistency of these datasets poses a challenge to apply ML approaches to off-target prediction as these datasets have been derived using different gRNAs and cell types. So, to judiciously use this cell-specific and species-specific information, more cell types need to be evaluated. Recently, a trend has been observed in using DL methods, which have the ability to learn on their own from the underlying sequence data and then use these features to predict off-targets. Deep learning algorithms outperform traditional machine learning algorithms. The majority of deep learning models, on the other hand, are still black boxes. Interpreting the value of model input features is one of the most difficult challenges for deep learning, giving room for the development of more sophisticated technologies like Explainable Artificial Intelligence (XAI).

In summary, significant progress has been made in developing experimental as well as computational tools for taking care of off-target effects. Further improvement in direction will aid in developing more precise therapeutic and other GE applications, covering a wide range of organisms in addition to currently dominating human and mouse species.

Acknowledgments

We acknowledge the institutional funding from the Department of Science and Technology, Government of India ("Fund for Improvement of Science and Technology Infrastructure (FIST)" grant D. O. No. SR/FST/PG College/2009), for the development of bioinformatics/ computational resources.

We are extremely thankful to Mr. Rishab Dhir, Assistant Professor, Department of Multimedia, Hans Raj Mahila Maha Vidyalaya, Jalandhar, for his help in designing the image showing the timeline of the emergence of various genome editing technologies.

References

[1] Hsu, P. D., Lander, E. S., & Zhang, F. (2014). Development and applications of CRISPR-Cas9 for genome engineering. *Cell*, *157*(6), 1262-1278.

[2] Komor, A. C., Badran, A. H., & Liu, D. R. (2017). CRISPR-based technologies for the manipulation of eukaryotic genomes. *Cell*, *168*(1-2), 20-36.

[3] Rothstein, R. J. (1983). One-step gene disruption in yeast. In *Methods in Enzymology* (Vol. 101, pp. 202-211). Academic Press.

[4] Gaj, T., Gersbach, C. A., & Barbas III, C. F. (2013). ZFN, TALEN, and CRISPR/Cas-based methods for genome engineering. *Trends in Biotechnology*, *31*(7), 397-405.

[5] Ghosh, D., Venkataramani, P., Nandi, S., & Bhattacharjee, S. (2019). CRISPR–Cas9 a boon or bane: the bumpy road ahead to cancer therapeutics. *Cancer Cell International*, *19*(1), 1-10.

[6] Cornu, T. I., Mussolino, C., & Cathomen, T. (2017). Refining strategies to translate genome editing to the clinic. *Nature Medicine*, *23*(4), 415-423.

[7] Barrangou, R., & Doudna, J. A. (2016). Applications of CRISPR technologies in research and beyond. *Nature Biotechnology*, *34*(9), 933-941.

[8] Wang, J., Zhang, X., Cheng, L., & Luo, Y. (2020). An overview and metanalysis of machine and deep learning-based CRISPR gRNA design tools. *RNA Biology*, *17*(1), 13-22.

[9] Hanna, K. E. (1991). Asilomar and Recombinant DNA: The End of the Beginning. In *Biomedical Politics*. National Academies Press (US).

[10] Nemudryi, A. A., Valetdinova, K. R., Medvedev, S. P., & Zakian, S. M. (2014). TALEN and CRISPR/Cas Genome Editing Systems: Tools of Discovery. *Acta Naturae*, *6*(3), 19–40.

[11] Smithies, O., Gregg, R. G., Boggs, S. S., Koralewski, M. A., & Kucherlapati, R. S. (1985). Insertion of DNA sequences into the human chromosomal β-globin locus by homologous recombination. *Nature*, *317*(6034), 230-234.

[12] Thomas, K. R., Folger, K. R., & Capecchi, M. R. (1986). High frequency targeting of genes to specific sites in the mammalian genome. *Cell*, *44*(3), 419-428.

[13] Thomas, K. R., & Capecchi, M. R. (1987). Site-directed mutagenesis by gene targeting in mouse embryo-derived stem cells. *Cell*, *51*(3), 503-512.

[14] Capecchi, M. R. (1989). Altering the genome by homologous recombination. *Science*, *244*(4910), 1288-1292.

[15] Lin, F. L., Sperle, K., & Sternberg, N. (1985). Recombination in mouse L cells between DNA introduced into cells and homologous chromosomal sequences. *Proceedings of the National Academy of Sciences of the United States of America*, *82*(5), 1391–1395. https://doi.org/10.1073/pnas.82.5.1391.

[16] Qiu, S., Adema, C. M., & Lane, T. (2005). A computational study of off-target effects of RNA interference. *Nucleic Acids Research*, *33*(6), 1834-1847.

[17] Wang, L., Zhao, Z., Meyer, M. B., Saha, S., Yu, M., Guo, A., ...& Xu, W. (2014). CARM1 methylates chromatin remodeling factor BAF155 to enhance tumor progression and metastasis. *Cancer Cell*, *25*(1), 21-36.

[18] Gupta, R. M., & Musunuru, K. (2014). Expanding the genetic editing tool kit: ZFNs, TALENs, and CRISPR-Cas9. *The Journal of Clinical Investigation*, *124*(10), 4154-4161.

[19] Rudin, N., Sugarman, E., & Haber, J. E. (1989). Genetic and physical analysis of double-strand break repair and recombination in Saccharomyces cerevisiae. *Genetics*, *122*(3), 519-534.

[20] Rouet, P., Smih, F., & Jasin, M. (1994a). Introduction of double-strand breaks into the genome of mouse cells by expression of a rare-cutting endonuclease. *Molecular and Cellular Biology*, *14*(12), 8096-8106.

[21] Rouet, P., Smih, F., & Jasin, M. (1994b). Expression of a site-specific endonuclease stimulates homologous recombination in mammalian cells. *Proceedings of the National Academy of Sciences*, *91*(13), 6064-6068.

[22] Kosicki, M., Tomberg, K., & Bradley, A. (2018). Repair of double-strand breaks induced by CRISPR–Cas9 leads to large deletions and complex rearrangements. *Nature Biotechnology*, *36*(8), 765-771.

[23] Choulika, A., Perrin, A., Dujon, B., & Nicolas, J. F. (1995). Induction of homologous recombination in mammalian chromosomes by using the I-SceI system of Saccharomyces cerevisiae. *Molecular and Cellular Biology*, *15*(4), 1968-1973.

[24] Jasin, M. (1996). Genetic manipulation of genomes with rare-cutting endonucleases. *Trends in Genetics*, *12*(6), 224-228.

[25] O'Driscoll, M., & Jeggo, P. A. (2006). The role of double-strand break repair—insights from human genetics. *Nature Reviews Genetics*, *7*(1), 45-54.

[26] Bibikova, M., Golic, M., Golic, K. G., & Carroll, D. (2002). Targeted chromosomal cleavage and mutagenesis in Drosophila using zinc-finger nucleases. *Genetics*, *161*(3), 1169–1175. https://doi.org/10.1093/genetics/161.3.1169.

[27] Bibikova, M., Beumer, K., Trautman, J. K., & Carroll, D. (2003). Enhancing gene targeting with designed zinc finger nucleases. *Science (New York, N.Y.)*, *300*(5620), 764. https://doi.org/10.1126/science.1079512.

[28] Porteus, M. H., & Baltimore, D. (2003). Chimeric nucleases stimulate gene targeting in human cells. *Science (New York, N.Y.)*, *300*(5620), 763. https://doi.org/10.1126/science.1078395.

[29] Urnov, F. D., Rebar, E. J., Holmes, M. C., Zhang, H. S., & Gregory, P. D. (2010). Genome editing with engineered zinc finger nucleases. *Nature Reviews Genetics*, *11*(9), 636-646.

[30] Silva, G., Poirot, L., Galetto, R., Smith, J., Montoya, G., Duchateau, P., & Pâques, F. (2011). Meganucleases and other tools for targeted genome engineering: perspectives and challenges for gene therapy. *Current gene therapy*, *11*(1), 11-27.

[31] Cathomen, T., & Joung, J. K. (2008). Zinc-finger nucleases: the next generation emerges. *Molecular Therapy*, *16*(7), 1200-1207.

[32] Boch, J., Scholze, H., Schornack, S., Landgraf, A., Hahn, S., Kay, S., & Bonas, U. (2009). Breaking the code of DNA binding specificity of TAL-type III effectors. *Science*, *326*(5959), 1509-1512.

[33] Zhang, F., Cong, L., Lodato, S., Kosuri, S., Church, G. M., & Arlotta, P. (2011). Efficient construction of sequence-specific TAL effectors for modulating mammalian transcription. *Nature Biotechnology*, *29*(2), 149-153.

[34] Joung, J. K., & Sander, J. D. (2013). TALENs: a widely applicable technology for targeted genome editing. *Nature Reviews Molecular Cell Biology*, *14*(1), 49-55.

[35] Al-Attar, S., Westra, E. R., Van Der Oost, J., & Brouns, S. J. (2011). Clustered regularly interspaced short palindromic repeats (CRISPRs): the hallmark of an ingenious antiviral defense mechanism in prokaryotes. *Biol Chem.* 2011 Apr;392(4):277-89.

[36] Cox, D. B. T., Platt, R. J., & Zhang, F. (2015). Therapeutic genome editing: prospects and challenges. *Nature Medicine*, *21*(2), 121-131.

[37] Ishino, Y., Shinagawa, H., Makino, K., Amemura, M., & Nakata, A. (1987). Nucleotide sequence of the iap gene, responsible for alkaline phosphatase isozyme conversion in *E. coli*, and identification of the gene product. *Journal of Bacteriology*, *169*(12), 5429-5433.

[38] Mojica, F. J., Díez-Villaseñor, C., García-Martínez, J., & Soria, E. (2005). Intervening sequences of regularly spaced prokaryotic repeats derive from foreign genetic elements. *Journal of Molecular Evolution*, *60*(2), 174-182.

[39] Jinek, M., Chylinski, K., Fonfara, I., Hauer, M., Doudna, J. A., & Charpentier, E. (2012). A programmable dual-RNA–guided DNA endonuclease in adaptive bacterial immunity. *Science*, *337* (6096), 816-821.

[40] Cong, L., Ran, F. A., Cox, D., Lin, S., Barretto, R., Habib, N., Hsu, P. D., Wu, X., Jiang, W., Marraffini, L. A. and Zhang, F. (2013). Multiplex genome engineering using CRISPR/Cas systems. *Science*, *339*(6121), pp.819-823.

[41] Mali, P., Yang, L., Esvelt, K. M., Aach, J., Guell, M., DiCarlo, J. E., Norville, J. E. and Church, G. M. (2013). RNA-guided human genome engineering via Cas9. *Science*, *339*(6121), pp.823-826.

[42] Barrangou, R., Fremaux, C., Deveau, H., Richards, M., Boyaval, P., Moineau, S., Romero, D. A. and Horvath, P. (2007). CRISPR provides acquired resistance against viruses in prokaryotes. *Science*, *315*(5819), pp.1709-1712.

[43] Garneau, J. E., Dupuis, M. È., Villion, M., Romero, D. A., Barrangou, R., Boyaval, P., Fremaux, C., Horvath, P., Magadán, A. H. and Moineau, S. (2010). The CRISPR/Cas bacterial immune system cleaves bacteriophage and plasmid DNA. *Nature*, *468*(7320), pp.67-71.

[44] Deltcheva, E., Chylinski, K., Sharma, C. M., Gonzales, K., Chao, Y., Pirzada, Z. A., Eckert, M. R., Vogel, J. and Charpentier, E. (2011). CRISPR RNA maturation by trans-encoded small RNA and host factor RNase III. *Nature*, *471*(7340), pp.602-607.

[45] Cho, S. W., Kim, S., Kim, J. M., & Kim, J. S. (2013). Targeted genome engineering in human cells with the Cas9 RNA-guided endonuclease. *Nature Biotechnology*, *31*(3), 230-232.

[46] Li, H., Yang, Y., Hong, W., Huang, M., Wu, M., & Zhao, X. (2020). Applications of genome editing technology in the targeted therapy of human diseases: mechanisms, advances and prospects. *Signal Transduction and Targeted Therapy*, *5*(1), 1-23.

[47] Bao, X. R., Pan, Y., Lee, C. M., Davis, T. H., & Bao, G. (2021). Tools for experimental and computational analyses of off-target editing by programmable nucleases. *Nature Protocols*, *16*(1), 10-26.

[48] Manghwar, H., Li, B., Ding, X., Hussain, A., Lindsey, K., Zhang, X., & Jin, S. (2020). CRISPR/Cas systems in genome editing: methodologies and tools for sgRNA design, off-target evaluation, and strategies to mitigate off-target effects. *Advanced Science*, *7*(6), 1902312.

[49] Kim, Y. G., Cha, J., & Chandrasegaran, S. (1996). Hybrid restriction enzymes: zinc finger fusions to Fok I cleavage domain. *Proceedings of the National Academy of Sciences*, *93*(3), 1156-1160.

[50] Beumer, K. J., Trautman, J. K., Bozas, A., Liu, J. L., Rutter, J., Gall, J. G., & Carroll, D. (2008). Efficient gene targeting in Drosophila by direct embryo injection with zinc-finger nucleases. *Proceedings of the National Academy of Sciences*, *105*(50), 19821-19826.

[51] Ramirez, C. L., Foley, J. E., Wright, D. A., Müller-Lerch, F., Rahman, S. H., Cornu, T. I., Winfrey, R. J., Sander, J. D., Fu, F., Townsend, J. A., Cathomen, T., Voytas, D. F., & Joung, J. K. (2008). Unexpected failure rates for modular assembly of engineered zinc fingers. *Nature Methods*, *5*(5), 374–375. https://doi.org/10.1038/nmeth0508-374.

[52] Szurek, B., Rossier, O., Hause, G., & Bonas, U. (2002). Type III-dependent translocation of the Xanthomonas AvrBs3 protein into the plant cell. *Molecular Microbiology*, *46*(1), 13-23.

[53] Kay, S., Hahn, S., Marois, E., Hause, G., & Bonas, U. (2007). A bacterial effector acts as a plant transcription factor and induces a cell size regulator. *Science*, *318*(5850), 648-651.

[54] Gao, H., Wu, X., Chai, J., & Han, Z. (2012). Crystal structure of a TALE protein reveals an extended N-terminal DNA binding region. *Cell Research*, *22*(12), 1716-1720.

[55] Zhu, W., Yang, B., Chittoor, J. M., Johnson, L. B., & White, F. F. (1998). AvrXa10 contains an acidic transcriptional activation domain in the functionally conserved C terminus. *Molecular Plant-Microbe Interactions*, *11*(8), 824-832.

[56] Yang, J., Zhang, Y., Yuan, P., Zhou, Y., Cai, C., Ren, Q., Wen, D., Chu, C., Qi, H. and Wei, W. (2014). Complete decoding of TAL effectors for DNA recognition. *Cell Research*, *24*(5), pp.628-631.

[57] Miller, J. C., Zhang, L., Xia, D. F., Campo, J. J., Ankoudinova, I. V., Guschin, D. Y., ...& Rebar, E. J. (2015). Improved specificity of TALE-based genome editing using an expanded RVD repertoire. *Nature Methods*, *12*(5), 465-471.

[58] Richter, A., Streubel, J., Blücher, C., Szurek, B., Reschke, M., Grau, J., & Boch, J. (2014). A TAL effector repeat architecture for frameshift binding. *Nature Communications*, *5*(1), 1-10.

[59] Tsuji, S., & Imanishi, M. (2019). Modified nucleobase-specific gene regulation using engineered transcription activator-like effectors. *Advanced Drug Delivery Reviews*, *147*, 59-65.

[60] Kubik, G., Schmidt, M. J., Penner, J. E., & Summerer, D. (2014). Programmable and highly resolved *In vitro* detection of 5-methylcytosine by TALEs. *Angewandte Chemie International Edition*, *53*(23), 6002-6006.

[61] Kubik, G., & Summerer, D. (2015). Achieving Single-Nucleotide Resolution of 5-Methylcytosine Detection with TALEs. *Chem Bio Chem*, *16*(2), 228-231.

[62] Qasim, W., Zhan, H., Samarasinghe, S., Adams, S., Amrolia, P., Stafford, S., ...& Veys, P. (2017). Molecular remission of infant B-ALL after infusion of universal TALEN gene-edited CAR T cells. *Science Translational Medicine*, *9*(374), eaaj2013.

[63] Menz, J., Modrzejewski, D., Hartung, F., Wilhelm, R., & Sprink, T. (2020). Genome edited crops touch the market: a view on the global development and regulatory environment. *Frontiers in Plant Science*, 1525.

[64] Doudna, J. A., & Charpentier, E. (2014). The new frontier of genome engineering with CRISPR-Cas9. *Science*, *346*(6213), 1258096.

[65] Gasiunas, G., Barrangou, R., Horvath, P., & Siksnys, V. (2012). Cas9–crRNA ribonucleoprotein complex mediates specific DNA cleavage for adaptive immunity in bacteria. *Proceedings of the National Academy of Sciences*, *109*(39), E2579-E2586.

[66] Barrangou, R. (2013). CRISPR-Cas systems and RNA-guided interference. *Wiley Interdisciplinary Reviews: RNA*, *4*(3), 267-278.

[67] Ledford, H., & Callaway, E. (2020). Pioneers of Crispr Gene Editing Win Chemistry Nobel. *Nature*, *586*(7829), 346-347.

[68] Smith, C., Abalde-Atristain, L., He, C., Brodsky, B. R., Braunstein, E. M., Chaudhari, P., ...& Ye, Z. (2015). Efficient and allele-specific genome editing of disease loci in human iPSCs. *Molecular Therapy*, *23*(3), 570-577.

[69] Lee, H. J., Kweon, J., Kim, E., Kim, S., & Kim, J. S. (2012). Targeted chromosomal duplications and inversions in the human genome using zinc finger nucleases. *Genome research*, *22*(3), 539-548.

[70] Park, C. Y., Kim, J., Kweon, J., Son, J. S., Lee, J. S., Yoo, J. E., Cho, S. R., Kim, J. H., Kim, J. S. and Kim, D. W. (2014). Targeted inversion and reversion of the blood coagulation factor 8 gene in human iPS cells using TALENs. *Proceedings of the National Academy of Sciences*, *111*(25), pp.9253-9258.

[71] Brunet, E., Simsek, D., Tomishima, M., DeKelver, R., Choi, V. M., Gregory, P., Urnov, F., Weinstock, D. M. and Jasin, M. (2009). Chromosomal translocations induced at specified loci in human stem cells. *Proceedings of the National Academy of Sciences*, *106*(26), pp.10620-10625.

[72] Cho, S. W., Kim, S., Kim, Y., Kweon, J., Kim, H. S., Bae, S., & Kim, J. S. (2014). Analysis of off-target effects of CRISPR/Cas-derived RNA-guided endonucleases and nickases. *Genome Research*, *24*(1), 132-141.

[73] Lee, H. J., Kim, E., & Kim, J. S. (2010). Targeted chromosomal deletions in human cells using zinc finger nucleases. *Genome Research*, *20*(1), 81-89.

[74] Koo, T., Lee, J., & Kim, J. S. (2015). Measuring and reducing off-target activities of programmable nucleases including CRISPR-Cas9. *Molecules and Cells*, *38*(6), 475.

[75] Naeem, M., Majeed, S., Hoque, M. Z., & Ahmad, I. (2020). Latest Developed Strategies to Minimize the Off-Target Effects in CRISPR-Cas-Mediated Genome Editing. *Cells*, *9*(7), 1608. https://doi.org/10.3390/cells9071608.

[76] Yee, J. K. (2016). Off-target effects of engineered nucleases. *The FEBS Journal*, *283*(17), 3239-3248.

[77] Hsu, P. D., Scott, D. A., Weinstein, J. A., Ran, F. A., Konermann, S., Agarwala, V., Li, Y., Fine, E. J., Wu, X., Shalem, O. and Cradick, T. J. (2013). DNA targeting specificity of RNA-guided Cas9 nucleases. *Nature Biotechnology*, *31*(9), pp.827-832.

[78] Fine, E. J., Cradick, T. J., Zhao, C. L., Lin, Y., & Bao, G. (2014). An online bioinformatics tool predicts zinc finger and TALE nuclease off-target cleavage. *Nucleic Acids Research*, *42*(6), e42-e42.

[79] Pattanayak, V., Lin, S., Guilinger, J. P., Ma, E., Doudna, J. A., & Liu, D. R. (2013). High-throughput profiling of off-target DNA cleavage reveals RNA-programmed Cas9 nuclease specificity. *Nature Biotechnology*, *31*(9), 839-843.

[80] Guilinger, J. P., Pattanayak, V., Reyon, D., Tsai, S. Q., Sander, J. D., Joung, J. K., & Liu, D. R. (2014). Broad specificity profiling of TALENs results in engineered nucleases with improved DNA-cleavage specificity. *Nature Methods*, *11*(4), 429-435.

[81] Fu, Y., Foden, J. A., Khayter, C., Maeder, M. L., Reyon, D., Joung, J. K., & Sander, J. D. (2013). High-frequency off-target mutagenesis induced by CRISPR-Cas nucleases in human cells. *Nature Biotechnology*, *31*(9), 822-826.

[82] Singh, D., Sternberg, S. H., Fei, J., Doudna, J. A., & Ha, T. (2016). Real-time observation of DNA recognition and rejection by the RNA-guided endonuclease Cas9. *Nature Communications*, *7*(1), 1-8.

[83] Modrzejewski D, Hartung F, Lehnert H, Sprink T, Kohl C, Keilwagen J and Wilhelm R (2020) Which Factors Affect the Occurrence of Off-Target Effects Caused by the Use of CRISPR/Cas: A Systematic Review in Plants. *Front. Plant Sci.* 11:574959. doi: 10.3389/fpls.2020.574959.

[84] Gupta, A., Meng, X., Zhu, L. J., Lawson, N. D., & Wolfe, S. A. (2011). Zinc finger protein-dependent and-independent contributions to the *in vivo* off-target activity of zinc finger nucleases. *Nucleic Acids research*, *39*(1), 381-392.

[85] Bulyk, M. L., Huang, X., Choo, Y., & Church, G. M. (2001). Exploring the DNA-binding specificities of zinc fingers with DNA microarrays. *Proceedings of the National Academy of Sciences*, *98*(13), 7158-7163.

[86] Segal, D. J., Dreier, B., Beerli, R. R., & Barbas, C. F. (1999). Toward controlling gene expression at will: selection and design of zinc finger domains recognizing each of the 5′-GNN-3′ DNA target sequences. *Proceedings of the National Academy of Sciences*, *96*(6), 2758-2763.

[87] Segal, D. J., Beerli, R. R., Blancafort, P., Dreier, B., Effertz, K., Huber, A., Koksch, B., Lund, C. V., Magnenat, L., Valente, D. and Barbas, C. F. (2003). Evaluation of a modular strategy for the construction of novel polydactyl zinc finger DNA-binding proteins. *Biochemistry*, *42*(7), pp.2137-2148.

[88] Perez, E. E., Wang, J., Miller, J. C., Jouvenot, Y., Kim, K. A., Liu, O., Wang, N., Lee, G., Bartsevich, V. V., Lee, Y. L. and Guschin, D. Y. (2008). Establishment of HIV-1 resistance in CD4+ T cells by genome editing using zinc-finger nucleases. *Nature Biotechnology*, *26*(7), pp.808-816.

[89] Zykovich, A., Korf, I., & Segal, D. J. (2009). Bind-n-Seq: high-throughput analysis of *in vitro* protein–DNA interactions using massively parallel sequencing. *Nucleic Acids Research*, *37*(22), e151-e151.

[90] Kuscu, C., Arslan, S., Singh, R., Thorpe, J., & Adli, M. (2014). Genome-wide analysis reveals characteristics of off-target sites bound by the Cas9 endonuclease. *Nature Biotechnology*, *32*(7), 677-683.

[91] Wu, X., Scott, D. A., Kriz, A. J., Chiu, A. C., Hsu, P. D., Dadon, D. B., Cheng, A. W., Trevino, A. E., Konermann, S., Chen, S. and Jaenisch, R. (2014). Genome-wide binding of the CRISPR endonuclease Cas9 in mammalian cells. *Nature Biotechnology*, *32*(7), pp.670-676.

[92] Zhang, L., Rube, H. T., Vakulskas, C. A., Behlke, M. A., Bussemaker, H. J., & Pufall, M. A. (2020). Systematic *in vitro* profiling of off-target affinity, cleavage and efficiency for CRISPR enzymes. *Nucleic Acids Research*, *48*(9), 5037-5053.

[93] Pattanayak, V., Ramirez, C. L., Joung, J. K., & Liu, D. R. (2011). Revealing off-target cleavage specificities of zinc-finger nucleases by *in vitro* selection. *Nature Methods*, *8*(9), 765-770.

[94] Fu, B. X., St. Onge, R. P., Fire, A. Z., & Smith, J. D. (2016). Distinct patterns of Cas9 mismatch tolerance *in vitro* and *in vivo*. *Nucleic Acids Research*, *44*(11), 5365-5377.

[95] Huston, N. C., Tycko, J., Tillotson, E. L., Wilson, C. J., Myer, V. E., Jayaram, H., & Steinberg, B. E. (2019). Identification of guide-intrinsic determinants of Cas9 specificity. *The CRISPR Journal*, *2*(3), 172-185.

[96] Kim

[97] Kim, D., Kang, B. C., & Kim, J. S. (2021). Identifying genome-wide off-target sites of CRISPR RNA–guided nucleases and deaminases with Digenome-seq. *Nature Protocols*, *16*(2), 1170-1192.

[98] Atkins, A., Chung, C. H., Allen, A. G., Dampier, W., Gurrola, T. E., Sariyer, I. K., Nonnemacher, M. R. and Wigdahl, B. (2021). Off-Target Analysis in Gene Editing and Applications for Clinical Translation of CRISPR/Cas9 in HIV-1 Therapy. *Frontiers in Genome Editing*, p.16.

[99] Kim, D., Kim, S., Kim, S., Park, J., & Kim, J. S. (2016). Genome-wide target specificities of CRISPR-Cas9 nucleases revealed by multiplex Digenome-seq. *Genome Research*, *26*(3), 406-415.

[100] Kim, H. K., Min, S., Song, M., Jung, S., Choi, J. W., Kim, Y., Lee, S., Yoon, S. and Kim, H. H. (2018). Deep learning improves prediction of CRISPR–Cpf1 guide RNA activity. *Nature Biotechnology*, *36*(3), pp.239-241.

[101] Tsai, S. Q., Nguyen, N. T., Malagon-Lopez, J., Topkar, V. V., Aryee, M. J., & Joung, J. K. (2017). CIRCLE-seq: a highly sensitive *in vitro* screen for genome-wide CRISPR–Cas9 nuclease off-targets. *Nature Methods*, *14*(6), 607-614.

[102] Lazzarotto, C. R., Nguyen, N. T., Tang, X., Malagon-Lopez, J., Guo, J. A., Aryee, M. J., Joung, J. K. and Tsai, S. Q. (2018). Defining CRISPR–Cas9 genome-wide nuclease activities with CIRCLE-seq. *Nature Protocols*, *13*(11), pp.2615-2642.

[103] Lazzarotto, C. R., Malinin, N. L., Li, Y., Zhang, R., Yang, Y., Lee, G., Cowley, E., He, Y., Lan, X., Jividen, K. and Katta, V. (2020). CHANGE-seq reveals genetic and epigenetic effects on CRISPR–Cas9 genome-wide activity. *Nature Biotechnology*, *38*(11), pp.1317-1327.

[104] Cameron, P., Fuller, C. K., Donohoue, P. D., Jones, B. N., Thompson, M. S., Carter, M. M., Gradia, S., Vidal, B., Garner, E., Slorach, E. M. and Lau, E. (2017). Mapping the genomic landscape of CRISPR–Cas9 cleavage. *Nature Methods*, *14*(6), pp.600-606.

[105] Crosetto, N., Mitra, A., Silva, M. J., Bienko, M., Dojer, N., Wang, Q., Karaca, E., Chiarle, R., Skrzypczak, M., Ginalski, K. and Pasero, P. (2013). Nucleotide-resolution DNA double-strand break mapping by next-generation sequencing. *Nature Methods*, *10*(4), pp.361-365.

[106] Mirzazadeh, R. (2019). *Novel methods to study genomic fragility and structural variation* (Doctoral dissertation, Karolinska Institutet (Sweden)).

[107] Lensing, S. V., Marsico, G., Hänsel-Hertsch, R., Lam, E. Y., Tannahill, D., & Balasubramanian, S. (2016). DSBCapture: in situ capture and sequencing of DNA breaks. *Nature Methods*, *13*(10), 855-857.

[108] Canela, A., Sridharan, S., Sciascia, N., Tubbs, A., Meltzer, P., Sleckman, B. P., & Nussenzweig, A. (2016). DNA breaks and end resection measured genome-wide by end sequencing. *Molecular Cell*, *63*(5), 898-911.

[109] Wong, N., John, S., Nussenzweig, A., & Canela, A. (2021). END-seq: an unbiased, high-resolution, and genome-wide approach to map DNA double-strand breaks and resection in human cells. In *Homologous Recombination* (pp. 9-31). Humana, New York, NY.

[110] Yan, W. X., Mirzazadeh, R., Garnerone, S., Scott, D., Schneider, M. W., Kallas, T., ...& Crosetto, N. (2017). BLISS is a versatile and quantitative method for genome-wide profiling of DNA double-strand breaks. *Nature Communications*, *8*(1), 1-9.

[111] Wang, X., Wang, Y., Wu, X., Wang, J., Wang, Y., Qiu, Z., Chang, T., Huang, H., Lin, R. J. and Yee, J. K. (2015). Unbiased detection of off-target cleavage by CRISPR-Cas9 and TALENs using integrase-defective lentiviral vectors. *Nature Biotechnology*, *33*(2), pp.175-178.

[112] Gabriel, R., Lombardo, A., Arens, A., Miller, J. C., Genovese, P., Kaeppel, C., Nowrouzi, A., Bartholomae, C. C., Wang, J., Friedman, G. and Holmes, M. C. (2011). An unbiased genome-wide analysis of zinc-finger nuclease specificity. *Nature Biotechnology*, *29*(9), pp.816-823.

[113] Tsai, S. Q., Zheng, Z., Nguyen, N. T., Liebers, M., Topkar, V. V., Thapar, V., Wyvekens, N., Khayter, C., Iafrate, A. J., Le, L. P. and Aryee, M. J. (2015). GUIDE-seq enables genome-wide profiling of off-target cleavage by CRISPR-Cas nucleases. *Nature Biotechnology*, *33*(2), pp.187-197.

[114] Zheng, Z., Liebers, M., Zhelyazkova, B., Cao, Y., Panditi, D., Lynch, K. D., Chen, J., Robinson, H. E., Shim, H. S., Chmielecki, J. and Pao, W. (2014). Anchored multiplex PCR for targeted next-generation sequencing. *Nature Medicine*, *20*(12), pp.1479-1484.

[115] Chaudhari, H. G., Penterman, J., Whitton, H. J., Spencer, S. J., Flanagan, N., Lei Zhang, M. C., Huang, E., Khedkar, A. S., Toomey, J. M., Shearer, C. A. and Needham, A. W. (2020). Evaluation of homology-independent CRISPR-Cas9 off-target assessment methods. *The CRISPR Journal*, *3*(6), pp.440-453.

[116] Nobles, C. L., Reddy, S., Salas-McKee, J., Liu, X., June, C. H., Melenhorst, J. J., Davis, M. M., Zhao, Y. and Bushman, F. D. (2019). iGUIDE: an improved pipeline for analyzing CRISPR cleavage specificity. *Genome Biology*, *20*(1), pp.1-6.

[117] Giannoukos, G., Ciulla, D. M., Marco, E., Abdulkerim, H. S., Barrera, L. A., Bothmer, A., Dhanapal, V., Gloskowski, S. W., Jayaram, H., Maeder, M. L. and Skor, M. N. (2018). UDiTaS™, a genome editing detection method for indels and genome rearrangements. *BMC Genomics*, *19*(1), pp.1-10.

[118] Costello, M., Pugh, T. J., Fennell, T. J., Stewart, C., Lichtenstein, L., Meldrim, J. C., Fostel, J. L., Friedrich, D. C., Perrin, D., Dionne, D. and Kim, S. (2013). Discovery and characterization of artifactual mutations in deep coverage targeted capture sequencing data due to oxidative DNA damage during sample preparation. *Nucleic Acids Research*, *41*(6), pp.e67-e67.

[119] Chen, L., Liu, P., Evans Jr, T. C., & Ettwiller, L. M. (2017). DNA damage is a pervasive cause of sequencing errors, directly confounding variant identification. *Science*, *355*(6326), 752-756.

[120] Chen, L., Liu, P., Evans, T. C. Jr., and Ettwiller, L. M. (2018). Response to comment on "DNA damage is a pervasive cause of sequencing errors, directly confounding variant identification." *Science* 361 (6409), eaat0958.

[121] Schmid-Burgk, J. L., Gao, L., Li, D., Gardner, Z., Strecker, J., Lash, B., & Zhang, F. (2020). Highly parallel profiling of Cas9 variant specificity. *Molecular Cell*, *78*(4), 794-800.

[122] Picelli, S., Björklund, Å. K., Reinius, B., Sagasser, S., Winberg, G., & Sandberg, R. (2014). Tn5 transposase and tagmentation procedures for massively scaled sequencing projects. *Genome Research*, *24*(12), 2033-2040.

[123] Chiarle, R., Zhang, Y., Frock, R. L., Lewis, S. M., Molinie, B., Ho, Y. J., Myers, D. R., Choi, V. W., Compagno, M., Malkin, D. J. and Neuberg, D. (2011). Genome-wide translocation sequencing reveals mechanisms of chromosome breaks and rearrangements in B cells. *Cell*, *147*(1), pp.107-119.

[124] Frock, R. L., Hu, J., Meyers, R. M., Ho, Y. J., Kii, E., & Alt, F. W. (2015). Genome-wide detection of DNA double-stranded breaks induced by engineered nucleases. *Nature Biotechnology*, *33*(2), 179-186.

[125] Zhou, Z. X., Zhang, M. J., Peng, X., Takayama, Y., Xu, X. Y., Huang, L. Z., & Du, L. L. (2013). Mapping genomic hotspots of DNA damage by a single-strand-DNA-compatible and strand-specific ChIP-seq method. *Genome Research*, *23*(4), 705-715.

[126] Hu, J., Meyers, R. M., Dong, J., Panchakshari, R. A., Alt, F. W., & Frock, R. L. (2016). Detecting DNA double-stranded breaks in mammalian genomes by linear amplification–mediated high-throughput genome-wide translocation sequencing. *Nature Protocols*, *11*(5), 853-871.

[127] Hoffman, E. A., Frey, B. L., Smith, L. M., & Auble, D. T. (2015). Formaldehyde crosslinking: a tool for the study of chromatin complexes. *Journal of Biological Chemistry*, *290*(44), 26404-26411.

[128] Kim, T. H., & Ren, B. (2006). Genome-wide analysis of protein-DNA interactions. *Annu. Rev. Genomics Hum. Genet.*, *7*, 81-102.

[129] Wienert, B., Wyman, S. K., Richardson, C. D., Yeh, C. D., Akcakaya, P., Porritt, M. J., Morlock, M., Vu, J. T., Kazane, K. R., Watry, H. L. and Judge, L. M. (2019). Unbiased detection of CRISPR off-targets *in vivo* using DISCOVER-Seq. *Science*, *364*(6437), pp.286-289.

[130] Shibata, A., Moiani, D., Arvai, A. S., Perry, J., Harding, S. M., Genois, M. M., Maity, R., van Rossum-Fikkert, S., Kertokalio, A., Romoli, F. and Ismail, A. (2014). DNA double-strand break repair pathway choice is directed by distinct MRE11 nuclease activities. *Molecular Cell*, *53*(1), pp.7-18.

[131] Liao, S., Tammaro, M., & Yan, H. (2016). The structure of ends determines the pathway choice and Mre11 nuclease dependency of DNA double-strand break repair. *Nucleic Acids Research*, *44*(12), 5689-5701.

[132] Akcakaya, P., Bobbin, M. L., Guo, J. A., Malagon-Lopez, J., Clement, K., Garcia, S. P., Fellows, M. D., Porritt, M. J., Firth, M. A., Carreras, A. and Baccega, T. (2018). *In vivo* CRISPR editing with no detectable genome-wide off-target mutations. *Nature*, *561*(7723), pp.416-419.

[133] Altschul, S. F., Gish, W., Miller, W., Myers, E. W., & Lipman, D. J. (1990). Basic local alignment search tool. *Journal of Molecular Biology*, *215*(3), 403-410.

[134] Langmead, B., Trapnell, C., Pop, M., & Salzberg, S. L. (2009). Ultrafast and memory-efficient alignment of short DNA sequences to the human genome. *Genome Biology*, *10*(3), 1-10. Langdon, W. B. (2015).

[135] Langdon, W. B. (2015). Performance of genetic programming optimised Bowtie2 on genome comparison and analytic testing (GCAT) benchmarks. *Bio Data Mining*, *8*(1), 1-7.

[136] Mandell, J. G., & Barbas, C. F. (2006). Zinc Finger Tools: custom DNA-binding domains for transcription factors and nucleases. *Nucleic Acids Research*, *34*(suppl_2), W516-W523.

[137] Cradick, T. J., Ambrosini, G., Iseli, C., Bucher, P., & McCaffrey, A. P. (2011). ZFN-site searches genomes for zinc finger nuclease target sites and off-target sites. *BMC Bioinformatics*, *12*(1), 1-10.

[138] Iseli, C., Ambrosini, G., Bucher, P., & Jongeneel, C. V. (2007). Indexing strategies for rapid searches of short words in genome sequences. *PloS One*, *2*(6), e579.

[139] Cermak, T., Doyle, E. L., Christian, M., Wang, L., Zhang, Y., Schmidt, C., Baller, J. A., Somia, N. V., Bogdanove, A. J. and Voytas, D. F. (2011). Efficient design and assembly of custom TALEN and other TAL effector-based constructs for DNA targeting. *Nucleic Acids Research*, *39*(12), pp.e82-e82.

[140] Doyle, E. L., Booher, N. J., Standage, D. S., Voytas, D. F., Brendel, V. P., VanDyk, J. K., & Bogdanove, A. J. (2012). TAL Effector-Nucleotide Targeter (TALE-NT) 2.0: tools for TAL effector design and target prediction. *Nucleic Acids Research*, *40*(W1), W117-W122.

[141] Grau, J., Boch, J., & Posch, S. (2013). TALENoffer: genome-wide TALEN off-target prediction. *Bioinformatics*, *29*(22), 2931-2932.

[142] Montague, T. G., Cruz, J. M., Gagnon, J. A., Church, G. M., & Valen, E. (2014). CHOPCHOP: a CRISPR/Cas9 and TALEN web tool for genome editing. *Nucleic Acids Research*, *42*(W1), W401-W407.

[143] Hruscha, A., Krawitz, P., Rechenberg, A., Heinrich, V., Hecht, J., Haass, C., & Schmid, B. (2013). Efficient CRISPR/Cas9 genome editing with low off-target effects in zebrafish. *Development*, *140*(24), 4982-4987.

[144] Zhu, X., Xu, Y., Yu, S., Lu, L., Ding, M., Cheng, J., Song, G., Gao, X., Yao, L., Fan, D. and Meng, S. (2014). An efficient genotyping method for genome-modified animals and human cells generated with CRISPR/Cas9 system. *Scientific Reports*, *4*(1), pp.1-8.

[145] Cancellieri, S., Canver, M. C., Bombieri, N., Giugno, R., & Pinello, L. (2020). CRISPRitz: rapid, high-throughput and variant-aware in silico off-target site identification for CRISPR genome editing. *Bioinformatics*, *36*(7), 2001-2008.

[146] McKenna, A., & Shendure, J. (2018). FlashFry: a fast and flexible tool for large-scale CRISPR target design. *BMC Biology*, *16*(1), 1-6.

[147] Listgarten, J., Weinstein, M., Kleinstiver, B. P., Sousa, A. A., Joung, J. K., Crawford, J., Gao, K., Hoang, L., Elibol, M., Doench, J. G. and Fusi, N. (2018). Prediction of off-target activities for the end-to-end design of CRISPR guide RNAs. *Nature Biomedical Engineering*, *2*(1), pp.38-47.

[148] Pliatsika, V., & Rigoutsos, I. (2015). "Off-Spotter": very fast and exhaustive enumeration of genomic lookalikes for designing CRISPR/Cas guide RNAs. *Biology Direct*, *10*(1), 1-10.

[149] Bae, S., Park, J., & Kim, J. S. (2014). Cas-OFFinder: a fast and versatile algorithm that searches for potential off-target sites of Cas9 RNA-guided endonucleases. *Bioinformatics*, *30*(10), 1473-1475.

[150] Li, H., & Durbin, R. (2009). Fast and accurate short read alignment with Burrows–Wheeler transform. *Bioinformatics*, *25*(14), 1754-1760.

[151] Xiao, A., Cheng, Z., Kong, L., Zhu, Z., Lin, S., Gao, G., & Zhang, B. (2014). CasOT: a genome-wide Cas9/gRNA off-target searching tool. *Bioinformatics*, *30*(8), 1180-1182.

[152] Jacquin, A. L., Odom, D. T., & Lukk, M. (2019). Crisflash: open-source software to generate CRISPR guide RNAs against genomes annotated with individual variation. *Bioinformatics*, *35*(17), 3146-3147.

[153] Zhang, Z. R., & Jiang, Z. R. (2022). Effective use of sequence information to predict CRISPR-Cas9 off-target. *Computational and Structural Biotechnology Journal*.

[154] Stemmer, M., Thumberger, T., del Sol Keyer, M., Wittbrodt, J., & Mateo, J. L. (2017). Correction: CCTop: an intuitive, flexible and reliable CRISPR/Cas9 target prediction tool. *PLoS One*, *12*(4), e0176619.

[155] Cradick, T. J., Qiu, P., Lee, C. M., Fine, E. J., & Bao, G. (2014). COSMID: a web-based tool for identifying and validating CRISPR/Cas off-target sites. *Molecular Therapy-Nucleic Acids*, *3*, e214.

[156] Singh, R., Kuscu, C., Quinlan, A., Qi, Y., & Adli, M. (2015). Cas9-chromatin binding information enables more accurate CRISPR off-target prediction. *Nucleic Acids Research, 43*(18), e118-e118.

[157] Labun, K., Montague, T. G., Gagnon, J. A., Thyme, S. B., & Valen, E. (2016). CHOPCHOP v2: a web tool for the next generation of CRISPR genome engineering. *Nucleic Acids Research*, *44*(W1), W272-W276.

[158] Labun, K., Montague, T. G., Krause, M., Torres Cleuren, Y. N., Tjeldnes, H., & Valen, E. (2019). CHOPCHOP v3: expanding the CRISPR web toolbox beyond genome editing. *Nucleic Acids Research*, *47*(W1), W171-W174.

[159] Xie, X., Ma, X., Zhu, Q., Zeng, D., Li, G., & Liu, Y. G. (2017). CRISPR-GE: a convenient software toolkit for CRISPR-based genome editing. *Molecular Plant*, *10*(9), 1246-1249.

[160] Doench, J. G., Fusi, N., Sullender, M., Hegde, M., Vaimberg, E. W., Donovan, K. F., Smith, I., Tothova, Z., Wilen, C., Orchard, R. and Virgin, H. W. (2016). Optimized sgRNA design to maximize activity and minimize off-target effects of CRISPR-Cas9. *Nature Biotechnology*, *34*(2), pp.184-191.

[161] Liu, G., Zhang, Y., & Zhang, T. (2020). Computational approaches for effective CRISPR guide RNA design and evaluation. *Computational and Structural Biotechnology Journal*, *18*, 35-44.

[162] Concordet, J. P., & Haeussler, M. (2018). CRISPOR: intuitive guide selection for CRISPR/Cas9 genome editing experiments and screens. *Nucleic Acids Research*, *46*(W1), W242-W245.

[163] Moreno-Mateos, M. A., Vejnar, C. E., Beaudoin, J. D., Fernandez, J. P., Mis, E. K., Khokha, M. K., & Giraldez, A. J. (2015). CRISPRscan: designing highly efficient sgRNAs for CRISPR-Cas9 targeting *in vivo*. *Nature Methods*, *12*(10), 982-988.

[164] Aprilyanto, V., Aditama, R., Tanjung, Z. A., Utomo, C., & Liwang, T. (2021). CROP: a CRISPR/Cas9 guide selection program based on mapping guide variants. *Scientific Reports*, *11*(1), 1-7.

[165] Corsi, G. I., Gadekar, V. P., Gorodkin, J., & Seemann, S. E. (2022). CRISPRroots: on-and off-target assessment of RNA-seq data in CRISPR–Cas9 edited cells. *Nucleic Acids Research*, *50*(4), e20-e20.

[166] Xie, K., Zhang, J., & Yang, Y. (2014). Genome-wide prediction of highly specific guide RNA spacers for CRISPR–Cas9-mediated genome editing in model plants and major crops. *Molecular Plant*, *7*(5), 923-926.

[167] Minkenberg, B., Zhang, J., Xie, K., & Yang, Y. (2019). CRISPR-PLANT v2: An online resource for highly specific guide RNA spacers based on improved off-target analysis. *Plant Biotechnology Journal*, *17*(1), 5.

[168] Lei, Y., Lu, L., Liu, H. Y., Li, S., Xing, F., & Chen, L. L. (2014). CRISPR-P: a web tool for synthetic single-guide RNA design of CRISPR-system in plants. *Molecular Plant*, *7*(9), 1494-1496.

[169] Liu, H., Ding, Y., Zhou, Y., Jin, W., Xie, K., & Chen, L. L. (2017). CRISPR-P 2.0: an improved CRISPR-Cas9 tool for genome editing in plants. *Molecular Plant*, *10*(3), 530-532.

[170] Alkan, F., Wenzel, A., Anthon, C., Havgaard, J. H., & Gorodkin, J. (2018). CRISPR-Cas9 off-targeting assessment with nucleic acid duplex energy parameters. *Genome Biology*, *19*(1), 1-13.

[171] Zhang, D., Hurst, T., Duan, D., & Chen, S. J. (2019). Unified energetics analysis unravels SpCas9 cleavage activity for optimal gRNA design. *Proceedings of the National Academy of Sciences*, *116*(18), 8693-8698.

[172] Heigwer, F., Kerr, G., & Boutros, M. (2014). E-CRISP: fast CRISPR target site identification. *Nature Methods*, *11*(2), 122-123.

[173] Oliveros, J. C., Franch, M., Tabas-Madrid, D., San-León, D., Montoliu, L., Cubas, P., & Pazos, F. (2016). Breaking-Cas—interactive design of guide RNAs for CRISPR-Cas experiments for ENSEMBL genomes. *Nucleic Acids Research*, *44*(W1), W267-W271.

[174] Haeussler, M., Schönig, K., Eckert, H., Eschstruth, A., Mianné, J., Renaud, J. B., Schneider-Maunoury, S., Shkumatava, A., Teboul, L., Kent, J. and Joly, J. S. (2016). Evaluation of off-target and on-target scoring algorithms and integration into the guide RNA selection tool CRISPOR. *Genome Biology*, *17*(1), pp.1-12.

[175] Doench, J. G., Hartenian, E., Graham, D. B., Tothova, Z., Hegde, M., Smith, I., Sullender, M., Ebert, B. L., Xavier, R. J. and Root, D. E. (2014). Rational design of highly active sgRNAs for CRISPR-Cas9–mediated gene inactivation. *Nature Biotechnology*, *32*(12), pp.1262-1267.

[176] Fusi, N., Smith, I., Doench, J., & Listgarten, J. (2015). In silico predictive modeling of CRISPR/Cas9 guide efficiency. *BioRxiv*, 021568.

[177] Sanson, K. R., Hanna, R. E., Hegde, M., Donovan, K. F., Strand, C., Sullender, M. E., Vaimberg, E. W., Goodale, A., Root, D. E., Piccioni, F. and Doench, J. G. (2018). Optimized libraries for CRISPR-Cas9 genetic screens with multiple modalities. *Nat Commun* 9: 5416.

[178] Chari, R., Yeo, N. C., Chavez, A., & Church, G. M. (2017). sgRNA Scorer 2.0: a species-independent model to predict CRISPR/Cas9 activity. *ACS Synthetic Biology*, *6*(5), 902-904.

[179] Chari, R., Mali, P., Moosburner, M., & Church, G. M. (2015). Unraveling CRISPR-Cas9 genome engineering parameters via a library-on-library approach. *Nature Methods*, *12*(9), 823-826.

[180] Wilson, L. O., Reti, D., O'Brien, A. R., Dunne, R. A., & Bauer, D. C. (2018). High activity target-site identification using phenotypic independent CRISPR-Cas9 core functionality. *The CRISPR Journal*, *1*(2), 182-190.

[181] Abadi, S., Yan, W. X., Amar, D., & Mayrose, I. (2017). A machine learning approach for predicting CRISPR-Cas9 cleavage efficiencies and patterns underlying its mechanism of action. *PLoS Computational Biology*, *13*(10), e1005807.

[182] Peng, H., Zheng, Y., Zhao, Z., Liu, T., & Li, J. (2018a). Recognition of CRISPR/Cas9 off-target sites through ensemble learning of uneven mismatch distributions. *Bioinformatics*, *34*(17), i757-i765.

[183] Peng, H., Zheng, Y., Blumenstein, M., Tao, D., & Li, J. (2018b). CRISPR/Cas9 cleavage efficiency regression through boosting algorithms and Markov sequence profiling. *Bioinformatics*, *34*(18), 3069-3077.

[184] Trivedi, T. B., Boger, R., Kamath, G. M., Evangelopoulos, G., Cate, J., Doudna, J., & Hidary, J. (2020). Crispr2vec: Machine Learning Model Predicts Off-Target Cuts of CRISPR systems. *BioRxiv*.

[185] Lin, J., & Wong, K. C. (2018). Off-target predictions in CRISPR-Cas9 gene editing using deep learning. *Bioinformatics*, *34*(17), i656-i663.

[186] Chuai, G., Ma, H., Yan, J., Chen, M., Hong, N., Xue, D., Zhou, C., Zhu, C., Chen, K., Duan, B. and Gu, F. (2018). DeepCRISPR: optimized CRISPR guide RNA design by deep learning. *Genome Biology*, *19*(1), pp.1-18.

[187] Kim, H. K., Min, S., Song, M., Jung, S., Choi, J. W., Kim, Y., Lee, S., Yoon, S. and Kim, H. H. (2018). Deep learning improves prediction of CRISPR–Cpf1 guide RNA activity. *Nature Biotechnology*, *36*(3), pp.239-241.

[188] Zhang, S., Li, X., Lin, Q., & Wong, K. C. (2019). Synergizing CRISPR/Cas9 off-target predictions for ensemble insights and practical applications. *Bioinformatics*, *35*(7), 1108-1115.

[189] Lin, J., Zhang, Z., Zhang, S., Chen, J., & Wong, K. C. (2020). CRISPR-Net: A Recurrent Convolutional Network Quantifies CRISPR Off-Target Activities with Mismatches and Indels. *Advanced Science*, *7*(13), 1903562.

[190] Zhang, G., Zeng, T., Dai, Z., & Dai, X. (2021). Prediction of CRISPR/Cas9 single guide RNA cleavage efficiency and specificity by attention-based convolutional neural networks. *Computational and Structural Biotechnology Journal*, *19*, 1445-1457.

[191] Liu, Q., He, D., & Xie, L. (2019). Prediction of off-target specificity and cell-specific fitness of CRISPR-Cas System using attention boosted deep learning and network-based gene feature. *PLoS Computational Biology*, *15*(10), e1007480.

[192] Dimauro, G., Colagrande, P., Carlucci, R., Ventura, M., Bevilacqua, V., & Caivano, D. (2019). CRISPRLearner: A deep learning-based system to predict CRISPR/Cas9 sgRNA on-target cleavage efficiency. *Electronics*, *8*(12), 1478.

[193] Zhang, Y., Long, Y., Yin, R., & Kwoh, C. K. (2020). DL-CRISPR: A Deep Learning Method for Off-Target Activity Prediction in CRISPR/Cas9 With Data Augmentation. *IEEE Access*, *8*, 76610-76617.

[194] Niu, R., Peng, J., Zhang, Z., & Shang, X. (2021). R-CRISPR: A Deep Learning Network to Predict Off-Target Activities with Mismatch, Insertion and Deletion in CRISPR-Cas9 System. *Genes*, *12*(12), 1878.

[195] Abadi, M., Agarwal, A., Barham, P., Brevdo, E., Chen, Z., Citro, C., Corrado, G. S., Davis, A., Dean, J., Devin, M. and Ghemawat, S. (2015). TensorFlow: Large-scale machine learning on heterogeneous systems. *ArXiv Preprint ArXiv:1603.04467*.

[196] Xiang, X., Corsi, G. I., Anthon, C., Qu, K., Pan, X., Liang, X., Han, P., Dong, Z., Liu, L., Zhong, J. and Ma, T. (2021). Enhancing CRISPR-Cas9 gRNA efficiency prediction by data integration and deep learning. *Nature Communications*, *12*(1), pp.1-9.

[197] Chen, X., Liu, J., Janssen, J. M., & Gonçalves, M. A. (2017). The chromatin structure differentially impacts high-specificity CRISPR-Cas9 nuclease strategies. *Molecular Therapy-Nucleic Acids*, *8*, 558-563.

[198] Dagdas, Y. S., Chen, J. S., Sternberg, S. H., Doudna, J. A., & Yildiz, A. (2017). A conformational checkpoint between DNA binding and cleavage by CRISPR-Cas9. *Science Advances*, *3*(8), eaao0027.

[199] Ricci, C. G., Chen, J. S., Miao, Y., Jinek, M., Doudna, J. A., McCammon, J. A., & Palermo, G. (2019). Deciphering off-target effects in CRISPR-Cas9 through accelerated molecular dynamics. *ACS Central Science*, *5*(4), 651-662.

[200] Palermo, G. (2019). Structure and dynamics of the CRISPR–Cas9 catalytic complex. *Journal of Chemical Information and Modeling*, *59*(5), 2394-2406.

[201] Zheng, L., Shi, J., & Mu, Y. (2018). Dynamics changes of CRISPR-Cas9 systems induced by high fidelity mutations. *Physical Chemistry Chemical Physics*, *20*(43), 27439-27448.

Chapter 6

Regulations and Policies of Genome Editing in Plants

Sukhdeep Kumar*
Department of Medical Laboratory Sciences, School of Allied Medical Sciences, Lovely Professional University, Phagwara, Punjab, India

Abstract

Genome editing techniques have empowered scientists to introduce desired point mutations and site-specific DNA sequence addition or removal. Such genome editing activities are aimed at developing plants with enhanced productivity. The problems of the ever-increasing magnitude of an increasing human population, decreasing area of agricultural land, and rapid climate change are looming over the human population. Developing plant varieties with increased productivity and stress tolerance is the only plausible solution. While notable advancement has been made by using traditional plant breeding methods to produce the plants with desired traits, it lacks specificity, making it a time-consuming process. Alternatively, plant breeders can quickly use modern genome editing tools to make rapid specific changes in the crop genome to speed up the crop improvement. The regulatory landscape is unclear for categorising the genome-edited plants (GE plants). Some nations categorise these plants as genetically modified organisms (GMOs) and have established elaborate regulatory frameworks, whereas others are uncertain about the categorisation and regulations. This chapter will overview the crop improvement methods, the similarities and differences between the plants generated through conventional breeding methods, and gene editing to provide insights into the inclusion, and exclusion of GE plants in different regulatory frameworks. This chapter discusses a detailed analysis of prevailing regulations, and policies on plant genome editing across different countries.

Keywords: genome-edited plants, CRISPR, GMOs, site-directed nuclease, crop improvement, mutational breeding

Introduction

Human survival revolves around various plant-based products. The transition of the human population from hunter-gatherers to agriculturalists was instrumental in supporting the thriving human population. The agricultural sector faces constant pressure to increase production due

* Corresponding Author's Email: sukhdeep28@gmail.com.

In: Genome Editing for Crop Improvement
Editors: Rahul Kumar, Ashutosh Sharma, Yogesh Kumar et al.
ISBN: 979-8-88697-740-0

to the growing human population, which is projected to cross nine billion by 2050 [1]. Particularly in developing and emerging nations, there is a lingering problem of rising food costs, and dwindling food security while rapid urbanisation occurs at the cost of agricultural land. Additionally, agricultural production is under intense pressure due to global climate change, and escalating biotic and abiotic pressures. Agriculture requires freshwater for irrigation, and globally, freshwater supplies are depleting. Therefore, agricultural practices should adapt to lower water requirements to conserve valuable water supplies and cut carbon emissions. There is a growing need for future crops to be nutrient-rich, climate-resilient, high-productivity, and resistant to biotic and abiotic stresses. There is a looming need to undertake swift action to develop crops with the desired traits.

Traditional Methods for Crop Improvement

Agriculturists have been improving crops since the start of agriculture by selecting desirable traits, and improving crop varieties through conventional plant breeding methods of cross-pollination and self-pollination. These traditional methods suffer from the disadvantage of transferring unwanted traits into the resulting crop variety, which requires years of extensive breeding programmes that segregate and eliminate the unwanted traits and generate a crop variety with only desired traits. Induced mutagenesis is another method perceived as a conventional method of the crop improvement. Plant breeders had utilised mutant breeding since the 1920s, when it was discovered that irradiation and chemical mutagenesis could produce heritable changes. In the period between the 1950s and 1960s, extensive use of mutational breeding resulted in the release of many mutant crop varieties [2-6]. In 1934, tobacco became the first commercially available mutant cultivar, and by 1995, 484 types of mutant cultivars were available [7, 8]. Since then, the number has been increasing, with new mutant strains being reported regularly around the globe. Fruit trees (peach, apple, and citrus), ornamentals (dahlia, and chrysanthemum), food crops (barley, rice, corn, and wheat), and other plants are among the plants available. Mutational breeding has yielded crop varieties with desirable agronomic traits such as improved product quality, early maturity, and winter hardiness. Irradiation and chemical mutagen exposure are the two most common ways of inducing mutations in plants. High energy radiations introduce a combination of point mutations, and the deletion of large DNA fragments, whereas chemical mutagens introduce single base changes in the genome. Radiation mutagenesis has created over 2,500 plant species (including rice, wheat, grapefruit, lettuce, and many fruits) [9, 10]. Chemical mutagens such as sodium azide, ethyl methanesulfonate (EMS), N-methyl methanesulfonate (MMS), methyl-N-nitrosourea (MNU), and hydrogen fluoride (HF) have been extensively explored to create valuable cultivars [11-14].

The traditional crop breeding methods are considered natural, and the crop varieties generated through these methods are not regulated. The traditional methods of crop improvement are slow and lack specificity, and they are heavily dependent on random combinations and mutations that might yield improved crop varieties. Nevertheless, the present demographic needs cannot be met by these methods.

Transgenics Methods for Crop Improvement

A giant leap in the field of crop improvement came with advancements in the field of recombinant DNA technology. These advancements equipped plant biologists to produce transgenic plants by late 1970. Genetic engineering was used to integrate novel genes into plant genomes to produce transgenic plants, ushering in an era of genetically engineered plants. The genetically engineered plants are also known as genetically-modified (GM) plants. The desired trait is introduced in these plants by inserting the gene responsible for the trait from other species into the plant genome. The inserted gene may be derived from another plant, animal, or bacteria. These gene inserts are codon-optimized for optimal expression in the host plant. Most transgenic plants are generated through *Agrobacterium tumefaciens* mediated transformation or particle gun methods [15-18]. A transgenic cultivar is generated to improve the host plant by conferring traits for insect resistance, drought resistance, herbicide resistance, salinity tolerance, and bioremediation. Transgenic plants with insect resistance have been created by inserting the cry protein gene from *Bacillus thuringiensis* [19]. Another well-known example of transgenic plants is herbicide glyphosate-tolerant plants with the transgenic gene for CP4 EPSP synthase from *Agrobacterium* sp. [20]. The presence of transgenes in GM plants classifies them as genetically modified organisms (GMOs). GM crops have increased productivity with a reduced cost of production, thereby contributing to the agricultural economy, sustainability, and human welfare at large. The cultivation of GM crops has steadily increased in 28 major GM crop growing nations, and the area under GM crops increased to 190 million hectares in 2019 [21]. GM plants are extensively cultivated in countries like the United States, Brazil, India, Canada, Australia, Argentina, Spain, and Portugal.

Genome Editing for Crop Improvement

The crops generated through some traditional and transgenic methods involve random changes in the plant genome. These random, although stable, changes may adversely impact the plant gene regulatory elements or change the open reading frame of a gene, resulting in the production of novel peptides, proteins, and metabolites with unknown biological activity. Therefore, the cultivars generated through these methods require extensive testing for compliance with biosafety. Genome editing, also known as New Breeding Techniques (NBT), is a relatively recent methodology for crop improvement, with CRISPR Cas9 being looked at as a significant breakthrough in crop improvement.

Genome editing utilises different types of site-directed nucleases (SDNs). These include Meganucleases, Zinc Finger Nucleases (ZFNs), Transcription Activator Like Effector Nucleases (TALENs), and the latest addition to the genome editing toolkit "CRISPR/Cas" [22]. Meganucleases are known as homing endonucleases. These endonucleases are capable of generating double-stranded breaks at specific DNA sequences through large cleavage sites of about 14-25 bp. Several meganucleases are expressed endogenously inside living organisms [23]. The use of meganucleases in gene editing is limited due to their non-modular structure and difficulty in designing the meganuclease to target the desired genomic sequence. ZFNs consist of a DNA binding domain with zinc finger motifs and a nuclease domain to cleave the target DNA. Intrinsically, the zinc finger domain is capable of interacting with three base-pair sites on target DNA which can be increased by the addition of more DNA binding motifs [24]. TALENs' mechanism of action is similar to that of ZFNs and utilises the nuclease activity to introduce double-strand breaks and use dimerization to bind the target site on the DNA. The DNA binding domain in TALENs consists of a tandem array of 33-35 amino acid repeats and has single nucleotide recognition. When contrasted to ZFNs, the amino acid repeats have single nucleotide recognition, which increases targeting capabilities and specificity. Even at a single nucleotide precision, employing TALENs as a gene-editing method is time-consuming and expensive. TALEN-based genome editing suffers from design constraints due to TALENs' structure, which explicitly needs a 5' thymine and a 3' adenine at the target site, which restricts target customizability. Furthermore, TALENs have worse editing effectiveness in extensively methylated areas. Delivery of TALENs into cells is complicated since TALENs are substantially bigger than ZFNs, requiring more time and money to generate a successful edit. While TALENs demonstrated advancements in genome editing technologies, the high labour and monetary costs restricted their extensive use [25].

SDNs create precise changes to a genome at predetermined sites, lowering the risk of random mutagenesis. SDNs provide an affordable and efficient method of introducing targeted genome alterations. Some SDN-created plants are devoid of foreign genetic material and may contain alterations similar to traditional mutagenesis methods. SDN1 systems cause gene knockout, gene disruption, or gene silencing through the introduction of a point mutation. To implement the appropriate improvements, double-strand break (DSB) repair in SDN1 does not require the use of a repair template. Scientists are arguing whether or not off-targets in SDN1 should be investigated further. The results of SDN1 are indistinguishable from those of spontaneous mutations. SDN2 systems utilisea template to alter one or more nucleotides at DSB. The genetic changes introduced by this system are indistinguishable from the changes introduced through the use of traditional plant breeding methods. In this method, the bases to be changed are specified in the repair template. SDN2 systems are capable of altering six to twenty nucleotides at the DSB with the help of microhomology end joining (MHEJ). In the case of the SDN3 system, the repair template contains 500 or more base complementarity in the regions flanking the target site. SDN3 systems are used to introduce a new genetic element at a specific site, unlike the transgenic approach, which introduces the new genetic element at a random site in the genome. Because of the limited efficiency of HDR in plants, SDN3 applications are wasteful and burdensome [26].

There are three types of plants that may be created with SDN:

(i) Plants with genomic alterations resulting in the introduction of a new gene.
(ii) Plants with genomic alterations resulting in the introduction of point mutation but no new genetic element.
(iii) Plants without any genome change.

Most of the shortcomings of Meganucleases, ZFNs, and TALEN-based genome editing were rectified in 2012, when scientists discovered a new method for gene editing based on the bacterial defense system CRISPR/Cas [27]. This defense system functions in bacteria by recognising and destroying the invading viral DNA. CRISPR is a component of the bacterial defence mechanism that specifically destroys invading viral DNA. Jennifer Doudna and Emmanuelle Charpentier performed the paradigm-changing pioneering work on CRISPR characterization, which fascinated the scientific community. Feng Zhang published a paper in

2013 describing the CRISPR may be used to modify eukaryotic DNA [28]. The CRISPR toolkit consists of a guide RNA and a Cas9 nuclease. The guide RNA identifies the 20-nucleotide sequence in the target DNA, which is subsequently broken down by Cas9 nuclease. Scientists may personalise their guide RNAs with CRISPR, and algorithms have been created to determine the likelihood of off-target consequences. On the other hand, CRISPR is considerably more adaptable and cost-effective, making it more accessible to scientists with limited budgets and time. Figure 1 explains various gene editing methodolies that are employed in the plant genome editing.

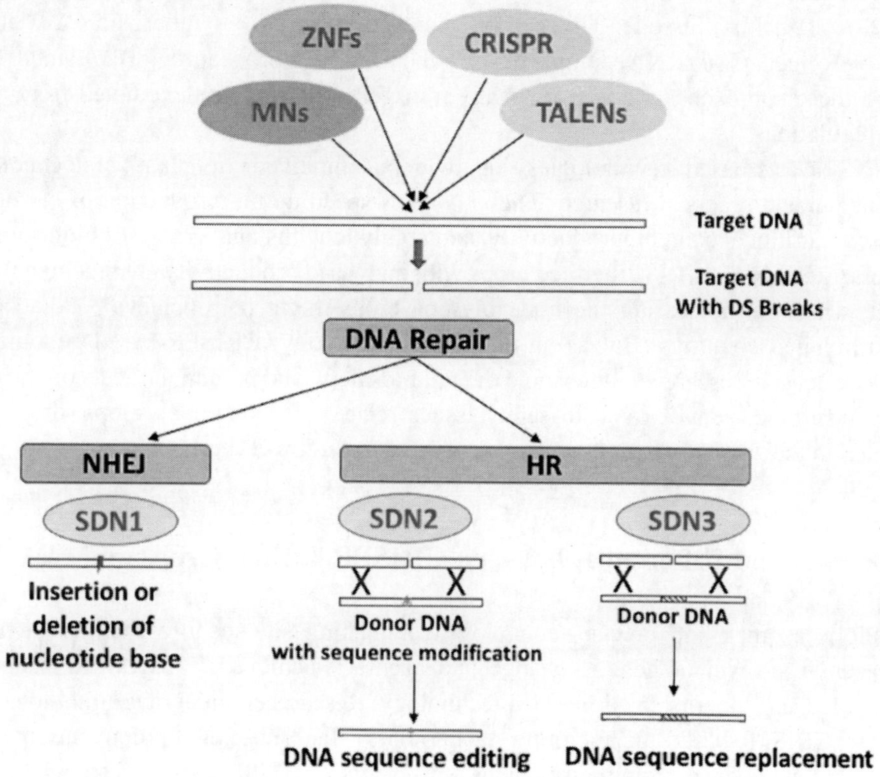

Figure 1. Overview of gene editing methodologies used in plant genome editing.

CRISPR/Cas is commonly used to solve crop plant issues. This technique can quickly deliver desired genetic changes in crops for increasing nutritional content, productivity, and resilience to biotic and abiotic challenges. The novel CRISPR/Cas technologies and applications for crop improvement that are currently accessible have far-reaching consequences for ensuring food availability in poorer populations. CRISPR/Cas is a relatively new technology in the field of biotechnology that holds promising prospects in agriculture, and by far, it is the most accurate and efficient of all the crop improvement methods. The regulatory frameworks will decide the thrust areas and extent to which they will be used. Although the CRISPR technique is pretty accurate, but the associated unwanted genetic changes in the form of point mutations, translocations, deletions, and inversions cannot be completely ruled out. Off-target genetic changes can impact the human health environment and may contribute to ecological instability. There is a remarkable similarity in the generation of gene-edited plants and GM plants, which involves the transfer of DNA cassettes in the target cells [29]. In this manner, the genome-editing procedure is similar to transgenics procedures, which involve the transfer of a DNA construct into the host cell. The CRISPR/Cas9 DNA cassette, once inside the target cell, produces the gRNA and proteins that result in the creation of DSBs at the target sites in the target genome. The experts in this field are skeptical about the consequences of long-term integration into the target genome and its long-term impacts on plant diversity, biosafety, and human health. In order to address the persistence of the Cas9 gene in the host plant, the techniques like backcrossing with the offspring might result in the deletion Cas9 gene. However, other approaches entirely omit the exposure of the target genome to the CRISPR/Cas9 DNA Cassette. This can be achieved by *in vitro* synthesized Cas9 mRNA or using preformed Cas-gRNA complexes for the CRISPR procedure [30]. Plants produced through these approaches are free from any transgene, and may be exempted from the strict GMO regulations.

CRISPR/Cas-based technologies are widely employed in plant biotechnology for fundamental and practical research. These systems speed up the production of cultivars with increased nutritional content, productivity, and ability to withstand a range of biotic and abiotic stressors. Steadily the CRISPR edited crops with increased productivity are reaching the global market, where further easing the trade of such crops might help achieve food security for underdeveloped countries. But as an emerging technology, CRISPR-based genome editing requires extensive research to determine overall biosafety, and potential effects on the ecology. Due to insufficient data related to safety issues related to gene-edited crops, they might be subjected to the same restrictions in some countries faced by GM crops.

Similarities and Differences between CRISPR-Edited Crops and GM Crops

CRISPR is a quick and precise technology for making site-specific changes in the DNA sequences. The level of genetic change and accuracy achieved by CRISPR technology is unparalleled and superior to earlier GM technology. Researchers in agricultural biotechnology consider CRISPR-based gene-editing technologies far superior to conventional genetic engineering methods of crop improvement. Most of the CRISPR crops lack transgenes; hence, these can be categorized along with the crops generated through the conventional crossing and mutagenesis methods; thereby, such crops have more potential for acceptance by different stakeholders. GMOs such as transgenic plants are generated by inserting a useful transgene into the plant genome. The most common example is the insertion of the Bt toxin gene from the *Bacillus thuringiensis* gene into the cotton plant. The Bt toxin is then expressed in different plant tissues, and provides protection against different insect pests. Overall, the insertion of a toxic transgene in some GM crops has negatively impacted their acceptance by farmers and consumers alike. Environmentally, these crops have the potential to destroy the population of harmless insects, and there is also a remote possibility of the transfer of transgene into unintended plants. The economic costs of developing and large-scale testing of GMOs are very high, resulting in the creation of monopolies of a few wealthy multinational corporations [31].

In contrast, CRISPR technology is relatively low-cost and readily available to small companies in underdeveloped countries. Transgenic technology suffers from the element of random incorporation of a transgene into the plant genome and lacks specificity, which can adversely influence the fitness of the resulting GMO. On the contrary, CRISPR techniques are used to create precise genome editing allowing for better control of the final product. CRISPR-based base editing can be used to change a specific base in the plant genome resulting in a plant with changes equivalent to the traditional mutagenesis process. DNA-free methods of using CRISPR RNPs can be used to produce a cultivar devoid of any foreign DNA [30]. Technically, some CRISPR/Cas approaches can obtain transgene-free gene editing, leading to the categorization of such gene-edited plants as non-GMOs. The CRISPR DNA cassette-based approach, like GMOs, utilizes biolistics or *Agrobacterium* to deliver into the cell, which results in random insertion into the genome. Another common problem for the CRISPR/Cas DNA cassette-based approach and GMOs is the use of a selection marker to select the plants with desired changes. Although different breeding strategies may get rid of these selection markers, some residual markers may persist in CRISPR edited crops, which raises biosafety concerns similar to GMOs.

Environmental Issues with CRISPR Crops

CRISPR-based plant genome editing makes it possible to generate tree varieties with a robust response to abiotic and biotic stress, thereby augmenting environmental protection in rapid climate change. But there are some potentially negative environmental impacts of CRISPR plants, mainly when used to produce traits like herbicide resistance. The possibility of genetic spillover to other plant varieties and problems with herbicide residues in crop products further raise concerns about the use of these techniques. This is similar to the negative environmental impacts of GM crops. When CRISPR is used to create cultivars similar to existing GMOs, it can lead to decreased biodiversity and the elimination of friendly insects [32].

Global Guiding Policies for the Regulation of Genome Editing in Plants

The broader perspectives for designing regulatory frameworks for genome-edited plants can be derived from the Convention on Biological Diversity (CBD) [33]. The convention is a multilateral treaty signed by 196 countries and came into force in 1993. The nations participating in the Convention on Biological Diversity (CBD) are required to "establish or maintain means to regulate, manage, or control the risks associated with the use and release of living modified organisms (LMOs)." Article 1 of the Cartagena Protocol on Biosafety (CPB) states "to contribute to ensuring an adequate level of protection in the field of the safe transfer, handling, and use of living modified organisms resulting from modern biotechnology that may have adverse effects on the conservation, and sustainable use of biological diversity, also taking into account risks to human health, and specifically focusing on transboundary movements" [34].

Regulations and Policies for Genome Editing in Plants in Continents-Wise Select Countries

This section describes the regulations and policies for genome editing in select counties of the world. These countries are already engaged in the development and cultivation of GM crops, and have a regulatory framework in place for regulations of GM crops. Generally, SDN1 and SDN2 technologies are classified as non-GMOs in most significant GM crop growing countries, whereas plants produced through SDN3 technologies of genome editing are classified as GMOs and are subjected to strict regulatory control. Table 1 highlights the regulatory status of GM and genome-edited plants in various countries.

Table 1. Regulatory status of GM and genome-edited plants in various countries

S.No	Country	GM Plants	Genome-edited plants
1.	India	Regulated	Unregulated for SDN1 and SDN2 methodologies
2.	China	Regulated	Regulations under development
3.	Japan	Regulated	Regulated
4.	Eswatini	Regulated	Regulations under development
5.	Kenya	Regulated	Mostly unregulated
6.	South Africa	Regulated	Regulated
7.	United States	Regulated	Regulated
8.	Argentina	Regulated	Mostly unregulated/Case by case
9.	Brazil	Regulated	Mostly unregulated/Case by case
10.	Canada	Regulated	Regulated
11.	Australia	Regulated	Mostly unregulated
12.	New Zealand	Regulated	Regulated
13.	European Union	Regulated/Prohibited	Regulated/Prohibited
14.	Switzerland	Regulated	Regulated
15.	Russia	Regulated	Regulated

Asia

The Asian continent occupies approximately 30% of earth's total land area and is home to 59.76% of the total human population. Bangladesh, China, India, Indonesia, Myanmar, Pakistan, Philippines, and Vietnam are involved in the commercial farming of GM crops. China and India are the major GM crop producers in Asia, taking active steps to design frameworks for regulating plant genome editing.

India

India has the world's fourth-largest acreage of genetically modified (GM) crops at 11.4 million hectares in 2017. However, unlike other major producers, the entire GM acreage is under a single cotton crop that contains genes from the *Bacillus thuringiensis*, or Bt soil bacterium, which encodes resistance to Heliothis bollworm insect pests. In India, the Department of Biotechnology (DBT) is engaged in producing policies for the regulation of genome-edited plants. DBT has notified the guidelines for assessing the biosafety and efficacy of GE crops in 2022 [35]. This regulatory framework utilizes a hierarchical categorization of GE plants based on the process. GE crops with verified SDN1 status are placed in Group I. GE crops categorized under these regulations are subject to only local supervision by the institutional biosafety committee. The GE plants generated through the SDN2 process are placed in Group II, and the proposed guidelines require extensive field testing and validation of such plants. The hierarchy of regulatory bodies for testing and validation of this category has not been finalised yet. The SDN3-based plants are placed in Group III. The GE plants in this group have either external DNA inserts or extensive DNA rearrangements. Plants categorized in this group require intensive regulation and testing before release in the market.

China

China has the sixth position in terms of area under GM crops, with 3.70 million hectares of arable land used to cultivate Bt cotton and Bt papaya. China was the pioneering country to start commercial cultivation of GMO crops with the cultivation of virus-resistant tobacco in 1988. China is also the biggest importer of GM soybean and canola. China adopts a GMO regulatory policy that amalgamates the two extreme regulatory approaches followed by the European Union, and the USA. However, extensive research is going on in plant genome editing, as evident from the number of papers published in China on CRISPR technologies [36]. China strictly controls the import and local cultivation of genetically engineered plants. However, many plant scientists expect China to follow the United States' lead and regulate most gene-editing methods in the same way as traditional plants. China carefully governs genetically engineered crops under the "Regulations on Administration of Agricultural Genetically Modified Organisms Safety," which the Ministry of Agriculture issued in 2001. The assessment, regulation, and inspection of new crops is the responsibility of an organization made up of numerous departments from several ministries. The state council evaluates the issues of safety.

In contrast to the United States, China governs the method of creating GM seeds rather than the features of the end goods and covers goods made from GMOs under GM regulations [37]. The Chinese Ministry of Agriculture (MOA) released a Guide for Biosafety Assessment of Gene-edited Crops on 24 January 2022, to govern gene-edited crops. It simplifies the regulatory procedures for gene-edited crops compared to GM crops, thereby allowing speedy commercialization of the gene-edited crops. Maize, rice, and soybeans have been successfully gene-edited in China. The publication of the Guide is anticipated to accelerate the commercial production of gene-edited crops [38].

Japan

Japan introduced the regulation for GMOs in 2003, which follows a product-based regulation of GMOs. The introduction of external RNA and DNA inside an organism is the critical trigger for the regulatory regime's application and categorization of an organism as a GMO. The regulatory approach may be described as "product-based" without considering the process of obtaining the GE plant [39]. If nucleic acids are found in plants of identical species, or other species that interchange nucleic acids with genetically altered organisms, they are not considered alien. In the case of SDN methodology, the persistence of guide RNA in GE plants leads to its categorization as GMO. Japan has four ministries to regulate crops and foods that have been produced through genetic engineering: the Ministry of Education, Culture, Sports, Science and Technology (MEXT), the Ministry of Health, Labor and Welfare (MHLW), the Ministry of Agriculture, Forestry and Fisheries (MAFF), and the Ministry of Environment (MOE). Apart from these ministries, there is another autonomous body, the Food Safety Commission (FSC), which assessment of risks associated with feed and food.

Overall, Japan has a permissible regulatory environment for genome-edited crops and GMOs. Japan has become the major importer of GM crops globally by approving the use of

133 GM crops for import, distribution and cultivation [40]. Although there is no restriction on GMOs in Japan, and a comparably large number of GM cultivars have been permitted for farming, there is no actual farming of such crops. Non-regulation-related elements can help to explain this strange scenario. Companies in Japan pay a significant level of consideration to customers' satisfaction and are pretty responsive to expressed apprehensions or wishes. Local farmers and manufacturers are hesitant to utilize GMOs because of the prevalent adverse attitude towards GMOs in Japanese society. The lack of acceptance for GM crops also stems from the existing agronomical practices that will not benefit from the use of GM crops. However, local administrative regulations obstruct the use of GM crops. Local governments enforce onerous application criteria, impose difficult-to-follow coexistence measures, or levy application costs that make the growing of GM crops prohibitively expensive [41]. While evaluating GMOs, the legislative requirements are considered in an economic, political, and social context.

Africa

The African continent is the second-largest continent with an area of 11.7 million square miles and the second most populated continent, with the population projected to be 1.39 billion in 2022. Only three countries on the African continent, namely, Eswatini, Kenya, and Nigeria, are actively working on laying down policies and regulatory frameworks for plant genome editing.

Eswatini

GM crops in Eswatini are regulated by the Biosafety Act of 2012, which is now being redesigned to accommodate NBTs, including genome editing [42]. Eswatini Environmental Authority accepts the applications from plant breeders for the introduction of new GM crops and forwards the application to the National Biosafety Advisory Committee for review and recommendations. Finally, under the supervision of the Ministry of Tourism and Environmental Affairs, the Biosafety Board provides the final decision based on the report and recommendations of the National Biosafety Advisory Committee.

Nigeria

Nigeria published its first draft of National biosafety guidelines for the regulation of gene editing in 2020 [43]. The Nigerian regulatory framework is similar to the framework adopted by Argentina, Brazil, and India.

Kenya

National biosafety authority of Kenya issued the guidelines for the regulation of genome-edited plants in February 2022 [44]. These guidelines exempt genome-edited plants without incorporating heterologous DNA from the GMO regulations.

South Africa

South Africa was the pioneering country on the African continent to cultivate GM crops. It started with the cultivation of GM maize in 1996, followed gradually by GM cotton and GM soybean. The cultivation area for these GM crops was 2.73 million hectares in 2017. South

Africa is slow to design a framework for NBTs, including gene editing. However, the Academy of Sciences of South Africa published a consensus study on "The Regulatory Implications of New Breeding Techniques (NBTs)" in 2017 [45]. Still, no formal regulatory frameworks have been adopted by South Africa for gene editing in plants.

North America

North America is the third largest continent, with a human population of 596 million. A lot of climatic diversity is observed across North America than on other continents. North America can be divided into the cool temperate zone, tropical zone, subtropical zone, and dry zone, with each zone contributing a different variety of agricultural produce.

United States

The United States is a leading nation in the research and commercialization of genetically modified crops. The regulatory framework for GMOs revolves around the Coordinated Framework for the Regulation of Biotechnology 1984 [46]. This framework is product-oriented and categorizes the organism on the basis of introduced traits without any consideration for the process involved in modifying the organism. In 2016, the Working Group on Biotechnology modified the existing regulatory framework to accommodate new technologies like CRISPR/Cas [47]. This has been performed to provide a conducive environment for innovation without ignoring the risk assessment of such innovations. The United States' approach to GMO regulation varies from others in that it employs distinct but selective triggers. As a result, if GMOs do not match any of the conditions that lead to the initiation of regulatory provisions for GMOs, they might theoretically avoid the regulatory framework entirely. For example, if a GE plant does not contain any external DNA sequence, it avoids categorization as a GMO. Another distinguishing element of the US regulatory framework is that it uses an informal consultation method to regulate food using a product-based trigger. There is much legal bemusement regarding the statutory position of GEOs under the existing regulatory framework.

Canada

Canada began commercial cultivation of GM canola, GM corn and GM soybean in 1995, 1996, and 1997, respectively. Canada follows an entirely product-based regulatory approach. The presence of a new agronomic feature in a plant variety, rather than some genetic modification or any other traditional or current breeding strategy, is the fundamental trigger for a stricter licensing process in this case. According to the definition of a novel trait variety, "it is a plant containing a trait not present in plants of the same species already existing as stable, cultivated populations in Canada or present at a level significantly outside the range of that trait in stable, cultivated populations of that plant species in Canada" [48]. As a result, genome altered plants, crops developed from "traditional" genetic engineering, and conventional varieties are all subject to the same restrictions. Only the cultivation clearance procedure is product-based. The gene-edited crops and foods are regulated jointly by Health Canada, and the Canadian Food Inspection Agency (CFIA). A framework consisting of three regulations, namely, the Food and Drugs Act and Regulations, the Feeds Act and Regulations, and the Seeds Act and Regulations, is used to evaluate the gene-edited organisms.

South America

The continent of South America has an area of 17,840,000 square kilometers and is home to approximately 437 million people. Brazil and Argentina are the major GM crop producers, ranking after the US in second and third place, respectively.

Argentina

Argentina was the first nation to pass a new regulation in 2015 addressing GE crops. According to this, the GE crops without any foreign DNA are categorized and regulated as the plants generated from conventional breeding methods. The regulation and assessment of gene-edited plants are performed by the Argentine Biosafety Commission (CONABIA). The GE plant regulation defines the procedural details for the assessment of GE plants as GMO or non-GMO. The GE crops without any foreign DNA are categorized and regulated as the plants generated from conventional breeding methods. If the GE plant is determined to be GMO, then it is subjected to an extensive Argentine regulatory framework for GMOs [49]. This regulation allows the crop developers to submit their cultivars in the early stage of research development to the competent authorities to evaluate non-GMO or GMO status. This evaluation step is required to be completed within sixty days. This early-stage and fast-paced category assessment helps the plant breeder in proper categorization and process change at an early stage of crop development.

The regulatory status of the cultivar is retained after a thorough assessment with the help of molecular biology tools that determine the stipulated changes incorporated in the final cultivar. If competent authorities find some potential risk associated with GE crops, then plant breeders may be provided consultation on safety. If a plant or plant variety does not fall under the GMO category, it must follow the same criteria as 'conventional' plants or plant variations. Nevertheless, suppose the authorized entity in charge of the (advance) consultation procedure finds concerns associated with an NBT-derived crop variant. In that case, it may give safety-related advice to the plant breeder. Furthermore, the authorities are required to communicate with the plant variety registration agency about any (potential) dangers. If commercial production and use of the plant variety offer intolerable hygiene or hazards to other plants, the registration could be refused in the end. Chile and Columbia follow gene-editing regulations similar to Argentina. Similar steps have been taken by other South American nations such as Brazil, Ecuador, Guatemala, Honduras, and Paraguay. Ecuador exempted SDN1 and cisgenesis from GMO regulations in 2019 [50].

Brazil

Brazil is the third-largest producer of GM crops, with a cultivation area of 51.3 million hectares for GM crops. The National Biosafety Technical Commission (CTNBio) is empowered by Brazilian Biosafety Law 2005 to oversee the biotechnological progress in the country and recommend the enactment of new regulations to balance technological advances with biosafety standards. The Normative Resolution No. 16 (2018) of CTNBio allows non-GMO classification of the plant varieties obtained through SDN1 and SDN2 technologies of gene editing [51]. The Brazilian regulatory framework places the SDN3 products as GMOs and requires a thorough risk assessment and biosafety compliance from CTNBio before being released into the country.

Australia/Oceania

Australia is the smallest continent, with an area of 7,692,000 square kilometers, and it is home to approximately 39 million humans.

Australia and New Zealand

Both countries have a unique bi-national body, Food Standards Australia New Zealand (FSANZ), which separates the Australian regulatory system for GMOs from the other nations. Prior to their commercial distribution in Australia and New Zealand, FSANZ is in charge of assessing and approving GM foods. FSANZ does not conduct separate assessments for Australia and New Zealand but grants both nations a unified market authorization [52]. As a result, both jurisdictions must follow the same regulations and processes. When it is considered that GMOs are not grown in New Zealand but are grown by Australian farmers, the idea of a shared sanctioning body is very intriguing. Although Australia and New Zealand take quite different approaches to GMO agriculture, they cooperate on GM product consumption. Australia also has a separate regulatory framework for the farming of GM crops.

Australia has adopted a process-based approach to regulating gene-edited crops. The Gene Technology Regulator (GTR) regulates the gene-edited crops under Gene Technology Regulations 2001 [53]. The Gene Technology Regulations 2019 amendments do not regulate the plants generated through SDN1 technologies involving only the genome nicking at specific sequence locations. Such plants are considered equivalent to cultivars generated by conventional mutagenesis techniques and do not need licensing. However, the plants generated through SDN2 and SDN3 methodologies require licensing. Australia and 12 other countries stated the World Trade Organization advocating moderate regulations for genome-edited plants without prejudice between plants generated through genome editing and plants produced through conventional methods. Because of the various legal definitions, the regulatory standing of GEOs in Australia is now determined by whether they are utilized for culture or food production; consequently, GE crops may be subjected to a rigorous GMO regime instead of a GM food regulation framework. FSANZ has been working with various stakeholders to include and regulate food produced through gene editing. In New Zealand, genome-edited plants are treated as GMOs, and these are regulated through an elaborate process-driven regulatory framework under "The hazardous substance and New Organisms Act 1996". Further, being an agriculture export-based economy, New Zealand has adopted a wait and watch policy in accepting gene-edited crops in the context of acceptance in the importing countries [54].

Europe

Europe is the second smallest continent of the world with an area of 9,938,000 square kilometers and is inhabited by approximately 751 million humans.

European Union

Spain, Portugal, the Czech Republic, Romania, and Slovakia are the five EU nations that cultivate GM maize. Spain has the highest arable area of 0.1 Mha under GM maize cultivation. Genome-edited plants are considered GMOs and regulated through the EU regulatory framework for GMOs. The European regulatory framework is presently the only one that permits a prior assessment of the regulatory categorization of GEOs in farming and the trading of produce generated from GE crops. The European Court of Justice (CJEU) adjudged the regulatory equivalence of genome-edited organisms and genetically modified organisms, thereby subjecting genome-edited plants also to the existing GMO regulations [55]. These regulations also cover the plants generated through precision techniques like oligonucleotide-directed mutagenesis and SDN methodology. The coverage of the European GMO definition can be determined from the GMO statement "GMO categorization is dependent on if the breeding method employed was natural." The European definition of GMOs extends to mutagenic and non-mutagenic gene editing procedures. Overall, the European regulatory framework has clear criteria for the inclusion of GE crops in GMOs, which results in more significant legal and regulatory uniformity [56].

Switzerland

The Swiss Federal Council published its intentions to change present gene technology legislation in General Secretariat UVEK (2018) to accommodate the newest breakthroughs in new breeding technologies [57]. A comprehensive study by the Economic Affairs, Education, and Research (EAER), the existing GMO legislation from 2004 is insufficient to regulate the GE plants, and subsequent classification of GE plants into the existing GMO framework is not possible. Switzerland has prohibited GMO production until the end of 2021; however, it is unclear if that includes plants developed via genome editing [58]. Any new rule is expected to incorporate a risk classification system for items and technologies. It should be emphasized that Switzerland is positioned within EU territory, and transboundary trade is impacted by the regulations in the EU.

Russia

The farming and breeding of genetically engineered crops have been prohibited in Russia since 2016, except for scientific research. The Russian Government published Resolution No. 770 in June 2017, which modifies Russia's legal framework for licensing GMOs, and goods produced from or consisting of such organisms to maintain the ban on creating and breeding GMOs [59]. There is insufficient information on the progress of genome editing in Russia. The majority of agricultural biotechnology research is focused on biological biofertilizers, plant protection and growth inducers. Genome-edited plants have an unresolved position in existing GMO regulations. Russian Government changed its anti-GMO stance with the publication of Decree No: 479 by the Russian Ministry of Education and Science to gain rapid prowess in the field of emerging genetic technologies [60]. This regulation categorizes the plant products obtained through a "particular type" of genome editing are comparable to those obtained through traditional plant breeding. Furthermore, obstructive rules were identified, as well as suggestions for modification. As a result, revising existing Russian rules can promote genome editing in the coming years.

Conclusion

There is significant heterogeneity in the perception and use of GE plants. Some nations are unsure of the regulatory status of GE plants and have adopted a wait and watch policy in framing a regulatory framework for these plants. In contrast, other nations consider GE plants equivalent to GM plants and strictly impose regulations on such plants. Still, countries like Argentina have been proactive in accommodating the GE plants by designing a regulatory framework. The field of GE plants is still in its infancy and requires more research to evaluate the biosafety, biodiversity, and environmental impacts.

Genome editing holds a great potential to act as a method of achieving agricultural sustainability in rapidly changing environmental scenarios. The biggest roadblock to the commercialization of CRISPR-edited crops is the regulatory classification as GMOs. Especially the methods of delivery of the CRISPR reagents, which might cause the integration of DNA corresponding to sgRNA and Cas9 into the plant genome, resulting in the regulatory classification of such plants as GMOs. DNA-free approaches involving RNPs can generate transgene-free plants but suffers from poor gene editing efficiency.

The way forward is to develop novel strategies and methodologies to enhance gene editing efficiency without incorporating heterologous DNA. Regulatory trends in nations already involved in the cultivation of GM crops points toward acceptance of plant genome editing with SDN1 and SDN2 methodologies involving no heterologous DNA insertion. Overall, the present global regulatory landscape lacks uniformity. Greater cooperation at the global level is required for timely framing of the regulatory framework for evaluating potential concerns associated with the cultivation and consumption of GE plants-based products.

Acknowledgments

The author gratefully thanks the Lovely Professional University in Phagwara for all the infrastructural support during the preparation of this manuscript.

References

[1] Cohen, Joel E. (2003). "*Human Population: The Next Half Century.*" 302 (5648), 1172-1175. doi:10.11 26/science.1088665.

[2] Stadler, Lewis J. (1928). "Genetic effects of X-rays in maize." *Proceedings of the National Academy of Sciences of the United States of America*, 14, no. 1, 69.

[3] Tollenaar, D. (1934). "Untersuchungen ueber Mutation bei Tabak [*Studies on mutation in tobacco*]." *Genetica*, 16, no. 1, 111-152.

[4] Tollenaar, D. (1938). "Untersuchungen ueber Mutation bei Tabak. [*Studies on mutation in tobacco*]" *Genetica*, 20, no. 3, 285-294.

[5] Auerbach, Ch. (1950). "Differences between effects of chemical and physical mutagens." *Differences between Effects of Chemical and Physical Mutagens.* 22, no. Suppl.

[6] Smith, Harold H. (1958). "Radiation in the production of useful mutations." *The Botanical Review*, 24, no. 1, 1-24.

[7] Tollenaar, D. (1934). "Untersuchungen ueber Mutation bei Tabak. [*Studies on mutation in tobacco*]" *Genetica*, 16, no. 1, 111-152.

[8] Tollenaar, D. (1938). "Untersuchungen ueber Mutation bei Tabak. [*Studies on mutation in tobacco*]" *Genetica*, 20, no. 3, 285-294.

[9] Tanaka, S. (1972). *Radiation Application for Improvement of Plant Breeding. Breeding Plants with High Nutrients*. Inst. of Radiation Breeding, Omiya, Japan.

[10] Yang, T. C., and Tobias, C. A. (1980). "Potential use of heavy-ion radiation in crop improvement." In *Gamma Field Symposia*, pp. 141-154.

[11] Joint, F. A. O. (1986). "*Chemical Mutagenesis for Crop Improvement*."

[12] Leont'ev, S. I., and Abramov, A. G. (1990). "Effect of the mutagen ethyl methanesulfonate (EMS) on yield from the main ear in spring wheat hybrids." *Effect of the Mutagen Ethyl Methanesulfonate (EMS) on Yield from the Main Ear in Spring Wheat Hybrids.* 50-52.

[13] Rines, H. W. (1985). "Sodium azide mutagenesis in diploid and hexaploid oats and comparison with ethyl methanesulfonate treatments." *Environmental and Experimental Botany*, 25, no. 1, 7-16.

[14] Szarejko, I., Szurman-Zubrzycka, M., Nawrot, M., Marzec, M., Gruszka, D., and Kurowska. M. (2017). "Biotechnologies for plant mutation breeding: protocols." *Creation of a TILLING Population in Barley After Chemical Mutagenesis with Sodium Azide and MNU*, 91-111.

[15] Ishida, Yuji, Hideaki Saito, Shozo Ohta, Yukoh Hiei, Toshihiko Komari, and Takashi Kumashiro. (1996). "High efficiency transformation of maize (Zea mays L.) mediated by Agrobacterium tumefaciens." *Nature Biotechnology*, 14, no. 6, 745-750.

[16] Cheng, Ming, Joyce E. Fry, Shengzhi Pang, Huaping Zhou, Catherine M. Hironaka, David R. Duncan, Timothy W. Conner, and Yuechun Wan. (1997). "Genetic transformation of wheat mediated by Agrobacterium tumefaciens." *Plant Physiology*, 115, no. 3, 971-980.

[17] Toki, Seiichi. (1997). "Rapid and efficientAgrobacterium-mediated transformation in rice." *Plant Molecular Biology Reporter*, 15, no. 1, 16-21.

[18] Anand, Ajith, and Todd J. Jones. (2018). "Advancing Agrobacterium-based crop transformation and genome modification technology for agricultural biotechnology." *Agrobacterium Biology*, 489-507.

[19] Meeusen, Ronald L., and Yousef Atallah. (1990). "Recombinant proteins in food: the Bt crystal protein example." In *Biotechnology and Food Safety*, pp. 267-273. Butterworth-Heinemann.

[20] Funke, Todd, Huijong Han, Martha L. Healy-Fried, Markus Fischer, and Ernst Schönbrunn. (2006). "Molecular basis for the herbicide resistance of Roundup Ready crops." *Proceedings of the National Academy of Sciences*, 103, no. 35, 13010-13015.

[21] Turnbull, Crystal, Morten Lillemo, and Trine A. K. Hvoslef-Eide. (2021). "Global regulation of genetically modified crops amid the gene edited crop boom–a review." *Frontiers in Plant Science*, 12,258.

[22] Coates, Craig J., Joseph M. Kaminski, James B. Summers, David J. Segal, Andrew D. Miller, and Andreas F. Kolb. (2005). "Site-directed genome modification: derivatives of DNA-modifying enzymes as targeting tools." *Trends in Biotechnology*, 23, no. 8, 407-419.

[23] Stoddard, Barry L. (2005). "Homing endonuclease structure and function." *Quarterly Reviews of Biophysics*, 38, no. 1, 49-95.

[24] Mandell, Jeffrey G., and Carlos F. Barbas. (2006). "Zinc Finger Tools: custom DNA-binding domains for transcription factors and nucleases." *Nucleic Acids Research*, 34, no. suppl_2, W516-W523.

[25] Hubbard, Basil P., Ahmed H. Badran, John A. Zuris, John P. Guilinger, Kevin M. Davis, Liwei Chen, Shengdar Q. Tsai, Jeffry D. Sander, J. Keith Joung, and David R. Liu. (2015). "Continuous directed evolution of DNA-binding proteins to improve TALEN specificity." *Nature Methods*, 12, no. 10, 939-942.

[26] Osakabe, Yuriko, and Keishi Osakabe. (2015). "Genome editing with engineered nucleases in plants." *Plant and Cell Physiology*, 56, no. 3, 389-400.

[27] Jinek, Martin, Krzysztof Chylinski, Ines Fonfara, Michael Hauer, Jennifer A. Doudna, and Emmanuelle Charpentier. (2012). "A programmable dual-RNA–guided DNA endonuclease in adaptive bacterial immunity." *Science*, 337, no. 6096, 816-821.

[28] Ran, F. A. F. A., Patrick D. Hsu, Jason Wright, Vineeta Agarwala, David A. Scott, and Feng Zhang. (2013). "Genome engineering using the CRISPR-Cas9 system." *Nature Protocols*, 8, no. 11, 2281-2308.

[29] Zhang, Zhengzhi, Lei Hua, Ajay Gupta, David Tricoli, Keith J. Edwards, Bing Yang, and Wanlong Li. (2019). "Development of an Agrobacterium-delivered CRISPR/Cas9 system for wheat genome editing." *Plant biotechnology Journal*, 17, no. 8, 1623-1635.

[30] Park, Jongjin, and Sunghwa Choe. (2019). "DNA-free genome editing with preassembled CRISPR/Cas9 ribonucleoproteins in plants." In *Transgenic Research*, vol. 28, no. 2, pp. 61-64. Springer International Publishing.

[31] Shiva, Vandana, and Tom Crompton. (1998). "Monopoly and monoculture: trends in Indian seed industry." *Economic and Political Weekly*, A137-A151.

[32] Kuiken, Todd, Rodolphe Barrangou, and Khara Grieger. "(Broken) Promises of Sustainable Food and Agriculture through New Biotechnologies: The CRISPR Case." *The CRISPR Journal*, 4, no. 1 (2021): 25-31.

[33] Stock, Taking. *The Convention on Biological Diversity*. (1992).

[34] Unit, Biosafety. "Text of the Cartagena Protocol on Biosafety." *The Biosafety Clearing-House (BCH). Secretariat of the Convention on Biological Diversity*, April 5, 2021. https://bch.cbd.int/protocol/text/. (Accessed: 22 August 2022).

[35] Barse, Brajesh, and Syed Shams Yazdani. "India—GMOs/Synthetic Biology Rules/Regulations and Biodiversity–A Legal Perspective from India." In *GMOs*, pp. 559-576. Springer, Cham, 2020.

[36] Cohen, Jon. "China's CRISPR revolution." (2019): 420-421.

[37] Chen, Zhu, Hong-Guang Wang, Zhao-Jun Wen, and Yihuang Wang. (2007). "Life sciences and biotechnology in China." *Philosophical Transactions of the Royal Society B: Biological Sciences*, 362, no. 1482, 947-957.

[38] Jian-Kang Zhu, (2022). The future of gene-edited crops in China, *National Science Review*, Volume 9, Issue 4, April.

[39] Ishii, Tetsuya. (2019). "Regulation of genome editing in plant biotechnology: Japan." In *Regulation of Genome Editing in Plant Biotechnology*, pp. 239-262. Springer, Cham.

[40] The Japanese Policy For Genetically Modified Crops (2017). https://ap.fftc.org.tw/article/1169. (Accessed: 22 August 2022).

[41] Hokkaido Prefecture. (2007). *Minutes of 2006 Hokkaido Food Safety and Comfort Committee*. http://www.pref.hokkaido.lg.jp/ns/shs/shokuan/gm-bukai.htm. (Accessed: 22 August 2022).

[42] The Biosafety Act, (2012). https://bch.cbd.int/api/v2013/documents/58db27a7-110b-43dd-b85c-fae0daa0f4a9/attachments/14851/BIOSAFETY%20ACT.pdf(Accessed: 22 August 2022).

[43] The guidelines for the regulation of genome-edited plants. https://www.aatf-africa.org/wp-content/uploads/2021/02/Government-of-Nigeria-approved-National-Biosafety-Guideline-on-Gene-Editing_Lagos_Nigeria_02-08-2021.pdf(Accessed: 22 August 2022).

[44] National biosafety authority of Kenya guidelines for the regulation of genome-edited plants https://www.biosafetykenya.go.ke/images/GENOME-EDITING-GUIDELINES-FINAL-VERSION-25th-Feb-2022-03.pdf(Accessed: 22 August 2022).

[45] Academy of Science of South Africa (ASSAf), (2016). The regulatory implications of new breeding techniques. [Online] Available at: DOI http://dx.doi.org/10.17159/assaf.2016/0011(Accessed: 22 August 2022).

[46] of Science, US Office, and Technology Policy. (1985). "Coordinated framework for regulation of biotechnology; establishment of the Biotechnology Science Coordinating Committee; notice." *Federal register*, 50, no. 220, 47174-47195.

[47] Kuzma, Jennifer. (2016). "A missed opportunity for US biotechnology regulation." *Science*, 353, no. 6305, 1211-1213.

[48] Smyth, Stuart J. (2019). "Regulation of genome editing in plant biotechnology: Canada." In *Regulation of Genome Editing in Plant Biotechnology*, pp. 111-135. Springer, Cham.

[49] Whelan, Agustina I., and Martin A. Lema. (2015). "Regulatory framework for gene editing and other new breeding techniques (NBTs) in Argentina." *GM Crops & Food*, 6, no. 4, 253-265.

[50] Gatica-Arias, Andrés. (2020). "The regulatory current status of plant breeding technologies in some Latin American and the Caribbean countries." *Plant Cell, Tissue and Organ Culture (PCTOC)*, 141, no. 2, 229-242.

[51] Vieira, L. R., Freitas, N. C., Justen, F., MIRANDA, V. de J., GARCIA, B. de O., Nepomuceno, A. L., Fuganti-Pagliarini, R., et al. (2021). "Regulatory framework of genome editing in Brazil and worldwide." *Embrapa Soja-Capítulo em livro científico (ALICE)*.

[52] Martin, Tania, Elizabeth Dean, Brigid Hardy, Tony Johnson, Fiona Jolly, Fiona Matthews, Ian McKay, Richard Souness, and Jenny Williams. (2003). "A new era for food safety regulation in Australia." *Food Control*, 14, no. 6, 429-438.

[53] Commonwealth of Australia (2001). Gene Technology Regulations 2001. *The Federal Register of Legislation* at https:// www.legislation.gov.au/Details/F2016C00615.(Accessed: 22 August 2022).

[54] The New Zealand Government (2016). *GMO Regulations Clarified*. Available online at: https://www.beehive.govt.nz/release/gmo-regulations-clarified. (Accessed: 22 August 2022).

[55] Broll, Hermann, Albert Braeuning, and Alfonso Lampen. (2019). "European Court of Justice decision for genome editing: Consequences on food/feed risk assessment and detection." *Food Control*, 104, 288-291.

[56] Custers, René. (2017). "The regulatory status of gene-edited agricultural products in the EU and beyond." *Emerging Topics in Life Sciences*, 1, no. 2, 221-229.

[57] Generalsekretariat UVEK (2018). Neue gentechnische Verfahren: bundesrat prüft Anpassung der rechtlichen Regelung [*New genetic engineering procedures: Federal Council examines amendment of the legal regulation*]. https://www.uvek.admin.ch/uvek/de/home/uvek/ medien/medienmitteilungen.msg-id-73173.html(Accessed: 22 August 2022).

[58] Chandrasekhar, A. (2016). Government approves GMO ban extension. *SWI*. Available Online at: https://www.swissinfo.ch/eng/society/genetically-modified-organisms_government-approves-gmo-ban-extension/42260828 (Accessed: 22August 2022).

[59] Chokheli, Vasiliy, Vishnu Rajput, Pavel Dmitriev, Tatiana Varduny, Tatiana Minkina, Rupesh Kumar Singh, and Abhishek Singh. (2021). "Status and policies of GM crops in Russia." In *Policy Issues in Genetically Modified Crops*, pp. 57-74. Academic Press.

[60] Chetvertakov, Sergey. (2016). "Welfare Analysis of Lifting the GM Ban in Russia." *AGRIS on-line Papers in Economics and Informatics*, 8, no. 665-2016-45120, 49-56.

Section 2. Genome Editing for Stress Tolerance

Chapter 7

CRISPR-Cas-Mediated Genome Editing Strategies for Abiotic Stress Tolerance in Crops

Meghna Patial[1]
Ram Kumar Sharma[1,2]
and Rohit Joshi[1,2*]
Division of Biotechnology, CSIR-Institute of Himalayan Bioresource Technology, Palampur, Himachal Pradesh, India
Academy of Scientific and Innovative Research (AcSIR), Ghaziabad, Uttar Pradesh, India

Abstract

Field crops are continuously exposed to various abiotic stresses that pose serious challenges for crop productivity, and cause substantial reduction in yield and global economy. Although better varieties with desired traits have been developed using conventional and mutation breeding techniques, still there is a huge gap between the demand and supply of food in the present era. Genome editing technology is diversifying the platform for the production of customized plants. Recent advancements in genome editing technologies have expanded the potential for crop improvement through precise genetic engineering for specified crop traits. However, identification of potential genes is vital to achieving tolerance against stresses.

CRISPR/Cas-based genome editing is an emerging technique that utilizes artificially engineered nucleases following the direction of specific guide RNA, for digesting DNA at targeted locations in the genome for high-throughput biotechnology-based breeding. Due to the versatility of this user-friendly CRISPR/Cas system and its wide applications in gene-editing, transcriptional regulation, high efficiency and easy manipulation, several plant specific CRISPR-Cas9 vector systems have been designed for developing climate resilient crops. Genome editing combined with speed breeding has revolutionized the pace and accuracy for the improvement of crop species. Recently, advanced genome-editing tools i.e., CRISPR/Cas12a, Cas13, Cas14 and base editors have also been introduced to prevent "off target" variations. Several reports are available for the development of tolerance against various abiotic stresses in plants through this technology. Recombinases and prime editing tools are more advantageous for precise and efficient editing.

---

* Corresponding Author's Email: rohitjoshi@ihbt.res.in.

In: Genome Editing for Crop Improvement
Editors: Rahul Kumar, Ashutosh Sharma, Yogesh Kumar et al.
ISBN: 979-8-88697-740-0

Briefly, the current chapter is an attempt to garner knowledge on the CRISPR/Cas system, their associated strategies, and application for developing crops with improved tolerance to abiotic stresses. Further, we discuss the limitations, biosafety issues and concerns to obtain transgene-free plants for their commercialization. This new system will further open new gates to improve plant abiotic stress tolerance and accelerate basic plant research.

Keywords: applications, climate change, limitations, mutants, strategies, transgene-free plants

Introduction

Climatic vagaries are a threat to agricultural productivity and global food security [1, 2]. Extreme climatic conditions such as salinity, drought and temperature substantially reduce crop yield, resulting in insufficient food supply to feed the growing world population [3, 4]. Thus, engineering cultivars that can tolerate abiotic stresses along with improved yield and quality would be essential to increase agricultural productivity by at least 70-85% by the year 2050 [1, 5, 6]. Although conventional cross-breeding and mutation breeding has significantly contributed towards abiotic stress tolerance and yield improvement, these untargeted breeding approaches are prone to human biases, require large mutants besides being tedious and time-consuming [7]. Further, these traditional techniques lead to undesired gene insertions along with the targeted genes, and it makes these even more non-specific in function. The primary target of plant biologists at present is to improve tolerance against abiotic stresses in crops using high-throughput targeted gene manipulation and its commercial applications in agriculture. With the advent of genetic engineering techniques, the ease and ability to develop genetically modified crops have revolutionized agriculture [8]. Therefore, efficient, robust, fast and precise crop breeding techniques i.e., 'genome editing,' offer excellent opportunities to potentially address these challenges of conventional mutational breeding [9, 10]. Recent advancements in gene-editing technologies have modernised crop improvement strategies by enabling precise and efficient editing of even a single base [11]. This would further improve the efficiency for targeted improvement for abiotic stress tolerance [12]. However, detailed information regarding gene function, structure, sequence and QTLs for specific traits is vital for genome editing applications.

Site-specific endonucleases (SSE), such as, clustered regularly interspaced short palindromic repeats (CRISPR)-CRISPR-associated protein (Cas) system, transcription activator-like effector nucleases (TALEN) and zinc finger nucleases (ZFN) are being widely used for gene editing [13]. Further, various modified forms of Cas proteins are available, i.e., Cas8, Cas9, Cas12a, or Cpf1, Cas12b, Cas13a, Cas13b, Cas14, nCas9 (nickase Cas9) and dCas9 (endonuclease deficient or catalytically dead Cas9). Among these, the CRISPR/Cas9 tool is the most versatile, reliable and advanced system for multiplex genome editing to develop stress resilience in a broad range of plant species [14]. This technology offers extended potential for crop editing because of specific *in vivo* alterations in DNA sequences [15]. Few countries have already approved genome-edited crops, whereas others are still assessing their public acceptance [16]. Due to the direct consumption of various crops, the ethics and regulatory issues of genome-edited crops need to be discussed for crop improvement [13]. Keeping in view with the rapid expansion of gene editing technologies, we discuss the use of CRISPR/Cas system and its potential in utilization for rapid and precise genome editing for stress tolerance in crops. The safety concerns and limitations of genome editing have also been discussed, that concern the societal acceptance of transgene-free crops. Rapidly improving modern technologies to edit plant genomes will help in developing new approaches for adding economically valuable traits in plants to produce improved varieties of essential crops.

Overview of CRISPR/Cas Technology

At present, the CRISPR/Cas system is the most consistent tool for precise genome editing [17]. Various genes encoding DNA repair proteins are reported to be associated with CRISPR were named Cas (CRISPR-associated) [14]. CRISPR-Cas9 is based on the acquired immune response of microbes that helps bacteria to defend themselves from invading phages, offering an excellent alternative to first-generation site-directed nucleases [18]. Despite identification of CRISPRs in bacterial DNA in 1987, their role in providing immunity with Cas protein could be demonstrated in 2007. Gene editing using CRISPR-Cas was first done using the cell-free system by Jinek et al. [19]. Later, gene editing using CRISPR-Cas9 was reported in *Nicotiana benthamiana*, *Arabidopsis* and *Oryza sativa* as model plants [20, 21]. CRISPR-Cas9 system has three key modules i.e., CRISPR RNA (crRNA) for matching and identifying target DNA, Cas9 protein and trans-activating CRISPR RNA (tracrRNA) [19]. Cas9 acts as "molecular scissors" and consists of two domains i.e., His-Asn-His (HNH) and RuvC which creates a double-stranded break at the target location [22]. The repair machinery of the host finally repairs these DSBs through either non-homologous end joining (NHEJ) or homology-directed repair (HDR) for insertions, knockouts and genomic alterations [23].

The Cas proteins comprise of four functional modules: (a) ancillary module for modulatory and other CRISPR-associated roles, (b) adaptation module for spacer insertion, (c) expression module for pre-CRISPR-RNA programming and (d) interference module which binds CRISPR-RNA with the target and its cleavage [24]. Figure 1 gives an overview of the CRISPR/Cas9 technology. A Cas9 nuclease complex forms the Cas9 protein and the single guide RNA (sgRNA) [25]. Cas9/sgRNA complex searches for the target sequence of DNA next to protospacer adjacent motif (PAM) that is required for effective detection of the target [26]. After identifying the target site, Cas9 cleaves the desired sequence allowing sgRNA to anneal with the complementary target strand. The expression of sgRNAs is driven by U3/U6 small nuclear RNA (snRNA) promoters in plants. RNA polymerase III synthesizes sgRNA [27]. R-loop formation activates RuvC and HNH [28], which then breaks double-strand DNA (dsDNA) 3 bp upstream of PAM of target DNA [29]. After that a chimeric module of guide RNA was formed, consisting of tracrRNA and crRNA ahead of the spacer sequence, complementary to the target DNA sequence [17]. In NHEJ, during repair donor template is not required, whereas it is required in HDR [30]. However, due to the low frequency of HDR than NHEJ, still, NHEJ is preferentially utilized for the repair of DSB [31].

Cas9-VirD2 chimeric protein (Cas9 fused with *Agrobacterium* VirD2) creates DSBs through Cas9 and brings DRT nearer to the DSB site using VirD2 relaxase, which facilitates HDR at the edited site [31]. For multiplex genome editing, polycistronic tRNA-gRNA (PTG) is constructed and every gRNA is combined with the target spacer sequence. After cleavage by the endogenous RNaseP and RNaseZ, numerous gRNAs are released which direct Cas9 to target multiple sites of the genome [32]. Furthermore, the base editing (BE) technique, which also improves editing by enhancing their accuracy and proficiency, is introduced and is successfully applicable in plant biology.

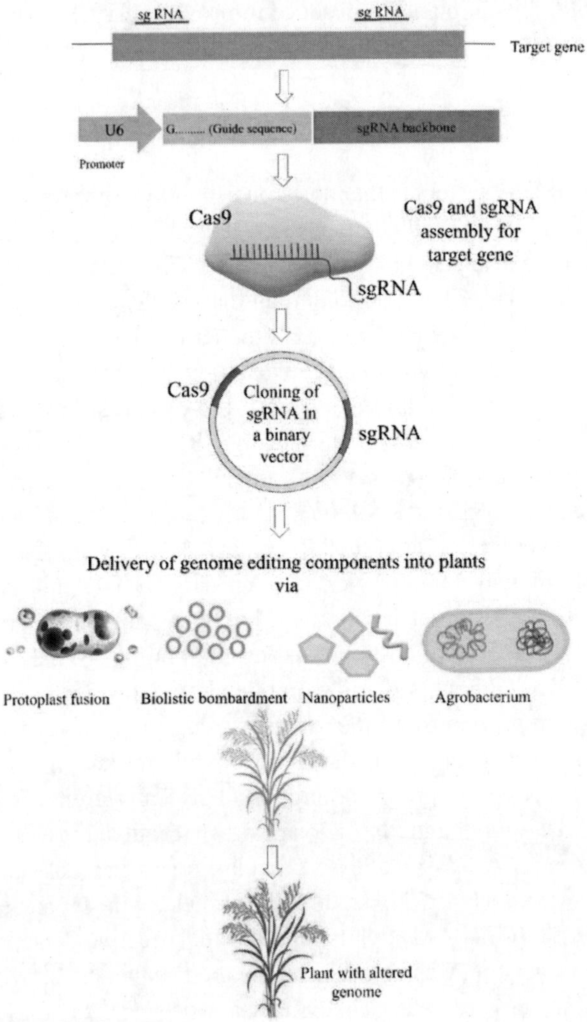

Figure 1. Overview of CRISPR/Cas technology.

BE is of two types, cytosine base editors (CBEs) and adenine base editors (ABEs) [16]. In addition, BE system includes dCpf1-BE [33], Targeted-AID [34] and BE4 [35]. All these BEs use Cas12a or Cas9 nickase attached with cytidine deaminase, causing precise C-to-T substitutions [14]. Engineered SaCas9 and SpCas9 variants, developed by fusing Cas9 nickase with human APOBEC3A, were used to modify the editing efficiency of BE [36]. CRISPR prime editing (PE) system utilizes reformed Cas9 protein/s and PE guide RNA (pegRNA) that confirms single strand cutting and improves the precision of editing. PE tool creates 12 different point mutations, containing conversions, deletions, insertions and substitutions, making it a powerful system [13]. A single RNA-guided endonuclease, Cas12a, recognizes T-rich PAMs, in contrast to Cas9, and eases strong DNA interference and creates staggered cuts [37].

Moreover, Cas14 is one-third in size than Cas9 protein and targets ssDNA without a PAM constraint [38]. In dCas9, the activity of endonuclease is removed through point mutations in its endonuclease domain.

The CRISPR-dCas9 system is composed of three major subunits: a nuclease-dead Cas9 (also known as dcas9 null mutant), sgRNA and transcriptional activators. The Cas9 nuclease domains were mutated in *S. pyrogens* by producing H840A mutation in the HNH domain, and D10A mutation in the RuvC domain [19]. This dead version of the Cas9 cannot bind DNA anymore, but it still targets and connects DNA to sgRNA with the same precision. Rather than permanently changing the genome, linking dCas9 disrupts transcription at the selected location, causing temporary gene silencing. Instead, standard CRISPR/Cas9 relies on DSBs by means of Cas9 endonuclease action, which is then accompanied by manipulating gene-editing DNA repair mechanisms, dCas9 activation systems integrate transcriptional N and C-terminal protein enabler [39]. Thus, dCas9 can modulate gene expression without permanently modifying the DNA [40]. Further, for callus-based transformations, promoters such as ubiquitin of *Arabidopsis, Zea mays* and *Oryza sativa,* and Cauliflower mosaic virus (CaMV) 35S promoter can be applied to induce the Cas9 gene in plants for successful genome editing [41]. Cas9 and sgRNA vectors are induced in plant cells through *Agrobacterium*-mediated transformation as it is the most successful method of transformation in plants [42].

Genome Editing for Abiotic Stress Tolerance

Under different abiotic stresses, metabolic, signaling and other regulatory pathways in plants are modulated by various genes. Classical breeding showed limited success in accumulating stress regulatory genes and production of abiotic stress-tolerant plants. Thus, researchers are now interested in improving stress adaptation in plants by targeting these genes using CRISPR/Cas9 technology. Identification of appropriate promoter for Cas9 expression, guide RNA, identification of abiotic stress-responsive genes and gene delivery system are essential for CRISPR/Cas9 gene editing [43]. CRISPR/Cas9 system is an efficient system to edit any gene of an organism, and even multiple genes can successfully be targeted within a single organism to produce climate-resilient plants, as demonstrated in *Physcomitrella patens, Arabidopsis thaliana*, wheat, tomato, rice, corn, and cotton (Table 1; 16, 44, 45]. Thus, CRISPR/Cas9 holds great potential for the development of multiple stress-tolerant crops by simultaneously targeting multiple genes in a sensitive and high-yielding cultivar. Several *T* genes encoding different antioxidant enzymes, which scavenge ROS molecules and are concomitant with abiotic stress tolerance, have been identified and targeted in plants to attain tolerance [14]. Similarly, *S* genes cause ROS production, a decrease in antioxidant activity and an increase in programmed cell death, making plants sensitive towards abiotic stresses [14]. Recently, the *S* gene, *SlAGAMOUS-LIKE 6* (*SlAGL6*) has been targeted using CRISPR/Cas9-mediated genome editing to improve fruit set under high temperature in tomato [46]. Gene knock-out mutants of *PYL1/4/6* (abscisic acid receptor genes) in rice showed heat stress tolerance [47].

Table 1. Application of CRISPR-Cas in the development of abiotic stress tolerant plants

S. No	Crop	Cas protein	Abiotic stress	Target sequence	Type of mutation	Result	Reference
1	*Oryza sativa*	Cas 9	Salt and oxidative stress	*PQT3*	Gene knockout	Knockout mutants show enhanced resistance to abiotic stresses and higher yield	[66]
2	*Oryza sativa*	Cas9	Salt stress	*OsRR22*	Gene knockout	Mutant plants showed enhanced salt tolerance at seedling stage	[62]
3	*Oryza sativa*	Cas9	Salt stress	*BBS1*	Gene knockout	Mutant plants exhibited hypersensitivity towards salt stress	[67]
4	*Solanum lycopersicum*	Cas9	Salt stress	*HyPRP1*	Gene knockout	Enhanced salinity tolerance at the germination and vegetative stages	[68]
5	*Zea mays*	Cas9	Drought stress	*ARGOS8*	Gene knockout	Variants showed increased grain yield under flowering stress conditions	[49]
6	*Arabidopsis*	Cas9	Drought stress	*HAT1*	Gene knockout	Mutants showed higher chlorophyll content and faster stomatal aperture under water deficit	[56]
7	*Oryza sativa*	Cas9	Drought stress	*ERA1*	Gene knockout	Mutant plants showed increased primary root growth	[54]
8	*Solanum lycopersicum*	Cas9	Drought stress	*LBD40*	Gene knockout	Transgenic plants showed enhanced drought tolerance	[52]
9	*Oryza sativa*	Cas9	Drought and salt stress	*DST*	Gene knockout	Mutants have broader leaves and reduced stomatal density, thus enhanced leaf water retention under dehydration stress. Seedlings exhibited higher salt tolerance	[60]
10	*Oryza sativa*	Cas9	Heat stress	*PYL1/4/6*	Multiple gene silencing	Improved growth and increased grain yield	[47]
11	*Oryza sativa*	Cas9	Cold stress	*PIN5b; GS3; MYB30*	Multiple gene editing	Mutant plants exhibited enhanced panicle length, grain size and cold tolerance	[63]
12	*Solanum lycopersicum*	Cas9	Cold stress	*MAPK3*	Gene knockout	Mutant plants exhibited enhanced wilting, hydrogen peroxide content, membrane damage and reduced antioxidant enzymes activity under drought stress	[48]
13	*Oryza sativa*	Cas9	Cold stress	*MYB30*	Gene knockout	Mutants showed increased cold tolerance	[59]
14	*Oryza sativa*	Cas9	Cadmium stress	*Nramp5*	Gene knockout	Mutant plants showed low cadmium accumulation	[67]
10	*Oryza sativa*	Cas9	Cesium stress	*HAK1*	Gene knockout	Reduced Cs^{+} uptake by rice plants	[65]

Moreover, CRISPR/Cas9 has been used to generate mutants to understand abiotic stress mechanisms in different crops such as mitogen-activated protein kinases3 (*slmapk3*) in tomato [48], *ARGOS8* gene (negatively regulates ethylene response) in maize [49] to increase yield under water deficiency and C-repeat binding factors (*cbfs*), double and triple mutants in *Arabidopsis* [50].

Similarly, targeted modifications of *TMS5* gene resulted in development of temperature-sensitive rice [51]. Knocking out of *OsSAPK2* (osmotic stress/ABA–activated protein kinase 2) and *OsPQT3* (*PARAQUAT TOLERANCE 3*) in rice revealed its positive role in osmotic stress, salinity, drought and seed dormancy through ABA signaling cascade [52]. Similarly, mutant rice plants of *OsPYL9* [53], *ERA1* (Enhanced Response to ABA1) [54], *OsSRL1/SRL2* (SEMI-ROLLED LEAF 1 and 2) [55], *ABRE1/ABF2* (abscisic acid-responsive element-binding protein 1/ABRE binding factor 2) [56], *OsEPFL9* [57] and *OST2* (regulating stomatal movement) showed higher drought tolerance than WT plants due to enhanced ROS scavenging enzyme activity, abscisic acid content, panicle number, and cuticle wax and reduced malondialdehyde content, transpiration rate, stomatal conductance and number of stomata. Osakabe et al. [58] applied tru-guide RNA (truncated guide RNA) for site-specific editing in *Arabidopsis* by targeting *OST2/AHA1* gene under regulation of a tissue-specific promoter for improving drought stress tolerance.

However, knockout mutants of *R2R3-type MYB TF* (*OsMYB30*) in rice [59] increased cold tolerance and yield in comparison to WT plants. CRISPR/Cas9-mediated editing of a zinc finger TF, *DST* [60] and *OsmiR535* [61], showed higher tolerance against osmotic, salinity and drought stress than WT by improving plant architecture, leaf water retention and stomatal conductance. In addition, knockout mutants of *OsRR22* (amino acid B-type response regulator) TF improves shoot architecture and salinity tolerance in rice [62]. Zeng et al. [63] have developed triple mutants by simultaneously knocking-out an endogenous *MYB30* gene along with two other yield-related genes to improve cold stress tolerance in rice. Previous researchers have also reduced mineral toxicity by developing rice knockout mutants of *OsNramp5* (metal transporter) for low cadmium [64], *OsARM1* for arsenic stress [48] and *OsHAK1* for cesium stress [65]. The above examples specify the efficiency of CRISPR/Cas9 technology to modify/eliminate genes by generating novel allelic variants to increase tolerance against various abiotic stresses, e.g., nutrient deficiency, heavy metals, extreme temperatures, drought and salinity [12]. In addition, CRISPR/Cas9 base editing also shows remarkable potential for targeted programming for creating novel traits for speed breeding [14].

Transgene-Free Genome-Edited Abiotic Stress-Tolerant Cultivars

Undoubtedly, with declining cultivable lands and resources, and changing climatic conditions, advancements in CRISPR/Cas9 technology have unlocked new possibilities for speed breeding of crop plants towards complex abiotic stress traits i.e., salinity, drought, and high-temperature tolerance [14]. However, despite being too promising for genetic engineers engaged in stress tolerance, developed transgenic mutants still face serious issues in their public and societal acceptance. However, to adopt these new technologies, various countries are developing legislation and regulatory frameworks. However, the EU Court of Justice (ECJ) and New Zealand have classified all genome-edited plants as GMOs, and they regulate these through

GM biosafety rules [13]. Few countries like Argentina, Brazil, Chile and Columbia have classified crops generated through SDN-3 under GMOs, whereas plants edited through SDN-1 as non-GMOs. In addition, plants edited through SDN-2 do not fall under any regulatory norms [13]. In contrast, China has not set any rules for gene-edited crops, and is focusing on processes or techniques with huge financial support to create new crop varieties. Thus, genome-edited plants must be transgene-free to obtain regulatory approval and commercialization [14]. Thus, to avoid any transgene, off-target mutations and gene integration are the most critical challenges for utilizing this robust tool.

Recently, transgene-free genome editing using CRISPR/Cas9 ribonucleoproteins (RNPs) was developed, and transgene was removed using piggyback, *Cre/loxP*, and *FLP/FRT* transposon systems in lettuce, tobacco, rice, and *Arabidopsis* [69]. Previous studies have demonstrated the production of Cas9-free lines after gene-editing by backcrossing or selfing or through segregation [70, 71]. In addition, for the development of Cas9-free genome-edited plants, mCherry cassette was fused with CRISPR/Cas9 vector for visually isolating Cas9 free plants. The seeds without any fluorescence contain Cas9-free plants [72], and are tested in tomato, rice and *Arabidopsis* [73]. Further, RNPs complexes of sgRNA and Cas9 protein are incorporated in the plant by transfecting or biolistic particle delivery to develop transgene-free plants [74]. In addition, the transient transformation of CRISPR/Cas9 RNA or DNA (TECCRNA or TECCDNA) is used as an alternate genome-editing method through callus regeneration. This system is more precise and effective for developing transgene-free crops such as wheat [75]. These strategies are extremely appropriate for the development of transgene-free genome-edited crops for commercialization with reduced time and labour [14].

Limitations and Proposed Solutions

Gene-editing technologies have vast applications over conventional breeding in developing abiotic stress-tolerant plants due to their ability for multiplex editing, robustness, precision and high efficiency. However, still few restrictions obstruct their utilization in crop improvement. Smaller proteins are to be targeted for gene editing, as larger proteins were reported to decrease the effectiveness of editing and are not appropriate for vector-based transformation in somatic cells. Similarly, Cas9 variants need multiple PAM sites recognition and broad compatibility with PAM, as inadequate PAM sites further reduce its efficiency and increase off-target mutations. Even though the CRISPR/Cas system is developing transgene-free plants, edited plants are acceptable only in a few countries [13]. In addition, disruption/mutation of the edited gene may reduce fitness, which can hamper any pathway or product, or element linked with other genes and regulating pathways. Several mutated *cis*-regulatory sequences contributing towards abiotic stress tolerance can serve as a potential binding location for TFs. regulating metabolic pathways and abiotic stress tolerance. For example, editing W-box (TTGAC) binding cis-element *GmWRKY13, GmWRKY21* and *GmWRKY54* improves tolerance against salinity stress in soybean [76]. To overcome this limitation, base editing is applied through escaping interference with other genes, single nucleotide mutation or targeting promoter for allele development [77].

However, limited PAM sites, huge editing windows and high off-target activity limit the efficiency of BE. Undesired or unintended editing generated by gRNA- independent or deceptive gRNA technology or non-specific sites are termed as "off-target mutants" [78]. To cope with off-target mutations, target specificity of Cas9 proteins is applied, including Sniper Cas9 [79], HypaCas9 [80], HF-Cas9 [4] and eSpCas9 [81]. Moreover, prime editing is also applied to reduce off-target mutations.

Further, the RNP approach is also applied to remove the obstacles of BE and to increase its efficiency. In addition, tissue culture protocols are also labor-intensive, create somaclonal variations and reduce regeneration efficiency. To surpass this obstacle, novel gene delivery methods such as direct delivery in pollen were developed especially for recalcitrant species [14, 82]. dCas9 can interrupt undesirable gene function by processing through CRISPR interference (CRISPRi). Blending transcriptional activation domains VP16/VP64 with dCas9 activates target genes and permits transmission for gain-of-function mutants for tolerance against abiotic stresses [83]. Advancements in molecular breeding using essential modifications in CRISPR/Cas9 system makes it more efficient for the development of new genotypes with improved abiotic stress tolerance.

Prospects and Conclusion

Rapid changes in the climatic conditions continuously impose different abiotic stresses to plants, which challenge global food security. Twenty-first century is witnessing the advancement of CRISPR-Cas9 technique as a robust, simple, efficient, flexible and target-specific gene-editing method for introducing abiotic stress tolerance in crops. In future, large-scale applications with the huge potential of CRISPR/Cas9 technology are anticipated to enable improvement in abiotic stress-regulated traits. Detailed knowledge of various aspects of the gene-editing process, such as genome sequencing, designing of guide RNA, tissue transformation and regeneration protocols can improve the applications in plant research. This technique will be used for the development of precise, programmable gene trails, which can provide tolerance against individual and combined abiotic stresses.

This technology can even be implemented in the domestication of wild genotypes with their ability to grow under adverse climatic conditions without compromising with the yield. In terms to its application in agriculture, this technology can potentially be used for abiotic stress-tolerant crop development via key gene targeting of different regulatory pathways. Further, by enabling multiplex genome editing, plants can be produced with enhanced tolerance against combined abiotic stresses. In addition, core promoter editing will also be a dependable approach to fine-tune any desired gene transcript, unravelling new possibilities for breeding improved cultivars. In addition, potential and novel delivery protocols to develop transgene-free products will help in overcoming the ethical and regulatory issues related to non-GMOs for social acceptance and commercialization.

In conclusion, this newly emerged gene-editing technology will act as a promising tool in the future for precise and efficient gene-editing in plants related to abiotic stresses to tackle sustainable developmental goals. We hope that these genome editing tools will significantly contribute towards the development of crops varieties tolerant against multiple abiotic stresses with wider public acceptance in the future.

Acknowledgments

Meghna Patial, Ram Kumar Sharma and Rohit Joshi gratefully acknowledge the director, CSIR-Institute of Himalayan Bioresource Technology, Palampur, for providing the facilities to carry out this work. CSIR support in the form of project MLP0201, MLP0170 (CSIR-CIMAP Project Ref No. FC2020-Z31MLP-07), and MLP0172 (No. 34/1/TD-AgriNutriBiotech./NCP-FBR 2020-RPPBDD-TMD-SeMI), DBT (No. BT/PR45280/NER/95/1918/2022) and Agriculture Department, Himachal Pradesh in the form of project [No. Agr.H.(Project Cell) F(2)10/2004.Vol.-1 and No. Agr KgrTech(F)NBM VoII(2019-20)1100032] for this study is highly acknowledged. This manuscript represents CSIR-IHBT communication no 5013.

References

[1] Anwar, K., Joshi, R., Dhankher, O. P., Singla-Pareek, S. L. and Pareek, A. (2021). Elucidating the response of crop plants towards individual, combined and sequentially occurring abiotic stresses. *Intnl. J. Mol. Sci.* 22: 6119.

[2] Joshi, R., Sahoo, K. K., Singh, A. K., Anwar, K., Pundir, P., Gautam, R. K., Krishnamurthy, S. L., Sopory, S. K., Pareek, A. and Singla-Pareek, S. L. (2020). Enhancing trehalose biosynthesis improves yield potential in marker-free transgenic rice under drought, saline, and sodic conditions. *J. Expt. Bot.* 71: 653-668.

[3] Joshi, R. and Chinnusamy, V. (2014). Antioxidant enzymes: Defense against high temperature stress. In: Ahmed, P. (Ed.) *Oxidative Damage to Plants* (pp. 369-396). Academic Press.

[4] Wang, D., Niu, C., Han, J., Ma, D. and Xi, Z. (2019). Target DNA mutagenesis-based fluorescence assessment of off-target activity of the CRISPR/Cas9 system. *RSC Adv.*, 9: 9067-9074.

[5] Pareek, A., Joshi, R., Gupta, K. J., Singla-Pareek, S. L. and Foyer, C. (2020). Sensing and signalling in plant stress responses: Ensuring sustainable food security in an era of climate change. *New Phytol.*, 228: 823–827.

[6] Shailani, A., Joshi, R., Singla-Pareek, S. L. and Pareek, A. (2021a). Stacking for future: Pyramiding genes to improve drought and salinity tolerance in rice. *Physiol. Plant.*, 172: 1352-1362.

[7] Bado, S., Forster, B. P., Nielen, S., Ali, A. M., Lagoda, P. J., Till, B. J. and Laimer, M. (2015). Plant mutation breeding: current progress and future assessment. In: Janick, J. (Ed.), *Plant Breeding Reviews* (pp. 23-88). Wiley-Blackwell.

[8] Naeem, M., Majeed, S., Hoque, M.Z. and Ahmad, I. (2020). Latest developed strategies to minimize the off-target effects in CRISPR/Cas-mediated genome editing. *Cells* 9: 1608.

[9] Anders, S., Cowling, W., Pareek, A., Gupta, K. J., Singla-Pareek, S. L. and Foyer, C. H. (2021). Gaining acceptance of novel plant breeding technologies. *Trends Plant Sci.* 26: 575-587.

[10] Shailani, A., Wungrampha, S., Dkhar, J., Singla-Pareek, S. L., Pareek, A. (2021b) Genetic Improvement of Rice for Food and Nutritional Security. In: Kavi Kishor, P. B., Rajam, M. V., Pullaiah, T. (Eds.) *Genetically Modified Crops*. (pp 13-32) Springer, Singapore.

[11] Wang, T., Zhang, H. and Zhu, H. (2019). CRISPR technology is revolutionizing the improvement of tomato and other fruit crops. *Hort. Res.* 6: 77.

[12] Razzaq, M. K., Aleem, M., Mansoor, S., Khan, M. A., Rauf, S., Iqbal, S. and Siddique, K. H. (2021). Omics and CRISPR/Cas9 approaches for molecular insight, functional gene analysis, and stress tolerance development in crops. *Intl. J. Mol. Sci*. 22: 1292.

[13] Tabassum, J., Ahmad, S., Hussain, B., Mawia, A. M., Zeb, A. and Ju, L. (2021). Applications and potential of genome-editing systems in rice improvement: current and future perspectives. *Agronomy*, 11: 1359.

[14] Patial, M., Devi, K. and Joshi, R. (2022). CRISPR/Cas9-mediated targeted mutagenesis in medicinal plants. In: Wani SH and Hensel G (Eds.) *Genome Editing*. (pp 55-70) Springer Nature Switzerland.

[15] Zafar, S. A., Zaidi, S. S. E. A., Gaba, Y., Singla-Pareek, S. L., Dhankher, O. P., Li, X., Mansoor, S. and Pareek, A. (2020). Engineering abiotic stress tolerance via CRISPR/Cas-mediated genome editing. *J. Expt. Bot.*, 71: 470-479.

[16] Shukla, M., Al-Busaidi, K. T., Trivedi, M. and Tiwari, R. K. (2018). Status of research, regulations and challenges for genetically modified crops in India. *GM Crops Food*, 9: 173-188.

[17] Dey, A. (2021). CRISPR/Cas genome editing to optimize pharmacologically active plant natural products. *Pharmacol. Res.* 164: 105359.

[18] Terns, M. P. and Terns, R. M. (2011). CRISPR-based adaptive immune systems. *Curr. Opin. Microbiol.* 14: 321-327.

[19] Jinek, M., Chylinski, K., Fonfara, I., Hauer, M., Doudna, J. A. and Charpentier, E. (2012). A programmable dual-RNA–guided DNA endonuclease in adaptive bacterial immunity. *Science*, 337: 816-821.

[20] Jaganathan, D., Ramasamy, K., Sellamuthu, G., Jayabalan, S. and Venkataraman, G. (2018). CRISPR for crop improvement: an update review. *Front. Plant Sci.* 9: 985.

[21] Kuluev, B. R., Gumerova, G. R., Mikhaylova, E. V., Gerashchenkov, G. A., Rozhnova, N. A., Vershinina, Z. R., Khyazev, A. V., Matniyazov, R. T., Baymiev, A., Baymiev, A. and Chemeris, A. V. (2019). Delivery of CRISPR/Cas components into higher plant cells for genome editing. *Russ. J. Plant Physiol.* 66: 694-706.

[22] Cong, L., Ran, F. A., Cox, D., Lin, S., Barretto, R., Habib, N., Hsu, P. D., Wu, X., Jiang, W., Marraffini, L. A. and Zhang, F. (2013). Multiplex genome engineering using CRISPR/Cas systems. *Science*. 339: 819-23.

[23] Manghwar, H., Lindsey, K., Zhang, X. and Jin, S. (2019). CRISPR/Cas system: recent advances and future prospects for genome editing. *Trends Plant Sci.* 24: 1102-1125.

[24] Makarova, K. S., Wolf, Y. I., Alkhnbashi, O. S., Costa, F., Shah, S. A., Saunders, S. J., Barrangou, R., Brouns, S. J., Charpentier, E., Haft, D. H. and Horvath, P. (2015). An updated evolutionary classification of CRISPR–Cas systems. *Nat. Rev. Microbiol.* 13: 722-736.

[25] Ma, X., Zhang, Q., Zhu, Q., Liu, W., Chen, Y., Qiu, R., Wang, B., Yang, Z., Li, H., Lin, Y. and Xie, Y. (2015). A robust CRISPR/Cas9 system for convenient, high-efficiency multiplex genome editing in monocot and dicot plants. *Mol. Plant*, 8: 1274-1284.

[26] Sternberg, S. H., Redding, S., Jinek, M., Greene, E. C. and Doudna, J. A. (2014). DNA interrogation by the CRISPR RNA-guided endonuclease Cas9. *Nature*, 507: 62-67.

[27] Wang, M., Wang, S., Liang, Z., Shi, W., Gao, C. and Xia, G. (2018). From genetic stock to genome editing: gene exploitation in wheat. *Trends Biotechnol.*, 36: 160-172.

[28] Jiang, F., Taylor, D. W., Chen, J. S., Kornfeld, J. E., Zhou, K., Thompson, A. J., Nogales, E. and Doudna, J. A. (2016). Structures of a CRISPR/Cas9 R-loop complex primed for DNA cleavage. *Science*, 351: 867-871.

[29] Jiang, F. and Doudna, J. A. (2017). CRISPR–Cas9 structures and mechanisms. *Ann. Rev. Biophys.* 46: 505-529.

[30] Char, S. N., Neelakandan, A. K., Nahampun, H., Frame, B., Main, M., Spalding, M. H., Becraft, P. W., Meyers, B. C., Walbot, V., Wang, K. and Yang, B. (2017). An *Agrobacterium*-delivered CRISPR/Cas9 system for high-frequency targeted mutagenesis in maize. *Plant Biotechnol. J.* 15: 257-68.

[31] Ali, Z., Shami, A., Sedeek, K., Kamel, R., Alhabsi, A., Tehseen, M., Hassan, N., Butt, H., Kababji, A., Hamdan, S. M. and Mahfouz, M. M. (2020). Fusion of the Cas9 endonuclease and the VirD2 relaxase facilitates homology-directed repair for precise genome engineering in rice. *Commun. Biol.*, 3: 44.

[32] McCarty, N. S., Graham, A. E., Studená, L. and Ledesma-Amaro, R. (2020). Multiplexed CRISPR technologies for gene editing and transcriptional regulation. *Nat. Comm.* 11: 1281.

[33] Li, C., Zong, Y., Wang, Y., Jin, S., Zhang, D., Song, Q., Zhang, R. and Gao, C. (2018). Expanded base editing in rice and wheat using a Cas9-adenosine deaminase fusion. *Gen. Biol.* 19: 59.

[34] Hunziker, J., Nishida, K., Kondo, A., Kishimoto, S., Ariizumi, T. and Ezura, H., 2020. Multiple gene substitution by Target-AID base-editing technology in tomato. *Sci. Rep.* 10: 20471.

[35] Komor, A. C., Zhao, K. T., Packer, M. S., Gaudelli, N. M., Waterbury, A. L., Koblan, L. W., Kim, Y.B., Badran, A. H. and Liu, D. R. (2017). Improved base excision repair inhibition and bacteriophage Mu

Gam protein yields C: G-to-T: A base editors with higher efficiency and product purity. *Sci. Adv.* 3: p.eaao4774.

[36] Zong, Y., Wang, Y., Li, C., Zhang, R., Chen, K., Ran, Y., Qiu, J. L., Wang, D. and Gao, C. (2017). Precise base editing in rice, wheat and maize with a Cas9-cytidine deaminase fusion. *Nat. Biotechnol.*, 35: 438-440.

[37] Alok, A., Sandhya, D., Jogam, P., Rodrigues, V., Bhati, K. K., Sharma, H. and Kumar, J. (2020). The rise of the CRISPR/Cpf1 system for efficient genome editing in plants. *Front. Plant Sci.,* 11: 264.

[38] Xu, X., Chemparathy, A., Zeng, L., Kempton, H.R., Shang, S., Nakamura, M. and Qi, L. S. (2021). Engineered miniature CRISPR/Cas system for mammalian genome regulation and editing. *Mol. Cell*, 81: 4333-4345.

[39] Li, Z., Zhang, D., Xiong, X., Yan, B., Xie, W., Sheen, J. and Li, J.F. (2017). A potent Cas9-derived gene activator for plant and mammalian cells. *Nat. Plants,* 3: 930-936.

[40] Moradpour, M. and Abdulah, S. N. A. (2020). CRISPR/dC as9 platforms in plants: strategies and applications beyond genome editing. *Plant Biotech. J.* 18: 32-44.

[41] Ren, Q., Sretenovic, S., Liu, S., Tang, X., Huang, L., He, Y., Liu, L., Guo, Y., Zhong, Z., Liu, G. and Cheng, Y. (2021). PAM-less plant genome editing using a CRISPR–SpRY toolbox. *Nat. Plants*, 7: 25-33.

[42] Schmitz, D.J., Ali, Z., Wang, C., Aljedaani, F., Hooykaas, P. J., Mahfouz, M. and de Pater, S. (2020). CRISPR/Cas9 mutagenesis by translocation of Cas9 protein into plant cells via the *Agrobacterium* type IV secretion system. *Front. Genom.* Ed., 2: 6.

[43] Osakabe, Y. and Osakabe, K. (2017). Genome editing to improve abiotic stress responses in plants. *Prog. Mol. Biol. Transl. Sci.* 149: 99-109.

[44] Arora, L. and Narula, A. (2017). Gene editing and crop improvement using CRISPR/Cas9 system. *Front. Plant Sci.,* 8:1932.

[45] Zhang, D., Hussain, A., Manghwar, H., Xie, K., Xie, S., Zhao, S., Larkin, R. M., Qing, P., Jin, S. and Ding, F. (2020). Genome editing with the CRISPR-Cas system: an art, ethics and global regulatory perspective. *Plant Biotechnol. J.*, 18: 1651-1669.

[46] Klap, C., Yeshayahou, E., Bolger, A. M., Arazi, T., Gupta, S. K., Shabtai, S., Usadel, B., Salts, Y. and Barg, R. (2017). Tomato facultative parthenocarpy results from Sl AGAMOUS-LIKE 6 loss of function. *Plant Biotechnol. J.* 15: 634-647.

[47] Miao, C., Xiao, L., Hua, K., Zou, C., Zhao, Y., Bressan, R. A. and Zhu, J. K. (2018). Mutations in a subfamily of abscisic acid receptor genes promote rice growth and productivity. *Proc. Natl. Acad. Sci.* 115: 6058-6063.

[48] Wang, L., Chen, L., Li, R., Zhao, R., Yang, M., Sheng, J. and Shen, L. (2017). Reduced drought tolerance by CRISPR/Cas9-mediated SlMAPK3 mutagenesis in tomato plants. *J. Agric. Food Chem.,* 65: 8674-8682.

[49] Shi, J., Gao, H., Wang, H., Lafitte, H. R., Archibald, R. L., Yang, M., Hakimi, S. M., Mo, H. and Habben, J. E. (2017). ARGOS 8 variants generated by CRISPR-Cas9 improve maize grain yield under field drought stress conditions. *Plant Biotech. J.* 15: 207-216.

[50] Zhao, C., Zhang, Z., Xie, S., Si, T., Li, Y. and Zhu, J.K. (2016). Mutational evidence for the critical role of CBF transcription factors in cold acclimation in *Arabidopsis*. *Plant Physiol.*, 171: 2744-2759.

[51] Barman, H.N., Sheng, Z., Fiaz, S., Zhong, M., Wu, Y., Cai, Y., Wang, W., Jiao, G., Tang, S., Wei, X. and Hu, P. (2019). Generation of a new thermo-sensitive genic male sterile rice line by targeted mutagenesis of TMS5 gene through CRISPR/Cas9 system. *BMC Plant Biol.* 19: 109.

[52] Lou, D., Wang, H., Liang, G. and Yu, D. (2017). OsSAPK2 confers abscisic acid sensitivity and tolerance to drought stress in rice. *Front. Plant Sci.* 8: 993.

[53] Usman, B., Nawaz, G., Zhao, N., Liao, S., Liu, Y. and Li, R. (2020). Precise editing of the ospyl9 gene by rna-guided Cas9 nuclease confers enhanced drought tolerance and grain yield in rice (Oryza sativa l.) by regulating circadian rhythm and abiotic stress responsive proteins. *Intl. J. Mol. Sci.* 21: 7854.

[54] Ogata, T., Ishizaki, T., Fujita, M. and Fujita, Y. (2020). CRISPR/Cas9-targeted mutagenesis of OsERA1 confers enhanced responses to abscisic acid and drought stress and increased primary root growth under nonstressed conditions in rice. *PloS one,* 15: e0243376.

[55] Liao, S., Qin, X., Luo, L., Han, Y., Wang, X., Usman, B., Nawaz, G., Zhao, N., Liu, Y. and Li, R. (2019). CRISPR/Cas9-induced mutagenesis of semi-rolled leaf1, 2 confers curled leaf phenotype and drought tolerance by influencing protein expression patterns and ROS scavenging in rice (*Oryza sativa* L.). *Agronomy*, 9: 728.

[56] Paixão, J. F. R., Gillet, F. X., Ribeiro, T. P., Bournaud, C., Lourenço-Tessutti, I. T., Noriega, D. D., Melo, B. P. D., de Almeida-Engler, J. and Grossi-de-Sa, M. F. (2019). Improved drought stress tolerance in Arabidopsis by CRISPR/dCas9 fusion with a Histone AcetylTransferase. *Sci. Rep.* 9: 8080.

[57] Yin, X., Anand, A., Quick, P. and Bandyopadhyay, A. (2019). Editing a stomatal developmental gene in rice with CRISPR/Cpf1. In: Yiping, Q. (Ed.) *Plant Genome Editing with CRISPR Systems* (pp. 257-268). Humana Press, New York.

[58] Osakabe, Y., Watanabe, T., Sugano, S. S., Ueta, R., Ishihara, R., Shinozaki, K. and Osakabe, K. (2016). Optimization of CRISPR/Cas9 genome editing to modify abiotic stress responses in plants. *Sci. Rep.* 6: 26685.

[59] Lv, Y., Yang, M., Hu, D., Yang, Z., Ma, S., Li, X. and Xiong, L. (2017). The OsMYB30 transcription factor suppresses cold tolerance by interacting with a JAZ protein and suppressing β-amylase expression. *Plant Physiol.* 173: 1475-1491.

[60] Santosh Kumar, V. V., Verma, R. K., Yadav, S. K., Yadav, P., Watts, A., Rao, M. V. and Chinnusamy, V. (2020). CRISPR/Cas9 mediated genome editing of drought and salt tolerance (OsDST) gene in indica mega rice cultivar MTU1010. *Physiol. Mol. Biol. Plants.* 26: 1099-1110.

[61] Yue, E., Cao, H. and Liu, B. (2020). OsmiR535, a potential genetic editing target for drought and salinity stress tolerance in *Oryza sativa*. *Plants*, 9: 1337.

[62] Zhang, A., Liu, Y., Wang, F., Li, T., Chen, Z., Kong, D., Bi, J., Zhang, F., Luo, X., Wang, J. and Tang, J. (2019). Enhanced rice salinity tolerance via CRISPR/Cas9-targeted mutagenesis of the *OsRR2* gene. *Mol. Breed.*, 39: 47.

[63] Zeng, Y., Wen, J., Zhao, W., Wang, Q. and Huang, W. (2020). Rational improvement of rice yield and cold tolerance by editing the three genes OsPIN5b, GS3, and OsMYB30 with the CRISPR–Cas9 system. *Front. Plant Sci.*, 10: 1663.

[64] Tang, L., Mao, B., Li, Y., Lv, Q., Zhang, L., Chen, C., He, H., Wang, W., Zeng, X., Shao, Y. and Pan, Y. (2017). Knockout of OsNramp5 using the CRISPR/Cas9 system produces low Cd-accumulating indica rice without compromising yield. *Sci. Rep.* 7: 14438.

[65] Nieves-Cordones, M., Mohamed, S., Tanoi, K., Kobayashi, N. I., Takagi, K., Vernet, A., Guiderdoni, E., Périn, C., Sentenac, H. and Véry, A. A. (2017). Production of low-Cs^+ rice plants by inactivation of the K^+ transporter *OsHAK 1* with the CRISPR-Cas system. *Plant J.*, 92: 43-56.

[66] Alfatih, A., Wu, J., Jan, S. U., Zhang, Z. S., Xia, J. Q. and Xiang, C. B. (2020). Loss of rice PARAQUAT TOLERANCE 3 confers enhanced resistance to abiotic stresses and increases grain yield in field. *Plant Cell Environ.*, 43: 2743-54.

[67] Zeng, D. D., Yang, C. C., Qin, R., Alamin, M., Yue, E. K., Jin, X. L. and Shi, C. H. (2018). A guanine insert in *OsBBS1* leads to early leaf senescence and salt stress sensitivity in rice (*Oryza sativa* L.). *Plant Cell Rep.*, 37: 933-946.

[68] Tran, M. T., Doan, D. T. H., Kim, J., Song, Y. J., Sung, Y. W., Das, S., Kim, E. J., Son, G. H., Kim, S. H., Van Vu, T. and Kim, J. Y. (2021). CRISPR/Cas9-based precise excision of SlHyPRP1 domain (s) to obtain salt stress-tolerant tomato. *Plant Cell Rep.*, 40: 999-1011.

[69] Woo, J. W., Kim, J., Kwon, S. I., Corvalán, C., Cho, S. W., Kim, H., Kim, S. G., Kim, S. T., Choe, S. and Kim, J. S. (2015). DNA-free genome editing in plants with preassembled CRISPR/Cas9 ribonucleoproteins. *Nat. Biotechnol.* 33: 1162-1164.

[70] Romero, F. M. and Gatica-Arias, A. (2019). CRISPR/Cas9: development and application in rice breeding. *Rice Sci.* 26: 265-281.

[71] Yang, H., Wu, J. J., Tang, T., Liu, K. D. and Dai, C. (2017). CRISPR/Cas9-mediated genome editing efficiently creates specific mutations at multiple loci using one sgRNA in *Brassica napus*. *Sci. Rep.*, 7: 7489.

[72] Gao, F., Yao, H., Zhao, H., Zhou, J., Luo, X., Huang, Y., Li, C. and Chen, H. (2016). Tartary buckwheat FtMYB10 encodes an R2R3-MYB transcription factor that acts as a novel negative regulator of salt and drought response in transgenic Arabidopsis. *Plant Physiol. Biochem.* 109: 387-396.

[73] Aliaga-Franco, N., Zhang, C., Presa, S., Srivastava, A. K., Granell, A., Alabadí, D., Sadanandom, A., Blázquez, M. A. and Minguet, E. G. (2019). Identification of transgene-free CRISPR-edited plants of rice, tomato, and arabidopsis by monitoring DsRED fluorescence in dry seeds. *Front. Plant Sci.,* 10: 1150.

[74] Zhang, Y., Iaffaldano, B. and Qi, Y. (2021). CRISPR ribonucleoprotein-mediated genetic engineering in plants. *Plant Commun.*, 2: 100168.

[75] Zhang, Y., Liang, Z., Zong, Y., Wang, Y., Liu, J., Chen, K., Qiu, J. L. and Gao, C. (2016). Efficient and transgene-free genome editing in wheat through transient expression of CRISPR/Cas9 DNA or RNA. *Nat. Commun.*, 7: 12617.

[76] Zhou, Q. Y., Tian, A. G., Zou, H. F., Xie, Z. M., Lei, G., Huang, J., Wang, C. M., Wang, H. W., Zhang, J. S. and Chen, S. Y. (2008). Soybean WRKY-type transcription factor genes, *GmWRKY13*, *GmWRKY21*, and *GmWRKY54*, confer differential tolerance to abiotic stresses in transgenic Arabidopsis plants. *Plant Biotechnol. J.*, 6: 486-503.

[77] Rees, H. A. and Liu, D. R. (2018). Base editing: precision chemistry on the genome and transcriptome of living cells. *Nat. Rev. Genet*. 19: 770-88.

[78] Napier, J. A., Haslam, R. P., Tsalavouta, M. and Sayanova, O. (2019). The challenges of delivering genetically modified crops with nutritional enhancement traits. *Nat. Plants*, 5: 563-567.

[79] Lee, J., Jung, M. H., Jeong, E. and Lee, J. K. (2019). Using Sniper-Cas9 to minimize off-target effects of CRISPR/Cas9 without the loss of on-target activity via directed evolution. *J. Visual. Expt*. 144: e59202.

[80] Ikeda, A., Fujii, W., Sugiura, K. and Naito, K. (2019). High-fidelity endonuclease variant HypaCas9 facilitates accurate allele-specific gene modification in mouse zygotes. *Commun. Biol*. 2: 371.

[81] Fan, R., Chai, Z., Xing, S., Chen, K., Qiu, F., Chai, T., Qiu, J. L., Zhang, Z., Zhang, H., Gao, C. (2020). Shortening the sgRNA-DNA interface enables SpCas9 and eSpCas9 (1.1) to nick the target DNA strand. *Sci. China Life Sci.* 63: 1619-30.

[82] Gao, C. (2018). The future of CRISPR technologies in agriculture. *Nat. Rev. Mol. Cell Biol*. 19: 275-276.

[83] Perez-Pinera, P., Kocak, D. D., Vockley, C. M., Adler, A. F., Kabadi, A. M., Polstein, L. R., Thakore, P. I., Glass, K. A., Ousterout, D. G., Leong, K. W. and Guilak, F. (2013). RNA-guided gene activation by CRISPR/Cas9–based transcription factors. *Nat. Meth*. 10: 973-976.

Chapter 8

CRISPR/Cas: A Promising Implement for Empowering Plants against Viral Stress

Asha Rani[1]
Rashmi Suman[1]
Taruna Gupta[1]
Narayan Rishi[1]
Yashika Walia[2]
Sunny Dhir[2,*]
and Vanita Chandel[1,†]
[1] Amity Institute of Virology and Immunology,
Amity University Uttar Pradesh, Noida, U.P., India
[2] Department of Biotechnology,
Maharishi Markandeshwar (Deemed to be University), Mullana, Ambala [Haryana], India

Abstract

Clustered Regularly Interspaced Short Palindromic Repeats (CRISPR) is carrying genetic engineering to a new world. Numerous gene editing techniques are available but CRISPR/Cas is more specific and robust. With the increasing necessity of sustainable agriculture development and improvement, the tools and techniques are also getting improved. Virus infection usually has negative impact on host plants which later causes disease. Biotic stresses get influenced on attack of viruses on plants. The infection makes plants deprive of nutrients which causes their death. It ultimately affects productivity. The stress problem can be solved by studying the genetic mechanism of virus infection. Genetically modified plants are doing well against biotic stresses. But this is very time consuming. The book chapter aims at providing brief overview of CRISPR/Cas technology and its implementation in plant stress management.

Keywords: CRISPR, viruses, viral stress, gene editing, genetically modified crops

---

* Corresponding Author's Email: sunny72155@gmail.com.
† Corresponding Author's Email: vchandel@amity.edu.

In: Genome Editing for Crop Improvement
Editors: Rahul Kumar, Ashutosh Sharma, Yogesh Kumar et al.
ISBN: 979-8-88697-740-0

Introduction

Earlier, people were not aware of the root cause of the reduction in the crop yield. Genome editing is a gene-specific method used for crop improvement. CRISPR/Cas is among the four families of restriction nucleases that can recognize the genome as well as cut the specific sequences. The function of CRISPR/Cas was first identified in bacterial defence mechanism thus it is a naturally occurring defense mechanism.

Plants infected with various infectious agents tries to evade the condition by various mechanisms such as accumulation of antiviral components, cells and tissues death responses, oxidative stress development etc. CRISPR/Cas system has different number of classes, types, and subtypes i.e., 2, 6 and 33 respectively. Different components of CRISPR/Cas system include crRNA that guides Cas nuclease to the target sequence, PAM is the signal for binding of Cas nuclease (PAM is required adjacent to the target sequence), tracrRNA (trans activating crRNA) helps in binding of Cas enzyme, and a 20-nucleotide long spacer is required. It should be complementary to the site that needs to be targeted. In class 1 (rarely used), there occurs multiple Cas proteins while there is single Cas protein in class 2 (3 types and 17 subtypes) CRISPR/Cas system. Different Cas enzymes recognize different PAM sequences as in case of 2nd class of CRISPR/Cas system while Cas 13 and 14 are the enzymes that target RNA and ssDNA sequences respectively [1].

CRISPR/Cas systems can use either plant or virus-mediated resistance. Plant viruses depend on their hosts for their requirement of the factors essential to complete their infection cycle [2]. Many host factors involved in their life cycle have been recognized. Modification in these factors results in resistant varieties production [3]. Requirements for this methodology involve gRNA and Cas9 nuclease (have two domains i.e., RuvC-like domain, HNH domain). Among all systems, CRISPR/Cas9 is well elucidated, in which crRNA maturation occurs with the help of tracrRNA. TracrRNA pairs with precrRNA which generate dsRNAs, which is further cleaved by RNase III and production of mature crRNA occurs. Now these crRNA and tracrRNA associate with Cas9 endonuclease [4].

When Cas9 associates with single-stranded or double-stranded target DNA, it retains companionship with guide RNA also known as sgRNA. sgRNA helps Cas9 for specific binding with the DNA (target). This results in formation of blunt double-stranded DNA break. These breaks get repaired by either Non-Homologous End Joining (NHEJ) or Homology Directed Repair (HDR) as shown in Figure 1.

Cas9 is preferred because of its high effectiveness in genome editing. However, off-target cleavage at undesired genome regions has been a drawback *in vivo*. Cas9 recognizes the PAM for its binding to the target DNA. *Streptococcus pyogenes* Cas9 or SpCas9 detects 5' NGA 3' as the PAM sequence, whereas *Francisella novicida* Cas9 or FnCas9 identifies 5' NGG 3' [5].

Cas12 (Cpf1) is another competent enzyme which belongs to class 2 type V Cas protein class and produces staggered cuts in double-stranded DNA. It produces individual crRNAs, allowing greater multiplexing. Cas12 has provided an epigenome editing platform. When Cas12a is activated by a targeted DNA fragment with the same spacer sequence as its spacer sequence, it can slice ssDNA [6]. Along with this, Cas12a identifies T-rich PAM sequence rather than G-rich and generates staggered dsDNA breaks [7, 8]. CRISPR/Cas12a complex performs cis-trans activity of cleavage on single-stranded DNA, which is utilised to diagnose a variety of viral diseases [6]. Cas13 (C2C2) is the effector associated with CRISPR that recognises complementary RNA and performs cleavage activity. It involves adjacent RNA collateral as well as trans cleavage [9]. It helps in diagnosis of viruses nowadays [10]. Cas13a, Cas13b, Cas13c, and Cas13d are proteins that target RNA strands cleavage [11].

Viral diseases are among other diseases affecting agriculture, accounting for over half of all the microorganisms. Plant viruses endanger farmed plants by limiting their development and vitality as well as lowering gross production through lowering plant quality [12]. Crop losses have economically affected most people whose livelihoods are entirely dependent on agriculture. The emergence and recurrence of viruses, as well as their potential to infect number of hosts, has exacerbated the viral problem [13]. Currently, a new methodology, CRISPR/Cas technique, that involves genetic engineering has paved a way to empower plants against viral stress. Despite its infancy stage, implementation of the CRISPR/Cas9 system-based editing in the genome for crop enhancement is becoming popular. In fact, multiple studies have already stated that this technique has been successfully applied to a wide range of plant species [14, 15].

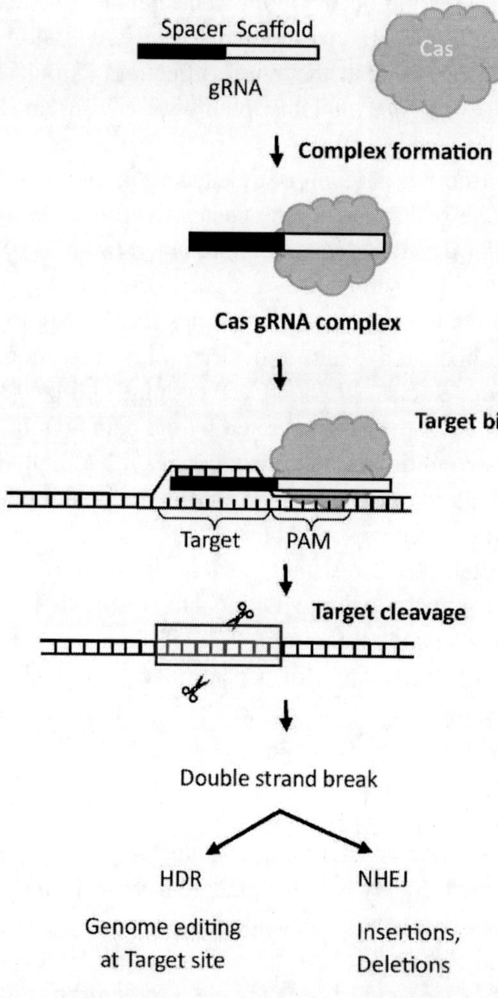

Figure 1. Two methods of gene repair: non-homologous end joining (NHEJ) and homology directed repair (HDR).

There are two ways through which CRISPR/Cas acts as a tool for improvement in resistance of plants against viruses as it can help in the introduction of mutations in the host genes that help the viruses for their establishment, and targeting the viral genome directly [16]. CRISPR provides a suitable way of achieving genome editing in plants. It offers several advantages over other methods. It is a simple, cost effective, and robust technique that is used for genome editing in many organisms. This chapter will help in gaining more information about how CRISPR/Cas is empowering plants against viral stress.

Viral Stress on Plants and CRISPR/Cas

Viruses are intracellular obligate parasites that make many possibilities to arise because of the interaction between their proteins with the host factors. Accumulation of viruses in host cell to an extent of damaging the plant growth and development indicates the potential control thereof viruses can have on host cell machinery. As studied in *Arabidopsis* and *Nicotiana*, major changes in response to viral infection occurs in the genes related to stress and defense in plant cell [17]. In the case of *Tobacco mosaic virus* inoculated on *N. glutinosa,* a gene N, was found to have a role in resistance against the viral infection. Similarly, tomato (*Solanum lycopersicum*) and pepper (*Capsicum annuum*) plants were virus inoculated for identifying resistance gene against viral infection [18].

During infection, changes occur at levels of defence hormone also like salicylic acid, nitric oxide and jasmonic acid. The reactive oxygen species like O^{2-} and hydrogen peroxide also accumulates not only in the infected but also in noninfected tissues [19]. Proteins involved in signalling of defence cascades get activated Expression of glucanase, defensin and chitinase gets upregulated. These are the from the protein families that belong to the pathogenesis [20]. Banana is a vegetatively propagated plant, and viruses are most commonly transferred by mechanical means and insect vectors. CRISPR/Cas9 system is utilized in disruption of coding region of Banana streak virus which was integrated by the pathogen in the plant host genome to get activated under stress conditions. The transgenic plant had mutation in the targeted site [21]. Genome editing has made progress in banana by phytoene desaturase gene knock-out in one of the researches and multiple genes in another [22].

Earlier, the mutant alleles for resistant variety were introduced through breeding procedures, but these were uncertain, time taking and resulted in larger genome regions insertion rather than just one gene. Requirement of PAM sequences become a challenge sometimes but not in case of inactivation of a particular gene at the target by introducing mutation. Modification in Cas nucleases is being carried out as in the case of xCas9, which can recognise wide range of PAM sequences. Multiple genes can also be targeted for different viral infections in a single host plant but in that case off-target events may increase.

Viral latency is also a threat to the plants as it involves integration of a part or full genome of a virus to the genome of the host plant due to some environmental stress or during any developmental stage of the host. It may help the plant to have resistance properties to that virus but in any span of time, these complete genome integration events of viruses inside host genome can lead to severe infection that may cause high loss in the production and thus economy. CRISPR/Cas9 editing has successfully solved the problem in research on endogenous (integrated) Banana streak virus, because of which, the activation process of this integrated genome of BSV stopped, that could turn it into an infectious particle. Although, other examples of are also there where viruses (endornaviruses) cause persistent infections. This stress (seed germinates faster, pod length of beans becomes higher, weight of the seed goes higher) to the plants because of infection helps in breeding [23].

Abiotic stresses make plant susceptible to infection by the biotic factors such as virus infection [24]. CRISPR/Cas9 technique has been used for many crop plants such as rice, maize, tomato, wheat, citrus, cotton, soybean, potato, alfalfa, watermelon, grass, legumes, cassava, ipomoea, lettuce, carrot, cacao, salvia etc [25].

Plant viruses have differences based on their genome i.e., dsDNA, ssDNA, dsRNA, ssRNA-, and ssRNA+. The most common type of virus with CRISPR application involves ssDNA viruses (G*eminiviridae* (family) e.g., *Begomovirus* (genus)]. The members of *Begomovirus* infect dicots and in the phloem part of the infected host. BSCTV (Beet severe curly top virus) was found to have forty-three locations for sgRNA or Cas9 in its genome. Virus accumulation was reduced in the target leaves Nicotiana and Arabidopsis on all the sites. In the same way, another research with eleven sites as target in *Bean yellow dwarf virus* infection, the technique was also successful in reduction of the virus by 87% in *Nicotiana benthamiana* [26].

CRISPR/Cas9 system was exploited for targeting viral genomes at the time of their replication. sgRNAs specific for different regions of yellow leaf curl virus were delivered on *Nicotiana benthamiana* plant. The most effective resistance was shown where the target site was sequence of stem-loop invariant present in the intergenic region of the virus. As similar sequences were present in other members of same family i.e., *Beet curly top virus* as well as *Merremia mosaic virus*, the plant with their mixed infection showed immunity against both of them [27].

SpCas9 can recognise dsDNA only and not the RNA viruses, which makes the resistance against RNA viruses more difficult. However, with time, FnCas9 nuclease from *Francisella novicida* and LwaCas13a nuclease from *Leptotrichia wadei* were able to cut RNA. FnCas9 has been reported to cut sgRNAs which were specific to *Tobacco mosaic virus* as well as Cucumber mosaic virus. *N. benthamiana* and *Arabidopsis thaliana* showed up to 80% resistance against these two viruses [7].

Cas13a has also been used in manipulation of Turnip mosaic virus (RNA genome). Four sites from its genome were targeted. HC-Pro and GFP genes were showing good results as it affected the viral replication, thus preventing the spread of the virus [29]. Viruses establish themselves using host machinery in the host plant cells, that includes the host translation factors like eIF4E and eIF4G [30]. Cucumber plants were made resistant to *Potyvirus* by mutating these genes. Other viruses from the same family were also targeted, and as a result, the plants showed immunity against *Zucchini yellow mosaic virus* (ZYMV), *Cucumber vein yellowing virus* (CVYV) and *Papaya ring spot mosaic virus* [41]. In *Arabidopsis* plant, eIF(iso)4E was mutated with single base pair insertion and deletion resulted in resistance against *Potyvirus* (ssRNA) [31].

Recessive resistance can be created by knocking out the genes related to host system which helps in life cycle of the viruses. Important genes of the host machinery can make host weaker, and the virus can overcome resistant. This has not proved to be a good practice for the development of resistant varieties, and further investigations need to be carried out.

Applications of CRISPR in Genome Editing

Over the last few decades, CRISPR/Cas technology has revolutionized different fields of life sciences especially in plant sciences, as it offers a powerful and novel approach towards classical plant breeding for trait improvement and domestication of wild species. Besides these, it has several other applications in plant biotechnology, such as novel delivery methods, modulation of gene-expression, high-throughput and multiplexed gene editing in plants, and *in-situ* directed evolution. Since plant viruses cause devastating diseases in crops with 90-100% loss in productivity, hence the application of this latest technology has not only successfully prevented crops from virus infections but also helped in crop improvement with increased productivity.

Inhibition of DNA and RNA Virus Replication

The scientists have used CRISPR approach to directly inhibit plant virus genome replication. The experiments on *Geminiviruses* having monopartite ssDNA genome, which is *Beet severe curly top virus* as well as *Bean yellow dwarf virus* were conducted in two plants i.e., *N. benthamiana* and *A. thaliana* [26]. Using CRISPR/Cas9 and sgRNA, they targeted Rep gene and non-coding region conserved in all *Geminiviruses,* that is nano nucleotide [TAATTATTAC] present in the intergenic region (IR) to develop transgenic plants with virus resistance. Results were significant as the virus accumulation was reduced by 87% and symptoms were also reduced in these transgenic plants [32][26]. In another study on *Tomato yellow leaf curl virus* (TYLCV), a bipartite *Geminivirus*, the scientist engineered sgRNA targeting RCRII motif of Rep gene, CP gene, and conserved stem-loop region to produce TYLCV resistant transgenic *N. benthamiana* plants [27]. In another report by Yin et al. in 2019, transgenic *N. benthamiana* plants were produced expressing Cas9 and dual sgRNA that targets rep gene and IR simultaneously, that were completely resistant to *Cotton leaf curl Multan virus* [33]. Further mutational deactivation of endonuclease Cas9 was still effective against inhibition of CLCV replication and accumulation [34].

Although development of CRISPR-mediated virus resistance was successful in model plants, it was not successful in many crop plants, and in few cases it resulted in development of a new viral strain. Overexpression of Cas9 and sgRNA in cassava plants, targeting genes for transcription activator protein (AC2) and replication enhancer protein (AC3) of *African cassava mosaic virus*, results in generation of new ACMV variant. CRISPR-mediated virus *Wheat dwarf virus*-resistant barley plants [35], and TYLCV-resistant tomato plants have also been produced [36]. This approach has also been used to develop resistance against *Cauliflower mosaic virus* (dsDNA virus) by targeting its CP gene via Cas9 and multiple sgRNA in *Arabidopsis* plants [37], and endogenous *Banana streak virus* by targeting its coding sequence in its genome in *Musa* spp. with 75% reduction in symptoms [21].

RNA viruses frequently infect crop plants with wide range of hosts, hence CRISPR-mediated resistance was developed in plants with the discovery of such systems namely Cas13(a-d), FnCas9 and SpCas9 was of huge advantage that enabled targeting of RNA viruses directly. Using these as a tool (FnCas9 and sgRNA), *Tobacco mosaic virus* and *Cucumber mosaic virus*-resistant plants (*Arabidopsis thaliana* and *N. benthamiana*) were developed which reduced the accumulation of virus by 80% [28]. Similarly, Cas13a was used to generate *Turnip mosaic virus* resistant *Nicotiana benthamiana*, potato with *Potato virus Y* resistance as well as rice was made resistant to *Rice stripe mosaic virus* and *Southern rice black streaked dwarf virus* [38].

Although CRISPR/Cas has efficiently been used to generate virus resistant plants, but there are certain limitations that must not be ignored because it must go through strict GMO (genetically modified crops) regulations. CRISPR/Cas technology requires selection of most advantageous sgRNA target sites so that the virus does not develop mutations after escaping cleavage by CRISPR/Cas system and give rise to severe novel strains [39]. Targeting multiplex regions and noncoding IR regions in the genome of virus have reduced mutation [39]. Further to decrease the RNA intermediates of DNA, different RNA binding Cas system can be used along with Cas9 because it will provide even more long-term resistance than nucleases which targets DNA directly [40].

Host Genome Editing

CRISPR/Cas tools can also be implemented on host genes that transcribe various transcription factors required during virus replication and movement within cell. Several plant genes were discovered as critical factors for virus infection, and are being focused by different scientists to use them for transgenic plant development. CRISPR allows mutation by deletion or insertion of nucleotides within recessive genes conferring durable resistance by plants. Some of the most used transcription initiation factors by plants viruses for replication are eIF4E, eIF(iso)4E and eIF4G whose inactivation has rendered resistance to plants [30]. Using these as a target for CRISPR/Cas tools, successful transgenic plants were developed in previous years. First experiment was conducted on cucumber plants whose eIF4E gene was knocked out and successfully achieved resistance against several viruses in *Potyviridae* members like *Papaya ringspot mosaic virus*, *Zucchini yellow mosaic virus*, and *Cucumber vein yellowing virus* [41]. Similarly, CRISPR/Cas system was applied to *A. thaliana* plants to knock out eIF(iso)4E genes to produce plants resistant to Turnip mosaic virus [31]. The other eIF4G gene was knocked for development of resistance against *Rice tungro spherical virus* in rice [31].

Besides knocking out, mutations can be introduced in these transcription initiation factors by CRISPR/Cas9. Using Cas9, the catalytic site of Cas9, namely nickase D10, was inactivated and fused to cytidine deaminase for the conversion of cytosine into thymidine on target site in tomato and rice [42]. This technique was considered later and using this methodology the scientists generated *Clover yellow vein virus*-resistant *Arabidopsis thaliana* plants to introduce single substitution mutation of N176K in eIF4E of host genome [43]. A recent report describes CRISPR/Cas9 usage in engineering double mutations on the nCBP-1 and nCBP-2 (novel cap-binding protein 1 and 2 that belongs to the eIF4E family, had successfully increased the resistance in cassava against *Cassava brown streak virus* [44].

Alternative approach using CRISPR/Cas9 as ribonucleo-proteins (RNP) which has been recently used to produce virus resistant plants [45]. Coilin gene is one of the most important proteins in Cazal body, a subnuclear structure that modulates virus-host interaction is the new target for genome editing [46]. CRISPR/Cas9 delivery in the form of RNPs using gold nanoparticles via biolistic bombardment method to target coilin gene into apical meristematic tissue of potato plants, caused resistance to PVY by deletion and editing allele of coilin gene [47].

Delivery of CRISPR/Cas Components via Different Methods

CRISPR/Cas components can be delivered to plants via agroinfiltration or physical methods such as biolistic, vacuum infiltration, protoplast transfection. However, these methods require mediators like RNPs, plasmid or virus as a vector. For stable expression, one must notice that if CRISPR/Cas components are delivered to germline cell, so that positive mutations might be transmitted to next generation saving time and labour.

Agrobacterium has the capability to transmit T-DNA (a part of its genome) to the cell nucleus and indiscriminately integrates in plant genomic DNA sequence [48]. Using this nature of *Agrobacterium*, expression cassette comprising of Cas9 and sgRNA could be cloned into Ti-plasmid and transformed into *Agrobacterium* so that they can be easily introduced into plants. This approach has been intensively used in various studies to achieve targeted mutations in plants. Editing efficiency reported in different studies are as follows: citrus with 20% [49], *N. benthamiana*, rice [50] and *A. thaliana* with up to 85% [51], soybean and some other plant species with 14.7% to 95% [52, 53].

Digestion of plant cell walls cellulases and lysosome enzymes helps in obtaining protoplasts along with other editing reagents, which could be delivered by two different techniques like electroporation or chemical treatment by polyethylene glycol. Using this plant protoplast, CRISPR/Cas system were delivered in plants to develop mutations and subsequently regenerated plants with either stable or transient expression with editing efficiency ranged from 3% to 46% in several crops including soybean, potato, grapevine, wheat, *A. thaliana*, rice, tobacco, and lettuce [54-57]. Delivery of preassembled Cas9-sgRNA RNPs was successful by this protoplast delivery technique which was not possible by *Agrobacterium* mediated delivery.

Another delivery method used in CRISPR/Cas system includes biolistic bombardment, where DNA coated with microprojectiles such as gold, silver, or tungsten particles are bombarded using biolistic gun with high pressure for the penetration into the cell wall of plant cells. This technique was used to introduce target mutations by CRISPR tools in plants such as rice, wheat, and soybean [58-60] genomes, which showed editing efficiencies within range of 14.5% to 76%, in maize and wheat [61, 62] ranging from 21.8% to 47%. Another report has mentioned the used of biolistic delivery of Cas9 and sgRNA via *Wheat dwarf virus*-based vector, (pWDV2) in wheat, resulted in 12-fold increase in editing efficiency [57]. Biolistic bombardment is usually efficient because it can simultaneously deliver multiple constructs and used for many plant species and produce multiple copies of introduced genes.

Virus-Mediated Delivery of CRISPR/Cas System

Plant viruses (both DNA and RNA) are being exploited for CRISPR/Cas delivery and expression in plants. Transient expression of Cas or sgRNA or even both in host genome is possible through RNPs or plasmid or other virus vectors that could effectively deliver Cas system [63]. This would reduce off-target effects and avoid hurdles associated with transgenic plants in GMO regulations. Use of viruses as autonomously replicating vectors since many years to express foreign RNAs and protein [64]. Some of commonly used viruses like DNA viruses are *Wheat dwarf virus*, *Bean yellow dwarf virus*, and *Cabbage leaf curl virus*, and RNA virus *Tobacco rattle virus* are used as delivery agent for CRISPR/Cas components [65, 64].

TRV is used as a delivery system for analysing gene function mainly by silencing mechanism [77]. However due to difficulty in carrying large gene like Cas9, another approach was used like split-gene where two different fragments of Cas9 was expressed by TMV vector and *Agrobacterium* separately and assembled to express active Cas9 protein. TMV derived vectors for transient expression of sgRNA and Cas9 in plants was efficiently used by Cody et al., for integration free genome editing [66]. Later a group of scientists designed an expression vector based on TMV in a more advanced form for the expression of target miR398 in tobacco. When this was transformed in plants, expression of miR398 eliminated the viral RNA in the regenerated plants during shoot development. They also used split-protein approach for the transient expression of Cas9 from two different fragments obtained via *Agrobacterium* and TMV separately and further the re-assembling of this Cas9 protein which induced a targeted mutation [67]. Effectiveness of virus-mediated delivery system as a medium to deliver CRISPR/Cas9 has evolved as an innovative approach in plant biotechnology and functional genomics.

CRISPR/Cas in Biosensing and Imaging

CRISPR/Cas-based systems has been effectively utilized to develop overly sensitive, cost-effective and virus specific diagnostic purposes. The remarkable collateral or non-specific activity of Cas12a to cleave ssDNA and Cas13 to cleave RNA during binding with target DNA/RNA has been explored completely for the development of new tools for diagnosis. One such diagnostic tool is SHERLOCK (specific high-sensitivity enzymatic reporter unlocking) that utilizes Cas13 for RNA virus detection, where RNA undergoes reverse transcription and then amplified by RPA and after that, they are transcribed into RNA by T7 RNA polymerase, for the cleavage by Cas13 which is coupled with fluorescent reporter and release signal upon quenching with Cas13 collateral cleavage [68]. This technique was developed as lateral flow test strips for rapid, extremely sensitive, and on-field plant virus detection [69].

Another diagnostic tool is DETECTR (DNA endonuclease targeted CRISPR trans-reporter) that has been employed in diagnosis of DNA viruses where Cas12a or Cas14 in combination with RPA for the subsequent cleavage of ssDNA sequences that is coupled with fluorescent reporters for visual detection of target DNA. Another technique that utilizes Cas12b are named as HOLMES (1-h low-cost multipurpose highly efficient system) which detect DNA/RNA viruses and its modified version HOLMESv2 which can detect even SNPs [71].

Later an advanced form of SHERLOCK technique was developed as HUDSON (heating unextracted diagnostic samples to obliterate nucleases) which can be used for the detection of even extremely low titre of virus in any sample [72]. SHERLOCKv2, another advanced version, utilizes Cas12 along with Cas13 in combination for multiplex detection. Further, the sensitivity of this technique was enhanced by adding another nuclease Csm6 to magnify the signal released during collateral cleavage by Cas13 to develop FAM biotin reporter kit [69]. In agriculture it has multiple application such as quantification PR (pathogen resistance) genes and genotyping [73], detection of significant traits and inspection of insect pests and disease.

For localization of nucleic acids in plants, live imaging technique was developed by Ma et al. who utilized Cas9 in catalytically inactive form which is fused with fluorescent marker and complexed with sgRNA to visualize target DNA in plant genome [74]. Similarly inactive Cas13 was used for live imaging of RNA transcripts in plant cells [75]. These approaches were used for detection of DNA/RNA viruses in virus-resistant plants.

Conclusion

CRISPR/Cas has brought an innovative approach to manipulate the plant genes in a way to establish sustainable agriculture and development. CRISPR enables live imaging technique that helps in localizing DNA or RNA in plant cells. CRISPR is applicable to produce heritable mutations and can target the existence of foreign gene present in the host plant genome. The other advantage is that there is no need of transgene plant production. Even in the horticulture crops, the technology is applied to increase shelf life, production and resistant from disease also. CRISPR opens a wide range of applications that includes exposure of different metabolic pathways, the various networks that regulate different genes and control stress response. While CRISPR/Cas systems are capable to control infection of plant viruses along with its overall enhancement but there is a limitation in their delivery methods [76].

To control these constraints, researchers must investigate in what form and how the delivery of these reagents of CRISPR/Cas can be successful in plant tissues, as well as appropriate techniques to ensure maximal and ecologically acceptable utility as many crops and production depends on technology that can be applied beyond virus free germplasm propagation especially in plants where vegetative propagation is the only option for their production in bulk. CRISPR applications have recently expanded to encompass screening of genome, multiplexing of genome editing, gene therapy, transcriptional regulation, and antiviral defence, thanks to greater understanding of different CRISPR/Cas systems. The recent discovery about Cas proteins has opened the way to novel tools for nucleic acid screening to provide cost-effective, portable diagnostics. The chapter briefs the implementation of CRISPR/Cas system for crop improvement. Although, the off-target effects and DNA damage needs improvement, but the technology has brought us much closer to the virus free plants production.

References

[1] Makarova, Kira S, and Eugene v Koonin. 2015. "Annotation and Classification of CRISPR/Cas Systems." *Methods in Molecular Biology* 1311. Humana Press Inc.: 47–75. doi:10.1007/978-1-4939-2687-9_4.

[2] Roger Hull. 2002. *Matthews' Plant Virology*. Academic Press. San Diego: Elsevier. doi:10.1016/b978-0-12-361160-4.x5050-6.

[3] Wittmann, Sylvie, Henriette Chatel, Marc G Fortin, and Jean François Laliberté. 1997. "Interaction of the Viral Protein Genome Linked of Turnip Mosaic *Potyvirus* with the Translational Eukaryotic Initiation Factor [Iso] 4E Of *Arabidopsis thaliana* Using the Yeast Two-Hybrid System." *Virology* 234 [1]. Academic Press: 84–92. doi:10.1006/VIRO.1997.8634.

[4] Kalinina, Natalia O, Andrey Khromov, Andrew J Love, and Michael E Taliansky. 2020. "CRISPR Applications in Plant Virology: Virus Resistance and Beyond." *Phytopathology* 110 [1]. *American Phytopathological Society*: 18–28. doi:10.1094/PHYTO-07-19-0267-IA.

[5] Wiedenheft, Blake, Samuel H Sternberg, and Jennifer A Doudna. 2012. "RNA-Guided Genetic Silencing Systems in Bacteria and Archaea." *Nature* 482 [7385]. England: 331–38. doi:10.1038/nature10886.

[6] Chen, Janice S, Enbo Ma, Lucas B Harrington, Maria Da Costa, Xinran Tian, Joel M Palefsky, and Jennifer A Doudna. 2018. "CRISPR/Cas12a Target Binding Unleashes Indiscriminate Single-Stranded DNase Activity." *Science [New York, N.Y.]* 360 [6387]: 436–39. doi:10.1126/science. aar6245.

[7] Stella, Stefano, Pablo Mesa, Johannes Thomsen, Bijoya Paul, Pablo Alcón, Simon B Jensen, Bhargav Saligram, Matias E Moses, Nikos S Hatzakis, and Guillermo Montoya. 2018. "Conformational Activation Promotes CRISPR/Cas12a Catalysis and Resetting of the Endonuclease Activity." *Cell* 175 [7]: 1856-1871.e21. doi: https://doi.org/10.1016/j.cell.2018.10.045.

[8] Swarts, D.C., van der Oost, J. and Jinek, M., 2017. Structural basis for guide RNA processing and seed-dependent DNA targeting by CRISPR-Cas12a. *Molecular Cell*, *66*[2], pp.221-233.

[9] Abudayyeh, Omar O, Jonathan S Gootenberg, Silvana Konermann, Julia Joung, Ian M Slaymaker, David B T Cox, Shmakov, S, Makarova KS, Semenova E, Minakhin, L, Severinov, K, Regev, A, Lander, ES, Koonin, EV, & Zhang, F. 2016. "C2c2 Is a Single-Component Programmable RNA-Guided RNA-Targeting CRISPR Effector." *Science [New York, N.Y.]* 353 [6299]: aaf5573. doi:10.1126/science. aaf5573.

[10] Kellner, Max J, Jeremy G Koob, Jonathan S Gootenberg, Omar O Abudayyeh, and Feng Zhang. 2019. "SHERLOCK: Nucleic Acid Detection with CRISPR Nucleases." *Nature Protocols* 14 [10]: 2986–3012. doi:10.1038/s41596-019-0210-2.

[11] Smargon, Aaron A, Yilan J Shi, and Gene W Yeo. 2020. "RNA-Targeting CRISPR Systems from Metagenomic Discovery to Transcriptomic Engineering." *Nature Cell Biology* 22 [2]: 143–50. doi:10.1038/s41556-019-0454-7.

[12] Anderson, Pamela K, Andrew A Cunningham, Nikkita G Patel, Francisco J Morales, Paul R Epstein, and Peter Daszak. 2004. "Emerging Infectious Diseases of Plants: Pathogen Pollution, Climate Change and Agrotechnology Drivers." *Trends in Ecology and Evolution* 19 [10]: 535–44. doi: 10.1016/j.tree.2004.07.021.

[13] Savary, Serge, Laetitia Willocquet, Sarah Jane Pethybridge, Paul Esker, Neil McRoberts, and Andy Nelson. 2019. "The Global Burden of Pathogens and Pests on Major Food Crops." *Nature Ecology and Evolution* 3 [3]: 430–39. doi:10.1038/s41559-018-0793-y.

[14] Green, J C, and J S Hu. 2017. "Editing Plants for Virus Resistance Using CRISPR/Cas." *Acta Virologica* 61 [2]. Slovakia: 138–42. doi:10.4149/av_2017_02_02.

[15] Wada, Naoki, Risa Ueta, Yuriko Osakabe, and Keishi Osakabe. 2020. "Precision Genome Editing in Plants: State-of-the-Art in CRISPR/Cas9-Based Genome Engineering." *BMC Plant Biology* 20 [1]: 234. doi:10.1186/s12870-020-02385-5.

[16] Price, Aryn A, Timothy R Sampson, Hannah K Ratner, Arash Grakoui, and David S Weiss. 2015. "Cas9-Mediated Targeting of Viral RNA in Eukaryotic Cells." *Proceedings of the National Academy of Sciences of the United States of America* 112 [19]. National Academy of Sciences: 6164–69. doi:10.1073/pnas.1422340112.

[17] Whitham, Steven A, Chunling Yang, and Michael M Goodin. 2006. "Global Impact: Elucidating Plant Responses to Viral Infection."*/1207 MPMI* 19 [11]: 1207–15. doi:10.1094/MPMI.

[18] Holmes, Francis O. 1929. "Local Lesions in Tobacco Mosaic." *Botanical Gazette* 87 [1]. University of Chicago Press: 39–55. doi:10.1086/333923.

[19] Culver, James N, and Meenu S Padmanabhan. 2008. "Virus-Induced Disease: Altering Host Physiology One Interaction at a Time." *Annual Review of Phytopathology*. doi: 10.1146/annurev.phyto.45.062806.094422.

[20] Mur, Luis AJ, Paul Kenton, Amanda J Lloyd, Helen Ougham, and Elena Prats. 2008. "The Hypersensitive Response; The Centenary Is upon Us but How Much Do We Know?" In *Journal of Experimental Botany*, 59:501–20. doi:10.1093/jxb/erm239.

[21] Tripathi, Jaindra N, Valentine O Ntui, Mily Ron, Samwel K Muiruri, Anne Britt, and Leena Tripathi. 2019. "CRISPR/Cas9 Editing of Endogenous *Banana Streak Virus* in the B Genome of *Musa Spp*. Overcomes a Major Challenge in Banana Breeding." *Communications Biology*. Springer US, 1–11. doi:10.1038/s42003-019-0288-7.

[22] Ntui, Valentine Otang, Jaindra Nath Tripathi, and Leena Tripathi. 2020. "Robust CRISPR/Cas9 Mediated Genome Editing Tool for Banana and Plantain [*Musa Spp*.]." *Current Plant Biology* 21 [January]. Elsevier B.V. doi: 10.1016/j.cpb.2019.100128.

[23] Herschlag, Rachel, Cesar Escalante, Eliezer Rodrigues de Souto, Surasak Khankhum, Ryo Okada, and Rodrigo A. Valverde. 2019. "Occurrence of Putative *Endornaviruses* in Non-Cultivated Plant Species in South Louisiana." *Archives of Virology* 164 [7]. Springer-Verlag Wien: 1863–68. doi:10.1007/s00705-019-04270-5.

[24] Gorovits, Rena, Iris Sobol, Miasser Altaleb, Henryk Czosnek, and Ghandi Anfoka. 2019. "Taking Advantage of a Pathogen: Understanding How a Virus Alleviates Plant Stress Response." *Phytopathology Research*. BioMed Central Ltd. doi:10.1186/s42483-019-0028-4.

[25] Peng, Junhua, Anna-Maria Botha-Oberholster, Deepa Jaganathan, Gayatri Venkataraman, Karthikeyan Ramasamy, Gothandapani Sellamuthu, and Shilpha Jayabalan. 2018. *"CRISPR for Crop Improvement: An Update Review*." doi:10.3389/fpls.2018.00985.

[26] Baltes, Nicholas J, Aaron W Hummel, Eva Konecna, Radim Cegan, Aaron N Bruns, David M Bisaro, and Daniel F Voytas. 2015. "Conferring Resistance to *Geminiviruses* with the CRISPR/Cas Prokaryotic Immune System." *Nature Plants* 1 [10]. doi:10.1038/nplants.2015.145.

[27] Ali, Zahir, Aala Abul-faraj, Lixin Li, Neha Ghosh, Marek Piatek, Ali Mahjoub, Mustapha Aouida, Piatek, A, Baltes, NJ, Voytas, DF, Dinesh-Kumar, S, & Mahfouz, MM. 2015. "Efficient Virus-Mediated Genome Editing in Plants Using the CRISPR/Cas9 System." *Molecular Plant*. England. doi: 10.1016/j.molp.2015.02.011.

[28] Zhang, Tong, Qiufeng Zheng, Xin Yi, Hong An, Yaling Zhao, Siqi Ma, and Guohui Zhou. 2018. "Establishing RNA Virus Resistance in Plants by Harnessing CRISPR Immune System." *Plant Biotechnology Journal* 16 [8]: 1415–23. doi:10.1111/pbi.12881.

[29] Aman, Rashid, Zahir Ali, Haroon Butt, Ahmed Mahas, Fatima Aljedaani, Muhammad Zuhaib Khan, Shouwei Ding, and Magdy Mahfouz. 2017. "RNA Virus Interference via CRISPR/Cas13a System in Plants." *BioRxiv*. doi:10.1101/213645.

[30] Sanfaçon, Hélène. 2015a. "Plant Translation Factors and Virus Resistance." *Viruses* 7: 3392–3419. doi:10.3390/v7072778.

[31] Pyott, Douglas E, Emma Sheehan, and Attila Molnar. 2016. "Engineering of CRISPR/Cas9-Mediated *Potyvirus* Resistance in Transgene-Free Arabidopsis Plants." *Molecular Plant Pathology* 17 [8]. Blackwell Publishing Ltd: 1276–88. doi:10.1111/mpp.12417.

[32] Ji, Xiang, Huawei Zhang, Yi Zhang, Yanpeng Wang, and Caixia Gao. 2015. "Establishing a CRISPR – Cas-like Immune System Conferring DNA Virus Resistance in Plants." *Nature Plants*, no. September. Nature Publishing Group: 5–9. doi:10.1038/nplants.2015.144.

[33] Yin, Kangquan, Ting Han, Ke Xie, Jinping Zhao, Junqi Song, and Yule Liu. 2019. "Engineer Complete Resistance to *Cotton Leaf Curl Multan Virus* by the CRISPR/Cas9 System in Nicotiana Benthamiana." *Phytopathology Research*., 1–9.

[34] Khan, Zulqurnain, Sultan Habibullah Khan, Aftab Ahmad, Sabin Aslam, and Muhammad Salman Mubarik. 2019. "CRISPR/DCas9-Mediated Inhibition of Replication of *Begomoviruses*." *International Journal of Agriculture and Biology* 21: 711–18. doi:10.17957/IJAB/15.0948.

[35] Kis, András, Éva Hamar, Gergely Tholt, Rita Bán, and Zoltán Havelda. 2019. "Creating Highly Efficient Resistance against *Wheat Dwarf Virus* in Barley by Employing CRISPR/Cas9 System." *Plant Biotechnology Journal*. doi:10.1111/pbi.13077.

[36] Tashkandi, Manal, Zahir Ali, Fatimah Aljedaani, Ashwag Shami, and Magdy M Mahfouz. 2018. "Engineering Resistance against *Tomato Yellow Leaf Curl Virus* via the CRISPR/Cas9 System in Tomato." *Plant Signaling & Behavior* 13 [10]: e1525996. doi:10.1080/15592324.2018.1525996.

[37] Liu, Haijie, Cara L Soyars, Jianhui Li, Qili Fei, Guijuan He, Brenda A Peterson, Blake C Meyers, Zachary L Nimchuk, and Xiaofeng Wang. 2018. "CRISPR/Cas9-Mediated Resistance to *Cauliflower Mosaic Virus*." *Plant Direct* 2 [3]: e00047. doi:10.1002/pld3.47.

[38] Zhang, Tong, Yaling Zhao, Jiajie Ye, Xue Cao, Chenhui Xu, Biao Chen, Hong An, Jiao, Y, Zhang, F, Yang, X, & Zhou, G. 2019. "Establishing CRISPR/Cas13a Immune System Conferring RNA Virus Resistance in Both Dicot and Monocot Plants." *Plant Biotechnology Journal* 17: 1185–87. doi:10.1111/pbi.13095.

[39] Ali, Zahir, Shakila Ali, Manal Tashkandi, Syed Shan-e-ali Zaidi, and Magdy M Mahfouz. 2016. "CRISPR / Cas9-Mediated Immunity to *Geminiviruses*: Differential Interference and Evasion." *Nature Publishing Group*, no. May. Nature Publishing Group. doi:10.1038/srep26912.

[40] Zaidi, Syed Shan-E-Ali, Manal Tashkandi, Shahid Mansoor, and Magdy M Mahfouz. 2016. "Engineering Plant Immunity: Using CRISPR/Cas9 to Generate Virus Resistance." *Frontiers in Plant Science* 7: 1673. doi:10.3389/fpls.2016.01673.

[41] Chandrasekaran, Jeyabharathy, Marina Brumin, Dalia Wolf, Diana Leibman, Chen Klap, Mali Pearlsman, Amir Sherman, Tzahi Arazi, and Amit Gal-On. 2016a. "Development of Broad Virus Resistance in Non-Transgenic Cucumber Using CRISPR/Cas9 Technology." *Molecular Plant Pathology* 17 [7]: 1140–53. doi:10.1111/mpp.12375.

[42] Shimatani, Z., Kashojiya, S., Takayama, M., Terada, R., Arazoe, T., Ishii, H., Teramura, H., Yamamoto, T., Komatsu, H., Miura, K. and Ezura, H., 2017. Targeted base editing in rice and tomato using a CRISPR-Cas9 cytidine deaminase fusion. *Nature Biotechnology*, *35*[5], pp.441-443.

[43] Bastet, Anna, Delyan Zafirov, Nathalie Giovinazzo, Anouchka Guyon-Debast, Fabien Nogué, Christophe Robaglia, and Jean-Luc Gallois. 2019. "Mimicking Natural Polymorphism in EIF4E by CRISPR/Cas9 Base Editing Is Associated with Resistance to *Potyviruses*." *Plant Biotechnology Journal* 17 [9]: 1736–50. doi:10.1111/pbi.13096.

[44] Gomez, Michael A, Z Daniel Lin, Theodore Moll, Raj Deepika Chauhan, Luke Hayden, Kelley Renninger, Getu Beyene, Taylor, NJ, Carrington, JC, Staskawicz, BJ, & Bart, RS. 2019. "Simultaneous CRISPR/Cas9-Mediated Editing of Cassava EIF4E Isoforms NCBP-1 and NCBP-2 Reduces Cassava Brown Streak Disease Symptom Severity and Incidence." *Plant Biotechnology Journal* 17 [2]: 421–34. doi:10.1111/pbi.12987.

[45] Ran, Yidong, Zhen Liang, and Caixia Gao. 2017. "Current and Future Editing Reagent Delivery Systems for Plant Genome Editing." *Science China. Life Sciences* 60 [5]. China: 490–505. doi:10.1007/s11427-017-9022-1.

[46] Shaw, Jane, Andrew J Love, Svetlana S Makarova, Natalia O Kalinina, Bryan D Harrison, and Michael E Taliansky. 2014. "Coilin, the Signature Protein of Cajal Bodies, Differentially Modulates the Interactions of Plants with Viruses in Widely Different Taxa." *Nucleus [Austin, Tex.]* 5 [1]: 85–94. doi:10.4161/nucl.28315.

[47] Makhotenko, AV, Khromov, AV, Snigir, EA, Makarova, SS, Makarov, VV, Suprunova, TP, Kalinina, NO and Taliansky, ME. 2019. "Functional Analysis of Coilin in Virus Resistance and Stress Tolerance of Potato *Solanum Tuberosum* Using CRISPR/Cas9 Editing." *Doklady. Biochemistry and Biophysics* 484 [1]. Russia [Federation]: 88–91. doi:10.1134/S1607672919010241.

[48] Nester, Eugene W. 2015. "Agrobacterium: Nature's Genetic Engineer." *Frontiers in Microbiology* 5 [January]: 1–16. doi:10.3389/fpls.2014.00730.

[49] Jia, Hongge, and Nian Wang. 2014. "Targeted Genome Editing of Sweet Orange Using Cas9/SgRNA." *PloS One* 9 [4]: e93806. doi: 10.1371/journal.pone.0093806.

[50] Wang, Mugui, Yuming Lu, José Ramón Botella, Yanfei Mao, Kai Hua, and Jian-Kang Zhu. 2017. "Gene Targeting by Homology-Directed Repair in Rice Using a *Geminivirus*-Based CRISPR/Cas9 System." *Molecular Plant*. England. doi: 10.1016/j.molp.2017.03.002.

[51] Yin, Kangquan, Ting Han, Guang Liu, Tianyuan Chen, Ying Wang, Alice Yunzi L Yu, and Yule Liu. 2015. "A Geminivirus-Based Guide RNA Delivery System for CRISPR/Cas9 Mediated Plant Genome Editing." *Scientific Reports* 5 [1]: 14926. doi:10.1038/srep14926.

[52] Cai, Yupeng, Li Chen, Xiujie Liu, Shi Sun, Cunxiang Wu, Bingjun Jiang, Tianfu Han, and Wensheng Hou. 2015. "CRISPR/Cas9-Mediated Genome Editing in Soybean Hairy Roots." *PloS One* 10 [8]: e0136064. doi: 10.1371/journal.pone.0136064.

[53] Sun, Xianjun, Zheng Hu, Rui Chen, Qiyang Jiang, Guohua Song, Hui Zhang, and Yajun Xi. 2015. "Targeted Mutagenesis in Soybean Using the CRISPR/Cas9 System." *Scientific Reports* 5 [1]: 10342. doi:10.1038/srep10342.

[54] Woo, Je Wook, Jungeun Kim, Soon Il Kwon, Claudia Corvalán, Seung Woo Cho, Hyeran Kim, Sang-Gyu Kim, Sang-Tae Kim, Sunghwa Choe, and Jin-Soo Kim. 2015. "DNA-Free Genome Editing in Plants with Preassembled CRISPR/Cas9 Ribonucleoproteins." *Nature Biotechnology* 33 [11]: 1162–64. doi:10.1038/nbt.3389.

[55] Malnoy, Mickael, Roberto Viola, Min-hee Jung, Ok-jae Koo, Seokjoong Kim, Jin-soo Kim, Riccardo Velasco, and Chidananda Nagamangala Kanchiswamy. 2016. "DNA-Free Genetically Edited Grapevine and Apple Protoplast Using CRISPR/Cas9 Ribonucleoproteins." *Frontiers in Microbiology* 7 [December]: 1–9. doi:10.3389/fpls.2016.01904.

[56] Andersson, Mariette, Helle Turesson, Alessandro Nicolia, Ann-Sofie Fält, Mathias Samuelsson, and Per Hofvander. 2017. "Efficient Targeted Multiallelic Mutagenesis in Tetraploid Potato [*Solanum Tuberosum*] by Transient CRISPR/Cas9 Expression in Protoplasts." *Plant Cell Reports* 36 [1]: 117–28. doi:10.1007/s00299-016-2062-3.

[57] Gil-Humanes, Javier, Yanpeng Wang, Zhen Liang, Qiwei Shan, Carmen V Ozuna, Susana Sánchez-León, Nicholas J Baltes, Starker, C, Barro, F, Gao, C, & Voytas, DF. 2017. "High-Efficiency Gene Targeting in Hexaploid Wheat Using DNA Replicons and CRISPR/Cas9." *The Plant Journal: For Cell and Molecular Biology* 89 [6]: 1251–62. doi:10.1111/tpj.13446.

[58] Shan, Q., Wang, Y., Li, J. and Gao, C., 2014. Genome editing in rice and wheat using the CRISPR/Cas system. *Nature Protocols*, *9*[10], pp.2395-2410.

[59] Wang, Y., Cheng, X., Shan, Q., Zhang, Y., Liu, J., Gao, C. and Qiu, J.L., 2014. Simultaneous editing of three homoeoalleles in hexaploid bread wheat confers heritable resistance to powdery mildew. *Nature Biotechnology*, *32*[9], pp.947-951.

[60] Li, Zhongsen, Zhan-Bin Liu, Aiqiu Xing, Bryan P Moon, Jessica P Koellhoffer, Lingxia Huang, R Timothy Ward, Elizabeth Clifton, S Carl Falco, and A Mark Cigan. 2015. "Cas9-Guide RNA Directed Genome Editing in Soybean." *Plant Physiology* 169 [2]: 960–70. doi:10.1104/pp.15.00783.

[61] Svitashev, Sergei, Christine Schwartz, Brian Lenderts, Joshua K Young, and A Mark Cigan. 2016. "Genome Editing in Maize Directed by CRISPR–Cas9 Ribonucleoprotein Complexes." *Nature Communications* 7 [1]: 13274. doi:10.1038/ncomms13274.

[62] Liang, Zhen, Kunling Chen, Tingdong Li, Yi Zhang, Yanpeng Wang, Qian Zhao, Jinxing Liu, Zhang, H, Liu, C, Ran, Y, & Gao, C. 2017. "Efficient DNA-Free Genome Editing of Bread Wheat Using CRISPR/Cas9 Ribonucleoprotein Complexes." *Nature Communications* 8 [1]: 14261. doi:10.1038/ncomms14261.

[63] Kuluev, BR, Gumerova, GR, Mikhaylova, EV, Gerashchenkov, GA, Rozhnova, NA, Vershinina, ZR, Khyazev, AV, Matniyazov, RT, Baymiev, An. Kh.,. Baymiev Al. Kh & Chemeris. AV. 2019. "Delivery of CRISPR/Cas Components into Higher Plant Cells for Genome Editing." *Russian Journal of Plant Physiology* 66: 694–706.

[64] Zaidi, Syed Shan-e-ali, and Shahid Mansoor. 2017. "Viral Vectors for Plant Genome Engineering." *Frontiers in Plant Science* 8 [April]: 539. doi:10.3389/fpls.2017.00539.

[65] Tomáš Čermák, Nicholas J. Baltes, Radim Čegan, Yong Zhang & Daniel F. Voytas. 2015. "High-Frequency, Precise Modification of the Tomato Genome." *Genome Biology* 16: 232. doi:10.1186/s13059-015-0796-9.

[66] Cody, Will B, Herman B Scholthof, and T Erik Mirkov. 2017. "Multiplexed Gene Editing and Protein Overexpression Using a *Tobacco Mosaic Virus* Viral Vector." *Plant Physiology* 175 [1]: 23–35. doi:10.1104/pp.17.00411.

[67] Kaya, Hidetaka, Kazuhiro Ishibashi, and Seiichi Toki. 2017. "A Split Staphylococcus Aureus Cas9 as a Compact Genome-Editing Tool in Plants." *Plant and Cell Physiology* 58 [4]. Oxford University Press: 643–49. doi:10.1093/pcp/pcx034.

[68] Gootenberg, Jonathan S, Omar O Abudayyeh, Jeong Wook Lee, Patrick Essletzbichler, Aaron J Dy, Julia Joung, Vanessa Verdine, Donghia, N, Daringer, NM, Freije, CA, Myhrvold, C, Bhattacharyya, RP, Livny, J, Regev, A, Koonin, EV, Hung, DT, Sabeti, PC, Collins, JJ, & Zhang, F. 2017. "Nucleic Acid Detection with CRISPR/Cas13a/C2c2." *Science [New York, N.Y.]* 356 [6336]: 438–42. doi:10.1126/science. aam9321.

[69] Gootenberg, Jonathan S, Omar O Abudayyeh, Max J Kellner, Julia Joung, James J Collins, and Feng Zhang. 2018. "Multiplexed and Portable Nucleic Acid Detection Platform with Cas13, Cas12a, and Csm6." *Science [New York, N.Y.]* 360 [6387]: 439–44. doi:10.1126/science. aaq0179.

[70] Ai, Jing-Wen, Xian Zhou, Teng Xu, Minling Yang, Yuanyuan Chen, Gui-Qing He, Ningp Pan, Cai, Y, Li, Y, Wang, X, Su, H, Wang, T, Zeng, W, & Zhang, WH. 2019. "CRISPR-Based Rapid and Ultra-Sensitive Diagnostic Test for *Mycobacterium tuberculosis*." *Emerging Microbes & Infections* 8 [1]: 1361–69. doi:10.1080/22221751.2019.1664939.

[71] Li, Linxian, Shiyuan Li, Na Wu, Jiacheng Wu, Gang Wang, Guoping Zhao, and Jin Wang. 2019. "HOLMESv2: A CRISPR/Cas12b-Assisted Platform for Nucleic Acid Detection and DNA Methylation Quantitation." *ACS Synthetic Biology* 8 [10]. United States: 2228–37. doi:10.1021/acssynbio.9b00209.

[72] Myhrvold, Cameron, Catherine A Freije, Jonathan S Gootenberg, Omar O Abudayyeh, Hayden C Metsky, Ann F Durbin, Max J Kellner, Tan, AL, Paul, LM, Parham, LA, Garcia, KF, Barnes, KG, Chak, B, Mondini, A, Nogueira, ML, Isern, S, Michael, SF, Lorenzana, I, Yozwiak, NL, Sabeti, PC. 2018. "Field-Deployable Viral Diagnostics Using CRISPR/Cas13." *Science [New York, N.Y.]* 360 [6387]: 444–48. doi:10.1126/science. aas8836.

[73] Abudayyeh, Omar O, Jonathan S Gootenberg, Max J Kellner, and Feng Zhang. 2019. "Nucleic Acid Detection of Plant Genes Using CRISPR/Cas13." *The CRISPR Journal* 2 [3]: 165–71. doi:10.1089/crispr.2019.0011.

[74] Ma, Hanhui, Li-Chun Tu, Ardalan Naseri, Maximiliaan Huisman, Shaojie Zhang, David Grunwald, and Thoru Pederson. 2016. "Multiplexed Labeling of Genomic Loci with DCas9 and Engineered SgRNAs Using CRISPRainbow." *Nature Biotechnology* 34 [5]: 528–30. doi:10.1038/nbt.3526.

[75] Abudayyeh, Omar O, Jonathan S Gootenberg, Patrick Essletzbichler, Shuo Han, Julia Joung, Joseph J Belanto, Vanessa Verdine, Cox, DBT, Kellner, MJ, Regev, A, Lander, ES, Voytas, DF, Ting, AY, & Zhang, F. 2017. "RNA Targeting with CRISPR/Cas13." *Nature* 550 [7675]: 280–84. doi:10.1038/nature24049.

[76] Baltes, Nicholas J, Javier Gil-Humanes, Tomas Cermak, Paul A Atkins, and Daniel F Voytas. 2014. "DNA Replicons for Plant Genome Engineering." *Plant Cell* 26 [1]. American Society of Plant Biologists: 151–63. doi:10.1105/tpc.113.119792.

[77] Ali, Zahir, Aala Abulfaraj, Ali Idris, Shakila Ali, Manal Tashkandi, and Magdy M. Mahfouz. 2015. "CRISPR/Cas9-Mediated Viral Interference in Plants." *Genome Biology* 16 [1]. Genome Biology: 1–11. doi:10.1186/s13059-015-0799-6.

Chapter 9

CRISPR-Cas9-Based Genome Editing for Disease Resistant Plants

Poornima Pandey*
Mansi Srivastava
Preeti Sharma
Narendra Kumar Sharma
and Pracheta Janmeda
Department of Bioscience and Biotechnology, Banasthali Vidyapith, Tonk, Rajasthan, India

Abstract

Agricultural biotechnology is an extension of genetic engineering which enables us to make changes to the genome of the plant for wide ranges of application. From changing the colour, size and aroma of ornamental flowers to creating disease resistance in the plant, it is all possible through genome editing. Plant diseases are the biggest threat to crop yields which lead to food shortage, shooting up of costs and various other consequences. Various disease resistance genes in plants are found in its genome that protects it against various pathogens in natural state and other genes that pre-dispose the plant to pathogenic attack can be edited out of the genome to impart resistance.

With the rise in technology, use of CRISPR systems to edit the genome of the plant provides us with an easier alternate to knock out – knock in or base edit enzymes and other proteins. The CRISPR system allows for specific sequence dependent endonuclease activity causing double stranded DNA breaks which can be recombined by endogenous machineries of the cell while introducing mutations in the coding as well as non-coding regions of the genome. The advancements in the field make it possible to reduce the off-target sites for nuclease making it more efficient. In this chapter, techniques used for various plants to make them resistant to different types of diseases have been briefly outlined.

Keywords: agricultural biotechnology, genetic engineering, CRISPR, DNA sequences

---

* Corresponding Author's Email: purnimapandey@banasthali.in.

In: Genome Editing for Crop Improvement
Editors: Rahul Kumar, Ashutosh Sharma, Yogesh Kumar et al.
ISBN: 979-8-88697-740-0

Introduction

With the biggest hunger crisis of the century, the entire human population gravely depends on plants as a source of food. Plants constantly fall victim to environmental stresses like cold, drought, salt, and ROS etc. Various biotic elements like bacteria, viruses, fungi, oomycetes, nematodes, insects, and even some plants of parasitic nature are the causative agents of plant diseases. The filamentous growth of fungi and oomycetes are the prominent cause of host infection in plants (Dong et al., 2015). For many years, scientists have been trying to figure out how plants interact with populations of bacteria, fungus, and other microbes. A multilayer defense system is activated during pathogen attacks to combat invading micro-organisms. Defense mechanisms work at both exogenous levels and endogenous levels with the help of special molecular receptors. One of these set of receptors present on the surface of the cell are known as pattern recognition receptors (PRR) that recognize microbial surface glycoproteins conserved in pathogens known as PAMPs (Pathogen associated molecular patterns) to initiate the immune response. This response leads to the trigger of the PRR triggered immunity (PTI). A disease-resistant R gene encodes for intracellular nucleotide-binding domains in the cytoplasm of the plant cell. A leucine-rich domain, and NOD-like receptors (NLRs) are also present within the cell that mediate the pathogen effectors pathway leading to a diseased condition of the host by encouraging the growth of the pathogen using plant machinery (Langner et al., 2018).

Traditional breeding efforts in the pre-genomic era were predicated to identifying the mutant alleles. These alleles that showcased resistance were then imbibed in the crops using plant breeding technologies. However, there are several drawbacks, including high labour costs, genetic heterogeneity, and induction of unwanted genes along with desirable traits.

The modern era of edited genome tools has transformed the field of research for plant variety improvement. Precision DNA modification has been made possible because of the revelation of programmable nucleases. Use of nucleases that can mediate double stranded breaks in the DNA such as transcription activator life effector nucleases (TALENs), meganucleases, zinc finger nuclease (ZFNs) and, clustered regularly interspaced palindromic repeats (CRISPR) are proving to be the greatest discovery which has allowed us to overcome the constraints of plant breeding technology(Cermak et al., 2011; Gao et al., 2010; Li et al., 2012; Nekrasov et al., 2013; Shan et al., 2013; Townsend et al., 2009). After the breaks are made, host cell repair systems carry out their processes using indels of nucleotides. These indels can be near homologous end-joining (NHEJ), microhomology-mediated end joining (MMEJ), or a gene modification by homologous recombination (HR) (Mohanta et al., 2017).

The rise in CRISPR technology suggests that it is feasible to adopt because of the increased success rate and easy to use nature of the tool. Multiple positive features of plants have been modified using CRISPR-Cas9, including morphological changes and disease resistance (Borrelli et al., 2018; Langner et al., 2018; Yin & Qiu, 2019; Zaynab et al., 2020). This chapter will thus outline the mechanism of CRISPR-Cas9, advantages, applications in imparting resistance towards diseases caused by several pathogens and the limitations of this technique.

CRISPR-Cas9

CRISPR-Cas is found in microbes as an adaptive immunity mechanism that acts on foreign genomic elements by cleaving them using RNA-guided nucleases. Many CRISPR systems compatible with various bacterial and archaeal hosts have been reported in the literature. All these systems consist of Cas gene clusters, repetitive elements such as direct repeats and few RNAs (non-coding). CRISPR-RNA also termed crRNA is a compilation of repeats that are interspaced due to the presence of short variable sequences. A DNA target, called a protospacer, is also a part of the system, which is accompanied by a Protospacer Adjacent Motif (PAM) that acts as a cleavage site for CRISPR to act on. It varies vividly from one system to another (Ran et al., 2013). After the discovery of class II adaptive immunity in bacteria Streptococcus pyogenes, CRISPR-Cas nucleases are known to be highly accessible.(Barrangou et al., 2007; Bhaya et al., 2011; Sorek et al., 2013).

CRISPR-Cas9 was foremost used in human cells, followed by applications in varied species of plants and fungi for genome editing. Cas9 endonuclease and a single guide RNA known as sgRNA are the basic requirements for CRISPR-Cas9 editing. sgRNA binds the target DNA as per the base-pairing rules. Cas9 has a bi-lobed structure with nuclease domains and PAM binding domain. A large REC lobe acts as a recognition site and NUC segment with the nucleases occur as lobes of Cas9.

The complex of Cas9-sgRNA finds standard G-rich PAM (5-NGG-3) sequence in the DNA double helix, as well as possible complementarity in PAM sequences about guide RNA. The heteroduplex of gRNA-DNA is produced, then Cas9 cleaves the DNA target which allows for destabilization of the tDNA and forms a seed area of about 8 to 12 bp (Langner et al., 2018). The PAM sequences of Staphylococcus aureus (SaCas9), Staphylococcus thermophilus (StCas9), and Neisseria meningitides (NmCas9) are all orthologs of SpCas9. SaCas9 requires a 5-NNGRRT-3 PAM sequence (R = A or G) and StCas9 needs a 5-NNGGAA-3 PAM sequence (Nishimasu et al., 2014). NmCas9 necessitates a unique PAM: 5 -NNNNGATT-3 or 5'-NNNNGCTT-3 (Hou et al., 2013). A flowchart of the steps involved in genome editing using CRISPR is described in Figure 1.

Generating fertile plants from CRISPR edited variants is the biggest challenge in realizing the full potential of this technology(Altpeter et al., 2016). The effectiveness of the tool for plants is determined by species and genotype, which means that only a few lab variations exist in crop species; CRISPR-Cas9 technology is available only to them. The quality of the CRISPR-Cas9 design is the important facet, along with the approach used to introduce it into the cell of a plant. The introduction of CRISPR-Cas9 tools can be accomplished by:

(i) Cas along with sgRNA segments transformation into the targeted genome by Agrobacterium-mediated transformation (AMT).
(ii) Introducing a ribonucleoprotein (RNP) complex into plant cells via PEG-mediated protoplast transformation to ensure plant genome editing without the use of DNA (Figure 2).
(iii) RNP complex biolistic transport into plant embryo cells. Herbicide selective markers like bialaphos resistance gene (bar), kanamycin resistance gene (kanR), or hygromycin resistance gene (hph), could be used to select transformants carrying the expression cassette.

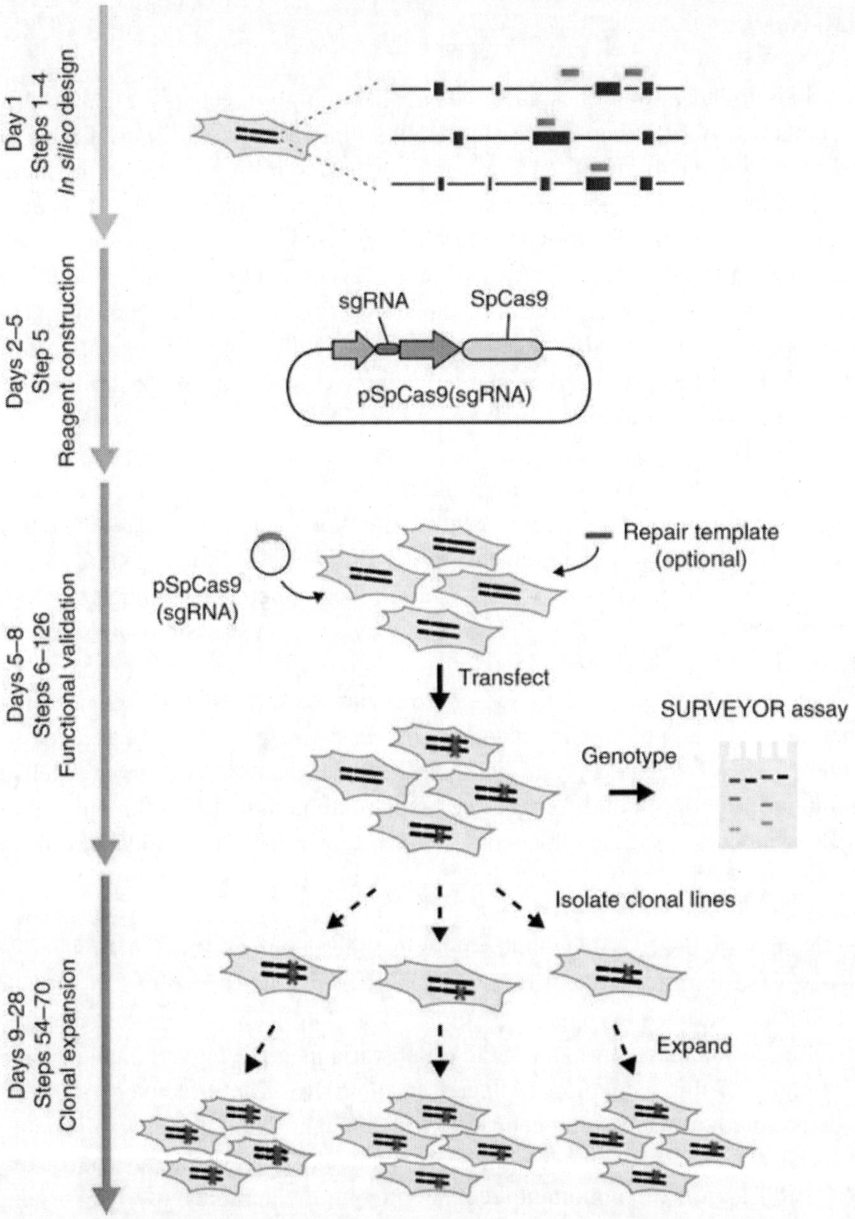

Figure 1. Flowchart of CRISPR-Cas steps of the procedure adopted from (Ran et al., 2013).

The last two systems provide an upper hand by transferring both-geneCas9 along with the sgRNA and allowing ribonucleoproteins (RNPs) to enter, which are assembled *in vitro* (Liang et al., 2017; Malnoy et al., 2016; Svitashev et al., 2016). Both the transfer techniques make it possible to easily raise rmDNA (repair matrix) to CRISPR DNA ratio (Borrelli et al., 2018).

When compared to mutagenesis induced chemically or via irradiation, it is found to be highly specific which has raised the question of whether guide RNAs can only target complementary DNA (genomic) sequences or it can also be applied to off-target region cleavage using CRISPR tools to avoid unwanted domains and traits. Scientists and regulatory bodies are concerned about two sorts of off-target effects: predicted off-targets in regions of the genome with substantial similarity in the sequences to the target, and unanticipated off-targets found in distinct genome regions. Predicted off-targets are dealt with using PCR amplification followed by sequencing of sections similar to the target, while the unexpected off-targets requires whole genome sequencing. CRISPR-Cas9 is preferred over mega nucleases, ZFNs and others because of its ease with which numerous genes are targeted with a single molecular constant simultaneously. For example, in a study on Arabidopsis species, a single construct of CRISPR-Cas9 was found to mutate 14 distinct genes (Peterson et al., 2016).

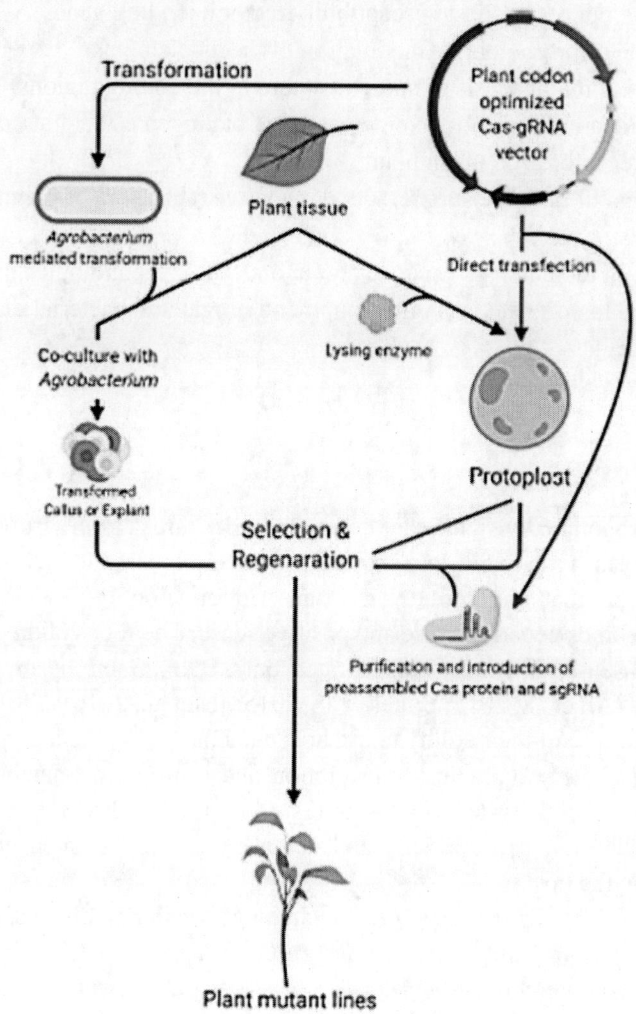

Figure 2. Modification of plant genome using CRISPR-Cas by *Agrobacterium*-mediated transformation adopted from (Paul et al., 2021).

Plant Disease Resistance Using CRISPR–Cas9

Plant diseases are mostly conferred by abiotic/biotic stress stimuli leading to a decrease in plant yields. Microorganisms act as a biotic stressors which account for 20-40% of global agricultural output losses. With such obstacles on the way, feeding the ever-rising population of the world can be done by transforming the plants into resistant variants; this will diminish the impact of disease on improved crops. Improvements in genome editing tools have created a path for enhancing crop resistance. The CRISPR-Cas system has helped address several agricultural obstacles in recent years, including boosting biotic stress resistance (Arora & Narula, 2017).

According to a study the CRISPR-Cas system can be used to study plant disease resistance in a variety of ways (Schenke & Cai, 2020). This includes:

(i) Knocking out of S genes (susceptibility factor encoding genes).
(ii) Alteration of promoters by modulation of thecis-elements.
(iii) Introducing mutagenesis of specific nature in the coding regions.
(iv) Modifying amino acids to avoid evasion of pathogen effectors secreted by surface receptor proteins in plants.
(v) Knocking out immune suppressors or negative regulators of plants defensive responses.
(vi) Changes in the defense response regulators. These technologies have primarily been used to combat virus infection, mend fungal and bacterial disease resistance.

Viruses

A large number of economically significant crops are under threat from viruses. Based on the type of genome, the plant viruses fall into six subgroups. Single-stranded DNA viruses are the most prevalent; viruses that undergo reverse transcription; double-stranded RNA, single-stranded RNA of two kinds negative sense and positive sense. The focus of the research studies for virus mediated diseases has been concentrated on ssDNA *Geminivirus* genome editing through CRISPR/Cas (Ali et al., 2015; Baltes et al., 2015; Jia et al., 2016). Viruses are the most dreaded pathogens because of their ability to use host machinery for translation; therefore, they hijack the plant's machinery including transcription and translational factors and use it to replicate themselves, and proliferate. As a result, resistance to viral diseases can be imparted to host plants by knocking out pathogenicity and defense regulation related genes or altering susceptibility genes in the host.

In 2015, the first-ever CRISPR-Cas based technology was implemented for viral disease resistance in plants. *Geminiviruses* like *Beet Severe Curly Top Virus* (BSCTC), *Tomato Leaf Curl Virus* (TYLCV) and *Bean Yellow Dwarf Virus* (BeYDV), were all studied by three independent research groups in model plants of Arabidopsis and Nicotiana. Both coding as well as non-coding portions of the viral genome were edited using the CRISPR-Cas system targeting nucleases and gRNA. This editing was done on capsid proteins, and replication protein, resulting in fewer symptoms and enhanced viral resistance (Ali et al., 2016; Baltes et al., 2015; Hirano et al., 2016). The researchers confirmed that the CRISPR technology is effective in inducing specific mutations in viruses, making them less able to multiply and improving plant resilience. Non-coding domains were found more efficient in inferring resistance than coding because of better adaptability of the virus to evolve by changing coding sequences in the coding areas, leaving a potential target of conserved non-coding regions (Tyagi et al., 2021).

Fungi

Fungal plant pathogens cause a variety of plant diseases around the world, including rot, smut, rust, mildew, and others, resulting in significant production losses in crop yields. In addition to illnesses, several fungal infections produce mycotoxins which are secondary metabolites that present major health hazards in both animals and humans, which can also be fatal in some cases. Response to fungal attacks mediated by effector-triggered immunity in plants; this ETI is usually activated in plant-fungal interactions when plant R proteins and pathogen effectors interact, which leads to activation of the plant defense mechanism. Conventionally R gene insertion in the plant genome has been the most promising approach of acquired immunity for plant disease resistance but it lacks lasting effects due to the absence of pathogen specificity. Fungi encase genetic plasticity as well, which allows for the pathogen to disrupt resistance genes altering the disease and making it more difficult to control pathogenicity. This results in short-lived resistance conferred via this technique which fungi easily overcome. However, modifying sensitive short genes like domains that contribute to disease development and genes that act as targets for pathogen mediated cascades, can be a more realistic method of increasing resistance.

To begin the infection process, fungi release enzymes targeting the cell wall proteins, which disrupt the cell wall, allowing the entry of the pathogen into the host. Plants release inhibitors of these enzymes in response to fungal pathogen invasion. Other processes, such as thickening of the cell wall due to callose production, also help in containing the infection. Overexpressing or modifying genes related to the above two mechanisms might be a promising target for the development of fungus resistance. This technique was used to edit the plants genome, and impart resistance to the powdery mildew pathogen Oidium neolycopersic. This resistance was attained by targeting the Powdery Mildew Resistance 4 (PMR4) gene ortholog SlPMR4 (PRR gene) responsible for callose deposition. The inversion mutation causes loss of function mutation of the PMR4 gene that enhanced the level of salicylic acid produced and led to activation of the hypersensitive response in plants upon pathogen exposure, conferring resistance. For long term resistance, cell-wall disintegrating enzymes can be inhibited, overexpression of PRR, and/or defective S genes can be the mediator of changes that increase disease resistance in plants towards fungal pathogens (Tyagi et al., 2021).

Bacteria

Bacteria are diverse, ubiquitous, multiply rapidly, and perform an important (beneficial or destructive) function in the biological system. Plant infection mediating bacteria are a huge concern for agricultural practices because they are the cause of a variety of diseases like spots, mosaics, and rot. They are difficult to contain, resulting in significant production losses. Pathogens of various kinds elicit varied reactions in their hosts, making resistance development difficult. Antibiotic resistance in bacteria is the most prominent concern while mediating resistance, as their genome is constantly evolving and transmitting the evolving factors via horizontal gene transfer. The entry of bacteria into the cell happens through openings of stomata, stigma and others, or artificial wounds, or by secreting bioactive chemicals. This entry elicits signals that are recognized by the plant defense system, which responds properly to eradicate the infection. Successful pathogens, on the other hand, use a complicated signaling cascade in which host plant genes, including several S genes, play a role, allowing pathogens to overcome plant protection and commence infection.

As previously stated, the resistance granted by the S genes are considered more lasting, hence acts as a great target for modification to provide bacterial disease resistance. However, progress in the production of bacterial disease-resistant cultivars has been slow due to the complexity of the plant genome and ever-evolving bacteria. Even then there has been less research using the CRISPR-Cas9 system that generates transgenic plants resistant to bacterial pathogens than there has been for fungal and viral pathogens.

Citrus canker disease, by the causative agent *Xanthomonas citri*, causes severe yield loss in citrus plants. To achieve resistance to *Xanthomonas citri*, the CRISPR-Cas9 technology was deployed. It works by modifying one of the ETI target genes. CsLOB1 gene promoter, the effector binding element (EBE) PthA4, was found to be involved in host susceptibility. Thus, by altering the promoter the potential to recognize and retaliate towards bacterial secretions was lost, resulting in enhanced resistance to infection. Peng et al., Peng et al., 2017 established a correlation between CsLOB1 promoter activity in Wanjincheng (orange) towards its susceptibility to citrus greening bacteria (CBC) disease (Peng et al., 2017). The knockout of the whole EBEPthA4 gene from CsLOB1 genes increased the CBC resistance. CsLOB1 promoter modified plants did not show any change in morphology (Borrelli et al., 2018). There is a dire need for more work in the area of bacterial resistance and CRISPR-Cas, for this has been an unexplored arena.

S Genes Edited with CRISPR-Cas9

Mildew Resistance Locus O

Mildew resistance locus O (MLO), is a conventional susceptibility factor, that has been found to play its role in the pathogenesis of powdery mildew pathogen *Blumeria graminis* f. sp. hordei found in barley, with wild-type alleles conferring sensitivity to all known isolates. The gene has been discovered in the genomes of higher plants, where it is found in several families of plants. MLO is a plasma membrane-associated protein that contains a cytoplasmic tail for calmodulin binding. A structural link to G-protein-coupled receptors has been found in metazoans. The role of this protein in powdery mildew susceptibility has been verified in both monocots and dicots.

In *Arabidopsis thaliana*, partial resistance is mediated against several powdery mildew species (Golovinomyces sp.) by Atmlo2 loss-of-function alleles. MLO-dependent susceptibility is also found in pepper. MLO2, here, has also been linked to bacterial pathogen *Xanthomonas campestris* susceptibility.

These conserved MLO genes led to the initiation of the use of endonucleases to mediate deletion of MLO. The conserved regions present in MLO exon 2 was targeted by the nucleases. Two guided RNAs were used to modify T0 plants. SIMlo1 locus was altered leading to a 48 bp deletion. Disease resistance studies on the Tomelo primary progeny (T1) indicated response by conferring resistance to the fungus, *Oidium neolycopersici*, with no pleiotropic implications on plant development or fruit output. Whole-genome sequencing was used to study the transgenic lines thus formed. It was found that the Tomelo segregants were transgene-free, without any off-target mismutations. This indicates the specificity and efficacy of CRISPR-Cas for disease resistance (Langner et al., 2018).

Rice Ethylene Response Factor 922

In rice, the transcription factor is encoded by OsERF922 which is an ethylene response factor (ERF) deriving from the APETALA2 superfamily. These factors play roles in the metabolic pathways of both plant development and disease resistance, like salinity, pathogen response, drought resistance etc. ERFs function as a modulator of diverse stress responses and are found engaged in balancing signaling pathways (agonistic and antagonistic) via hormonal crosstalk mechanisms involving abscisic acid, jasmonic acid, and ethylene. These factors are transcriptional regulators that are specific in plants. AP2/ERF domains were found to be acquired horizontally, from endonucleases of bacteria and viruses. There are 164 AP2/ERF genes in the rice genome, all of which belong to the ERF subclade. In monocots, the repeats and variations of AP2/ERFs vary from 53 in barley to over 180 in maize. In the japonica rice variety Kuiku131, CRISPR-Cas9 genome editing was used to create indels adjacent to the initiation codon of OsERF922 mRNA. Six homozygous T2 mutant lines with no transgene resistance to the M. oryzae were increased by several frameshift indels. Edited agronomic variables- flag leaf length, plant height, and leaf area were also investigated. No plants that were observed with OsERF922 mutations displayed significantly different features than wild-type plants, demonstrating that OsERF922 can be modified to generate plants with increased resilience without compromising plant growth (Langner et al., 2018).

Eukaryotic Initiation Factor 4E

The translation initiation complex contains the EIF4E (eukaryotic initiation factor), which has a binding site for 5' 7-methylguanylate caps of mRNAs and eIF4G together. This binding leads to the promotion of recruitment of the 40s ribosomal subunit at the mRNA 5' end to eventually cause initiation of translation. Plant eIF4Es has been discovered to be important components of viral infection susceptibility. Alleles with a natural resistance have since been discovered, and they now make up the biggest group of virus resistance genes in both monocotyledonous and dicotyledonous plants.

eIF4E-dependent resistance in cucumbers has been imparted with genetic disruption using CRISPR-Cas. eIF4E and eIF(iso)4E both are encoded in the cucumber genome, and they share 56% nucleotide identity. The use of specific sgRNAs to target eIF4E exons 1 and 3 resulted in modest deletions in both of the non-homologous targeted sites. To create homozygous T3 plant lines, it was crossed with heterozygous T1 plants. The tested lines showed an increase in resistance to a variety of viruses such as vein yellowing viruses of cucumber, the yellow mosaic virus of zucchini and ring spot mosaic virus of papaya. The heterozygous lines were highly susceptible, due to the recessive nature of the resistance mediated (Langner et al., 2018).

Pathogens Resistant Plants

The modified list of various pathogens resistant plants that are genetically modified by CRISPR-Cas9 tool (Borrelli et al., 2018) are as follows:

Rolling Circle Mechanism

Nicotiana benthamiana: BeYDV, TYLCV BCTV MEMV and CLCUKov MMV (target genes: CP. Rep. IR, and LIR).

Replication Mechanism

Nicotiana benthamiana: TUMV, CMV, and TMV (target genes: ORF, GFP1, GFP2, HC-Pro. CP).

Factors Found in Host for RNA Viruses Translation

- *Cucumis sativus:* CVYV ZYMV PRSV-W (target gene-1F4E)
- *Arabidopsis thaliana:* TUMV (elftarget gene- iso 4E)
- *Oryza sativa L. japonica*: RTSV (target gene- elf 4G).

Susceptibility (S) Gene Involved in Powdery Mildew Disease

Triticum aestivum, Solanum lycopersicum and Vitis vinifera: Powdery mildew (*Blumeria graminis*, *Oidium neolycopersici* and *Erysiphe necator*): (target gene: MLO-A1, MLO1, and MLO-7).

Stress Related Transcription Factors

- *Vitis vinifera:* Gray mold (*Botrytis cinerea*, target gene-WAKY52).
- *Oryza sativa L. japonica:* Rice blast disease (*Magnaporthe oryzae*, target gene-ERF922).

Regulator of the Immune System

Theobroma cacao: Black pod disease (*Phytophthora tropicalis*, target gene- NPR3).

Subunit of the Exocyst Complex

Oryza sativa L. japonica: Rice blast disease (*Magnaporthe oryzae*, target gene- SEC3A).

Sucrose Transporter Gene

Oryza sativa: Bacterial blight (*Xanthomonas oryzae* pv. oryzae, target gene- SWEET13).

Susceptibility (S) Gene Promoting Pathogen Growth and Pustule Formation

Citrus paradisi and *Citrus sinensis* Osbeck*:* Citrus canker (*Xanthomonas citri* subspecies *citric*, target gene- LOB1).

Factor Involved in Fire Blight Disease

Malus domestica: Fire blight (*Erwinia amylovora*, target genes- DIPM-1 DIPM-2 DIPM-4).

Conclusion

Revolution in the field of science like food production, crop improvement etc. is greatly influenced by genome editing, especially after the introduction of CRISPR-Cas9. Use of CRISPR systems for making plants tolerant and resistant to various stresses has made life easier for those in farming sector, but there is still a long way to go. Such advancements are very much essential to feed the need for an increasing global population. Using such a versatile method that can improve crop variety one should also keep in mind the various ethical, biosafety and societal apprehensions.

There is an urgent need to introduce of general awareness about ethical usage and regulated mechanisms for gene editing Recent advancement has highlighted the importance of CRISPR-Cas9 for genome engineering. Plant breeding has gained a lot more advantage as various Cas9-based applied techniques are being utilized in plants that help in the proper and clear understanding of Plant physiology and thus enhancing the commercial value of a plant. Undeniably, the CRISPR-Cas9 system is one of the most influential tools in crop genetic modification.

References

Ali, Z., Abulfaraj, A., Idris, A., Ali, S., Tashkandi, M., & Mahfouz, M. M. (2015). CRISPR/Cas9-mediated viral interference in plants. *Genome Biol.*, *16*, 238. https://doi.org/10.1186/s13059-015-0799-6.

Ali, Z., Ali, S., Tashkandi, M., Zaidi, S. S., & Mahfouz, M. M. (2016). CRISPR/Cas9-Mediated Immunity to Geminiviruses: Differential Interference and Evasion. *Sci Rep.*, *6*, 26912. https://doi.org/10.1038/srep26912.

Altpeter, F., Springer, N. M., Bartley, L. E., Blechl, A. E., Brutnell, T. P., Citovsky, V., Conrad, L. J., Gelvin, S. B., Jackson, D. P., Kausch, A. P., Lemaux, P. G., Medford, J. I., Orozco-Cardenas, M. L., Tricoli, D. M., Van Eck, J., Voytas, D. F., Walbot, V., Wang, K., Zhang, Z. J., & Stewart Jr., C. N. (2016). Advancing Crop Transformation in the Era of Genome Editing. *Plant Cell*, *28*(7), 1510–1520. https://doi.org/10.1105/tpc.16.00196.

Arora, L., & Narula, A. (2017). Gene Editing and Crop Improvement Using CRISPR-Cas9 System. *Front. Plant Sci.*, *8*, 1932. https://doi.org/10.3389/fpls.2017.01932.

Baltes, N. J., Hummel, A. W., Konecna, E., Cegan, R., Bruns, A. N., Bisaro, D. M., & Voytas, D. F. (2015). Conferring resistance to geminiviruses with the CRISPR-Cas prokaryotic immune system. *Nat. Plants*, *1*(10). https://doi.org/10.1038/nplants.2015.145.

Barrangou, R., Fremaux, C., Deveau, H., Richards, M., Boyaval, P., Moineau, S., Romero, D. A., & Horvath, P. (2007). CRISPR provides acquired resistance against viruses in prokaryotes. *Science*, *315*(5819), 1709–1712. https://doi.org/10.1126/science.1138140.

Bhaya, D., Davison, M., & Barrangou, R. (2011). CRISPR-Cas systems in bacteria and archaea: versatile small RNAs for adaptive defence and regulation. *Annu. Rev. Genet.*, *45*, 273–297. https://doi.org/10.1146/annurev-genet-110410-132430.

Borrelli, V. M. G., Brambilla, V., Rogowsky, P., Marocco, A., & Lanubile, A. (2018). The Enhancement of Plant Disease Resistance Using CRISPR/Cas9 Technology. *Front. Plant Sci.*, *9*, 1245. https://doi.org/10.3389/fpls.2018.01245.

Cermak, T., Doyle, E. L., Christian, M., Wang, L., Zhang, Y., Schmidt, C., Baller, J. A., Somia, N. V, Bogdanove, A. J., & Voytas, D. F. (2011). Efficient design and assembly of custom TALEN and other TAL effector-based constructs for DNA targeting. *Nucleic Acids Res.*, *39*(12), e82. https://doi.org/10.1093/ nar/gkr218.

Dong, S., Raffaele, S., & Kamoun, S. (2015). The two-speed genomes of filamentous pathogens: waltz with plants. *Curr. Opin. Genet. Dev.*, *35*, 57–65. https://doi.org/10.1016/j.gde.2015.09.001.

Gao, H., Smith, J., Yang, M., Jones, S., Djukanovic, V., Nicholson, M. G., West, A., Bidney, D., Falco, S. C., Jantz, D., & Lyznik, L. A. (2010). Heritable targeted mutagenesis in maize using a designed endonuclease. *Plant. J.*, *61*(1), 176–187. https://doi.org/10.1111/j.1365-313X.2009.04041.x.

Hirano, H., Gootenberg, J. S., Horii, T., Abudayyeh, O. O., Kimura, M., Hsu, P. D., Nakane, T., Ishitani, R., Hatada, I., Zhang, F., Nishimasu, H., & Nureki, O. (2016). Structure and Engineering of Francisella novicida Cas9. *Cell*, *164*(5), 950–961. https://doi.org/10.1016/j.cell.2016.01.039.

Hou, Z., Zhang, Y., Propson, N. E., Howden, S. E., Chu, L. F., Sontheimer, E. J., & Thomson, J. A. (2013). Efficient genome engineering in human pluripotent stem cells using Cas9 from Neisseria meningitidis. *Proc. Natl. Acad. Sci. U S A*, *110*(39), 15644–15649. https://doi.org/10.1073/pnas.1313587110.

Jia, H., Orbovic, V., Jones, J. B., & Wang, N. (2016). Modification of the PthA4 effector binding elements in Type I CsLOB1 promoter using Cas9/sgRNA to produce transgenic Duncan grapefruit alleviating XccDeltapthA4:dCsLOB1.3 infection. *Plant. Biotechnol. J.*, *14*(5), 1291–1301. https://doi.org/10.1111/pbi.12495.

Langner, T., Kamoun, S., & Belhaj, K. (2018). CRISPR Crops: Plant Genome Editing Toward Disease Resistance. *Annu. Rev. Phytopathol.*, *56*, 479–512. https://doi.org/10.1146/annurev-phyto-080417-050158.

Li, T., Liu, B., Spalding, M. H., Weeks, D. P., & Yang, B. (2012). High-efficiency TALEN-based gene editing produces disease-resistant rice. *Nat. Biotechnol.*, *30*(5), 390–392. https://doi.org/10.1038/nbt.2199.

Liang, Z., Chen, K., Li, T., Zhang, Y., Wang, Y., Zhao, Q., Liu, J., Zhang, H., Liu, C., Ran, Y., & Gao, C. (2017). Efficient DNA-free genome editing of bread wheat using CRISPR/Cas9 ribonucleoprotein complexes. *Nat. Commun.*, *8*, 14261. https://doi.org/10.1038/ncomms14261.

Malnoy, M., Viola, R., Jung, M. H., Koo, O. J., Kim, S., Kim, J. S., Velasco, R., & Nagamangala Kanchiswamy, C. (2016). DNA-Free Genetically Edited Grapevine and Apple Protoplast Using CRISPR/Cas9 Ribonucleoproteins. *Front. Plant. Sci.*, *7*, 1904. https://doi.org/10.3389/fpls.2016.01904.

Mohanta, T. K., Bashir, T., Hashem, A., Abd Allah, E. F., & Bae, H. (2017). Genome Editing Tools in Plants. *Genes (Basel)*, *8*(12). https://doi.org/10.3390/genes8120399.

Nekrasov, V., Staskawicz, B., Weigel, D., Jones, J. D., & Kamoun, S. (2013). Targeted mutagenesis in the model plant Nicotiana benthamiana using Cas9 RNA-guided endonuclease. *Nat. Biotechnol.*, *31*(8), 691–693. https://doi.org/10.1038/nbt.2655.

Nishimasu, H., Ran, F. A., Hsu, P. D., Konermann, S., Shehata, S. I., Dohmae, N., Ishitani, R., Zhang, F., & Nureki, O. (2014). Crystal structure of Cas9 in complex with guide RNA and target DNA. *Cell*, *156*(5), 935–949. https://doi.org/10.1016/j.cell.2014.02.001.

Paul, N. C., Park, S. W., Liu, H., Choi, S., Ma, J., MacCready, J. S., Chilvers, M. I., & Sang, H. (2021). Plant and Fungal Genome Editing to Enhance Plant Disease Resistance Using the CRISPR/Cas9 System. *Frontiers in Plant Science*, *12*(August), 1–10. https://doi.org/10.3389/fpls.2021.700925.

Peng, A., Chen, S., Lei, T., Xu, L., He, Y., Wu, L., Yao, L., & Zou, X. (2017). Engineering canker-resistant plants through CRISPR/Cas9-targeted editing of the susceptibility gene CsLOB1 promoter in citrus. *Plant. Biotechnol. J.*, *15*(12), 1509–1519. https://doi.org/10.1111/pbi.12733.

Peterson, B. A., Haak, D. C., Nishimura, M. T., Teixeira, P. J., James, S. R., Dangl, J. L., & Nimchuk, Z. L. (2016). Genome-Wide Assessment of Efficiency and Specificity in CRISPR/Cas9 Mediated Multiple Site Targeting in Arabidopsis. *PLoS One*, *11*(9), e0162169. https://doi.org/10.1371/journal.pone.0162169.

Ran, F. A., Hsu, P. D., Wright, J., Agarwala, V., Scott, D. A., & Zhang, F. (2013). Genome engineering using the CRISPR-Cas9 system. *Nat. Protoc.*, *8*(11), 2281–2308. https://doi.org/10.1038/nprot.2013.143.

Schenke, D., & Cai, D. (2020). Applications of CRISPR/Cas to Improve Crop Disease Resistance: Beyond Inactivation of Susceptibility Factors. *IScience*, *23*(9), 101478. https://doi.org/10.1016/j.isci.2020.101478.

Shan, Q., Wang, Y., Li, J., Zhang, Y., Chen, K., Liang, Z., Zhang, K., Liu, J., Xi, J. J., Qiu, J. L., & Gao, C. (2013). Targeted genome modification of crop plants using a CRISPR-Cas system. *Nat. Biotechnol.*, *31*(8), 686–688. https://doi.org/10.1038/nbt.2650.

Sorek, R., Lawrence, C. M., & Wiedenheft, B. (2013). CRISPR-mediated adaptive immune systems in bacteria and archaea. *Annu Rev Biochem*, *82*, 237–266. https://doi.org/10.1146/annurev-biochem-072911-172315.

Svitashev, S., Schwartz, C., Lenderts, B., Young, J. K., & Mark Cigan, A. (2016). Genome editing in maize directed by CRISPR-Cas9 ribonucleoprotein complexes. *Nat. Commun.*, *7*, 13274. https://doi.org/10.1038/ ncomms13274.

Townsend, J. A., Wright, D. A., Winfrey, R. J., Fu, F., Maeder, M. L., Joung, J. K., & Voytas, D. F. (2009). High-frequency modification of plant genes using engineered zinc-finger nucleases. *Nature*, *459*(7245), 442–445. https://doi.org/10.1038/nature07845.

Tyagi, S., Kumar, R., Kumar, V., Won, S. Y., & Shukla, P. (2021). Engineering disease resistant plants through CRISPR-Cas9 technology. *GM Crops Food*, *12*(1), 125–144. https://doi.org/10.1080/21645698. 2020.1831729.

Yin, K., & Qiu, J. L. (2019). Genome editing for plant disease resistance: applications and perspectives. *Philos Trans. R. Soc. Lond. B Biol. Sci.*, *374*(1767), 20180322. https://doi.org/10.1098/rstb.2018.0322.

Zaynab, M., Sharif, Y., Fatima, M., Afzal, M. Z., Aslam, M. M., Raza, M. F., Anwar, M., Raza, M. A., Sajjad, N., Yang, X., & Li, S. (2020). CRISPR/Cas9 to generate plant immunity against pathogen. *Microb. Pathog.*, *141*, 103996. https://doi.org/10.1016/j.micpath.2020.103996.

Chapter 10

A Genome Editing Approach to Enhance Host Resistance against Plant-Parasitic Weeds

Vinay Kumar Bari*

Department of Biochemistry, School of Basic Sciences,
Central University of Punjab, VPO-Ghudda, Bathinda, Punjab, India

Abstract

Root-specific plant-parasitic weeds *Phelipanche* and *Orobanche* spp. (broomrape) induce potential damage to numerous crops and lead to heavy loss to the economy due to a reduction in total yield, especially in the Mediterranean, Europe, Africa and Asian countries. The lack of effective control measures restricts our capabilities to avoid newly developing and more virulent races of parasitic weeds, hence, innovative solutions are urgently needed to control these parasites. Significant research towards deciphering the plant genome and its functions have been made over the years, including the genome of the root parasitic weed *Phelipanche aegyptiaca*. Recent biotechnological advancements in genome manipulation such as silencing RNA molecules and T-DNA insertions have been significantly used for introducing resistance against several parasitic weeds. However, the available strategies to generate host resistance to the plant-parasitic weed need further advancement.

Several different methods for the control of parasitic weeds (such as herbicide, and breeding resistant variety crops) have been adopted in attempts to reduce germination of *Orobanche* commonly known as broomrapes, but several difficulties arise in targeting specific plant–plants (host plant and parasitic plant) interaction systems. Moreover, most control strategies to manage parasitic weeds have considerable drawbacks and are less effective. Recently developed gene-editing technology CRISPR/Cas9 has been used in numerous crops to enhance desired plant phenotypes and the same technique can also be employed to target the essential key processes of host-parasite interactions, such as strigolactone (SL) biosynthesis, signalling, haustorium development, and penetration of the host cell wall. A major attempt has been made to edit host genes rather than parasite-specific genes using CRISPR/Cas9 since the germination of these parasites is itself induced by the host. In this chapter, we summarize and discuss the genome editing approach used

* Corresponding Author's Email: vinay.bari@cup.edu.in.

In: Genome Editing for Crop Improvement
Editors: Rahul Kumar, Ashutosh Sharma, Yogesh Kumar et al.
ISBN: 979-8-88697-740-0

to edit the host-specific genes which confer resistance to root-specific plant-parasitic weeds.

Introduction

A parasitic plant affects the growth and development of the host, which disturbs the balance between them in terms of competition [1]. Parasitic plants utilize a unique heterotrophic lifestyle and their existence depends on either a complete or incomplete host because they are unable to live autotrophically, either partially or fully [2]. It is possible to define them as obligate in the early stages of development, or as facultative at an advanced stage. Different forms of adaptation enable the parasitic weed to survive on its host plant. In addition to aerial stem parasitic weeds such as *Cuscuta* and *Arceuthobium*, there are also underground parasites such as *Orobanche* spp. and *P. aegyptiaca* which target mainly host roots [2]. Holoparasitic obligate plant parasites lack photosynthetic apparatus but can be hemiparasites with chlorophyllous organelles [3]. During parasitic infection, the vascular system (xylem and phloem) of host plants is penetrated by pathogens using a special root-like structure known as haustorium, which absorbs water and organic nutrients from the hosts [4]. From the tropical rainforests at the equator to the arctic at the North Pole, different types of parasitism occur, which comprise one percent of the angiosperm population including 270 genera [5]. These parasitic plants can appear as herbs (*Rhinanthus*), vines (*Cuscuta*), shrubs (*Mistletoes*), and even big trees (*Okoubaka aubrevillei*) [6]. The genera of parasitic weeds, *Orobanche spp.*(broomrapes) and *Phelipanche* belong to the *Orobanchaceae* family, consisting of more than a hundred species, and represent one of the most devastating weeds and pose biggest economic challenges in agriculture [7]. Members of *Orobanche* are obligate plant-parasitic weeds that attack mainly through the roots of almost all economically important crops belonging to Asteraceae, Solanaceae, Cucurbitaceae, Fabaceae, Apiaceae, and Brassicaceae plant families [8].

The Life Cycle of Plant-Parasitic Weeds

Plant-parasitic weeds have evolved with sophisticated systems for detecting the presence of host roots in the surrounding areas and regulating their development in coordination with the host [9–11]. The initial stages of growth are very crucial for parasitic weed survival because a germinated parasite seedling that fails to establish a connection with an appropriate host will die [12]. The life cycle of *P. aegyptiaca* is mainly divided into two different parts; first, a period of pre-conditioning that requires small amounts of specific germination inducers produced by the host plant roots (Figure 1). These root exudates are normally produced in small quantities by the host plants. In the next step, the parasite requires a haustorium initiation factor, which induces the radicle tip of the parasite to re-differentiate into a structure known as haustorium that penetrates the host root tissue [4, 13]. Haustorium is the characteristic feature of the parasitic species that separates it from non-parasitic plants [14]. This structure forms the physical and physiological connection between parasite and host tissues which is important for the translocation of organic solute molecules from the host to the parasite. Certain organic molecules (phenolics, quinones, flavonoids, and anthocyanins) are known to induce haustorium development when applied to the parasite radicals [15].

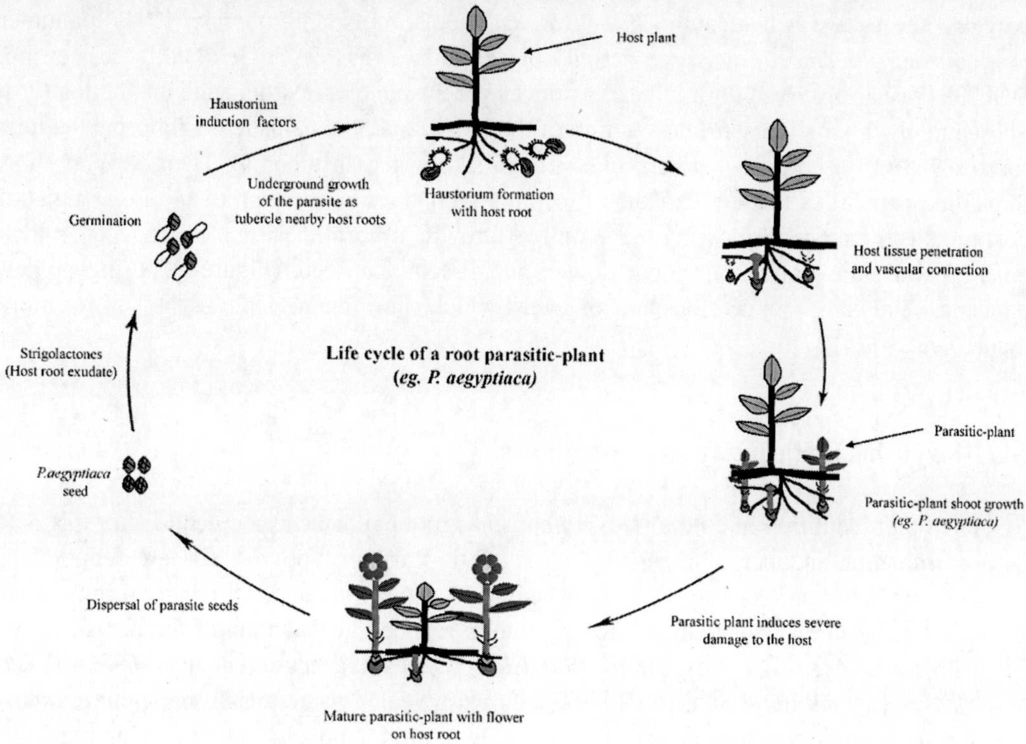

Figure 1. The life cycle of the root parasitic weed *P. aegyptiaca*. The root parasite seeds germinate only in presence of strigolactone (SL)stimulus from a neighboring host root. Subsequently, the root parasite grows as a tubercle that is attached to the host root, and in presence of haustorium-inducing factors, they make a network connection to the host vascular system and withdraw nutrients at the expense of the host. Eventually, a shoot develops from the tubercle, emerges above the soil, flowers, and disseminates seeds.

The first discovered natural haustorium-inducing molecules, 2, 6-dimethoxy-1, 4-benzoquinone (DMBQ), extracted from sorghum root, were found to induce haustoria in parasites within hours of application on their roots [16, 17]. Previous studies have reported two genes encoding a quinone oxidoreductase; TvQR1 (GenBank accession no. AF304461) and TvQR2 (GenBank accession no. AF304462) in *Triphysaria versicolor*, a facultative *Orobanchaceae* parasitic species induced by DMBQ [18]. A decrease in expression of TvQR1, but not TvQR2 significantly correlated with a decrease in the number of haustorium networks developed on the host, suggesting that TvQR1 is involved in haustorium induction [19]. Furthermore, to gain access to host tissue, parasitic plants must have to modify their biological characteristics such as modification in the cell wall, cell expansion, and division of the cortical and epidermal cell layers during interactions with the host plants [20]. Several parasite-specific genes which encode serine and cysteine protease, oxidoreductase, oligopeptide transporter, and glutamate receptors, alter their expression during haustorium induction [20, 21]. During the early stages of host tissue penetration, the parasite releases organic substances that facilitate its attachment to host cell walls, followed by the secretion of lytic enzymes which degrade host cell walls such as pectin methylesterase, polygalacturonase, and rhamnogalacturonase [22, 23].

Furthermore, during interaction with the host plant, increased activity of cellulase and polygalacturonase was detected in *P. aegyptiaca* shoots, roots, and tubercles [24]. In another root parasite, *P. ramosa,* increased activity of peroxidase (*PrPOX1*; GenBank accession no. DQ073917) was observed during the early development stage of host-parasite interaction [25]. Subsequently, after successful penetration of the host tissue, the parasite neighboring the host root grows into a bulbous network-like structure known as a tubercle. Then, several short thread-like structures that are capable of forming secondary haustorium to adjoining areas of host roots can emerge. Following one month of growth, a floral meristem is produced, which emerges above the ground to produce flowers and disseminate seeds (Figure 1). A single plant-parasitic weed can produce thousands of seeds, which can remain viable in the soil for more than 10 years [26, 27].

SL Biosynthesis Pathway in Host Plants

SLs have first identified in cotton (*Gossypium spp.*) root exudates as a crucial factor that acts as a germination inducer of *Striga lutea* [28, 29]. Various types of SLs are reported as germination stimulants for root parasitic weeds, and the decreased SL production in the host conferred resistance against parasitic weeds due to reduced germination of the parasite [28]. The *Carotenoid Cleavage Dioxygenases 7* (*CCD7*) (Genbank accession no. GQ468556.1) and 8 (*CCD8*; GenBank accession no. JF831532.1) wasrevealed to be the SL-deficient mutants, have been obtained in genetic studies that aimed to explore a novel shoot branching regulator [30, 31]. Subsequently, it was revealed that SLs are involved in the inhibition of plant shoot branching and utilized as a signalling molecule by the rhizosphere microflora [30-32].

Several recessive mutants of plants having increased shoot branching greatly helped to understand the SL biosynthesis pathway and reveal that the SL is derived from the plant carotenoids [33, 34]. Subsequently, the involvement of an iron-binding protein *D27* in rice (GenBank accession no. FJ641055) and *AtD27* in *Arabidopsis* was reported to be involved in SL biosynthesis [35, 36]. SLs biosynthesis occurs in all the plant kingdom and its biosynthesis starts with all-trans-*β*-carotene, a carotenoid molecule that produces 9-cis-β-carotene by the activity of enzyme all-trans-*9-cis-β-Carotene Isomerase* (*D27/MAX3*), after that *Carotenoid Cleavage Dioxygenases 7* (*CCD7/MAX3/DAD3/RMS5*) enzyme converts it into 9-cis β–apo 10′-carotenal, and final enzyme *Carotenoid Cleavage Dioxygenases 8 (CCD8/MAX4/ DAD1/RMS1)* leads to the production of carlactone. Carlactone is further oxidized by cytochrome P450 enzymes *(CYP) CYP711A-More Axillary Growth 1 (MAX1)* to yield carlactonoic acid (Figure 2).

Moreover, another cytochrome P450 *CYP722C* acts as an enzyme that catalyzes the direct conversion of carlactonic acid (CLA) to orobanchol in the cowpea and tomato. *CYP711A* and *CYP722C* family enzymes produce canonical SL orobanchol while *GaCYP722C* (*Gossypium arboretum*) and *LjCYP722C* (*Lotus japonicus*) produce strigol type SLs such as 5-deoxystrigol [37]. A recent study found a mutation in a gene, *LOW GERMINATION STIMULANT 1* (*LGS1*; NCBI SRA accession no. SRP098704), which encodes an enzyme that acts as a sulfotransferase [38]. The mutation in *LGS1* causes alteration of major SL in sorghum from 5-deoxystrigol to orobanchol, eventually leading to less induction of *Striga*, thus significantly reducing germination of *Striga* seed [38]. Non-canonical SL methyl-carlactonoic acid was synthesized from carlactonoic acid by unknown methyltransferase in *Arabidopsis* [39]. Methyl-carlactonoic acid was further converted into hydroxy-methyl carlactonoic acid by *LATERAL BRANCHING OXIDOREDUCTASE* (*LBO1;* TAIR accession no. AT3G21420), a gene that encodes an oxidoreductase-like enzyme of the family, belongs to 2-oxoglutarate and Fe(II)-dependent dioxygenase in tomato, *Arabidopsis,* maize, and sorghum [40, 41]. *LjLLD* in *Lotus japonicus* was involved in the synthesis of lotuslactone, a non-canonical SL, which strongly induces germination of *P. ramosa* and *O. minor*, but it is a less effective inducer to *Striga hermonthica* seeds [42].

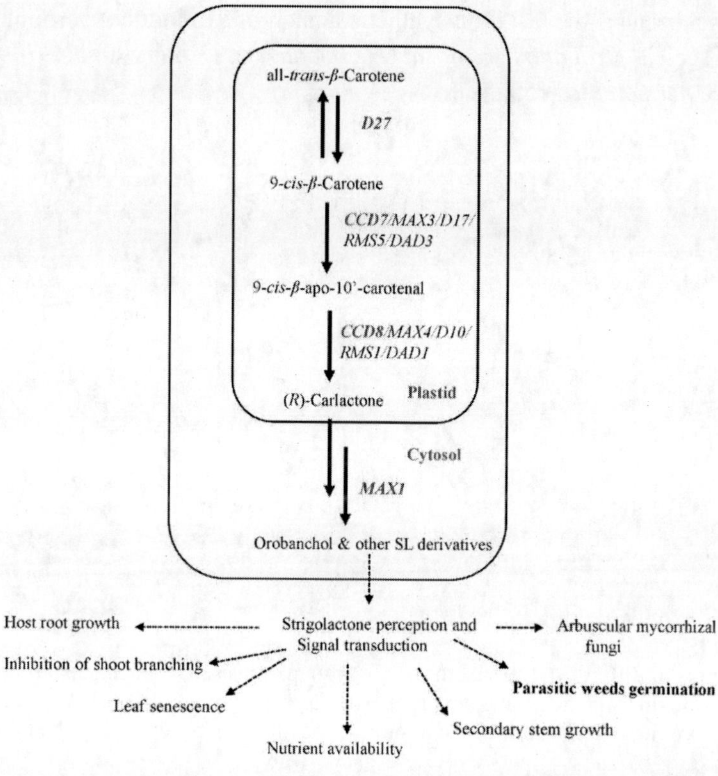

Figure 2. Summary of strigolactone biosynthetic and signalling pathways. The biosynthesis involves one novel iron-containing protein *(D27)* and two carotenoid cleavage dioxygenases, *CCD7 (MAX3, D17, RMS5, DAD3)* and *CCD8 (MAX4, D10, RMS1, DAD1),* one cytochrome P450 monooxygenase *(MAX1)* which is involved in the synthesis of orobanchol and other SL derivatives. The potential downstream components in the SL pathway led to the germination of parasitic weeds and arbuscular mycorrhiza symbiosis after exuding from host roots.

Recently, sphynolactone-7 (SPL7), and SL mimic were developed which exhibited highly potent activity similar to the natural stimulant 5-deoxystrigol [43]. The two fundamental SL biosynthetic pathway enzymes *CCD7* and *CCD8* have been characterized in tomato, however, there is only one *MAX1* homolog present and this is required for oxidation of the CL, an SL precursor molecule [44, 45]. In tomato, many different SLs, such as orobanchol, 4-deoxy orobanchol, solanacol, several didehydro-orobanchol (DDH) isomers, 7-hydroxyorobanchol, orobanchyl acetate, and 7-oxoorobanchol, were identified (Figure 3) [44–46]. SLs are essential to plant hormones responsible for normal plant growth and inhibiting branching and act as intercellular communication between plants and soil microbes. SLs also play a crucial role in the germination of plant-parasitic weeds [47]. However, the biochemical mechanism by which diverse SL is synthesized in tomatoes remains elusive. For the rhizosphere microbial flora, such as symbiotic arbuscular mycorrhizal fungi, SL molecules act as a signal to detect hosts, and also induce germination of plant-parasitic weeds [47, 48].

Host plant resistance to the *Striga,* an obligate root holoparasite, has been observed in crops with reduced SL production. In addition, previous studies found that the tomato plant gene *SL-ORT1* if mutagenized by the fast-neutron technique, displays resistance to *Phelipanche* and *Orobanche spp.*, which makes these plants unable to produce or secrete SLs to the rhizosphere [49]. Several lines of evidence suggest that these germination stimulants are species-dependent and race-specific, because the GR24, a synthetic analog of SL, induces germination in several parasitic weeds such as *Orobanche ramosa, O. cumana,* and *O. minor* up to ~70–90%, but only up to ~50% in *S. hermonthica* [50, 51].

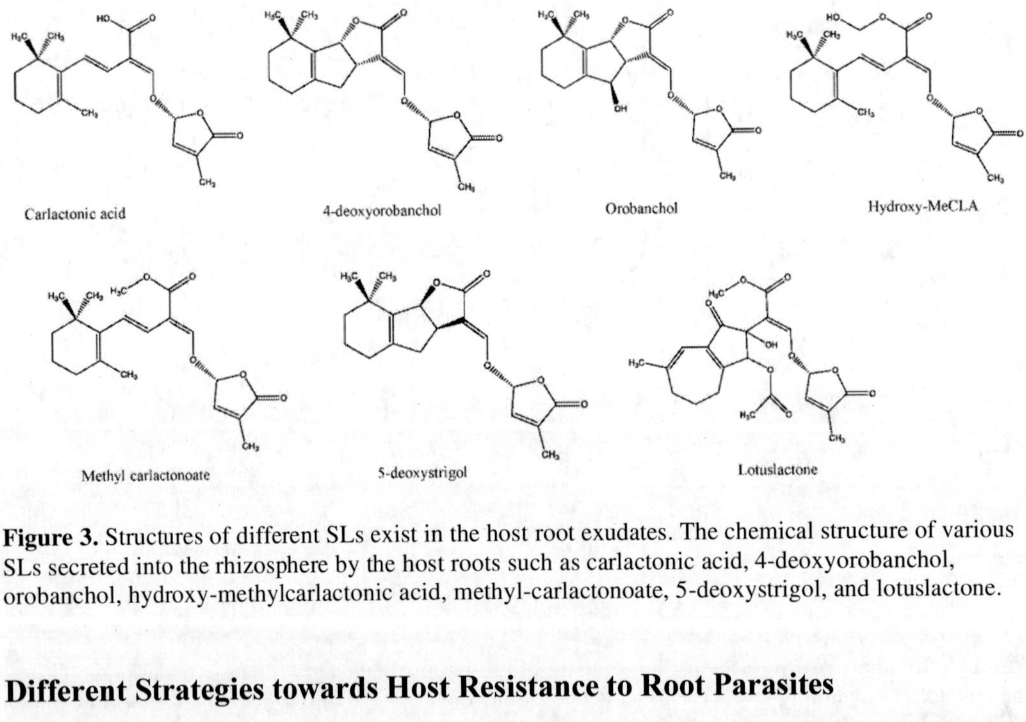

Figure 3. Structures of different SLs exist in the host root exudates. The chemical structure of various SLs secreted into the rhizosphere by the host roots such as carlactonic acid, 4-deoxyorobanchol, orobanchol, hydroxy-methylcarlactonic acid, methyl-carlactonoate, 5-deoxystrigol, and lotuslactone.

Different Strategies towards Host Resistance to Root Parasites

Host plants are continuously exposed to several biotic and abiotic stresses including parasitic weeds [52], and the host plant responds to them by utilizing various defense mechanisms such as synthesizing secondary metabolites, phenolic compounds, and suberization or lignification of their cell wall [53]. Previous studies revealed that invasion of *Orobanche crenata* into the host vascular system was stopped at the cortex layers before penetrating the vascular bundle, concomitantly increase in the accumulation of peroxidase and catalase [54]. Plant hormones such as jasmonic acid and salicylic acid have defense-related functions also known to contribute to host resistance in tomato against stem parasitic weeds such as *C. pentagona* [55]. Application of salicylic acid on red clover roots reduces the number of germinated tubercles of *O. minor via* lignification of the host endodermis cell layer [56]. Several salicylic-induced pathogenesis-related gene transcripts or proteins were shown to be increased in the resistance response of sunflower to *O. cumana* [57]. Moreover, transgenic tobacco host plants expressing antibacterial peptide sarcotoxin IA gene using HMG2 promoter showed increased resistance to *Phelipanche spp.* [58, 59].

Another strategy for long-term control of parasitic weeds is to breed resistant cultivars. However, using conventional methods for breeding to develop resistance against root-specific parasitic weeds in crops has faced lots of difficulties, and in many instances, resistance has proven to be unsuccessful [60, 61]. In addition, the herbicide-based chemical control of these parasitic weeds has been adopted through foliar applications [62]. Unfortunately, only selected herbicides are proven to be effective against these parasites in the fields. Successful weed control using herbicides is possible only when the host plant can sustain the effective amount of herbicide either through natural or genetic modification.

Earlier it was shown that the foliar application of the chlorsulfuron herbicide to transgenic plants with specific target-site mutation is efficient in controlling broomrapes [63]. Currently, two main categories of herbicides are used for the effective control of parasitic weeds: herbicides that target amino acid biosynthesis [64], and those that inhibit folic acid biosynthesis [65]. It was shown that glyphosate applied to infected tomato plants moves promptly to tomato meristematic tissues and further to underground *Phelipanche* and *Orobanche spp.* tubercles, where it accumulates, and inhibits the growth of the parasite [64]. The result demonstrates that the specific mutation allows migration of the herbicide towards parasites from the host, which are connected to its root tissues without any harmful effect on the host. Similarly, transgenic carrot resistance to the herbicide imazapyr allows the movement of non-metabolized herbicide through the host crop to abolish the germination of *P. aegyptiaca* [66]. Several other host plants such as carrot and tobacco cultivars, containing a mutation in (the aceto-hydroxy acid synthase gene (AHAS3R) have been shown to be resistant to the herbicide chlorsulfuron, hence translocation of the herbicide from plant leaves to roots, and successful control of the attached broomrape was observed [66, 67]. Another herbicide glyphosate is translocated in plants and mainly accumulates in young leaves and root meristems. This herbicide is known to inhibit the enzyme 5-enolpyruvylshikimate-3-phosphate synthase (EPSPS), involved in the synthesis of aromatic amino acids [68].

Plants have evolved a wide range of gene-silencing pathways based on siRNAs of 21–24 nucleotide length [69, 70]. siRNAs repress the expression of sequence-specific homologous genes during transcription, post-transcriptional modification, or translation [71]. Intimate physical contact between root parasitic weeds and their hosts is evident in their relationship with the vascular system. Translocation of RNA may also occur between some root parasitic plant species and their hosts, implying that a silencing signal (21–24 nucleotides) can move from host to parasite, and target a parasite gene [72]. In this basic approach, a plasmid encoding a double-stranded small hairpin RNA (ds hpRNA) was used to target key parasitic mRNA molecules for silencing [73]. It has been shown that molecular RNA trafficking between the parasite and host can be delineated [74]. A key gene in mannitol biosynthesis is M6PR (GenBank accession no. U83687), which encodes mannose-6-phosphate reductase. Mannitol formation during *Phelipanche* attachment is vital for nutrient and water absorption from the host plants [75, 76]. Hence, the *Phelipanche* M6PR gene appears to be a potential target for parasite control.

The parasite-specific M6PR gene was silenced in tomato plants using an hpRNA in a previous study and in transgenic plants, the endogenous M6PR mRNA from the tubercles and shoots of *P. aegyptiaca* was found to be drastically reduced due to siRNA movement from the host to the parasite tissues [77]. It has already been noted that more than 30 natural SLs have been discovered and their secretion by the roots of the members of Orobanchaceae is essential for their germination. These molecules are synthesized from plant carotenoids by involving CCD genes [78]. Antisense RNA gene silencing of *CCD7* (GenBank accession no. GQ468556.1) and *CCD8* (GenBank accession no. JF831532.1) transcripts in tomato causes a reduction of germination of the root parasitic weeds [44, 45]. Previous studies suggest that in transgenic tomato plants, silencing the host *CCD8* gene by hpRNA technique reduced infestation by 90% [45].

SL biosynthesis and exudation were reduced in transgenic plants, while arbuscular mycorrhizal symbiosis and fruit yields were barely affected. The tobacco rattle virus-VIGS system has been used to validate candidate genes in *S. hermonthica*, providing insight into parasite development and parasitism [79]. Using a transient knockdown of *CCD7* or *CCD8* in *P. aegyptiaca*, it was shown that parasite-tubercle development and infestation of *Nicotiana benthamiana* plants were significantly suppressed [80]. It has been recently demonstrated that trans-specific gene silencing of parasite genes can be used for the control of root parasitic plants [81]. DNA fragments (*ACS, M6PR,* and *Prx1*) of *P. aegyptiaca* genes were fused and cloned into hpRNA silencing vectors to pyramid the pathways targeted by RNAi. The results showed that exogenous siRNA mobile from the host was transmitted to the parasite, resulting in a reduction in the expression of targeted essential parasitic genes, leading to a significant reduction in parasite tubercles on the host plant [81].

CRISPR-Cas9-Mediated Genome Editing

The advent of CRISPR-Cas9-based genome editing technology has made it easier to manipulate plant genomes, understand gene functions, and improve growth, development, and other desired characteristics [82–84]. The genome editing platform has become a versatile tool for accurate and efficient gene editing in various plants [85, 86]. In comparison with other genome editing methods, it shows superiority because it is based on a specific DNA sequence required for genome editing. When it comes to manipulating traits, CRISPR/Cas9 is very promising and it has been widely used in agriculture with broad prospects [87].

CRISPR-mediated genome editing in host plants mainly includes:

(1) The first step to select desired genomic locus and design single guide (sg)RNA (18-22 nucleotide) to generate desired mutation.
(2) assemble the sgRNA and Cas9 expression components on suitable plant specific binary vector with an appropriate selectable marker.
(3) identification of the desired editing events in transgenics using suitable methods (PCR/RFLP/DNA sequencing) (Figure 4) [88].

There are standard guidelines that need to be considered when designing a guide RNA. First, the genomic loci where the desired mutation needs to be created using Cas9 nuclease must be selected, and second, the target region of the guide RNA (18-21 nucleotides) must be adjacent to a PAM site (5-NGG-3 in the DNA). For best results, a PAM site must be as close to the location of the desired target sequence as possible [89, 90]. To facilitate guide RNA design, several online web tools are available that should be used to determine the on-target specificity of a sgRNA sequence to avoid off-target mutagenesis [91]. If the targeted sgRNA sequences are non-specific, then Cas9 can bind to and create a break in DNA at different places than desired and result in off-target mutations [92].

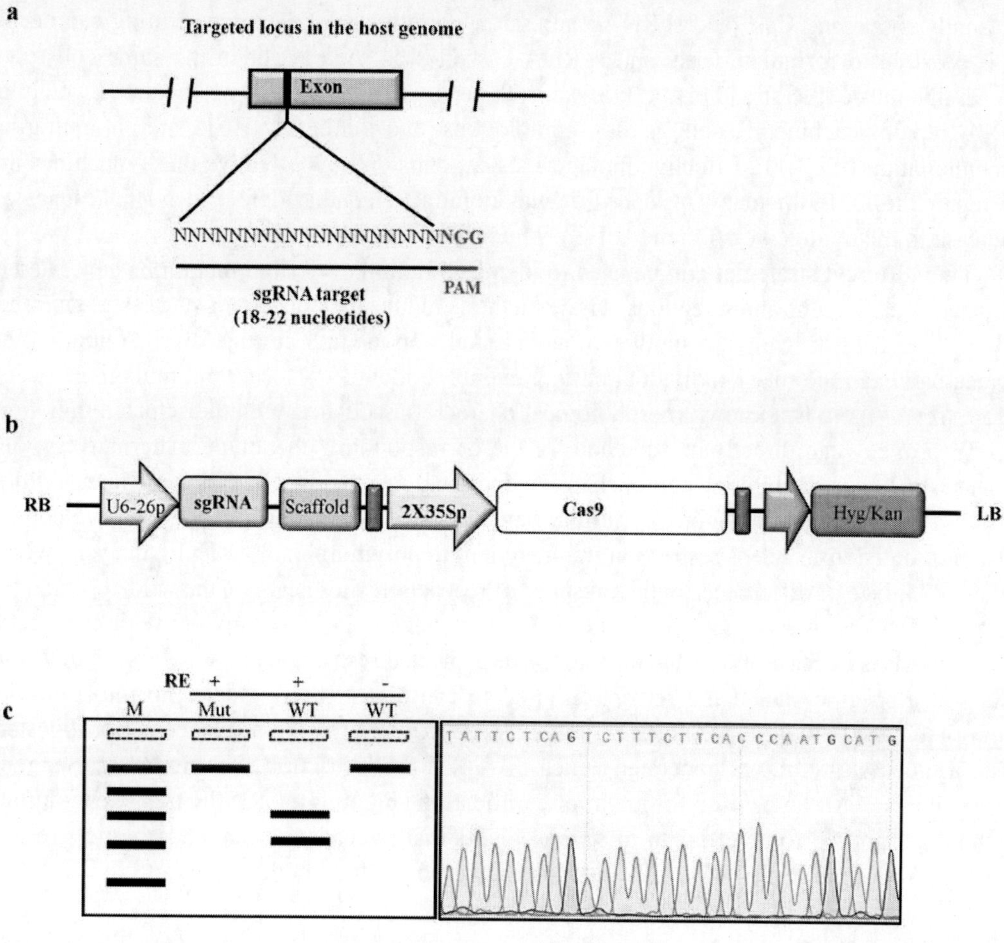

Figure 4. The steps involved in the CRISPR/Cas9-mediated genome editing. The CRISPR/Cas9 system contains two components; the Cas9 nuclease and, a single-guide RNA (sgRNA). (a) First, the gene-specific sgRNA sequence needs to be designed using online available tools. The sgRNA sequence is typically 19-22 nt in length and binds specifically to the target DNA sequence present in the gene of interest usually containing a 5'-NGG-3' PAM motif at the 3' end. (b) Assembly of targeted sgRNA, attached with the scaffold and the Cas9 nuclease on plant binary vector construct with suitable selection markers such as hygromycin or kanamycin which is needed to transform into host plant via different transformation techniques. (c) Screening of Cas9 edited events in plants using either diagnostic polymerase chain reaction or digesting PCR product to identify restriction enzyme site loss in the mutagenized plant, or next-generation/DNA sequencing (M, DNA ladder; WT, wild type control; mut, mutagenized; RE, restrictions enzyme).

Although it is not completely understood what affects guide RNA efficiency, several studies have established characteristics and properties of effective guide sequences, such as the presence of a purine (G /A) at the 3'-end of the target with increased efficiency [93].

CRISPR/Cas9 can introduce mutations directly even in the first generation of rice and tomato plants [94, 95]. CRISPR/Cas9 has proved to be a highly useful tool for gene editing and providing resistance to crops such as *Arabidopsis,* tobacco, rice, maize, tomato, and sorghum [96, 97]. Among plants, the most commonly used CRISPR/Cas9 system type II is made of Cas9 nuclease, tracer RNA, and crRNA. 3' end's NGG PAM target sequence is recognized by Cas9 protein [98]. For CRISPR/Cas9 mediated mutagenesis, plant cells must be incorporated with plasmids containing Cas9 and sgRNA-expression cassettes before genome editing can occur. It is possible to formulate Cas9 and sgRNA cassettes as separate or in the same construct. Generally, the double-strand break (DSB) introduced by Cas9 nuclease gets repaired by cellular DNA repair machinery such as non-homologous end-joining (NHEJ) and homologous recombination [99, 100]. Effective mutagenesis depends on how effective these machines are to repair the DSB. In many instances, it was demonstrated that defect in NHEJ causes an increase in the mutagenic efficiency of Cas9 nuclease [101, 102].

Two different strategies can be used to identify random or desired mutations generated at targeted sites: screens and selections. These methods reduce the concern of successful genome-editing event identification, a significant and crucial step in Cas9 editing [103]. A genotyping screen can examine many individual editing events to identify a successful mutagenic event. Moreover, if there is a known morphological character associated with the selected mutation, the T1 progeny can directly be screened for the corresponding phenotype. Alternatively, one can use the molecular biology approach to screen mutation generated by Cas9- mediated editing [104]. These methods are highly useful when the Cas9 editor generates a few nucleotides insertion/deletions, such as restriction fragment length polymorphisms (RFLP) analysis, which takes advantage of differences in the presence of restriction sites between the wild-type and the mutated DNA sequence generated due to Cas9 mutagenesis. The targeted region used for mutagenesis is PCR amplified and digested with a selected restriction endonuclease, which cuts DNA at a specific site. If the restriction enzyme can digest the fragment, no mutation was generated and the wild-type sequence still existed. However, if the sequence is not digested, then a successful edit was generated which disrupted the restriction site, moreover, Sanger's DNA sequencing can be used for mutation verification (Figure 4). CRISPR/Cas9 gene-editing technology proved to be efficient in several plants that overcome biotic and abiotic stresses, leading to successful plant research and increased crop production.

Cas9-Mediated Editing Approach to Generate Host Resistance to Parasitic Weeds

The selection of suitable host genes to edit plays a critical role in targeting the host-parasitic interactions system. Hence, the selection of target genes in the host should be carefully evaluated. The selected target host gene must be involved in a process that should be essential for the growth of the parasitic weed, and it needs to be lethal for the parasite at an early stage of their life cycle. For example, a gene encoding the SL biosynthesis the of host will reduce the germination parasites and abolish parasitic growth on the host roots [105].

To determine whether SL biosynthetic pathway contributes to host resistance against plant-parasitic weed *P. aegyptiaca*, a gene encoding *Carotenoid cleavage dioxygenase 8* (*Solyc08g066650*) in tomato is targeted using CRISPR/Cas9 system, in which the guide RNA is particularly targeted against the second exon of the host tomato (*Solanum lycopersicum* L.) plant (Table 1) [105]. Although Cas9 mutagenesis drastically affected the phenotypic morphology of the plants such as highly branched shoots due to a block in SL biosynthesis but disruption in the *CCD8* gene greatly enhances host resistance against the pathogenic weed *P. aegyptiaca* as determined by the growth of parasite tubercle on the host root [105]. Studies on engineering plant architecture using CRISPR/Cas9 suggest that the germination ability of parasitic weed *Striga* on *Carotenoid Cleavage Dioxygenase 7 (CCD7)* mutants of rice is reduced to a greater extent compared to the wild-type control plants [106]. Similarly, mutagenesis of *More axillary growth 1* (*MAX1*; *Solyc08g062950*), a gene that is involved in the SL biosynthesis pathway in tomatoes induces apical growth of axillary branches in contrast to *CCD8* which mainly reduces plants height. CRISPR/Cas9-induced mutagenesis in *MAX1* contributes to increased resistance to host tomato plants against the pathogenic weed *P. aegyptiaca* [107].

Table 1. CRISPR/Cas9 mediated targeted mutagenesis of SL biosynthesis pathway, used to generate resistance against plant-parasitic weeds

Host plant species	Gene-edited	Pathway targeted	Target parasitic weed	References
Rice	*CCD7 (LOC_Os04g46470)*	SL biosynthesis	*S. hermonthica*	[106]
Sorghum	*LGS1 (Sobic.005G213600)*	SL biosynthesis	*S. hermonthica*	[136]
Tomato	*CCD8 (Solyc08g066650)*	SL biosynthesis	*P. aegyptiaca*	[105]
	MAX1 (Solyc08g062950)	SL biosynthesis	*P. aegyptiaca*	[107]
	CYP722C (Solyc02g084930)	SL biosynthesis	*S. hermonthica* *O. crenata* *P. aegyptiaca*	[112]
	ABCG44 (Solyc08g067610) and *ABCG45 (Solyc08g067620)*	SL transport	*P. aegyptiaca*	[111]

A significant discovery of SL transport was made by identification of an ATP-binding cassette (ABC)-type transporter *PLEIOTROPIC DRUG RESISTANCE 1* (*PDR1*) in petunia (*Petunia hybrida*), which is responsible for exudation of SLs into the rhizosphere [108]. Recently, the involvement of *PDR1* homologs in SL transport was also reported in other host plants [109, 110]. However, detailed mechanisms of SL exudation from the host roots, translocation, and its role in parasitic weed resistance remain unclear. Recently, it has been demonstrated that Cas9 directed mutagenesis of two homologous genes of *Petunia hybrida pdr1* namely ATP binding cassette subfamily gene *ABCG44* (*Solyc08g067610*) and *ABCG45* (*Solyc08g067620*) in tomato eliminates the ability of germination of *P. aegyptiaca*as compared to the control plants [111]. Cytochrome-P450 (CYP) *CYP722C*, is another essential enzyme responsible for the direct synthesis of orobanchol from carlactonic acid in tomato and cotton [37, 112]. The *CYP722C*- mutated tomato plants were generated by CRISPR/Cas9 targeted genome editing system. Mutagenesis of *CYP722C* in tomato causes reduced growth of parasitic weeds such as *Striga hermonthica*, *Orobanche crenata,* and *P. aegyptiaca* suggesting that CRISPR/Cas9 can be used as an effective strategy to prevent parasitic weed growth on the host plants [112].

In conclusion, all these studies demonstrate that SL pathway is essential for pathogenic weed survival, and by targeting the suitable host gene(s), CRISPR/Cas9 system can prevent parasitic weed infection (Table 1). However, till now most of the studies focus only on SL biosynthesis. Further studies are needed to explore other suitable pathways and genes that could be used to achieve effective control of parasitic weeds.

Impact of SL Biosynthesis Inhibition on Host Plants

CRISPR/Cas9-mediated mutagenesis has been used to block SL biosynthesis in host plants, and to reduce or suppress the germination and infection of parasitic weeds [105, 107, 111]. However, SLs are also required for host plant growth and development such as inhibiting shoot branching and regulating the growth of primary and secondary roots [113]. It has been used in rice plants to create mutagenesis in the *CCD7* gene, which leads to reduced SL biosynthesis and inhibition of germination of *S. hermonthica* which exhibited a dwarf phenotype and an increase in tiller number [106]. Similarly, mutagenesis of the *CCD8* gene in tomatoes (*Solanum lycopersicum L.*) inhibits the germination of *P. aegyptiaca* [105]. Tomato *CCD8* mutants were dwarf in morphology with increased shoot branching and more growth of adventitious root, and similar phenotypes were also observed in other host plants defective in SL biosynthesis such as tobacco [114].

CRISPR/Cas9-induced mutagenesis of the *CCD8* gene in tobacco leads to a reduction in total shoot biomass however, editing of the *CCD8* gene in tomato resulted in an increased number of fruits, along with a decrease in size than wild-type. The best way to prevent a dwarf/shunted phenotype of the host is the use of grafting. CRISPR/Cas9 genome editing could be applied to host plant rootstock which is already resistant to various biotic stresses and could be grafted over a wild-type scion to achieve parasitic weed resistance and normal host growth simultaneously. Another strategy to avoid negative effects on host plants is the foliar application of the GR24, a synthetic SL analog, which can suppress the dwarf phenotype in *CCD7* and *CCD8* mutated host plants to wild-type plants levels. The reduced secretion of SL in the rhizosphere not only inhibits the germination of parasitic weeds *P. aegyptiaca* but also affects symbiotic relationship of host plants with arbuscular mycorrhizal fungi [44]. This is because SL regulates both branching and arbuscular mycorrhizal symbiosis. Therefore, thoughtful research is necessary to investigate how host SL's biosynthetic pathway can be manipulated to reduce the growth of parasitic weeds while retaining normal host growth characteristics and beneficial relationships with symbiotic arbuscular mycorrhizal fungi.

Factors Affecting the Efficiency of CRISPR/Cas9

How effectively CRISPR/Cas9 can edit a plant genome can vary greatly depending on various parameters [115]. Optimizing Cas9 codons, improving promoters for Cas9 expression, and inserting T-DNA in appropriate positions have been shown to alter the expression level of Cas9, which could affect Cas9 activity [116]. CRISPR/Cas9 induction efficiency will also be affected by GC content in the targeted DNA sequences and high GC content (50%-70%), makes Cas9-mediated editing relatively efficient. In preliminary studies, the first target nucleotide is limited to A or G, so the common form is AN(19)NGG which corresponds to U3 promoter, and GN(19)NGG which further corresponds to U6 promoter [117]. Despite this, one or more additional nucleotides (irregular sgRNA) can affect the genome-editing efficiency of the Cas9/sgRNA nuclease complex [118]. Moreover, if the targeted DNA sequence forms a stem-loop-like structure with the sgRNA backbone sequence (the part where sgRNA binds to Cas9), the editing efficiency will be reduced [119].

In addition, DNA repair capacity between different plants or different tissues of the same plant will also affect genome editing efficiency. A DSB in DNA caused by Cas9 can be successfully repaired in a species having high DNA repair capabilities which leads to the generation of fewer mutations that limit our ability. The effective activity of Cas9 is believed to be tightly controlled by the guide sequence and a PAM adjacent to the targeted sequence in the genome, however, potential off-target cleavage activity could still occur even with 3-4 base pairs mismatches in the sgRNA sequence [119]. Previous studies have demonstrated that differences in guide RNA structures can also affect the effective cleavage of target sites. The X-ray crystal structure studies on Cas9 nuclease suggest that the PAM site, the sequence corresponding to the 3′ ends of the crRNA sequence is crucial for the Cas9 binding and activation of Cas9 nuclease [120, 121]. In plants, Cas9 nuclease may generate unintended non-specific mutations.

The potential for off-target effect is an important question for genome-edited plants. Validated design approaches are required to discriminate between on-target and off-target edits. Using Cas9 nuclease to edit the genome has a few limitations that can cause the generation of non-specific off-target cleavage. Thus, these non-specific mutations may alter the normal cellular function of a gene which can cause genomic instability. Although diverse modifications in the Cas9 nuclease protein and sgRNA designing tools have been made to increase on-target site specificity [122, 123], the occurrence of off-target cleavage is still a major limiting factor associated with the Cas9 nuclease editing technology [124].

Conclusion

Although several resistant crop varieties were developed by breeding technology, several parasitic weeds have promptly evolved as new races to establish parasitic infestations [125]. Therefore, it is essential to understand different pathogenic infection mechanisms of parasitic weeds and their corresponding host plant defense mechanisms. Phenotypic analysis and quantification of the infection level are significant steps to identify the race or disease-causing variant of parasitic weeds at the initial stage of shoot emergence. Host plants have adopted several mechanisms of defense such as physical barriers (the cuticle and cell wall) and constitutive production of secondary metabolites [126], however, a major strategy to avoid parasitism is by reducing the exudation of germination stimulants, particularly SLs by the host roots.

CRISPR/Cas9 technology has been successfully used to enhance crop production by reducing biotic or abiotic stresses [127, 128]. Plant-parasitic weeds have characteristic properties that require the host to establish parasitic interaction, germination, and development. Plant parasite weeds are noxious weeds hampering the economic development and growth of agriculturally important crops throughout the world. CRISPR/Cas9- mediated editing of host genes has been successfully shown to get resistance from this parasitic plant. However, the morphological character of the host affected by this technology leads to several unwanted effects on the cellular physiology of the host. To avoid this, we need to develop and apply this technology in such a way that it will mutate the parasitic weed-specific gene instead of the host; this will be beneficial for the host plant. In the past decade, several RNA silencing technologies applied to shut down the parasite-specific gene, similarly, we can use the CRISPR methods to edit parasite-specific genes which will directly reduce the economic burden.

Future Prospects

Advancement of any new technology (including CRISPR/Cas9) demands progressive enhancement. Research in CRISPR/Cas9 has made a great impact on the evolution of gene-editing technology [129]. The development of novel tools is indispensable for scientific advancement. CRISPR/Cas9 technologies are among the major tools in the history of molecular biology, which enabled breakthrough discoveries in science. Gene editing in plants via Cas9 is a breakthrough technology that opens a new field for plant disease research through the ability to disrupt parasite-host interactions, and enhance plant defense [130, 131]. Currently, several transcriptome and genome-wide studies are in progress to identify the pathogenic effector molecules and corresponding signalling pathways in parasitic weeds [3, 132, 133], building a rich source of information for future studies [134, 135]. Genome editing technology using CRISPR/Cas9 has been utilized to reduce the parasitic weeds burden, but at the same time, it is also very important to ensure that the endogenous functions of host-specific SLs should not be disrupted while preserving the crosstalk with other endogenous hormone pathways.

Acknowledgments

The authors are thankful for the infrastructural support provided by the Central University of Punjab, VPO-Ghudda, Bathinda during the preparation of the manuscript.

References

[1] Press MC, Phoenix GK. Impacts of parasitic plants on natural communities. *New Phytologist.* 2005;166.

[2] Westwood JH, Yoder JI, Timko MP, dePamphilis CW. The evolution of parasitism in plants. *Trends in Plant Science*. 2010.

[3] Wickett, NJ, Honaas, LA, Wafula, EK, Das, M, Huang, K, Wu, B, Landherr, L, Timko, MP, Yoder, J, Westwood, JH, & dePamphilis, CW. Transcriptomes of the parasitic plant family orobanchaceae reveal surprising conservation of chlorophyll synthesis. *Current Biology.* 2011;21.

[4] Yoshida S, Cui S, Ichihashi Y, Shirasu K. The Haustorium, a Specialized Invasive Organ in Parasitic Plants. *Annual Review of Plant Biology*. 2016;67.

[5] Clarke CR, Timko MP, Yoder JI, Axtell MJ, Westwood JH. Molecular Dialog between Parasitic Plants and Their Hosts. *Annual Review of Phytopathology*. 2019;57.

[6] Nickrent DL, Musselman LJ. Introduction to Parasitic Flowering Plants. *The Plant Health Instructor.* 2004. https://doi.org/10.1094/phi-i-2004-0330-01.

[7] Joel DM, Hershenhorn J, Eizenberg H, Aly R, Ejeta G, Rich PJ, J. K. Ransom, J. Sauerborn, D. Rubiales. Biology and Management of Weedy Root Parasites. In: *Horticultural Reviews*. 2007.

[8] Habimana S, Nduwumuremyi A, Chinama R JD. Management of orobanche in field crops-a review. *Journal of Soil Science and Plant Nutrition*. 2014;14.

[9] Cardoso C, Ruyter-Spira C, Bouwmeester HJ. Strigolactones and root infestation by plant-parasitic Striga, Orobanche and Phelipanche spp. *Plant Science*. 2011.

[10] Bouwmeester HJ, Matusova R, Zhongkui S, Beale MH. Secondary metabolite signalling in host-parasitic plant interactions. *Curr Opin Plant Biol*. 2003.

[11] Yoder JI. Parasitic plant responses to host plant signals: A model for subterranean plant-plant interactions. *Current Opinion in Plant Biology*. 1999. https://doi.org/10.1016/S1369-5266(99)80013-2.

[12] Joel DM. The long-term approach to parasitic weeds control: Manipulation of specific developmental mechanisms of the parasite. In: *Crop Protection*. 2000.

[13] Kokla A, Melnyk CW. Developing a thief: Haustoria formation in parasitic plants. *Developmental Biology*. 2018;442.

[14] Govier RN, Pate JS, Kuijt J. The Biology of Parasitic Flowering Plants. *The Journal of Ecology*. 1970;58.

[15] Albrecht H, Yoder JI, Phillips DA. Flavonoids promote haustoria formation in the root parasite Triphysaria versicolor. *Plant Physiology*. 1999;119.

[16] Chang M, Lynn DG. The haustorium and the chemistry of host recognition in parasitic angiosperms. *Journal of Chemical Ecology*. 1986;12.

[17] Matvienko M, Torres MJ, Yoder JI. Transcriptional responses in the hemiparasitic plant Triphysaria versicolor to host plant signals. *Plant Physiology*. 2001;127.

[18] Matvienko M, Wojtowicz A, Wrobel R, Jamison D, Goldwasser Y, Yoder JI. Quinone oxidoreductase message levels are differentially regulated in parasitic and non-parasitic plants exposed to allelopathic quinones. *Plant Journal*. 2001;25.

[19] Bandaranayake, PCG, Filappova, T, Tomilov, A, Tomilova, NB, Jamison-McClung, D, Ngo, Q, Inoue, K, & Yoder, JI. A single-electron reducing quinone oxidoreductase is necessary to induce haustorium development in the root parasitic plant Triphysaria. *Plant Cell*. 2010;22.

[20] Mitsumasu K, Seto Y, Yoshida S. Apoplastic interactions between plants and plant root intruders. *Frontiers in Plant Science*. 2015;6 AUG.

[21] Mutuku JM, Cui S, Yoshida S, Shirasu K. Orobanchaceae parasite–host interactions. *New Phytologist*. 2021;230.

[22] Veronesi C, Bonnin E, Calvez S, Thalouarn P, Simier P. Activity of secreted cell wall-modifying enzymes and expression of peroxidase-encoding gene following germination of Orobanche ramosa. *Biologia Plantarum*. 2007;51.

[23] Losner-Goshen D, Portnoy VH, Mayer AM, Joel DM. Pectolytic activity by the haustorium of the parasitic plant Orobanche L. (Orobanchaceae) in host roots. *Annals of Botany*. 1998;81.

[24] Singh A, Singh M. Cell wall degrading enzymes in Orobanche aegyptiaca and its host Brassica campestris. *Physiologia Plantarum*. 1993;89.

[25] González-Verdejo CI, Barandiaran X, Moreno MT, Cubero JI, di Pietro A. A peroxidase gene expressed during early developmental stages of the parasitic plant Orobanche ramosa. In: *Journal of Experimental Botany*. 2006.

[26] Parker C. Parasitic Weeds: A World Challenge. *Weed Sci*. 2012. https://doi.org/10.1614/ws-d-11-00068.1.

[27] Musselman LJ. Parasitic Weeds of the World: Biology and Controls. *Economic Botany*. 1994;48.

[28] Yoneyama K, Awad AA, Xie X, Yoneyama K, Takeuchi Y. Strigolactones as germination stimulants for root parasitic plants. *Plant and Cell Physiology*. 2010;51.

[29] Sato D, Awad AA, Takeuchi Y, Yoneyama K. Confirmation and quantification of strigolactones, germination stimulants for root parasitic plants Striga and Orobanche, produced by cotton. *Bioscience, Biotechnology and Biochemistry*. 2005;69.

[30] Gomez-Roldan V, Fermas S, Brewer PB, Puech-Pagès V, Dun EA, Pillot JP, Fabien Letisse, Radoslava Matusova, Saida Danoun, Jean-Charles Portais, Harro Bouwmeester, Guillaume Bécard, Christine A

Beveridge, Catherine Rameau, Soizic F Rochange. Strigolactone inhibition of shoot branching. *Nature*. 2008. https://doi.org/10.1038/nature07271.

[31] Umehara, M, Hanada, A, Yoshida, S, Akiyama, K, Arite, T, Takeda-Kamiya, N, Magome, H, Kamiya, Y, Shirasu, K, Yoneyama, K, Kyozuka, J, & Yamaguchi, S. Inhibition of shoot branching by new terpenoid plant hormones. *Nature*. 2008. https://doi.org/10.1038/nature07272.

[32] Bouwmeester HJ, Roux C, Lopez-Raez JA, Bécard G. Rhizosphere communication of plants, parasitic plants and AM fungi. *Trends in Plant Science*. 2007.

[33] Dun EA, Brewer PB, Beveridge CA. Strigolactones: discovery of the elusive shoot branching hormone. *Trends in Plant Science*. 2009;14.

[34] Rameau C, Murfet IC, Laucou V, Floyd RS, Morris SE, Beveridge CA. Pea rms6 mutants exhibit increased basal branching. *Physiologia Plantarum*. 2002;115.

[35] Lin, H, Wang, R, Qian, Q, Yan, M, Meng, X, Fu, Z, Yan, C, Jiang, B, Su, Z, Li, J, & Wang, Y. DWARF27, an iron-containing protein required for the biosynthesis of strigolactones, regulates rice tiller bud outgrowth. *Plant Cell*. 2009;21.

[36] Waters MT, Brewer PB, Bussell JD, Smith SM, Beveridge CA. The Arabidopsis ortholog of rice DWARF27 acts upstream of MAX1 in the control of plant development by Strigolactones. *Plant Physiology*. 2012;159.

[37] Wakabayashi T, Shida K, Kitano Y, Takikawa H, Mizutani M, Sugimoto Y. CYP722C from Gossypium arboreum catalyzes the conversion of carlactonoic acid to 5-deoxystrigol. *Planta*. 2020;251.

[38] Gobena D, Shimels M, Rich PJ, Ruyter-Spira C, Bouwmeester H, Kanuganti S, Tesfaye Mengiste, and Gebisa Ejetaa. Mutation in sorghum *LOW GERMINATION STIMULANT 1* alters strigolactones and causes *Striga* resistance. *Proceedings of the National Academy of Sciences*. 2017. https://doi.org/10.1073/pnas.1618965114.

[39] Yoneyama, K, Mori, N, Sato, T, Yoda, A, Xie, X, Okamoto, M, Iwanaga, M, Ohnishi, T, Nishiwaki, H, Asami, T, Yokota, T, Akiyama, K, Yoneyama, K, & Nomura, T. Conversion of carlactone to carlactonoic acid is a conserved function of MAX1 homologs in strigolactone biosynthesis. *New Phytologist*. 2018;218.

[40] Yoneyama K, Akiyama K, Brewer PB, Mori N, Kawano-Kawada M, Haruta S, Nishiwaki, H, Yamauchi, S, Xie, X, Umehara, M, Beveridge, CA, Yoneyama, K, & Nomura, T. Hydroxyl carlactone derivatives are predominant strigolactones in Arabidopsis. *Plant Direct*. 2020;4.

[41] Brewer, PB, Yoneyama, K, Filardo, F, Meyers, E, Scaffidi, A, Frickey, T, Akiyama, K, Seto, Y, Dun, EA, Cremer, JE, Kerr, SC, Waters, MT, Flematti, GR, Mason, MG, Weiller, G, Yamaguchi, S, Nomura, T, Smith, SM, Yoneyama, K, & Beveridge, CA. Lateral branching oxidoreductase acts in the final stages of strigolactone biosynthesis in Arabidopsis. *Proc Natl Acad Sci USA*. 2016;113.

[42] Xie X, Mori N, Yoneyama K, Nomura T, Uchida K, Yoneyama K, & Akiyama, K. Lotuslactone, a non-canonical strigolactone from Lotus japonicus. *Phytochemistry*. 2019;157.

[43] Uraguchi, D, Kuwata, K, Hijikata, Y, Yamaguchi, R, Imaizumi, H, Am, S, Rakers, C, Mori, N, Akiyama, K, Irle, S, McCourt, P, Kinoshita, T, Ooi, T, & Tsuchiya, Y. A femtomolar-range suicide germination stimulant for the parasitic plant striga hermonthica. *Science* (1979). 2018;362.

[44] Vogel JT, Walter MH, Giavalisco P, Lytovchenko A, Kohlen W, Charnikhova T, Simkin, AJ, Goulet, C, Strack, D, Bouwmeester, HJ, Fernie, AR, & Klee, HJ. SlCCD7 controls strigolactone biosynthesis, shoot branching and mycorrhiza-induced apocarotenoid formation in tomato. *Plant Journal*. 2010. .

[45] Kohlen W, Charnikhova T, Lammers M, Pollina T, Tóth P, Haider I, Pozo, MJ, Maagd, RA, Ruyter-Spira, C, Bouwmeester, HJ, & López-Ráez, JA. The tomato carotenoid cleavage dioxygenase8 (SlCCD8) regulates rhizosphere signaling, plant architecture and affects reproductive development through strigolactone biosynthesis. *New Phytologist*. 2012. https://doi.org/10.1111/j.1469-8137.2012.04265.x.

[46] Kohlen W, Charnikhova T, Bours R, López-Ráez JA, Bouwmeester H. Tomato strigolactones: A more detailed look. *Plant Signaling and Behavior*. 2013;8.

[47] Akiyama K, Hayashi H. Strigolactones: Chemical signals for fungal symbionts and parasitic weeds in plant roots. *Handbook of Environmental Chemistry, Volume 5: Water Pollution*. 2006;97.

[48] Akiyama K, Matsuzaki KI, Hayashi H. Plant sesquiterpenes induce hyphal branching in arbuscular mycorrhizal fungi. *Nature*. 2005;435.

[49] Dor E, Yoneyama K, Wininger S, Kapulnik Y, Yoneyama K, Koltai H, Xie, X, & Hershenhorn, J. Strigolactone deficiency confers resistance in tomato line SL-ORT1 to the parasitic weeds Phelipanche and Orobanche spp. *Phytopathology*. 2011;101.

[50] Hu L, Wang J, Yang C, Islam F, Bouwmeester HJ, Muños S, & Zhou, W. The effect of virulence and resistance mechanisms on the interactions between parasitic plants and their hosts. *International Journal of Molecular Sciences*. 2020;21.

[51] Cohen M, Prandi C, Occhiato EG, Tabasso S, Wininger S, Resnick N, Steinberger, Y, Koltai, H, & Kapulnik, Y. Structure-function relations of strigolactone analogs: Activity as plant hormones and plant interactions. *Molecular Plant*. 2013;6.

[52] Hegenauer V, Körner M, Albert M. Plants under stress by parasitic plants. *Current Opinion in Plant Biology*. 2017;38.

[53] Pérez-De-Luque A, Lozano MD, Moreno MT, Testillano PS, Rubiales D. Resistance to broomrape (Orobanche crenata) in faba bean (Vicia faba): Cell wall changes associated with prehaustorial defensive mechanisms. *Annals of Applied Biology*. 2007;151.

[54] Pérez-de-Luque A, González-Verdejo CI, Lozano MD, Dita MA, Cubero JI, González-Melendi P, Risueño, MC, & Rubiales, D. Protein cross-linking, peroxidase and β-1,3-endoglucanase involved in resistance of pea against Orobanche crenata. *Journal of Experimental Botany*. 2006;57.

[55] Runyon JB, Mescher MC, Felton GW, de Moraes CM. Parasitism by Cuscuta pentagona sequentially induces JA and SA defence pathways in tomato. *Plant, Cell and Environment*. 2010;33.

[56] Kusumoto D, Goldwasser Y, Xie X, Yoneyama K, Takeuchi Y, Yoneyama K. Resistance of red clover (Trifolium pratense) to the root parasitic plant Orobanche minor is activated by salicylate but not by jasmonate. *Annals of Botany*. 2007;100.

[57] Letousey P, de Zélicourt A, Vieira Dos Santos C, Thoiron S, Monteau F, Simier P, Thalouarn, P, & Delavault, P. Molecular analysis of resistance mechanisms to Orobanche cumana in sunflower. *Plant Pathology*. 2007;56.

[58] Radi A, Dina P, Guy A. Expression of sarcotoxin IA gene via a root-specific tob promoter enhanced host resistance against parasitic weeds in tomato plants. *Plant Cell Reports*. 2006;25.

[59] Hamamouch N, Westwood JH, Banner I, Cramer CL, Gepstein S, Aly R. A peptide from insects protects transgenic tobacco from a parasitic weed. *Transgenic Research*. 2005;14.

[60] Aly R. Conventional and biotechnological approaches for control of parasitic weeds. In *Vitro Cellular and Developmental Biology - Plant*. 2007.

[61] Rubiales D, Pérez-de-Luque A, Fernández-Aparico M, Sillero JC, Román B, Kharrat M, Khalil, S, Joel, DM, & Riches, C. Screening techniques and sources of resistance against parasitic weeds in grain legumes. *Euphytica*. 2006;147.

[62] Okazawa A, Wakabayashi T. Chemical control of root parasitic weeds. In: *ACS Symposium Series*. 2015.

[63] Kaundun SS, Dale RP, Bailly GC. Molecular Basis of Resistance to Herbicides Inhibiting Acetolactate Synthase in Two Rigid Ryegrass (Lolium rigidum) Populations from Australia. *Weed Sci.* 2012;60.

[64] Nandula VK, Foy CL, Orcutt DM. Glyphosate for Orobanche aegyptiaca control in Vicia sativa and Brassica napus. *Weed Sci*. 1999;47.

[65] Surov T, Aviv D, Aly R, Joel DM, Goldman-Guez T, Gressel J. Generation of transgenic asulam-resistant potatoes to facilitate eradication of parasitic broomrapes (Orobanche spp.), with the sul gene as the selectable marker. *Theoretical and Applied Genetics*. 1998;96.

[66] Aviv D, Amsellem Z, Gressel J. Transformation of carrots with mutant acetolactate synthase for Orobanche (broomrape) control. *Pest Management Science*. 2002;58.

[67] Slavov S, Valkov V, Batchvarova R, Atanassova S, Alexandrova M, Atanassov A. Chlorsulfuron resistant transgenic tobacco as a tool for broomrape control. *Transgenic Research*. 2005;14.

[68] Nandula VK, Westwood JH, Foster JG, Foy CL. Influence of glyphosate on amino acid composition of Egyptian broomrape [Orobanche aegyptiaca (Pers.)] and selected hosts. *Journal of Agricultural and Food Chemistry*. 2001;49.

[69] Frizzi A, Huang S. Tapping RNA silencing pathways for plant biotechnology. *Plant Biotechnology Journal*. 2010;8.

[70] Mermigka G, Verret F, Kalantidis K. RNA silencing movement in plants. *Journal of Integrative Plant Biology*. 2016;58.

[71] Liu YX, Wang M, Wang XJ. Endogenous Small RNA Clusters in Plants. *Genomics, Proteomics and Bioinformatics*. 2014;12.

[72] LeBlanc M, Kim G, Westwood JH. RNA trafficking in parasitic plant systems. *Frontiers in Plant Science*. 2012;3 AUG.

[73] Yoder JI, Gunathilake P, Wu B, Tomilova N, Tomilov AA. Engineering host resistance against parasitic weeds with RNA interference. *Pest Management Science*. 2009;65.

[74] Tomilov AA, Tomilova NB, Wroblewski T, Michelmore R, Yoder JI. Trans-specific gene silencing between host and parasitic plants. *Plant Journal*. 2008;56.

[75] Everard JD, Cantini C, Grumet R, Plummer J, Loescher WH. Molecular cloning of mannose-6-phosphate reductase and its developmental expression in celery. *Plant Physiology*. 1997;113.

[76] Delavault P, Simier P, Thoiron S, Véronési C, Fer A, Thalouarn P. Isolation of mannose 6-phosphate reductase cDNA, changes in enzyme activity and mannitol content in broomrape (Orobanche ramosa) parasitic on tomato roots. *Physiologia Plantarum*. 2002;115.

[77] Aly R, Cholakh H, Joel DM, Leibman D, Steinitz B, Zelcer A, Naglis, A, Yarden, O, & Gal-On, A. Gene silencing of mannose 6-phosphate reductase in the parasitic weed Orobanche aegyptiaca through the production of homologous dsRNA sequences in the host plant. *Plant Biotechnology Journal*. 2009;7.

[78] Auldridge ME, Block A, Vogel JT, Dabney-Smith C, Mila I, Bouzayen M, Magallanes-Lundback, M, DellaPenna, D, McCarty, DR, & Klee, HJ. Characterization of three members of the Arabidopsis carotenoid cleavage dioxygenase family demonstrates the divergent roles of this multifunctional enzyme family. *Plant Journal*. 2006;45.

[79] Kirigia D, Runo S, Alakonya A. A virus-induced gene silencing (VIGS) system for functional genomics in the parasitic plant Striga hermonthica. *Plant Methods*. 2014;10.

[80] Aly R, Dubey NK, Yahyaa M, Abu-Nassar J, Ibdah M. Gene silencing of CCD7 and CCD8 in Phelipanche aegyptiaca by tobacco rattle virus system retarded the parasite development on the host. *Plant Signaling and Behavior*. 2014. https://doi.org/10.4161/psb.29376.

[81] Dubey NK, Eizenberg H, Leibman D, Wolf D, Edelstein M, Abu-Nassar J, Marzouk, S., Gal-On, A., & Aly, R. Enhanced Host-Parasite Resistance Based on Down-Regulation of Phelipanche aegyptiaca Target Genes Is Likely by Mobile Small RNA. *Frontiers in Plant Science*. 2017. https://doi.org/10.3389/fpls.2017.01574.

[82] Hsu PD, Lander ES, Zhang F. Development and applications of CRISPR/Cas9 for genome engineering. *Cell*. 2014.

[83] Song G, Jia M, Chen K, Kong X, Khattak B, Xie C, Li, A, & Mao, L. CRISPR/Cas9: A powerful tool for crop genome editing. *Crop Journal*. 2016.

[84] Liu X, Wu S, Xu J, Sui C, Wei J. Application of CRISPR/Cas9 in plant biology. *Acta Pharmaceutica Sinica B*. 2017.

[85] Jaganathan D, Ramasamy K, Sellamuthu G, Jayabalan S, Venkataraman G. CRISPR for Crop Improvement: An Update Review. *Frontiers in Plant Science*. 2018. https://doi.org/10.3389/fpls.2018.00985.

[86] Bortesi L, Fischer R. The CRISPR/Cas9 system for plant genome editing and beyond. *Biotechnology Advances*. 2015.

[87] Ricroch A, Clairand P, Harwood W. Use of CRISPR systems in plant genome editing: toward new opportunities in agriculture. *Emerging Topics in Life Sciences*. 2017. https://doi.org/10.1042/etls20170085.

[88] Ma X, Zhu Q, Chen Y, Liu YG. CRISPR/Cas9 Platforms for Genome Editing in Plants: Developments and Applications. *Molecular Plant*. 2016;9.

[89] Lowder L, Malzahn A, Qi Y. Rapid evolution of manifold CRISPR systems for plant genome editing. *Frontiers in Plant Science*. 2016;7.

[90] Wada N, Ueta R, Osakabe Y, Osakabe K. Precision genome editing in plants: State-of-the-art in CRISPR/Cas9-based genome engineering. *BMC Plant Biology*. 2020;20.

[91] Doench JG, Fusi N, Sullender M, Hegde M, Vaimberg EW, Donovan KF, Smith, I, Tothova, Z, Wilen, C, Orchard, R, Virgin, HW, Listgarten, J, & Root, DE. Optimized sgRNA design to maximize activity and minimize off-target effects of CRISPR/Cas9. *Nature Biotechnology*. 2016;34.

[92] Yan J, Xue D, Chuai G, Gao Y, Zhang G, Liu Q. Benchmarking and integrating genome-wide CRISPR off-target detection and prediction. *Nucleic Acids Research*. 2020;48.

[93] Xu H, Xiao T, Chen CH, Li W, Meyer CA, Wu Q, Wu, D, Cong, L, Zhang, F, Liu, JS, Brown, M, & Liu, XS. Sequence determinants of improved CRISPR sgRNA design. *Genome Research*. 2015;25.

[94] Pan C, Ye L, Qin L, Liu X, He Y, Wang J, Chen, L, & Lu, G. CRISPR/Cas9-mediated efficient and heritable targeted mutagenesis in tomato plants in the first and later generations. *Scientific Reports*. 2016. https://doi.org/10.1038/srep24765.

[95] Zhang H, Zhang J, Wei P, Zhang B, Gou F, Feng Z, Mao, Y, Yang, L, Zhang, H, Xu, N, & Zhu, JK. The CRISPR/Cas9 system produces specific and homozygous targeted gene editing in rice in one generation. *Plant Biotechnology Journal*. 2014. https://doi.org/10.1111/pbi.12200.

[96] Jiang W, Zhou H, Bi H, Fromm M, Yang B, Weeks DP. Demonstration of CRISPR/Cas9/sgRNA-mediated targeted gene modification in Arabidopsis, tobacco, sorghum and rice. *Nucleic Acids Research*. 2013;41.

[97] Brooks C, Nekrasov V, Lippman ZB, van Eck J. Efficient Gene Editing in Tomato in the First Generation CRISPR/Cas9 System. *Plant Physiolgy*. 2014;166.

[98] Adli M. The CRISPR tool kit for genome editing and beyond. *Nature Communications*. 2018;9.

[99] Kosicki M, Tomberg K, Bradley A. Repair of double-strand breaks induced by CRISPR–Cas9 leads to large deletions and complex rearrangements. *Nature Biotechnology*. 2018;36.

[100] Shen H, Strunks GD, Klemann BJPM, Hooykaas PJJ, de Pater S. CRISPR/Cas9-induced double-strand break repair in Arabidopsis nonhomologous end-joining mutants. *G3: Genes, Genomes, Genetics*. 2017;7.

[101] Lieber MR. The mechanism of double-strand DNA break repair by the nonhomologous DNA end-joining pathway. *Annual Review of Biochemistry*. 2010;79.

[102] Kanaar R, Hoeijmakers JHJ, van Gent DC. Molecular mechanisms of DNA double-strand break repair. *Trends in Cell Biology*. 1998;8.

[103] Zischewski J, Fischer R, Bortesi L. Detection of on-target and off-target mutations generated by CRISPR/Cas9 and other sequence-specific nucleases. *Biotechnology Advances*. 2017;35.

[104] Guo J, Li K, Jin L, Xu R, Miao K, Yang F, Qi, C, Zhang, L, Botella, JR, Wang, R, & Miao, Y. A simple and cost-effective method for screening of CRISPR/Cas9-induced homozygous/biallelic mutants. *Plant Methods*. 2018;14.

[105] Bari VK, Nassar JA, Kheredin SM, Gal-On A, Ron M, Britt A, Steele, D, Yoder, J, & Aly, R. CRISPR/Cas9-mediated mutagenesis of CAROTENOID CLEAVAGE DIOXYGENASE 8 in tomato provides resistance against the parasitic weed Phelipanche aegyptiaca. *Scientific Reports*. 2019;9.

[106] Butt H, Jamil M, Wang JY, Al-Babili S, Mahfouz M. Engineering plant architecture via CRISPR/Cas9-mediated alteration of strigolactone biosynthesis. *BMC Plant Biology*. 2018;18.

[107] Bari VK, Nassar JA, Aly R. CRISPR/Cas9 mediated mutagenesis of MORE AXILLARY GROWTH 1 in tomato confers resistance to root parasitic weed Phelipanche aegyptiaca. *Scientific Reports*. 2021;11.

[108] Kretzschmar T, Kohlen W, Sasse J, Borghi L, Schlegel M, Bachelier JB, Reinhardt, D, Bours, R, Bouwmeester, HJ, & Martinoia, E. A petunia ABC protein controls strigolactone-dependent symbiotic signalling and branching. *Nature*. 2012;483.

[109] Shiratake K, Notaguchi M, Makino H, Sawai Y, Borghi L. Petunia PLEIOTROPIC DRUG RESISTANCE 1 Is a Strigolactone Short-Distance Transporter with Long-Distance Outcomes. *Plant and Cell Physiology*. 2019;60.

[110] Banasiak J, Borghi L, Stec N, Martinoia E, Jasiński M. The Full-Size ABCG Transporter of Medicago truncatula Is Involved in Strigolactone Secretion, Affecting Arbuscular Mycorrhiza. *Frontiers in Plant Science*. 2020;11.

[111] Bari VK, Nassar JA, Meir A, Aly R. Targeted mutagenesis of two homologous ATP-binding cassette subfamily G (ABCG) genes in tomato confers resistance to parasitic weed Phelipanche aegyptiaca. *Journal of Plant Research*. 2021;134.

[112] Wakabayashi T, Hamana M, Mori A, Akiyama R, Ueno K, Osakabe K, Osakabe, Y, Suzuki, H, Takikawa, H, Mizutani, M, & Sugimoto, Y. Direct conversion of carlactonoic acid to orobanchol by cytochrome P450 CYP722C in strigolactone biosynthesis. *Science Advances*. 2019;5.

[113] Al-Babili S, Bouwmeester HJ. Strigolactones, a Novel Carotenoid-Derived Plant Hormone. *Annual Review of Plant Biology*. 2015. https://doi.org/10.1146/annurev-arplant-043014-114759.

[114] Gao J, Zhang T, Xu B, Jia L, Xiao B, Liu H, Liu, L, Yan, H, & Xia, Q. CRISPR/Cas9-mediated mutagenesis of Carotenoid cleavage dioxygenase 8 (CCD8) in tobacco affects shoot and root architecture. *International Journal of Molecular Sciences*. 2018;19.

[115] Mikami M, Toki S, Endo M. Parameters affecting frequency of CRISPR/Cas9 mediated targeted mutagenesis in rice. *Plant Cell Reports*. 2015;34.

[116] Li K, Wang G, Andersen T, Zhou P, Pu WT. Optimization of genome engineering approaches with the CRISPR/Cas9 system. *PLoS ONE*. 2014;9.

[117] Manghwar H, Li B, Ding X, Hussain A, Lindsey K, Zhang X, & Jin, S. CRISPR/Cas Systems in Genome Editing: Methodologies and Tools for sgRNA Design, Off-Target Evaluation, and Strategies to Mitigate Off-Target Effects. *Advanced Science*. 2020;7.

[118] Liu X, Yang J, Song Y, Zhang X, Wang X, Wang Z. Effects of sgRNA length and number on gene editing efficiency and predicted mutations generated in rice. *The Crop Journal*. 2021. https://doi.org/10.1016/j.cj.2021.05.015.

[119] Liang G, Zhang H, Lou D, Yu D. Selection of highly efficient sgRNAs for CRISPR/Cas9-based plant genome editing. *Scientific Reports*. 2016;6.

[120] Hirano S, Nishimasu H, Ishitani R, Nureki O. Structural Basis for the Altered PAM Specificities of Engineered CRISPR/Cas9. *Molecular Cell*. 2016;61.

[121] Anders C, Niewoehner O, Duerst A, Jinek M. Structural basis of PAM-dependent target DNA recognition by the Cas9 endonuclease. *Nature*. 2014;513.

[122] Scott T, Urak R, Soemardy C, Morris K v. Improved Cas9 activity by specific modifications of the tracrRNA. *Scientific Reports*. 2019;9.

[123] Filippova J, Matveeva A, Zhuravlev E, Stepanov G. Guide RNA modification as a way to improve CRISPR/Cas9-based genome-editing systems. *Biochimie*. 2019;167.

[124] Hajiahmadi Z, Movahedi A, Wei H, Li D, Orooji Y, Ruan H, & Zhuge, Q. Strategies to increase on-target and reduce off-target effects of the CRISPR/Cas9 system in plants. *International Journal of Molecular Sciences*. 2019;20.

[125] Hu L, Wang J, Yang C, Islam F, Bouwmeester HJ, Muños S, & Zhou, W. The effect of virulence and resistance mechanisms on the interactions between parasitic plants and their hosts. *International Journal of Molecular Sciences*. 2020;21.

[126] Pérez-De-Luque A, Moreno MT, Rubiales D. Host plant resistance against broomrapes (Orobanche spp.): Defence reactions and mechanisms of resistance. *Annals of Applied Biology*. 2008;152.

[127] Hyun TK. CRISPR/Cas-based genome editing to improve abiotic stress tolerance in plants. *Botanica Serbica*. 2020;44.

[128] Rao MJ, Wang L. CRISPR/Cas9 technology for improving agronomic traits and future prospective in agriculture. *Planta*. 2021;254.

[129] Randhawa S, Sengar S. The evolution and history of gene editing technologies. In: *Progress in Molecular Biology and Translational Science*. 2021.

[130] Rato C, Carvalho MF, Azevedo C, Oblessuc PR. Genome editing for resistance against plant pests and pathogens. *Transgenic Research*. 2021;30.

[131] Borrelli VMG, Brambilla V, Rogowsky P, Marocco A, Lanubile A. The enhancement of plant disease resistance using CRISPR/Cas9 technology. *Frontiers in Plant Science*. 2018;9.

[132] Yang C, Fu F, Zhang N, Wang J, Hu L, Islam F, Bai, Q, Yun, X, & Zhou, W. Transcriptional profiling of underground interaction of two contrasting sunflower cultivars with the root parasitic weed Orobanche cumana. *Plant and Soil*. 2020;450.

[133] Yoshida S, Kim S, Wafula EK, Tanskanen J, Kim YM, Honaas L, Yang, Z, Spallek, T, Conn, CE, Ichihashi, Y, Cheong, K, Cui, S, Der, JP, Gundlach, H, Jiao, Y, Hori, C, Ishida, JK, Kasahara, H, Kiba, T, … Shirasu, K. Genome Sequence of Striga asiatica Provides Insight into the Evolution of Plant Parasitism. *Current Biology*. 2019;29.

[134] Yang Z, Wafula EK, Honaas LA, Zhang H, Das M, Fernandez-Aparicio M, Huang, K, Bandaranayake, PCG, Wu, B, Der, JP, Clarke, CR, Ralph, PE, Landherr, L, Altman, NS, Timko, MP, Yoder, JI, Westwood, JH, & dePamphilis, CW. Comparative transcriptome analyses reveal core parasitism genes and suggest gene duplication and repurposing as sources of structural novelty. *Molecular Biology and Evolution*. 2015;32.

[135] Kösters LM, Wiechers S, Lyko P, Müller KF, Wicke S. WARPP-web application for the research of parasitic plants. *Plant Physiology*. 2021;185.

[136] Bellis ES, Kelly EA, Lorts CM, Gao H, DeLeo VL, Rouhan G, Budden, A, Bhaskara, GB, Hu, Z, Muscarella, R, Timko, MP, Nebie, B, Runo, SM, Chilcoat, ND, Juenger, TE, Morris, GP, dePamphilis, CW, & Lasky, JR. Genomics of sorghum local adaptation to a parasitic plant. *Proc Natl Acad Sci USA*. 2020;117.

Section 3. Genome Editing in Selected Plants/Crops

Chapter 11

CRISPR-Mediated Genome Editing in Model Plants

Vinita Sharma[1]
Ashita Bisht[2,3]
Ankita Mishra[1]
Himanshu Sharma[1]
Joy Roy[1]
and Abhishek Bhandawat[1,*]

[1] National Agri-Food Biotechnology Institute, Mohali, Punjab, India
[2] Department of Plant Breeding and Genetics,
Punjab Agricultural University, Ludhiana, Punjab, India
[3] CSK Himachal Pradesh Krishi Vishvavidyalaya,
Highland Agricultural Research and Extension Centre, Kukumseri, Lahaul and Spiti, India

Abstract

Model plants are most extensively studied organisms and are well-suited for understanding growth, development, stress responses and gene function. Success of various reverse genetics approaches to understand gene function relies mainly on genome sequence availability. With the advent of high throughput NGS technology, genome sequencing has been made rapid and more affordable. This has facilitated the emergence of CRISPR-Cas as a revolutionary tool for targeted genome editing to understand gene function and crop improvement. Increasing number of studies have replaced existing gene editing tools with robust, precise and more efficient CRISPR-based technology. This chapter sheds light on some of the most popular and interesting examples of gene knockout, gene activation, loss of function and other genetic modifications using CRISPR technology in four model plants namely *Arabidopsis*, tomato, rice and maize. Through this chapter, you will recognize the promises of CRISPR-Cas technology not only in investigating the function of genes but also development of varieties of crops and vegetables with improved yield, nutritional quality and/or stress resistance.

---

* Corresponding Author's Email: abhishek.bhandawat@gmail.com.

In: Genome Editing for Crop Improvement
Editors: Rahul Kumar, Ashutosh Sharma, Yogesh Kumar et al.
ISBN: 979-8-88697-740-0

CRISPR/Cas9 Editing in *Arabidopsis thaliana*

Arabidopsis thaliana (Brassicaceae) is a popular flowering plant for functional characterization studies and understanding plant development. Small diploid genome, availability of whole genome sequence, short lifecycle, easy laboratory handling, and robust transformation protocol make this a suitable model organism for functional genomics, and developing novel tools and approaches including CRISPR-based genome editing. In the following sections, applications of CRISPR-Cas system in biotic and abiotic stress tolerance, transcriptional activation, promoter methylation involving knockdown, overexpression and gene activation of target genes have been discussed along with few examples of *Arabidopsis thaliana* (Table 1).

Table 1. Examples of CRISPR-mediated gene editing in *Arabidopsis*

S. No.	Target gene	Genetic modification	Delivery approach	Observed phenotype	Reference
1	*HSFA6a, HSFA6b*	Knockout	*Agrobacterium*-mediated	Reduced ROS accumulation, ABA and osmotic stress insensitivity modulating ROS homeostasis	[1]
2	*eIF(iso)4E*	Knockout	*Agrobacterium*-mediated	*Potyvirus* resistance	[3]
3	43 coding or non-coding sites of the Beet severe curly top virus (BSCTV) genome	Knockout	*Agrobacterium*-mediated	Geminivirus-resistance	[4]
4	*UGT79B2, UGT79B3*	Knockout	*Agrobacterium*-mediated	Cold, salt and drought tolerance	[2]
5	*BAR,* GL1	Knockout, gene activation	*Agrobacterium*-mediated	Glufosinate resistance and reduced trichomes formation	[5]
6	*HOS1*	Knockout	*Agrobacterium*-mediated	Tolerance to cold and regulation of secondary metabolism	[6]
7	*WRKY30*	Gene activation	PEG-mediated protoplast transfection	Transcriptional activation of one or more target gene	[7]
8	*FWA* promoter	CG methylation	*Agrobacterium*-mediated	CG methylation of promoter resulting in delayed flowering	[8]

HSFA6a and *HSFA6b* Confers Abiotic Stress Tolerance

Using CRISPR/Cas9, single and double mutants of *hsfa6a* and *hsfa6b* showed varied tolerance to various abiotic stresses such as ABA, mannitol and NaCl. Double mutants were found more tolerant to these stresses than single mutants. Knockout mutations of these transcription factors resulted in reduced ROS accumulation, higher expression of ABA responsive genes including those involved in ROS level regulation [1].

In another study, RNAi and CRISPR-mediated knockout of UDP-glycosyltransferases *UGT79B2* and *UGT79B3*, increase the susceptibility of *Arabidopsis* plant to cold, salt and drought stresses. Additional studies show that CBF1 regulates *UGT79B2* and *UGT79B3* which act as anthocyanin rhamnosyltransferases and confer abiotic stress tolerance by altering anthocyanin accumulation [2].

Resistance to Turnip Mosaic Virus (TuMV)

Introduction of sequence-specific point mutations in eukaryotic elongation factor allele *eIF(iso)4E* in *Arabidopsis* using transgene-free CRISPR/Cas9 technology provided complete resistance to potyvirus (*Turnip mosaic virus*) which is detrimental for vegetable crops [3].

Understanding Viral Replication Mechanism

CRISPR/Cas9 was proved to be an efficient system for screening suitable sgRNA target sites for viral genome replication in *Arabidopsis*. The lines expressing CRISPR/Cas9 showed tolerance to *Beet severe curly top virus* (BSCTV), a *Geminivirus* [4].

Loss of Function and Gain of Function

BAR (*Bialaphos Resistance*) gene is commonly used as positive selection marker for conferring resistance to glufosinate herbicide. CRISPR-mediated mutation of mutant BAR allele resulted in gain of function, conferring partial tolerance to herbicide and rescues the plant. Additionally, CRISPR editing of *GL1* (GLABROUS1), which is required for the formation of trichomes, resulted in disruption of ORF and functional protein formation [5].

Regulation of Secondary Metabolism

HOS1 (High Expression of Osmotically Responsive Genes 1), an E3 ubiquitin ligase, is known to play important role in cold signalling. CRISPR-mediated disruption of this gene resulted in increased tolerance to cold and increase in expression of cold responsive genes [6]. This mutation also affected the secondary metabolism affecting phytolexin biosynthesis.

Cas9 for Transcriptional Activation

Cas9 recognizes and edits specific target using guide RNA. Li et al. (2017b) developed dCas9–TV (dCas9–TAD) transcriptional activation system for one or more genes [7]. For instance, using dCAS9-TV strongly induced expression of *WRKY30* up to 139 folds in T1 transgenic lines. In the study, dCas9–TV was found to be a potent transcriptional activation system for activation of one or more genes in plant and mammalian cells.

CRISPR-Based Targeting of DNA Methylation

Methylation of CG islands is an important epigenetic mechanism for gene silencing which is sustained through cell divisions. A novel application of CRISPR in CG-specific targeting of DNA methylation for plants was developed using bacterial variant of CG-specific DNA methyltransferase (MQ1). Mutant plant lacking methylation in *FWA* (*Flowering Wageningen)* promoter shows late flowering, which was used in the study [8]. Methylation of CG was specific to target *FWA* gene promoter, and it was heritably maintained in *Arabidopsis*. Such system can effectively be used to regulate expression of important genes to obtain desirable phenotypes in crop and other plants.

CRISPR/Cas9 Editing in Tomatoes

Tomatoes (*Solanum lycopersicum*) are a model fruit-bearing crop that has been bred for fruit quality, higher yield, and tolerance to stress. The applications of CRISPR/Cas9 in tomato first came into lime light in 2014, where the *Argonaute 7* gene was knocked-out resulting in the wiry phenotype having leaves without petioles and laminae [9]. Thereafter a huge number of research publications emerged, deploying CRISPR/Cas9 application for various traits in tomato that grouped into three categories: (i) resistance to biotic and abiotic stress, (ii) fruit nutrition quality improvement, and (iii) tomato domestication. In this section, we will discuss the applications of CRISPR/Cas9 technology for the above-mentioned categories (Table 2).

Table 2. Examples of CRISPR-mediated gene editing in *Solanum lycopersicum*

S. No.	Target gene	Genetic modification	Delivery approach	Phenotype (after modification)	Reference
1	*SGR1, LCY-E, LCY-B1, LCY-B2, Blc. LCY-E*	Knockdown	*Agrobacterium*-mediated	Enhanced lycopene content	[23]
2	*DCL2*	Knockout	*Agrobacterium*-mediated	Resistant against RNA viruses	[14]
3	SlPelo, SlMlo1	Knockout	*Agrobacterium*-mediated	Resistant against TYLCV	[13]
4	DMR6	Knockout	*Agrobacterium*-mediated	Resistant against downy mildew	[15]
5	Mlo1	Loss of function	*Agrobacterium*-mediated	Resistant against powdery mildew	[16]
6	*Solyc08g075770*	Gene complementation and knockout	*Agrobacterium*-mediated	Resistance to fusarium wilt	[30]
7	*SlJAZ2*	Overexpression	*Agrobacterium*-mediated	Resistance against *P. syringae*	[31]
8	*BZR1*	Overexpression	*Agrobacterium*-mediated	Thermotolerance	[17]
9	*CLV3 & SP*	Loss of function	*Agrobacterium*-mediated	An increase in floral organ number or fruit size, conferring enhanced tomato fruit yield	[19]
10	*RIN*	Knockout	*Agrobacterium*-mediated	Ripening associated changes	[32]
11	*ALC*	Knockout	*Agrobacterium*-mediated	Extended shelf life of fruit	[33]
12	*PL*	Knockout	*Agrobacterium*-mediated	Extended shelf life of fruit	[34]
13	*ANT1*	Overexpression	*Agrobacterium*-mediated	Increased anthocyanin content in fruit	[22]

S. No.	Target gene	Genetic modification	Delivery approach	Phenotype (after modification)	Reference
14	*SlGAD2* and *SlGAD3*	Overexpression	*Agrobacterium*-mediated	Fruits with increased GABA content	[24]
15	*Sl-ALMT9*	Overexpression	*Agrobacterium*-mediated	Decrease malate accumulation and Al resistant phenotype	[35]
16	*MYB12*	Knockout	*Agrobacterium*-mediated	Pink tomatoes	[20]
17	CRTISO and PSY1	Knockout	*Agrobacterium*-mediated	Orange and yellow fruits	[21]
18	*SlIAA9, SlAGL6, and ARF7*	Knockout	*Agrobacterium*-mediated	Induced Parthenocarpy	[28, 29]
19	*FW2.2, FAS, MULT* and *CycB*	Loss of function	*Agrobacterium*-mediated	Threefold increase in fruit size and tenfold increase in fruit number	[27]
20	*SP, SP5G* and *SlER*	Loss of function	*Agrobacterium*-mediated	Compactness and growth cycle of tomato plants were dramatically increased and shortened, respectively	[36]

Resistance to Biotic and Abiotic Stress

Biotic Stress

A huge damage to fruit crops caused by biotic stress that mainly includes: bacteria, fungi, viruses that attack plants at the time of fruiting and ultimately leading to yield reduction and crop loss. Since its successful application in 2014, CRISPR/Cas9 technology has been widely used to generate resistant plants against viral, fungal and bacterial pathogens causing severe loss to tomato [10, 11].

To resist the infection caused by viruses in tomato plants, two strategies have been adopted. One way is the sgRNA-based targeting of the viral genome through direct complementation, and other way is through modification of tomato genes inducing viral resistance. The tomato plants had been engineered using the CRISPR/Cas9 technology to generate resistance to *Tomato yellow leaf curl virus* (TYLCV). The transgenic plants were generated by targeting the coat protein and replicase loci of the viral genome. The transgenic plants thus generated showed lower viral DNA accumulation, and an efficient viral interference [12]. Another study by Pramanik et al. targeted *SlPelo* gene using CRISPR/Cas9 system to generate plants conferring resistance to TYLCV [13]. Wang et al. (2018) knocked-out *dicer-like 2* (*DCL2*) gene using CRISPR/Cas9 technology. The *dcl2* mutant conferred resistance against *Tomato mosaic virus* and other RNA viruses as well [14].

One of the major economic penalties from which tomato producers suffer is due to reduced yield and quality, caused by fungal infections including smut, rot, mildew and rust. Among the above-mentioned fungal diseases, powdery and downy mildew cause major loss to tomato producers. CRISPR/Cas9 system has been used to knockout the *Arabidopsis thaliana downy mildew resistant 6* (*DMR6*) ortholog in tomato, and found that the transgenic plants were resistant to various species such as *P. syringae*, *P. capsica*, and *Xanthomonas spp*. [15]. Loss of function of the *mildew resistant locus O 1* (*Mlo1*) using CRISPR/Cas9 technology developed mutant lines, completely resistant to powdery mildew disease caused by fungal pathogen *Oidium neolycopersici* [16].

Abiotic Stress

Increase in world's population and demand for food supply is a challenge in the present era. Main hurdles to cover the gap between demand and supply are posed by environmental conditions. Abiotic stresses such as heat, cold, draught and flood impose high risk to crops, therefore, the need of the hour is to improve crop production using advance biotechnological tools, and CRISPR/Cas9 technology is the most promising candidate. CRISPR/Cas9 technology was successfully employed to develop thermotolerant tomato by overexpressing the *Brassinazole resistant 1 (BZR1)* gene [17]. Tomato is cold sensitive, and chilling temperature can easily damage the fruit, reducing its quality. *C-repeat binding factor 1* (*CBF1*) is known to protect the plant from cold injury. Li et al. (2018) developed *cbf1* mutant and identified that the mutant exhibited extreme cold injury and electrolyte leakage compared to the wild counterpart [18].

Nutrition and Fruit Quality Improvement

Internal and external properties of a fruit ultimately determine its quality. The internal fruit properties include composition of nutrients and bioactive compounds, and external fruit properties can be assessed visually that include fruit size, colour and texture. The destruction of the *Clavata-Wuschel* (*CLV-WUS*) a stem cell circuit using CRISPR/Cas9 technology generated plants with larger fruits [19]. A *locule number* (*lc*) QTL was recreated in tomato using the CRISPR technology. The developed transgenic plants consisted of fruits with large number of locules [19]. Transgenic tomatoes with different colours such as pink, purple, orange and yellow have been engineered using CRISPR/Cas9 system, targeting the genes *MYB transcription factor 12* (*MYB12*), *anthocyanin 1* (*ANT1*), *carotenoid isomerase* (*CRTISO*) and *phytoene synthase 1* (*PSY1*) respectively [20–22]. Genetic engineering of multiple genes [*staygreen 1* (*SGR1*), *lycopene ε-cyclase* (*LCY-E*), *beta-lycopene* (*Blc*), *lycopene β-cyclase 1* (*LCY-B1*), and *LCY-B2*] from carotenoid biosynthesis pathway in tomato using the CRISPR/Cas9 technology was known to increase lycopene content in fruit up to 5.1-fold [23]. CRISPR/Cas9 induced overexpression of the two *glutamate decarboxylase* (GAD) genes (*SlGAD2* and *SlGAD3*), the key genes involved in GABA biosynthesis, significantly increased GABA accumulation up to 15-fold in tomato fruit [24].

Tomato Domestication

Crop domestication usually affects the genes regulating the plant architecture (size of seeds and its dispersal) and physiology (duration of seed germination, flowering and fruit ripening). Therefore, introducing alleles from elite cultivars or wild relatives into the modern cultivated species through classical breeding or modern rewilding is time consuming and labour intensive task [25, 26]. One of the smart strategies is genetic manipulation of the wild relatives to domesticate them *de novo* and study their adaptation under different environmental conditions [27]. The former has been accelerated using CRISPR/Cas9 technology. The *agamous-like 6* (*AGL6*) gene, which is responsible for parthenocarpic fruit production under stress conditions, was knocked-out in tomato along with *indole-3-acetic acid inducible 9* (*IAA9*) and *auxin response factor 7* (*ARF7*), whose involvement is in auxin signalling pathway, produced fruits without seeds [28, 29].

CRISPR/Cas-Mediated Gene Editing in Rice

Genome editing in model plants is critical to development of crops with desired traits by means of analysing the function of a gene. The model system, *Arabidopsis*, with fully decoded and relatively small genome provides a platform to understand plant biology, but the information is not always transferable to cereal crops [37]. Among the major cereal crops, maize, rice and wheat, transformation in rice have been routinely carried out in many labs, while transformation in wheat and maize still requires well-equipped laboratories and intensive expertise [38]. Therefore, rice has been successfully utilized to understand different traits and associated mechanisms of cereal crops. In the last decade, CRISPR/Cas technique has been employed to analyse gene function and crop improvement by biotic and abiotic stress tolerance, yield parameters and grain nutritional quality. Here, we discuss successful CRISPR/Cas-mediated targeting in rice that paved path for improvement of cereal crops (Table 3).

Table 3. Examples of CRISPR-mediated genome editing in rice

S. No.	Target gene	Knockout/ Mutation etc.	Delivery approach	Outcome/effect on phenotype/Trait targeted	Reference
1.	Fatty acid desaturase 2 (*OsFAD2-1*)	Knockout	*Agrobacterium*-mediated	Increased oleic acid content and decreased linoleic acid	[50]
2.	Betaine aldehyde dehydrogenase (*Badh2*)	T base insertion in first exon	*Agrobacterium*-mediated	Increase in 2AP contents (Fragrant rice)	[52]
3.	Grain Size 3 (*GS3*), Grain Width 2 (*GW2*), Grain Width 5 (*GW5*) and Thousand-Grain Weight 6 (*TGW6*)	Mutations	*Agrobacterium tumefaciens*-mediated	Improvement of grain weight	[40]
4.	Late flowering gene (*OsGhd7*)	Knockouts	*Agrobacterium tumefaciens*-mediated	Early flowering phenotype	[53]
5.	Phytochrome C (*PHYC*)	Knockout	*Agrobacterium- tumefaciens* mediated	Early flowering phenotype	[54]
6.	*GBSSI*	Promoter editing	*Agrobacterium tumefaciens*-mediated	Decrease in Amylose content	[55]
7.	*GBSSI*	Mutations in Exon-1	*Agrobacterium tumefaciens*-mediated	Decrease in Amylose content /glutinous rice	[56]
8.	Semi-dwarf 1 (*SD1*)	Mutations	*Agrobacterium tumefaciens*-mediated	Effects on height and yield depending on genetic back-grounds	[57]
9.	Drought and salt tolerance gene (*OsDST*)	Loss of function mutants	*Agrobacterium tumefaciens*-mediated	Moderate level tolerance to osmotic stress and high level of salt stress	[44]
10.	*OsSWEET14*	Promoter editing	Biolistic transformation	Resistance Against Bacterial Blight	[43]
11.	Annexin gene (*OsAnn3*)	Knockout	*Agrobacterium tumefaciens*-mediated	Decreased cold tolerance	[46]
12.	Orange (*Or*) gene	Direct gene disruption	*Agrobacterium tumefaciens*-mediated transformation	β-carotene fortification	[51]

Table 3. (Continued)

S. No.	Target gene	Knockout/ Mutation etc.	Delivery approach	Outcome/effect on phenotype/Trait targeted	Reference
13.	*FAD2* gene	mutations	Biolistic transformation	Increased oleic acid content in seeds	[58]
14.	Acetolactate Synthase (*ALS*)	Amino acid substitution	Particle bombardment	Herbicide-Resistant Rice	[59]
15.	Starch Branching Enzymes	Mutations	*Agrobacterium tumefaciens*-mediated	High amylose rice	[49]
16.	Semi-Dwarf1 (*SD1*)	Mutations	*Agrobacterium tumefaciens*-mediated	Semi-dwarf rice lines	[60]
17.	*Gn1a*, *DEP1*, *GS3*, and *IPA1*	Mutations	*Agrobacterium*-mediated	Reassessment of roles of yield-related genes	[40]
18	*OsSWEET13*	Knockout	*Agrobacterium*-mediated	Resistance for bacterial blight	[42]
19	*OsERF922*	InDel mutations	*Agrobacterium*-mediated	Resistance for rice blast	[41]
20	*eIF4G*	Mutations	*Agrobacterium*-mediated	Resistance to Rice tungro spherical virus	[61]
21	*OsSAPK2*	Loss-of-function mutants	*Agrobacterium*-mediated	Abscisic acid sensitivity and tolerance to drought stress	[45]

Genome Editing for High-Yielding Rice

Low-yield is the major constraint to ensure global food security. Green revolution has introduced many high-yielding varieties (HYVs) using breeding practices to enhance cereal production [39]. Advancement in genome sequencing and rice transformation technologies have enabled researchers to improve yield of several rice varieties by CRISPR/Cas-mediated targeting. Multiplex genome editing mediated by CRISPR/Cas9 system has enabled us to mutate QTLs; Grain Size 3 (GS3), Grain Width 2 (GW2), Grain Width 5 (GW5), and Thousand-Grain Weight 6 (TGW6) and null mutation of QTLs lead to increase in grain weight in rice. The double and triple mutants (*gw5tgw6* and *gw2gw5tgw6*) in this study showed 13% and 28% increase in thousand kernel weight respectively [40].

Genome Editing for Stress Tolerance

Plants have to withstand different types of biotic and abiotic stresses in their life cycle. Environmental factors influence plant growth and yield. Therefore, crops tolerant to different stresses is at the centre of research.

Biotic Stress Tolerance

Rice is susceptible to various viral, bacterial and fungal diseases, that results in reduction in crop yield. Genome editing methods have been successfully applied to develop pathogen-resistant rice varieties. Rice blast is one of the most destructive diseases of rice. ERF transcription factor, OsERF922, was targeted for developing resistance to blast disease in rice using CRISPR/Cas9 technology, and considerable resistance to the pathogen was achieved [41].

Bacterial leaf blight (BLB) caused by caused by *Xanthomonas oryzae* is another major threat to rice crops. Indica rice lines resistant to BLB was developed by CRISPR/Cas9-based knockout of OsSWEET13, that prevented its interaction with the effector [42]. Highly selective and promising immunity was achieved for Basmati rice by CRISPR/Cas9-based targeting of 4 EBEs located in the promoter of OsSWEET14 [43]. Resistance to *Rice tungro spherical virus* (RTSV) was introduced by application of CRISPR/Cas9-based editing of eIF4G. Findings suggested the revolutionary role of CRISPR/Cas9 technology for conferment of biotic stress tolerance to rice, and the information can be used for the improvement of other cereals.

Abiotic Stress Tolerance

Different abiotic stresses such as drought, salinity and cold, substantially affect rice cultivation. Genome editing has been used to develop rice varieties with improved stress tolerance. Recently, drought and salt tolerance-related gene (*OsDST*) was targeted by CRISPR-based editing and mutant lines were found to confer moderate level tolerance to osmotic stress and high level of salt stress [44]. CRISPR/Cas9-mediated loss of function mutants of Osmotic stress/ABA–activated protein kinase 2 (*SAPK2*) resulted in increased drought and salt tolerance in rice [45]. The study suggested that *SAPK2* is a good candidate for crop improvement.

CRISPR/Cas9-mediated knockouts of the annexin gene *OsAnn3* lead to decreased cold tolerance in rice, suggesting the role of *OsAnn3* to cold tolerance [46]. Recently, Zeng et al. (2020) reported improved cold tolerance in rice by targeting MYB30 and two other genes by CRISPR-based targeting [47].

Genome Editing for Nutritional Quality

The need of nutritional quality cannot be ignored in the shade of food security. Starch is a major constraint to nutritional quality of rice. CRISPR technology has been used to improve starch quality and aroma, altering oil content and biofortification. Starch with high amylose content is correlated with resistant starch and offers several health benefits [48]. CRISPR-Cas9-mediated editing of starch branching enzyme (*SBEII*) resulted in high amylose rice [49]. Rice rich in oleic acid content offers several health benefits. CRISPR/Cas9 was used to enhance oleic acid content in rice by targeting fatty acid desaturase 2 (OsFAD2-1) [50]. Biofortified rice with enhanced β-carotene was developed by genome editing of OSOr gene [51].

CRISPR/Cas Editing in Maize

Maize (*Zea mays* ssp. *mays*) ranks amongst the top three cereals in production globally. It plays a critical role as a staple crop, fuel and feed. It has numerous features making it an important model plant in research. Owing to its biology, even after being monoecious in nature, it has spatially separated female and male flowers, allowing easy and controlled crossing. This ease of creating hybrids that effectively utilizes the phenomenon of heterosis makes it a highly successful agricultural crop. Single cob bears hundreds of bold kernels arranged in proper rows eases the visualization of anomalies in colour or configuration to infer segregation ratios. Moreover, a large number of natural and induced mutants are available in maize, making it a perfect fit as a model organism in agriculture [62].

For production of superior hybrids, the availability of elite parental inbreds is a must. Conventional breeding requires almost a decade to incorporate a desirable allele in an agronomically superior line with extensive backcrossing and screening of lines in each generation. This process is laborious, time-consuming and involves unpredictable recombination events for the development of desirable phenotype. Sometimes the problem of linkage drag can also hinder this process [63]. Thus, creation of desirable phenotypes led to integration of fast and effective CRISPR-Cas technology in precise editing of maize genome. The first genome editing in maize using CRISPR-Cas was done to reduce the proportion of anti-nutritional factor (phytic acid inositol 1,2,3,4,5,6-hexakisphosphate) in maize kernels to improve digestibility by monogastric animals [64]. Since its discovery, this technology has been extensively utilized to improve many traits such as yield, modification in the architecture of the plant, tolerance to abiotic stress, herbicide resistance, induction of male sterility and haploidy, as documented in Table 4.

Table 4. Examples of CRISPR-mediated genome editing in maize

S. No.	Target gene	Genetic modification	Delivery approach	Phenotype (after modification)	Reference
1	*ZmIPK*	Base editing	*Agrobacterium*-mediated	Phytic acid biosynthesis	[64]
2	*ZmHKT1*	Base editing	*Agrobacterium*-mediated	Tolerance to salt stress	[69]
3	*ZmALS2*	Base editing	Biolistic	Resistance to herbicide (chlorsulfuron)	[70]
4	*MS26* and *MS45*	Base editing	Biolistic	Male sterility	[70]
5	*Psy1*	Base editing	*Agrobacterium*-mediated	Promotes variegation in leaves and kernels	[71]
6	*Zmzb7*	Base editing	*Agrobacterium*-mediated	Albinism	[72]
7	*MADS, MYBR* and *AP2*	Base editing	*Agrobacterium*-mediated	Decreased storage protein (zein) in kernels	[73]
8	*RPL* and *PPR*	Base editing	*Agrobacterium*-mediated	Improper plant development	[73]
9	*a1* and *a4*	Base editing	*Agrobacterium*-mediated	Anthocyaninless	[74]
10	*ZmAgo18a* and *ZmAgo18b*	Base editing	*Agrobacterium*-mediated	Sporogeneis	[74]
11	*LG1*	Knockout	*Agrobacterium*-mediated	Upright leaves and compact architecture	[75]
12	*ZmTMS5*	Knockout	Biolistic	Thermosensitive male-sterility	[76]
13	*ZmPLA1*	Knockout	Polyethylene glycol mediated	Haploid induction (endosperm abortion in kernels)	[77]
14	*ARGOS8*	Overexpression	Biolistic	Increased yield under drought stress	[78]
15	*MS8*	Knockout	*Agrobacterium*-mediated	Male sterility	[79]
16	*ZmMTL/ZmPLA1*	Base editing	*Agrobacterium*-mediated	Maternal haploid induction	[80]
17	*zb7*	Knockout	*Agrobacterium*-mediated	Albino seedlings	[81]
18	*zb7*	Base editing	*Agrobacterium*-mediated	Chlorotic seedlings	[81]
19	20 genes	Single or multiple knockouts	*Agrobacterium*-mediated	-	[82]
20	*gl2*	Base editing	*Agrobacterium*-mediated	Dull leaf surface	[83]
21	*GL2*	Base editing	*Agrobacterium*-mediated	Glossy leaf surface	[65]

S. No.	Target gene	Genetic modification	Delivery approach	Phenotype (after modification)	Reference
22	*LG1*	Base editing	-	Haploid induction	[68]
23	*UB2* and *UB3*	Base editing	-	Haploid induction	[68]
24	*Ms26* and *Ms45*	Base editing	*Agrobacterium*-mediated, biolistic and ribonucleotprotein	Male sterility	[66]
25	*Lg1*	Base editing	*Agrobacterium*-mediated, biolistic and ribonucleotprotein	Lack of ligules	[66]
26	*ZmDMP*	Knockout	*Agrobacterium*-mediated	Haploid induction	[84]
27	*Dek42*	Knockout	*Agrobacterium*-mediated	Defective kernels	[85]
28	Locations on chromosomes 1, 3 and 10	Base editing	*Agrobacterium*-mediated	No effect on grain yield, grain moisture, plant height, ear height, pollen shed and silking	[86]
29	*ZmALS1* and *ZmALS2*	Base editing	Polyethylene glycol mediated	Herbicide resistance	[87]
30	118 genes of agronomic and nutrional importance	Base editing	*Agrobacterium*-mediated	Various phenotypic changes	[67]
31	*ZmWx*	Base editing	-	Waxy kernels	[88]
32	*ZmGA20ox3*	Base editing	*Agrobacterium*-mediated	Semidwarf	[89]
33	*Opaque2*	Base editing	*Agrobacterium*-mediated	Opaque kernels	[90]
34	*ZmCLE7* and *ZmFCP1*	Knockout	*Agrobacterium*-mediated	Increased grain yield	[91]
35	*ZmPRO1* and *ZmPRO3*	Base editing	*Agrobacterium*-mediated	Modified pollen cells	[92]

A recent study reported the efficacy of LbCas12a in the production of a 100% mutation rate in T1 generation at 28°C in maize [65]. Additionally, various approaches are available to deliver CRISPR-Cas components in maize plants including *Agrobacterium*, biolistic and polyethylene glycol (PEG)-mediated transformation. *Agrobacterium*-mediated transformation is most common among these methods as it has demonstrated to have highest rate of on-target site genome editing and with the frequency of off-target cleavage when compared with other delivery approaches [66]. A large-scale experiment involving 118 genes with 412 edited sequences also proved the applicability of CRISPR/Cas9 in high-throughput genome editing of maize with clear phenotypic modifications [67].

Due to biosafety concerns over transgenic crops, researchers are encouraged to create genome-edited transgene-free plants. Wang and co-workers (2019) were able to overcome this problem and proposed the Haploid-Inducer Mediated Genome Editing (IMGE) approach for the development of transgene-free maize plants [68]. This approach has vast potential in accelerating breeding in different crops which will be readily accepted among the consumers.

Providing the advancement in next-generation sequencing technologies and fully sequenced genome of many of the crop plants, CRISPR/Cas technology has been proven successful for crop improvement and functional characterization. The use of functional knowledge acquired from model or reference species has further played a major role in detailed understanding of mechanisms and processes in different plant species. Unlike, the previous approaches like mutations and insertional lines, CRISPR/Cas technology is very successful in rice, maize, tomato and can be used to produce transgene-free plants, could be commercialized.

References

[1] Wenjing, W., Chen, Q., Singh, P. K., Huang, Y., and Pei, D. (2020) CRISPR/Cas9 edited HSFA6a and HSFA6b of Arabidopsis thaliana offers ABA and osmotic stress insensitivity by modulation of ROS homeostasis. *Plant Signaling & Behavior*. 15 (12), 1816321.

[2] Li, P., Li, Y. J., Zhang, F. J., Zhang, G. Z., Jiang, X. Y., Yu, H. M., and Hou, B. K. (2017) The Arabidopsis UDP-glycosyltransferases UGT79B2 and UGT79B3, contribute to cold, salt and drought stress tolerance via modulating anthocyanin accumulation. *The Plant Journal*. 89 (1), 85–103.

[3] Pyott, D. E., Sheehan, E., and Molnar, A. (2016) Engineering of CRISPR/Cas9-mediated potyvirus resistance in transgene-free Arabidopsis plants. *Molecular Plant Pathology*. 17 (8), 1276–1288.

[4] Ji, X., Zhang, H., Zhang, Y., Wang, Y., and Gao, C. (2015) Establishing a CRISPR–Cas-like immune system conferring DNA virus resistance in plants. *Nature Plants*. 1 (10), 1–4.

[5] Hahn, F., Mantegazza, O., Greiner, A., Hegemann, P., Eisenhut, M., and Weber, A. P. M. (2017) An efficient visual screen for CRISPR/Cas9 activity in Arabidopsis thaliana. *Frontiers in Plant Science*. 8 39.

[6] Shkryl, Y., Yugay, Y., Avramenko, T., Grigorchuk, V., Gorpenchenko, T., Grischenko, O., and Bulgakov, V. (2021) CRISPR/Cas9-Mediated Knockout of HOS1 Reveals Its Role in the Regulation of Secondary Metabolism in Arabidopsis thaliana. *Plants*. 10 (1), 104.

[7] Li, Z., Zhang, D., Xiong, X., Yan, B., Xie, W., Sheen, J., and Li, J. F. (2017) A potent Cas9-derived gene activator for plant and mammalian cells. *Nature Plants*. 3 (12), 930–936.

[8] Ghoshal, B., Picard, C. L., Vong, B., Feng, S., and Jacobsen, S. E. (2021) CRISPR-based targeting of DNA methylation in Arabidopsis thaliana by a bacterial CG-specific DNA methyltransferase. *Proceedings of the National Academy of Sciences*. 118 (23),.

[9] Brooks, C., Nekrasov, V., Lippman, Z. B., and Van Eck, J. (2014) Efficient gene editing in tomato in the first generation using the clustered regularly interspaced short palindromic repeats/CRISPR-associated9 system. *Plant Physiology*. 166 (3), 1292–1297.

[10] Chaudhary, R. and Atamian, H. (2017) Resistance-gene-mediated defense responses against biotic stresses in the crop model plant tomato. *Journal of Plant Pathology & Microbiology*. 8 (4), 1–11.

[11] Borrelli, V. M. G., Brambilla, V., Rogowsky, P., Marocco, A., and Lanubile, A. (2018) The enhancement of plant disease resistance using CRISPR/Cas9 technology. *Frontiers in Plant Science*. 1245.

[12] Tashkandi, M., Ali, Z., Aljedaani, F., Shami, A., and Mahfouz, M. M. (2018) Engineering resistance against Tomato yellow leaf curl virus via the CRISPR/Cas9CRISPR/Cas9 system in tomato. *Plant Signaling & Behavior*. 13 (10), e1525996.

[13] Pramanik, D., Shelake, R. M., Park, J., Kim, M. J., Hwang, I., Park, Y., and Kim, J. Y. (2021) CRISPR/Cas9-mediated generation of pathogen-resistant tomato against tomato yellow leaf curl virus and powdery mildew. *International Journal of Molecular Sciences*. 22 (4), 1878.

[14] Wang, Z., Hardcastle, T. J., Pastor, A. C., Yip, W. H., Tang, S., and Baulcombe, D. C. (2018) A novel DCL2-dependent miRNA pathway in tomato affects susceptibility to RNA viruses. *Genes & Development*. 32 (17–18), 1155–1160.

[15] de Toledo Thomazella, D. P., Brail, Q., Dahlbeck, D., and Staskawicz, B. (2016) CRISPR/Cas9 mediated mutagenesis of a DMR6 ortholog in tomato confers broad-spectrum disease resistance. *BioRxiv*. 64824.

[16] Nekrasov, V., Wang, C., Win, J., Lanz, C., Weigel, D., and Kamoun, S. (2017) Rapid generation of a transgene-free powdery mildew resistant tomato by genome deletion. *Scientific Reports*. 7 (1), 1–6.

[17] Yin, Y., Qin, K., Song, X., Zhang, Q., Zhou, Y., Xia, X., and Yu, J. (2018) BZR1 transcription factor regulates heat stress tolerance through FERONIA receptor-like kinase-mediated reactive oxygen species signaling in tomato. *Plant and Cell Physiology*. 59 (11), 2239–2254.

[18] Li, R., Zhang, L., Wang, L., Chen, L., Zhao, R., Sheng, J., and Shen, L. (2018) Reduction of tomato-plant chilling tolerance by CRISPR–Cas9-mediated SlCBF1 mutagenesis. *Journal of Agricultural and Food Chemistry*. 66 (34), 9042–9051.

[19] Rodríguez-Leal, D., Lemmon, Z. H., Man, J., Bartlett, M. E., and Lippman, Z. B. (2017) Engineering quantitative trait variation for crop improvement by genome editing. *Cell*. 171 (2), 470–480.

[20] Deng, L., Wang, H., Sun, C., Li, Q., Jiang, H., Du, M., Li, C. B. and Li, C. (2018) Efficient generation of pink-fruited tomatoes using CRISPR/Cas9 system. *Journal of Genetics and Genomics= Yi Chuan Xue Bao*. 45 (1), 51–54.

[21] Dahan-Meir, T., Filler-Hayut, S., Melamed-Bessudo, C., Bocobza, S., Czosnek, H., Aharoni, A. and Levy, A. A. (2018) Efficient in planta gene targeting in tomato using geminiviral replicons and the CRISPR/Cas9 system. *The Plant Journal*. 95 (1), 5–16.

[22] Čermák, T., Baltes, N. J., Čegan, R., Zhang, Y., and Voytas, D. F. (2015) High-frequency, precise modification of the tomato genome. *Genome Biology*. 16 (1), 1–15.

[23] Li, X., Wang, Y., Chen, S., Tian, H., Fu, D., Zhu, B., Luo, Y. and Zhu, H. (2018) Lycopene is enriched in tomato fruit by CRISPR/Cas9-mediated multiplex genome editing. *Frontiers in Plant Science*. 9 559.

[24] Nonaka, S., Arai, C., Takayama, M., Matsukura, C., and Ezura, H. (2017) Efficient increase of γ-aminobutyric acid (GABA) content in tomato fruits by targeted mutagenesis. *Scientific Reports*. 7 (1), 1–14.

[25] Zeder, M. A., Bradley, D. G., Smith, B. D., and Emshwiller, E. (2006) *Documenting Domestication: New Genetic and Archaeological Paradigms*. Univ of California Press, .

[26] Zeder, M. A. (2012) The domestication of animals. *Journal of Anthropological Research*. 68 (2), 161–190.

[27] Zsögön, A., Cermak, T., Voytas, D., and Peres, L. E. P. (2017) Genome editing as a tool to achieve the crop ideotype and de novo domestication of wild relatives: case study in tomato. *Plant Science*. 256 120–130.

[28] Ueta, R., Abe, C., Watanabe, T., Sugano, S. S., Ishihara, R., Ezura, H., Osakabe, Y. and Osakabe, K. (2017) Rapid breeding of parthenocarpic tomato plants using CRISPR/Cas9. *Scientific Reports*. 7 (1), 1–8.

[29] Hu, J., Israeli, A., Ori, N., and Sun, T. (2018) The interaction between DELLA and ARF/IAA mediates crosstalk between gibberellin and auxin signaling to control fruit initiation in tomato. *The Plant Cell*. 30 (8), 1710–1728.

[30] Prihatna, C., Barbetti, M. J., and Barker, S. J. (2018) A novel tomato fusarium wilt tolerance gene. *Frontiers in Microbiology*. 9 1226.

[31] Ortigosa, A., Gimenez-Ibanez, S., Leonhardt, N., and Solano, R. (2019) Design of a bacterial speck resistant tomato by CRISPR/Cas9-mediated editing of Sl JAZ 2. *Plant Biotechnology Journal*. 17 (3), 665–673.

[32] Ito, Y., Nishizawa-Yokoi, A., Endo, M., Mikami, M., and Toki, S. (2015) CRISPR/Cas9-mediated mutagenesis of the RIN locus that regulates tomato fruit ripening. *Biochemical and Biophysical Research Communications*. 467 (1), 76–82.

[33] Yu, Q. H., Wang, B., Li, N., Tang, Y., Yang, S., Yang, T., Xu, J., Guo, C., Yan, P., Wang, Q. and Asmutola, P. (2017) CRISPR/Cas9-induced targeted mutagenesis and gene replacement to generate long-shelf life tomato lines. *Scientific Reports*. 7 (1), 1–9.

[34] Uluisik, S., Chapman, N. H., Smith, R., Poole, M., Adams, G., Gillis, R. B., Besong, T. M., Sheldon, J., Stiegelmeyer, S., Perez, L. and Samsulrizal, N. (2016) Genetic improvement of tomato by targeted control of fruit softening. *Nature Biotechnology*. 34 (9), 950–952.

[35] Ye, J., Wang, X., Hu, T., Zhang, F., Wang, B., Li, C., Yang, T., Li, H., Lu, Y., Giovannoni, J. J. and Zhang, Y., 2017 (2017) An InDel in the promoter of Al-ACTIVATED MALATE TRANSPORTER9 selected during tomato domestication determines fruit malate contents and aluminum tolerance. *The Plant Cell*. 29 (9), 2249–2268.

[36] Kwon, C. T., Heo, J., Lemmon, Z. H., Capua, Y., Hutton, S. F., Van Eck, J., Park, S. J. and Lippman, Z. B. (2020) Rapid customization of Solanaceae fruit crops for urban agriculture. *Nature Biotechnology*. 38 (2),.

[37] Provart, N. J., Alonso, J., Assmann, S. M., Bergmann, D., Brady, S. M., Brkljacic, J., Browse, J., Chapple, C., Colot, V., Cutler, S. and Dangl, J. (2016) 50 years of Arabidopsis research: highlights and future directions. *New Phytologist*. 209 (3), 921–944.

[38] Hiei, Y., Ishida, Y., and Komari, T. (2014) Progress of cereal transformation technology mediated by Agrobacterium tumefaciens. *Frontiers in Plant Science*. 5 628.

[39] Conway, G. and Simmonds, N. W. (1998) The doubly Green Revolution: Food for all in the 21st century. *Nature*. 391 (6663), 139.

[40] Xu, R., Yang, Y., Qin, R., Li, H., Qiu, C., Li, L., Wei, P. and Yang, J. (2016) Rapid improvement of grain weight via highly efficient CRISPR/Cas9-mediated multiplex genome editing in rice. *Journal of Genetics and Genomics= Yi Chuan Xue Bao*. 43 (8), 529–532.

[41] Wang, F., Wang, C., Liu, P., Lei, C., Hao, W., Gao, Y., Liu, Y. G. and Zhao, K. (2016) Enhanced rice blast resistance by CRISPR/Cas9-targeted mutagenesis of the ERF transcription factor gene OsERF922. *PloS One*. 11 (4), e0154027.

[42] Zhou, J., Peng, Z., Long, J., Sosso, D., Liu, B. O., Eom, J. S., Huang, S., Liu, S., Vera Cruz, C., Frommer, W. B. and White, F. F. (2015) Gene targeting by the TAL effector PthXo2 reveals cryptic resistance gene for bacterial blight of rice. *The Plant Journal*. 82 (4), 632–643.

[43] Zafar, K., Khan, M. Z., Amin, I., Mukhtar, Z., Yasmin, S., Arif, M., Ejaz, K. and Mansoor, S. (2020) Precise CRISPR/Cas9 mediated genome editing in super basmati rice for resistance against bacterial blight by targeting the major susceptibility gene. *Frontiers in Plant Science*. 11 575.

[44] Santosh Kumar, V. V., Verma, R. K., Yadav, S. K., Yadav, P., Watts, A., Rao, M. V. and Chinnusamy, V. (2020) CRISPR/Cas9 mediated genome editing of drought and salt tolerance (OsDST) gene in indica mega rice cultivar MTU1010. *Physiology and Molecular Biology of Plants*. 26 (6), 1099–1110.

[45] Lou, D., Wang, H., Liang, G., and Yu, D. (2017) OsSAPK2 confers abscisic acid sensitivity and tolerance to drought stress in rice. *Frontiers in Plant Science*. 8 993.

[46] Shen, C., Que, Z., Xia, Y., Tang, N., Li, D., He, R. and Cao, M. (2017) Knock out of the annexin gene OsAnn3 via CRISPR/Cas9-mediated genome editing decreased cold tolerance in rice. *Journal of Plant Biology*. 60 (6), 539–547.

[47] Zeng, Y., Wen, J., Zhao, W., Wang, Q., and Huang, W. (2020) Rational Improvement of Rice Yield and Cold Tolerance by Editing the Three Genes OsPIN5b, GS3, and OsMYB30 With the CRISPR–Cas9 System. *Frontiers in Plant Science*. 10.

[48] Regina, A., Bird, A., Topping, D., Bowden, S., Freeman, J., Barsby, T., Kosar-Hashemi, B., Li, Z., Rahman, S. and Morell, M. (2006) High-amylose wheat generated by RNA interference improves indices of large-bowel health in rats. *Proceedings of the National Academy of Sciences of the United States of America*. 103 (10),.

[49] Sun, Y., Jiao, G., Liu, Z., Zhang, X., Li, J., Guo, X., Du, W., Du, J., Francis, F., Zhao, Y. and Xia, L. (2017) Generation of high-amylose rice through CRISPR/Cas9-mediated targeted mutagenesis of starch branching enzymes. *Frontiers in Plant Science*. 8.

[50] Abe, K., Araki, E., Suzuki, Y., Toki, S., and Saika, H. (2018) Production of high oleic/low linoleic rice by genome editing. *Plant Physiology and Biochemistry*. 131 58–62.

[51] Endo, A., Saika, H., Takemura, M., Misawa, N., and Toki, S. (2019) A novel approach to carotenoid accumulation in rice callus by mimicking the cauliflower Orange mutation via genome editing. *Rice*. 12 (1), 1–5.

[52] Shao, G., Xie, L., Jiao, G., Wei, X., Sheng, Z., Tang, S. and Hu, P. (2017) CRISPR/CAS9-mediated editing of the fragrant gene Badh2 in rice. *Chinese Journal of Rice Science*. 31 (2), 216–222.

[53] Wang, G., Wang, C., Lu, G., Wang, W., Mao, G., Habben, J. E., Song, C., Wang, J., Chen, J., Gao, Y. and Liu, J. (2020) Knockouts of a late flowering gene via CRISPR–Cas9 confer early maturity in rice at multiple field locations. *Plant Molecular Biology*. 104 (1), 137–150.

[54] Li, B., Du, X., Fei, Y., Wang, F., Xu, Y., Li, X., Li, W., Chen, Z., Fan, F., Wang, J. and Tao, Y. (2021) Efficient Breeding of Early-Maturing Rice Cultivar by Editing PHYC via CRISPR/Cas9. *Rice*. 14 (1), 1–4.

[55] Huang, L., Li, Q., Zhang, C., Chu, R., Gu, Z., Tan, H., Zhao, D., Fan, X. and Liu, Q. (2020) Creating novel Wx alleles with fine-tuned amylose levels and improved grain quality in rice by promoter editing using CRISPR/Cas9 system. *Plant Biotechnology Journal*. 18 (11), 2164.

[56] Zhang, J., Zhang, H., Botella, J. R., and Zhu, J. (2018) Generation of new glutinous rice by CRISPR/Cas9-targeted mutagenesis of the Waxy gene in elite rice varieties. *Journal of Integrative Plant Biology*. 60 (5), 369–375.

[57] Biswas, S., Tian, J., Li, R., Chen, X., Luo, Z., Chen, M., Zhao, X., Zhang, D., Persson, S., Yuan, Z. and Shi, J. (2020) Investigation of CRISPR/Cas9-induced SD1 rice mutants highlights the importance of

molecular characterization in plant molecular breeding. *Journal of Genetics and Genomics*. 47 (5), 273–280.

[58] Bahariah, B., Masani, M. Y. A., Rasid, O. A., and Parveez, G. K. A. (2021) Multiplex CRISPR/Cas9-mediated genome editing of the FAD2 gene in rice: a model genome editing system for oil palm. *Journal of Genetic Engineering and Biotechnology*. 19 (1), 1–13.

[59] Sun, Y., Zhang, X., Wu, C., He, Y., Ma, Y., Hou, H., Guo, X., Du, W., Zhao, Y. and Xia, L. (2016) Engineering herbicide-resistant rice plants through CRISPR/Cas9-mediated homologous recombination of acetolactate synthase. *Molecular Plant*. 9 (4), 628–631.

[60] Hu, X., Cui, Y., Dong, G., Feng, A., Wang, D., Zhao, C., Zhang, Y. U., Hu, J., Zeng, D., Guo, L. and Qian, Q. (2019) Using CRISPR/Cas9 to generate semi-dwarf rice lines in elite landraces. *Scientific Reports*. 9 (1), 1–7.

[61] Macovei, A., Sevilla, N. R., Cantos, C., Jonson, G. B., Slamet-Loedin, I., Čermák, T., Voytas, D. F., Choi, I. R. and Chadha-Mohanty, P. (2018) Novel alleles of rice eIF4G generated by CRISPR/Cas9-targeted mutagenesis confer resistance to Rice tungro spherical virus. *Plant Biotechnology Journal*. 16 (11), 1918–1927.

[62] Hake, S. and Ross-Ibarra, J. (2015) The natural history of model organisms: genetic, evolutionary and plant breeding insights from the domestication of maize. *Elife*. 4 e05861.

[63] Kumar, V., Vats, S., Kumawat, S., Bisht, A., Bhatt, V., Shivaraj, S. M., Padalkar, G., Goyal, V., Zargar, S., Gupta, S. and Kumawat, G. (2021) Omics advances and integrative approaches for the simultaneous improvement of seed oil and protein content in soybean (Glycine max L.). *Critical Reviews in Plant Sciences*. 40 (5), 398–421.

[64] Liang, Z., Zhang, K., Chen, K., and Gao, C. (2014) Targeted mutagenesis in Zea mays using TALENs and the CRISPR/Cas system. *Journal of Genetics and Genomics*. 41 (2), 63–68.

[65] Malzahn, A. A., Tang, X., Lee, K., Ren, Q., Sretenovic, S., Zhang, Y., Chen, H., Kang, M., Bao, Y., Zheng, X. and Deng, K. (2019) Application of CRISPR/Cas12a temperature sensitivity for improved genome editing in rice, maize, and Arabidopsis. *BMC Biology*. 17 (1), 1–14.

[66] Young, J., Zastrow-Hayes, G., Deschamps, S., Svitashev, S., Zaremba, M., Acharya, A., Paulraj, S., Peterson-Burch, B., Schwartz, C., Djukanovic, V. and Lenderts, B. (2019) CRISPR/Cas9 editing in maize: systematic evaluation of off-target activity and its relevance in crop improvement. *Scientific Reports*. 9 (1), 1–11.

[67] Liu, H. J., Jian, L., Xu, J., Zhang, Q., Zhang, M., Jin, M., Peng, Y., Yan, J., Han, B., Liu, J. and Gao, F. (2020) High-throughput CRISPR/Cas9 mutagenesis streamlines trait gene identification in maize. *The Plant Cell*. 32 (5), 1397–1413.

[68] Wang, B., Zhu, L., Zhao, B., Zhao, Y., Xie, Y., Zheng, Z., Li, Y., Sun, J. and Wang, H. (2019) Development of a haploid-inducer mediated genome editing system for accelerating maize breeding. *Molecular Plant*. 12 (4), 597–602.

[69] Xing, H. L., Dong, L., Wang, Z. P., Zhang, H. Y., Han, C. Y., Liu, B., Wang, X. C. and Chen, Q. J. (2014) A CRISPR/Cas9 toolkit for multiplex genome editing in plants. *BMC Plant Biology*. 14 (1), 1–12.

[70] Svitashev, S., Schwartz, C., Lenderts, B., Young, J. K., and Mark Cigan, A. (2016) Genome editing in maize directed by CRISPR–Cas9 ribonucleoprotein complexes. *Nature Communications*. 7 (1), 1–7.

[71] Zhu, J., Song, N., Sun, S., Yang, W., Zhao, H., Song, W. and Lai, J. (2016) Efficiency and inheritance of targeted mutagenesis in maize using CRISPR/Cas9. *Journal of Genetics and Genomics*. 43 (1), 25–36.

[72] Feng, C., Yuan, J., Wang, R., Liu, Y., Birchler, J. A., and Han, F. (2016) Efficient targeted genome modification in maize using CRISPR/Cas9 system. *Journal of Genetics and Genomics*. 43 (1), 37–43.

[73] Qi, W., Zhu, T., Tian, Z., Li, C., Zhang, W., and Song, R. (2016) High-efficiency CRISPR/Cas9 multiplex gene editing using the glycine tRNA-processing system-based strategy in maize. *BMC Biotechnology*. 16 (1), 1–8.

[74] Char, S. N., Neelakandan, A. K., Nahampun, H., Frame, B., Main, M., Spalding, M. H., Becraft, P. W., Meyers, B. C., Walbot, V., Wang, K. and Yang, B. (2017) An Agrobacterium-delivered CRISPR/Cas9 system for high-frequency targeted mutagenesis in maize. *Plant Biotechnology Journal*. 15 (2), 257–268.

[75] Li, C., Liu, C., Qi, X., Wu, Y., Fei, X., Mao, L., Cheng, B., Li, X. and Xie, C. (2017) RNA-guided Cas9 as an *in vivo* desired-target mutator in maize. *Plant Biotechnology Journal*. 15 (12), 1566–1576.

[76] Li, J., Zhang, H., Si, X., Tian, Y., Chen, K., Liu, J., Chen, H. and Gao, C. (2017) Generation of thermosensitive male-sterile maize by targeted knockout of the ZmTMS5 gene. *Journal of Genetics and Genomics= Yi Chuan Xue Bao*. 44 (9), 465–468.

[77] Liu, C., Li, X., Meng, D., Zhong, Y., Chen, C., Dong, X., Xu, X., Chen, B., Li, W., Li, L. and Tian, X. (2017) A 4-bp insertion at ZmPLA1 encoding a putative phospholipase A generates haploid induction in maize. *Molecular Plant*. 10 (3), 520–522.

[78] Shi, J., Gao, H., Wang, H., Lafitte, H. R., Archibald, R. L., Yang, M., Hakimi, S. M., Mo, H. and Habben, J. E. (2017) ARGOS 8 variants generated by CRISPR-Cas9 improve maize grain yield under field drought stress conditions. *Plant Biotechnology Journal*. 15 (2), 207–216.

[79] Chen, R., Xu, Q., Liu, Y., Zhang, J., Ren, D., Wang, G. and Liu, Y., 2018 (2018) Generation of transgene-free maize male sterile lines using the CRISPR/Cas9 system. *Frontiers in Plant Science*. 1180.

[80] Dong, L., Li, L., Liu, C., Liu, C., Geng, S., Li, X., Huang, C., Mao, L., Chen, S. and Xie, C. (2018) Genome editing and double-fluorescence proteins enable robust maternal haploid induction and identification in maize. *Molecular Plant*. 11 (9), 1214–1217.

[81] Feng, C., Su, H., Bai, H., Wang, R., Liu, Y., Guo, X., Liu, C., Zhang, J., Yuan, J., Birchler, J. A. and Han, F. (2018) High-efficiency genome editing using a dmc1 promoter-controlled CRISPR/Cas9 system in maize. *Plant Biotechnology Journal*. 16 (11), 1848–1857.

[82] Doll, N. M., Gilles, L. M., Gerentes, M. F., Richard, C., Just, J., Fierlej, Y., Borrelli, V. M., Gendrot, G., Ingram, G. C., Rogowsky, P. M. and Widiez, T. (2019) Single and multiple gene knockouts by CRISPR–Cas9 in maize. *Plant Cell Reports*. 38 (4), 487–501.

[83] Lee, K., Zhang, Y., Kleinstiver, B. P., Guo, J. A., Aryee, M. J., Miller, J., Malzahn, A., Zarecor, S., Lawrence-Dill, C. J., Joung, J. K. and Qi, Y. (2019) Activities and specificities of CRISPR/Cas9 and Cas12a nucleases for targeted mutagenesis in maize. *Plant Biotechnology Journal*. 17 (2), 362–372.

[84] Zhong, Y.U., Liu, C., Qi, X., Jiao, Y., Wang, D., Wang, Y., Liu, Z., Chen, C., Chen, B., Tian, X. and Li, J. (2019) Mutation of ZmDMP enhances haploid induction in maize. *Nature Plants*. 5 (6), 575–580.

[85] Zuo, Y., Feng, F., Qi, W., and Song, R. (2019) Dek42 encodes an RNA-binding protein that affects alternative pre-mRNA splicing and maize kernel development. *Journal of Integrative Plant Biology*. 61 (6), 728–748.

[86] Gao, H., Mutti, J., Young, J. K., Yang, M., Schroder, M., Lenderts, B., Wang, L., Peterson, D., St. Clair, G., Jones, S. and Feigenbutz, L. (2020) Complex trait loci in maize enabled by CRISPR/Cas9 mediated gene insertion. *Frontiers in Plant Science*. 11 535.

[87] Jiang, Y. Y., Chai, Y. P., Lu, M. H., Han, X. L., Lin, Q., Zhang, Y., Zhang, Q., Zhou, Y., Wang, X. C., Gao, C. and Chen, Q. J. (2020) Prime editing efficiently generates W542L and S621I double mutations in two ALS genes in maize. *Genome Biology*. 21 (1), 1–10.

[88] Qi, X., Wu, H., Jiang, H., Zhu, J., Huang, C., Zhang, X., Liu, C. and Cheng, B. (2020) Conversion of a normal maize hybrid into a waxy version using *in vivo* CRISPR/Cas9 targeted mutation activity. *The Crop Journal*. 8 (3), 440–448.

[89] Zhang, J., Zhang, X., Chen, R., Yang, L., Fan, K., Liu, Y., Wang, G., Ren, Z. and Liu, Y. (2020) Generation of transgene-free Semidwarf maize plants by gene editing of gibberellin-oxidase20-3 using CRISPR/Cas9. *Frontiers in Plant Science*. 1048.

[90] Gong, C., Huang, S., Song, R., and Qi, W. (2021) Comparative Study between the CRISPR/Cpf1 (Cas12a) and CRISPR/Cas9 Systems for Multiplex Gene Editing in Maize. *Agriculture*. 11 (5), 429.

[91] Liu, L., Gallagher, J., Arevalo, E. D., Chen, R., Skopelitis, T., Wu, Q., Bartlett, M. and Jackson, D. (2021) Enhancing grain-yield-related traits by CRISPR–Cas9 promoter editing of maize CLE genes. *Nature Plants*. 7 (3), 287–294.

[92] Jing, C., Wei, M., Fang, P., Song, R., and Qi, W. (2021) Pollen-specific CRISPR/Cas9 system to increase heritable gene mutations in maize. *Agriculture*. 11 (8), 751.

Chapter 12

Genome Editing in Some Cereal Crops

Priya Chaudhary[1]
Pracheta Janmeda[1,*]
Poornima Pandey[1]
Devendra Singh[2]
and Rahul Kumar[3]
[1] Department of Bioscience and Biotechnology, Banasthali Vidyapith, Rajasthan, India
[2] Department of Chemistry, Mohanlal Sukhadia University, Udaipur, Rajasthan, India
[3] Faculty of Agricultural Sciences, DAV University, Jalandhar, Punjab, India

Abstract

Improving the stability and yield of crops is one of the important challenges for agricultural sciences in the present time through the development of progressive superior cultivars. The editing of the genome provides an uncommon way to improve crops, with better speed and precision. Various techniques of genome editing have been utilized in cereals, e.g., CRISPR/Cas9, transcription activator-like effector nucleases (TALENs), meganucleases, and zinc finger nucleases (ZFNs). Other than these techniques, base editors and double-strand break (DSB) repair mechanisms are also applied for the precise editing of cereal crops. The use of complete genome sequences of various cereal crops in conjunction with genome editing increases the likelihood of meeting important quality and agronomic traits in cereals. This involves adaptive traits to improve the nutritional content and grain quality, higher yields, tolerance to biotic stresses and mitigating the effects of climate change. All traits cannot be improved through the editing of the genome. Therefore, different regulatory and technical challenges are required for the technology to achieve its full potential. However, genome editing has restructured the field of crop improvement and is shaping agricultural practices in combination with other breeding techniques. This chapter summarizes the latest status of genome editing technologies in some common cereal crops like barley, wheat, maize, and rice.

Keywords: agricultural sciences, agronomic traits, biotic stresses, cereal crops, climate change, CRISPR/Cas9, genome editing, yields

---

* Corresponding Author's Email: pracheta@banasthali.in.

In: Genome Editing for Crop Improvement
Editors: Rahul Kumar, Ashutosh Sharma, Yogesh Kumar et al.
ISBN: 979-8-88697-740-0

Introduction

Maize, wheat, rice, and barley are the four main cereal crops that are supplying 42% of all calorie-intake by the overall population across the globe. Maintenance of these staple crops while improving their quality of nutrition and addressing different climatic factors is quite challenging. Thus, it requires the application of various innovative breeding techniques [1]. Conventional techniques such as hybrid breeding, selection, and hybridization were found to be successful in exploring the information of quantitative trait loci and genes regulating the variable traits. The approach of molecular breeding is dependent on the recognition of molecular markers associated with the genes, and subsequently marker-assisted selection (MAS). However, these propagation techniques are majorly dependent on the natural variations that remain in the primary gene pool [2].

Genome editing is a modern tool with several uses in various sectors that includes agriculture for the development of better crops [1]. The effect of genome editing in plants was determined two decades back with low targeted integration frequencies and because of the declaration of novel nucleases inciting double-strand break (DSBs) at particular loci, the frequencies of genome editing were remarkably enhanced [3]. Genome editing emergence allows investigators to incorporate some sequence-specific alterations economically and rapidly into the genome of various cell types (Figure 1).

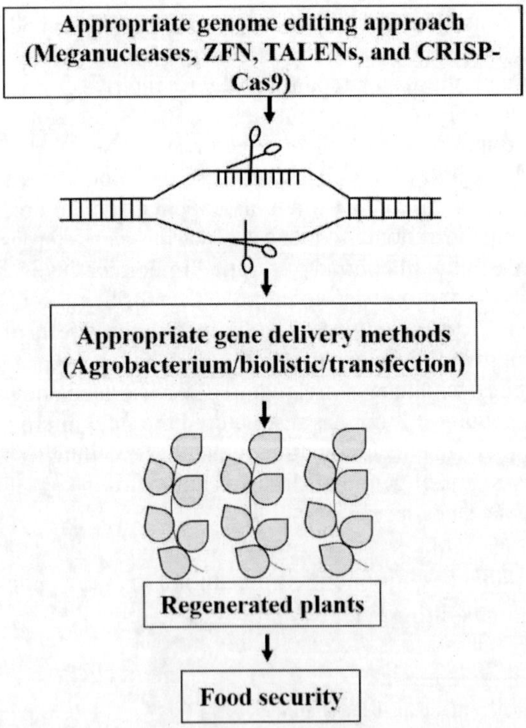

Figure 1. Genome editing technologies for global food security.

The main techniques used for editing genome are homing endonucleases or meganucleases, zinc-finger nucleases (ZFNs), transcription activator-like effector nucleases (TALENs), and clustered regularly interspaced short palindromic repeats (CRISPR)-CRISPR-associated protein 9 (Cas9). Particularly, the ease with TALENs and CRISPR-Cas9 can be customized to identify novel genome sequences that have speeded scientific discoveries and breakthroughs in different disciplines of agricultural sciences, neurosciences, drug discovery, disease modeling, human gene therapy, and synthetic biology [4, 5].

In the maize variety, the AHAS109 and AHAS108 (acetohydroxyacid synthase) genes were modified by applying the chimeric form of DNA/RNA oligonucleotides, and the chances of recurrence were 4-10 times. Moreover, genes that are more difficult to modify by the mutagenesis process can be effectively covered with the techniques of genome editing studied for their different roles. With the advancements in CRISPR/Cas, TALENs, and ZFNs, all these techniques have been implemented for the modification of specific loci or genes in various cereal crops [3].

Crop and trait improvement by modification of a pre-selected sequence of the plant genome in a precise way is the prime aim of plant breeding. It introduces genetic diversity that leads to the formation of better varieties, especially in the case of polyploid crops, which otherwise require longer programs of screening and crossing [6]. The application and development of genome-editing techniques in recent times have restructured the basic analysis in plant sciences and led to the production of new plant varieties.

Genome Editing Techniques

Zinc-Finger Nucleases

ZFNs, which are a fusion between the cleavage domain of the FokI restriction endonuclease and a custom-designed Cys2-His2 zinc-finger protein, were targeted first for widespread use [7, 8]. FokI cleaved domain dimerized the ZFN protein, which cleaves the DNA within a 5-7 bp spacer sequence that carries the separation of two flanking zinc-finger binding sites. Typically, ZFN is composed of 3 to 4 zinc-finger domains, with ~30 amino acid residues of each domain in the ββα motif [9]. The α-helical domain contains DNA recognition residues that interact with 3 bp DNA, with overlap from an adjacent domain. Many zinc-finger domains have been identified to recognize the distinct DNA triplet by using the method of phage display. These domains are tandemly fused using a canonical linker peptide for the generation of polydactyl zinc-finger proteins, which further target a broad range of DNA sequences [4].

TALENs

The genome editing mediated by TALEN has been utilized in various crop plants viz., maize, soybean, barley, rice, and Arabidopsis to recognize the function of different genes whose roles are still not revealed as well as to improve quantitative and qualitative traits [10].

The assembly of TALEN consists of a translocation and secretion signal at the N terminal, a highly conserved acidic transcription activation domain (AD) at C terminal, nuclear localization signal (NLS), and a DNA binding domain at the center. DNA binding domain consists of 33 to 35 tandem repeats and each repeat identifies the sole nucleotide in the sequence of targeted sites. Therefore, the TALEN design is more uncomplicated than ZFNs [11]. The specific nature of TALEN is controlled by the presence of repeat variable di-residue (RVD) at 12 and 13 positions. Different RVDs have been discovered; major RVDs are NG (Asn Gly), NK (Asn Lys), NN (Asn Asn), HD (His Asp), and NI (Asn Ile) that recognizes thymine (T), guanine (G), cytosine (C), and adenine (A) respectively [12]. Two TALENs get translocated to the nucleus through a type III secretion mechanism and attach to the target DNA sequence in opposite direction with 12 to 30 bp of spacing and then it imitates the transcription factor of the host organism to adjust the host gene expression. After the attaching step, FokI dimerized and cleaves at the region of spacer and leads to DSB in the targeted site [3].

CRISPR/Cas9

The methodology of CRISPR cleavage requires a Cas9 nuclease enzyme which cuts 3-4 bases after the protospacer adjacent motif and a 20-nucleotide short synthetic gRNA sequence that binds to the target site [13]. The Cas9 nuclease enzymes consist of an HNH domain and RuvC-like domain, each domain induced to cut on one DNA strand [14]. The implementation of the CRISPR system requires some basic steps viz. validation of edited lines, screening, introduction into host species or cell line, cloning of sgRNA into an appropriate binary vector, and synthesis of a single guide RNA (sgRNA). These basic steps even allow a small laboratory to carry out genome editing to transform a plant. Over the last half decades, the techniques of CRISPR/Cas9 have been utilized extensively as compared to other TALENs and ZFNs methods [15, 16, 17].

Meganucleases

Meganucleases (MegaN), which belong to the family endonucleases can cut and recognize a larger sequence of DNA from 12 to 40 bps, nearly so or unique in most genomes. They are recognized as an ideal tool for genome engineering due to the large recognition sites [18]. Different companies have modified meganucleases based on their application in genome editing. The area that interacts with a specific site of DNA and the area which is responsible for cleavage of DNA in meganucleases were altered by Cellectis Company to connect with variable sequences of DNA from original meganucleases while maintaining their greater degree of specificity and capacity to cut the DNA. A design process-directed nuclease editor (DNE) was progressed by Precision Biosciences, Inc. to create highly engineered meganucleases that strongly modify and target a user-directed location in a genome [19]. There is a disadvantage of meganucleases, in meganucleases, the construction of specific enzymes based on sequence is expensive and time-taken in contrast to other systems. For every genome-engineering target, there is a need for an initial stage of protein engineering to give rise to a custom meganuclease. On that account, these meganucleases are found to be difficult to work with and they are also hampered by disputes over patents [20].

Base Editing

Base editing allows agriculturalists and researchers to generate a novel mutant with better efficiency than conventional breeding methods. Production of gain-of-function and precise point mutation can improve different traits, such as a point mutation in acetolactate synthase

(ALS) gene providing resistance against imidazolinones and sulfonylureas herbicides [21]. Generation of point mutation through (DNA break) DSB-induced homology-directed repair (HDR) is quite challenging in plant cells. Various crop plants of agronomic interest have been modified with adenine base editors (ABEs) and cytosine base editors (CBEs).

In a study, BE3 results in an efficient point mutation in wheat, rice, and maize plants whereas the Target-AID (activation-induced deaminase) editor generates prominent editing in tomato and rice plants [22]. Recently in a report, researchers optimized the structure of ABE7.10 for the base editing of adenine in wheat and rice plants and utilized the resulting editor in the regeneration of plants through protoplast fusion [23]. These reports determined that base editing is a better approach for the genetic engineering of polyploid species of plants. From consumer and regulatory perspectives, the RNA delivery of BEs into target plants is important as the integration of transgene from plasmid delivery results in the generation of plants with the status of GMOs. The RNA delivery of base editors would allow DNA-free editing that inhibits the formation of GMO crops [24].

Double-Strand Break (DSB) Repair Mechanism

Double-strand break (DSB) has to be repaired before the replication of the genome as the repair of DSBs is found to be important for the survival of different organisms. Therefore, the repair of the DSBs can be done via two different pathways viz. non-homologous end-joining (NHEJ) and homologous recombination (HR). In the HR pathways, the sequence is linked into the region that is known to each other whereas in NHEJ the sequence does not involve the rejoining step of the two double strands [25].

In the last few years, a lot of information has been gathered regarding the DSB repair in crops. The function of various pathways of DSB repair and their respective frequencies make it possible to conclude the alterations that happen while the DSB is repaired in the genome of the plant. The sequenced genomes of different crops are accessible and allow the analysis of evolutionary changes at the complete genomic level. So, the latest information regarding the DSB repair can describe why the genomes progressed in organization and size that is in today's complexity. However, the information on DSB repair in crops is not complete, and the query regarding the DSB-directed translocation DSB-induced sister chromatid exchange has also not been answered yet. The main challenge is to reform and modify the genome manipulation techniques further. Meiotic recombination induction by site-directed DSBs may also be a crucial step along with the challenges in the coming years [26].

Genome Editing in Cereal Crops

Genome editing has been applied to attain major quality and agronomic traits in cereal crops as discussed below and presented in Figure 2.

Wheat

The first reported utilization of CRISPR-Cas9 was done in combination with TALENs to generate a stable wheat (*Triticum aestivum*) plant that includes targeted knockout of Mildew

Locus (Mlo) gene which confers resistance against *Blumeria graminis f.* sp. *tritici* [27]. Targeting of α-gliadin genes to decrease the content of gluten in grains [28], TaGW2 to increase the weight of grain [29], CENH3 and TaMTL for the induction of haploid plant [30], TaQsd1 to decrease sprouting while preharvest and TaZIP4-B2 to understand the meiotic homologous crossover [31].

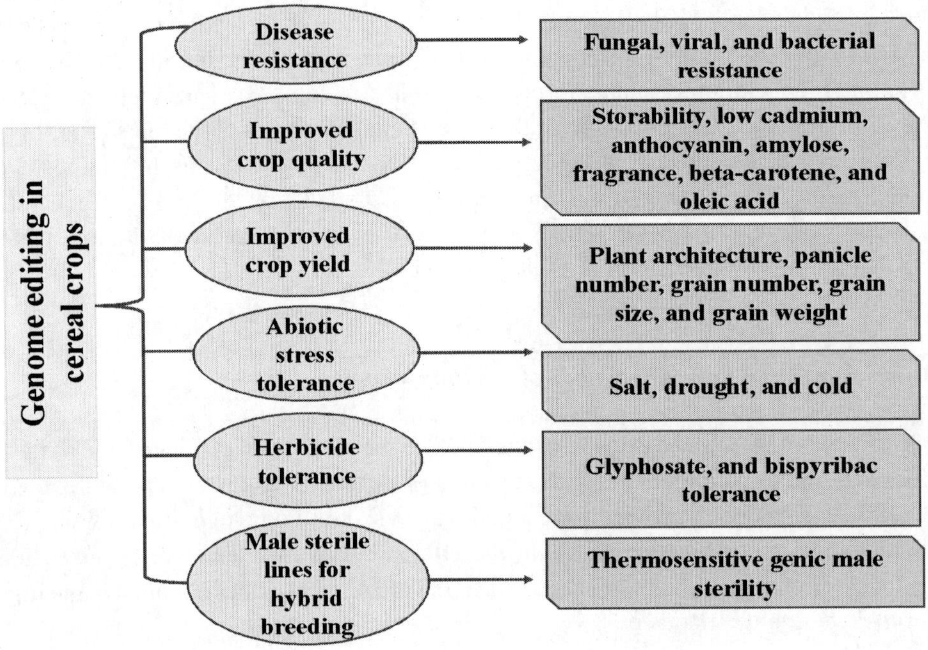

Figure 2. Genome editing in cereal crops.

Two modern approaches of biotechnology represent promising tools for the generation of wheat that is considered to be non-toxic for CD patients: CRISPR/Cas9 and RNA interference (RNAi). RNAi was utilized to downregulate the three gliadin gene types that result in the reduction of 92% of gliadin response, as determined utilizing the R5 monoclonal antibody (mAb) assay and reduction of DQ8 and DQ2 epitope by 10-100-fold in T-cell tests. Nowadays, RNAi was applied to silence the ω-gliadins. It indirectly targets these gluten genes through the help of RNA transcripts. However, this perspective needs stable transgene expression and genetic modifications. Moreover, government regulations for genetically modified (GM) products are time-consuming, expensive, and require food-safety evaluation before the product reaches the market [28].

CRISPR/Cas9 genome editing technique has the capacity to precisely and simultaneously alter multiple gliadin-encoded epitopes while maintaining the quality of gliadin proteins [28]. This technique has also successfully induced targeted mutation in a single copy of the gene across all three homologous loci in case of hexaploid wheat [32]. In CRISPR/Cas9 genome editing, a single guide RNA (sgRNA) directs the Cas9 endonuclease to induce a double-stranded break at the targeted DNA site. During the DNA repair mechanism, a plant cell undergoes mistakes which lead to the deletion of one or many nucleotides. In tandemly repeated gliadin genes, there is a concurrent break of double-stranded DNA in consecutive genes that results in the removal of larger fragments of DNA containing one or many gliadin genes [28].

Different techniques to modify certain genes and locus in the case of wheat are presented in Table 1.

Table 1. Genome editing in wheat crop with targeted gene and molecular functions

Genome editing strategy	Targeted gene	Molecular function	References
RNA interference	Vernalization gene, TaVRN2	Reduction of TaVRN1 by 60% led to earlier flowering	[33]
	TaVRN1	Reduction of TaVRN1 by 80% led to an increase in number of leaves and delay the flowering time by 14 to 19 days	[34]
RNAi silencing	ω-gliadins	Wheat variety deficient in ω-gliadins showed increased stability of protein with better dough properties under different growth conditions	[35]
RNAi silencing	γ-gliadins	Reduction of γ-gliadins of wheat results in elevated glutenins content, better dough, and resistance against over-mixing	[36]
RNAi expression plasmid	Selectively target ω-, α-, and γ-gliadins, and Low Molecular Weight (LMW) glutenins	Reduction of gluten epitopes related to coeliac diseases	[37]
RNAi	Targeting the myosin-5 gene (FaMyo5) from *F. asiaticum*	Results resistance against diseases in the case of wheat	[38]
RNAi	Fungal glucanosyltransferase genes (GTF1 and GTF2), and the virulence effector gene Avra10 silencing	Affects resistance of wheat to *Blumeria graminis*	[39]
CRISPR/Cas9	Transcription factor, TaNFXL1 gene	Confers resistance in wheat against *Fusarium graminearum*	[40]
CRISPR-Cas	TaEDR1	Resistance against *Blumeria graminis* f. sp. tritici	[41]
TALEN	TaMIo1	Resistance against *Blumeria graminis* f. sp. tritici	

Rice

Rice (*Oryza sativa* L.) is the food source for around 3.5 billion of the population across the globe [42, 90]. Over the last decades, molecular, mutational, and classical propagation methods have produced an enormous rise in the fertility of rice along with the generation of some new varieties of rice. In recent decades, the yield of rice has been stagnant due to various factors including climate change, phytopathogens, pests, and some other environmental problems which pose an ultimatum to worldwide food security. There is an immediate focus on this issue to produce more rice to manage the requirement of the still growing population. The emergence of different genome-editing techniques and advances in genomics has transformed the field of agriculture. Among these, CRISPR/Cas9 is the widely accepted and latest one due to its high efficiency, robustness, and simplicity respectively [43]. Some examples of trait improvement in the case of rice are presented in Table 2.

Table 2. Genome editing in rice with targeted gene and molecular functions

Genome editing strategy	Targeted gene	Molecular function	References
TALENs	Susceptibility gene Os11N3 of rice bacterial blight	Results in the production of disease-resistant cell lines in rice	[44]
TALENs	*Oryza sativa* betaine aldehyde dehydrogenase 2 (OsBADH2) gene	Enhances the rice fragrance	[45]
TALENs	OsSWEET13	Better resistance against bacterial blight	[44]
TALENs	Os09g29100	Improved resistance against bacterial leaf streak	[46]
CRISPR/Cas 9	Hs2, Hd4, and Hd5	Rice variety early maturity	[47]
CRISPR/Cas 9 system	OsHXK1 gene	Improving the rice yield and photosynthetic efficiency	[48]
CRISPR/Cas 9 system	CSA knockout	Reverse the photosensitive sterility	[49]
CRISPR/Cas 9 system	Gn1a knockout	Increases the grain number	
CRISPR/Cas 9 system	GW2/GW5/TGW6 knockout	Increases the grain weight	[50]
CRISPR/Cas 9 system	GS3 knockout	Increases the size of grains	[49]
CRISPR/Cas 9 system	SLR1 single-base editing	Provide resistance against herbicide	[60]
CRISPR/Cas 9 system	EPSPS replace	Provide resistance against herbicide	[49]
CRISPR/Cas 9 system	Ideal Plant Architecture 1 (IPA1) QTL	Plant height reduction and rise in grain yield and tiller number	[51]
CRISPR/Cas 9 system	Targeting ABA receptor gene Pyrabactin Resistance 9 (OsPYL9)	Increasing the yield of grain and further editing of PYL6, PYL4, and PYL1 increases the grain number by 31% which ultimately led to greater yield in case of mutants than the wild variety	[52]
CRISPR/Cas 9 system	OsMYB30	Provide tolerance against cold	
CRISPR/Cas 9 system	OsAnn3	Provide tolerance against cold	
CRISPR/Cas 9 system	OsmiR535	Provide tolerance against osmosis and salinity	
CRISPR/Cas 9 system	OsUBQ	Provide tolerance against drought	
CRISPR/Cas 9 system	OsPYL9	Provide tolerance against drought	
CRISPR/Cas 9 system	OsSAPK2	Reduced osmotic, drought, and salinity stress tolerance	
CRISPR/Cas 9 system	eIF4G	Resistance against Rice tungro spherical virus (RTSV)	[53]
CRISPR/Cas 9 system	OsPi21, OsALB1, OsERF922	Resistance against Rice blast (*Magnaporthe oryzae*)	[54]
CRISPR/Cas 9 system	OsSWEET11, OsSWEET13, OsSWEET14	Resistance against Bacterial leaf blight (*Xanthomonas oryzae* pv. Oryzae)	[55]
CRISPR/Cas 9 system	OsWaxy	Reduction in the content of amylose in XS134 and Zhonghua11	[56]
CRISPR/Cas 9 system	OsOr	Increase in the content of beta-carotene	[57]
CRISPR/Cas 9 system	OsSPL16/GW8	Increase in size of grain in Indica (VP4892)	[58]
Base editing	OsCDC48	Regulate death and senescence	[59]
Base editing	NRT1.1B gene	Enhances the efficiency of nitrogen use	[60]
Base editing	OsALS1, OsACC	Provide resistance against herbicide	[61] [62]
Base editing	SLR1, SPL14, SPL16, APL18, APL17	Improved yield	[63]
Base editing	OsCDC48, OsNRT1.1B, OsSPL14	Improved yield	[59]
Base editing	Wx or GBSSI	High content of amylose	[64]

Maize

Maize (*Zea mays* L.) is the other most consumed crop globally. It is easy to cultivate under varied soil and environmental conditions and myriad end uses have made this crop a desirable one across the globe. This crop is also utilized in the form of raw material for various industries, as feed for livestock, and as biofuel. However, for the improvement of its agronomic traits, scientists are persistently working to improve its genome with the help of various genetic techniques [65]. Before the era of precision gene editing and transgenesis, the genetic improvement of maize depended mainly on mutation breeding and traditional programs. For delivering CRISPR/Cas9 systems for any editing program in maize, *Agrobacterium* directed modification is the most suitable one. However, the low efficiency of transformation and recalcitrant nature of calli are considered to be the major factors that limit the prosperity of big modifications in maize varieties. To overcome this restriction, Zhang et al. [66] formed a new ternary vector in which morphogenic regulator genes have been incorporated into the CRISPR/Cas9 system. The delivery of this vector is directed by *Agrobacterium* and it was found to be quite favourable and resulted a positive effect in the ability of recalcitrant maize.

Table 3. Genome editing in maize with targeted gene and molecular functions

Genome editing strategy	Targeted gene	Molecular function	References
CRISPR/Cas9 system	Gibberellin-oxidase20-3 gene knock out	Gibberellin biosynthesis pathway blockage and generation of semi-dwarf maize plant with resistance to lodging	[72]
CRISPR/Cas9 system	BRACHYTIC2 gene	BRACHYTIC2 encodes an ATP binding cassette type B (ABCB) which is included in the transfer of auxin. BRACHYTIC2 targeting induced a one base pair frameshift mutation that give rise to a premature stop codon in the region of exon 5 and produced a semi-dwarf phenotype which is very much similar to natural br2 mutations	
CRISPR/Cas9 system	Wx1 gene	Granule-bound starch synthase encoded by Wx1 gene. Targeting of Wx1 gene results in the synthesis of amylose in maize	[73]
Zinc Finger Nucleases (ZFNs)	IPK1 gene	Targeting of IPK1 gene results in the decrease of phytate content in corn	[74]
CRISPR/Cas9 system	Phytic acid biosynthesis gene-ZmIPK	CRISPR/Cas9 induces targeted mutations in ZmIPK gene with 13.1% of efficiency in contrast to 9.1% of efficiency with TALENs	[65]
CRISPR/Cas 9	Acetolactate synthase (ALS2 gene) editing	Results in the production of mutant corn resistant to chlorsulfuron	[17]
ZFN	Targeted cleavage of Inositol-1,3,4,5,6-pentakisphosphate kinase 1 (IPK1)	Tolerance to herbicide and alterations in the profile of inositol phosphate	[3]

Various types of research have been done to develop a plan for doing genome-editing by applying the CRISPR/Cas9 system. Cas9 protein which was optimized in maize codon is found to be worthwhile for directing mutation in the phytoene synthase gene [67]. The direct delivery of CRISPR/Cas9 ribonucleoprotein complexes and the application of dmc-1 promoter-driven Cas9 [68] were improved further for precise editing in maize. Under drought conditions, the development of maize variety with an improved yield of grains is considered to be a feature of genome editing. According to a report, the correct insertion of GOS2 maize promoter within the region of 5'-UTR of ARGOS8 has raised the transcript of ARGOS8 and thereby increased the yield of grain at the flowering stage under drought conditions [69]. Nowadays, the need for baby, waxy, and sweet corn varieties is increasing. Keeping this thing in view, Dong et al. [70] have generated a waxy and super sweet corn type by mutating WX and SH2 genes that are included in the metabolism of sugar and starch.

Liu et al. [71] have utilized genetic mapping with CRISPR/Cas9 mutagenesis to confirm the role of underlying genes in quantitative trait locus (QTL). By the integration of these two perspectives, the function of 743 genes was surveyed in terms of nutritional and agronomic importance in maize. This collective knowledge for developing mutants will serve as a potential resource for generating a highly precise breeding system in maize varieties [69]. Different genome editing techniques have been identified to modify specific genes in the case of maize are presented in Table 3.

Barley

Barley is the fourth major cereal cultivated and utilized across the globe. It is majorly utilized as animal feed, and only a small part of this is being utilized for human consumption in the form of alcoholic beverages [75]. Recently, the completion of pan-genome sequencing of barley will contribute to the application and development of genomic-based tools for genome analysis, and gene discovery to support the breeding of barley, including CRISPR/Cas9 techniques that require a known genome to determine off-target and on-target activities [76]. An early effort for the establishment of genome editing includes TALENs which did not target the functional area but the phytase HvPAPhy-a promoter region. This study shows that the barley was manageable to edit without the requirement for primary transformants. Moreover, editing techniques in which the double-strand breaks in the targeted site were introduced by the *Agrobacterium*-directed delivery system of Cas9 confirm that the efficiency of editing is not a bottleneck [1].

The transformation of barley is majorly done with the help of immature embryo explant and the efficiency of modification is based on the genotype. The 'Golden Promise' cultivar is the highly efficient transformable genotype and the genome assembly for this genotype is also published recently. Till now, genome editing in the case of barley has been majorly done in the 'Golden Promise' cultivar, but there is a growing demand to utilize this editing technique directly for other superior varieties [77]. Gerasimova et al. [78] reported the cultivation of protoplasts obtained from a Siberian barley variety by modifying the protocols of tissue culture. Recently, T0 individuals of four commercial varieties of Australia were modified by using the *Agrobacterium* directed delivery system of CRISPR/Cas9. These genome-editing and transformation experiments may suffer from the low alteration possibilities of some accessions. However, the restricted receptors of barley have limited genetic background and agronomic value, resulting in constriction in the application of genome editing techniques for precise pyramiding and fine-tuning traits of interest in new varieties [20].

Hisano and Sato [79] have recognized loci that control transformation amenability in the chromosome regions: 3H and 2H in an F2 population which was obtained from a cross between varieties Haruna Nijo and Golden Promise. The introduction of these genomic regions at a target site in the target haplotype may raise the ability of transformation and capabilities of genome editing. A few studies that reported the improvement in barley traits through editing techniques are described below in Table 4.

Table 4. Genome editing in barley with targeted gene and molecular functions

Genome editing strategy	Targeted gene	Molecular function	References
CRISPR/Cas9	D hordein gene	Creation of new D hordein allele by using CRISPR/Cas9 that allows the propagation of new varieties with improved quality malt	[80]
Streptococcus pyrogenes Cas9 (SpCas9)-mediated knockout strategy	HvMORC1	Increased resistance against fungal pathogen	[81]
CRISPR/Cas9 gene editing	Site-specific mutations in Arabidopsis *At2OGO*	Provide resistance against *Fusarium* head blight in barley	[82]
CRISPR/Cas9 system	Inhibit geminivirus replication by targeting wheat dwarf virus strain	Provide resistance in response to wheat dwarf virus in case of barley	[83]
CRISPR/*Sp*Cas9	CRISPR/*Sp*Cas9-mediated double knockout of barley Microrchidia MORC1 and MORC6a	Participation in plant growth, silencing of transcriptional genes, and plant immunity	[84]

Challenges of Genome Editing in Cereals

Polyploidy

The presence of one or extra complete sets of chromosomes within a being is called polyploidy. It has been divided further into three classes viz. segmental allopolyploids, allopolyploids, and autopolyploids. The genome of polyploids is quite complicated in various ways [85]. They are hard to sequence because of their larger genomes and the presence of repetitive sequences [86].

Sometimes knockout and knockdown mutations of genes produce no phenotypic alterations as the result of paralogous copies of different genes. This type of issue is faced while dealing with editing technique in polyploidy crops [3]. Apart from this issue, the presence of various copies of genes makes it hard to attain precise editing in a certain copy of a gene. Sometimes manipulation of desired traits requires modifications in all paralogs which can greatly hamper the ability of transformation [87]. Recently, two ALCATRAZ (ALC) homologs of tetraploid rape oilseed were transformed using CRISPR-Cas9 constructs. Additionally, transgenic T1 plants with four alc mutant alleles were obtained using a single target sequence. The modification of multiple homologs through CRISPR-Cas9 without any mutation in the background will result in the utilization of mutant genotypes in breeding programs [3].

Transformation Efficiency

Modification of the genome for the improvement of crops is an innovative revolution with some constraints to its implementation. Distinguishing genetic regulations and genetic variations in different plant models and crop species has filled out our understanding to modify the specific gene in the crop improvement system. The exploration of genome editing for the development of certain crops largely depends on the efficiency of transformation [88].

Two consecutive and distinct types of transformation are stable transformation and transient transformation. Both these transformations are important for fundamental research and plant biotechnology. However, the stable transformation method is important in the development of transgenic plants with genetic variation. In this method, the regeneration step has a more restrictions in contrast to the stable transgene integration. From last few decades, the transformation directed through particle bombardment and *Agrobacterium* has been broadly utilized in agriculture field [89]. However, their application is still not effective for various crop varieties. The complexity that arise are: transfer of genes through the bombardment of particles exhibits less precision, DNA titer is low, lower frequency of stable transformation, and formation of transgene containing plants is highly time consuming. Advancement should be done to raises the intact single-copy expression cassette delivery with less damage to different tissues of the crops. Further study programs should be done for the magnification of regeneration and plant transformation responses by targeting a broad range of genotypes and tissues [3].

Conclusion and Future Prospects

Genome editing techniques are becoming a popular option for crop improvement and functional genomics. Many recent studies reported that these editing techniques are being utilized for their unmatched understanding of crop improvement and plant biology through targeted and rapid association with breeding and mutagenesis. Because of the various attractive features such as amenability, high specificity, efficiency, and simplicity, the genome editing techniques discussed here are revolutionizing the way by which crop improvements have been done and providing the way for the next generation of breeding. Still, the techniques of genome editing face various challenges in their application. So, there is a requirement to cope with these challenges and allow the successful utilization of editing techniques for the modification of different crops.

Out of various challenges, the efficiency of transformations is the major one that needs to be evaluated for the varied cultivars. For different cereal crops, the protocols of transformation are already streamlined albeit they are genotype-specific. In this case, genome editing cannot be applied effectively in the genome of commercial varieties. Another concern is related to the requirement of PAM which makes it difficult to undergo genome editing in the absence of a particular sequence of PAM. Despite the availability of various genome editing techniques, there are still restrictions in the utilization of CRISPR tools for the modification of varieties due to some global regulations. The production of healthier and new food can be increased by overcoming the restrictions related to CRISPR tools. Plants developed by the CRISPR/Cas9 system might be considered non-genetically modified organisms. So, for their acceptance at the level of the field, we can implement the technology of CRISPR/Cas9 in agriculture that will transform the fertility of crops and tend to be coherent with the wave of a green revolution that ensures the expanding population of the globe for their nutritional and food requirements.

Acknowledgments

The authors would like to warmly thank the Vice-Chancellor of Banasthali Vidyapith, Rajasthan, for providing excellent research facilities. We would also like to acknowledge the Bioinformatics Center, Banasthali Vidyapith, supported by DBT and DST, for providing computation and networking support through the FIST and CURIE programs at the Department of Bioscience and Biotechnology, Banasthali Vidyapith, Rajasthan, India.

References

[1] Matres, J. M., Hilscher, J., Datta, A., Armario-Nájera, V., Baysal, C., He, W., Huang, X., Zhu, C., Valizadeh-Kamran, R., Trijatmiko, K. R., Capell, T., Christou, P., Stoger, E., Slamet-Loedin, I. H. (2021). Genome editing in cereals crops: an overview. *Transgenic Res.*, 30: 461-498. https://doi.org/10.1007/s11248-021-00259-6.

[2] Ahmer, S., Gill, R. A., Jung, K.-H., Faheem, A., Qasim, M. U., Mubeen, M., Zhou, W. (2020). Conventional and molecular techniques from breeding to speed breeding in crop plants: advances and future outlook. *Int J Mol Sci.,* 21(7): 2590. https://doi.org/10.3390/ijms21072590.

[3] Ansari, W. A., Chandanshive, S. U., Bhatt, V., Nadaf, A. B., Vats, S., Katara, J. L., Sonah, H., Deshmukh, R. (2020). Genome editing in cereals: Approaches, applications, and challenges. *Int. J. Mol. Sci.*, 21(11), 4040. https://doi.org/10.3390/ijms21114040.

[4] Gaj, T., Sirk, S. J., Shui, S.-I., Liu, J. (2016). Genome-Editing technologies: principles and applications. *Cold Spring Harb Perspective Biol.*, 8(12): a023754. https://doi.org/10.1101/cshperspect.a023754.

[5] Guha, T. K., Wai, A., Hausner, G. (2017). Programmable genome editing tools and their regulation for efficient genome engineering. *Comput Struct Biotechnol J.*, 15: 146-160. https://doi.org/10.1016/j.csbj.2016.12.006.

[6] Jansing, J., Schiermeyer, A., Schillberg, S., Fischer, R., Bortesi, L. (2019). Genome editing in agriculture: technical and practical considerations. *Int. J. Mol. Sci.*, 20(12), 2888. https://doi.org/10.3390/ijms20122888.

[7] Carroll, D. (2011). Genome engineering with zinc-finger nucleases. *Genetics*, 188(4): 773-782. https://doi.org/10.1534/genetics.111.131433.

[8] Paschon, D. E., Lussier, S., Wangzor, T., Xia, D. F., Li, P. W., Hinkley, S. J., Scarlott, N. A., Lam, S. C., Waite, A. J., Truong, L. N., Gandhi, N., Kadam, B. N., Patil, D. P., Shivak, D. A., Lee, G. K., Holmes, M. C., Zhang, L., Miller, J. C., & Rebar, E. J. (2019). Diversifying the structure of zinc finger nucleases for high-precision genome editing. *Nat Commun.*, 10: 1133. https://doi.org/10.1038/s41467-019-08867-x.

[9] Chandrasegaran, S. and Carroll, D. (2016). Origins of programmable bucleases for genome engineering. *J Mol Biol.*, 428(5 Pt B): 963-989. https://doi.org/10.1016/j.jmb.2015.10.014.

[10] Aglawe, S. B., Barbadikar, K. M., Mangrauthia, S. K., Madhav, M. S. (2018). New breeding techniques "genome editing" for crop improvement: applications, potentials and challenges. *3 Biotech*, 8(8): 336. https://doi.org/10.1007/s13205-018-1355-3.

[11] Jankele, R. and Svoboda, P. (2014). TAL effectors: tools for DNA targeting. *Brief Funct Genomics.*, 13(5): 409-419. https://doi.org/10.1093/bfgp/elu013.

[12] Mak, A. N. S., Bradley, P., Bogdanove, A. J., & Stoddard, B. L. (2013). TAL effectors: function, structure, enginerring and applicatons. *Curr Opin Struct Biol.*, 23: 93-9. https://doi.org/10.1016/j.sbi.2012.11.001.

[13] Wu, X., Kriz, A. J., Sharp, P. A. (2014). Target specificity of the CRISPR-Cas9 system. *Quant Biol.*, 2(2): 59-70. https://doi.org/10.1007/s40484-014-0030-x.

[14] Hsu, P. D., Lander, E. S., Zhang, F. (2014). Development and applications of CRISPR-Cas9 for genome engineering. *Cell*, 157(6): 1262-1278. https://doi.org/10.1016/j.cell.2014.05.010.

[15] Jaganathan, D., Ramasamy, K., Sellamuthu, G., Jayabalan, S., Venkataraman, G. (2018). CRISPR for crop improvement: an update review. *Front. Plant Sci*. https://doi.org/10.3389/fpls.2018.00985.

[16] Khan, S. H. (2019). Genome-editing technologies: concept, pros, and cons of various genome-editing techniques and bioethical concerns for clinical application. *Mol Ther Nucleic Acids*., 16: 326-334. https://doi.org/10.1016/j.omtn.2019.02.027.

[17] Kamburova, V. S., Nikitina, E. V., Shermatov, S. E., Buriev, Z. T., Kumpatla, S. P., Emani, C., Abdurakhmonov, I. Y. (2017). Genome editing in plants: an overview of tools and applications. *Int. J. Agron*., 2017, 7315351. https://doi.org/10.1155/2017/7315351.

[18] Silva, G., Poirot, L., Galetto, R., Smith, J., Montoya, G., Duchateau, P., Pâques, F. (2011). Meganucleases and other tools for targeted genome engineering: perspectives and challenges for gene therapy. *Curr Gene Ther*., 11(1): 11-27. https://doi.org/10.2174/156652311794520111.

[19] Ashworth, J., Taylor, G. K., Havranek, J. J., Quadri, S. A., Stoddard, B. L., Baker, D. (2010). Computational reprogramming of homing endonuclease specificity at multiple adjacent base pairs. *Nucleic Acid Res*., 38(16): 5601-8. https://doi.org/10.1093/nar/gkq283.

[20] Abdallah, N. A., Prakash, C. S., McHughen, A. G. (2015). Genome editing for crop improvement: Challenges and opportunities. *GM Crops Food*., 6(4): 183-205. https://doi.org/10.1080/2165698.2015.1129937.

[21] Voytas, D. F. and Gao, C. (2014). Precision genome engineering and agriculture: opportunities and regulatory challenges. *PLos Biol*., 12, e1001877. https://doi.org/10.1371/journal.pbio.1001877.

[22] Molla, K. A., Sretenovic, S., Bansal, K. C., Qi, Y. (2021). Precise plant genome editing using base editors and prime editors. *Nat. Plants*, 7: 1166-1187. https://doi.org/10.1038/s41477-021-00991-1.

[23] Azameti, M. K. and Dauda, W. P. (2021). Base editing in plants: applications, challenges, and future prospects. *Front. Plant Sci*. https://doi.org/10.3389/fpls.2021.664997.

[24] Rees, H. A. and Liu, D. R. (2018). Base editing: precision chemistry on the genome and transcriptome of living cells. *Nat Rev Genet*., 19(12): 770-788. https://doi.org/10.1038/s41576-018-0059-1.

[25] Lieber, M. R. (2010). The mechanism of double-strand break repair by the nonhomologous DNA end joining pathway. *Annu Rev Bioschem*., 79: 181-211. https://doi.org/10.1146/annurev.biochem.052308.093131.

[26] Puchta, H. (2005). The repair of double-strand breaks in plants: mechanisms and consequences for genome evolution. *J. Exp. Bot*., 14. https://doi.org/10.1093/jxb/eri025.

[27] Borrelli, V. M. G., Brambilla, V., Rogowsky, P., Morocco, A., Lanubile, A. (2018). The enhancement of plant disease resistance using CRISPR/Cas9 technology. *Front Plant Sci*., 9: 1245. https://doi.org/10.3389/fpls.2018.01245.

[28] Jouanin, A., Gilissen, L. J. W. J., Schaart, J. G., Leigh, F. J., Cockram, J., Wallington, E. J., Boyd, L. A., van den Broeck, H. C., van der Meer, I. M., America, A. H. P., Visser, R. G. F., Smulders, M. J. M. (2020). CRISPR/Cas9 gene editing of gluten in wheat to reduce gluten content and exposure-reviewing methods to screen for coeliac safety. *Front. Nutr*., https://doi.org/10.3389/fnut.2020.00051.

[29] Zhang, Y., Li, D., Zhang, D., Zhao, X., Cao, X., Dong, L., Liu, J., Chen, K., Zhang, H., Gao, C., Wang, D. (2018). Analysis of the functions of TaGW2 homoeologs in wheat grain weight and protein content traits. *Plant J*., 94(5), 857–866. doi: 10.1111/tpj.13903.

[30] Wang, N., Gent, J. I., Dawe, R. K. (2021). Haploid induction by a maize cenh3 null mutant. *Sci Adv*., 7(4): eabe2299. https://doi.org/10.1126/sciadv.abe2299.

[31] Smedley, M. A., Hayta, S., Clarke, M., Harwood, W. A. (2021). CRISPR-Cas9 based genome editing in wheat. *Curr. Protoc*., Htpps://doi.org/10.1002/cpz1.65.

[32] Bratz, J., Harloff, H.-J., Mascher, M., Stein, N., Himmelbach, A., Jung, C. (2017). CRISPR-Cas9 targeted mutagenesis leads to simultaneous modifications of different homoeologous gene copies in polyploid oilseed rape (*Brassica napus*). *Plant Physiol*., 174(2): 935-942. https://doi.org/10.1104/pp.17.00426.

[33] Borisjuk, N., Kishchenko, O., Eliby, S., Schramm, C., Anderson, P., Jatayev, S., Kurishbayev, A., Shavrukov, Y. (2019). Genetic modifications for wheat improvement: from transgenesis to genome editing. *BioMed Res. Int*., 2019, 6216304. https://doi.org/10.1155/2019/6216304.

[34] Loukoianov, A., Yan, L. and Blechl, A. (2005). "Regulation of VRN-1 vernalization genes in normal and transgenic polyploid wheat,." *Plant Physiol*., 138(4): 2364–2373.

[35] Altenbach S. B. and Allen, P. V. (2011). "Transformation of the US bread wheat 'Butte 86' and silencing of omega-5 gliadin genes," *GM Crops*, 2: 66–73.

[36] Gil-Humanes, J., Wang, Y., Liang, Z., Shan, Q., Ozuna, C. V., Sánchez-León, S., Baltes, N. J., Starker, C., Barro, F., Gao, C., & Voytas, D. F. (2017). High-efficiency gene targeting in hexaploid wheat using DNA replicons and CRISPR/Cas9. *Plant J.*, 89: 1251–1262.

[37] Barro, F., Iehisa, J. C. M., Giménez, M. J., García-Molina, M. D., Ozuna, C. V., Comino, I., Sousa, C., & Gil-Humanes, J. (2016). "Targeting of prolamins by RNAi in bread wheat: effectiveness of seven silencing-fragment combinations for obtaining lines devoid of coeliac disease epitopes from highly immunogenic gliadins,." *Plant Biotechnol. J.*, 14(3): 986–996.

[38] Song, X. S., Gu, K. X., Duan, X. X., Xiao, X. M., Hou, Y. P., Duan, Y. B., Wang, J. X., & Zhou, M. G. (2018). "A myosin5 dsRNA that reduces the fungicide resistance and pathogenicity of *fusarium asiaticum*,." *Pestic Biochem Physiol.*, 150: 1–9.

[39] Nowara, D., Gay, A., Lacomme, C., Shaw, J., Ridout, C., Douchkov, D., Hensel, G., Kumlehn, J., & Schweizer, P. (2012). "HIGS: host-induced gene silencing in the obligate biotrophic fungal pathogen *Blumeria graminis*," *Plant Cell*, 22(9): 3130–3141.

[40] Brauer, E. K., Balcerzak, M., Rocheleau, H., Leung, W., Schernthaner, J., Subramaniam, R., Ouellet, T. (2020). Genome editing of a Deoxynivalenol-induced transcription factor confers resistance to *Fusarium graminearum* in wheat. *Mol. Plant-Microbe Interact.*, 33(3). https://doi.org/10.1094/MPMI-11-19-0332-R.

[41] Langner, T., Kamoun, S., Belhaj, K. (2018). CRISPR Crops: plant genome editing towards disease resistance. *Annu. Rev. Phytopathol.*, 56: 479-512. https://doi.org/10.1146/annurev-phyto-080417-050158.

[42] Muthayya, S., Sugimoto, J. D., Montgomery, S., Maberly, G. F. (2014). An overview of global rice production, supply, trade, and consumption. *Ann. N. Y. Acad. Sci.*, 1324(1): 7-14. https://doi.org/10.1111/nyas.12450.

[43] Mishra, R., Joshi, R. J., Zhao, K. (2018). Genome editing in rice: recent advances, challenges, and future implications. *Front. Plant Sci.* https://doi.org/10.3389/fpls.2018.01361.

[44] Li, T., Liu, B., Spalding, M. H., Weeks, D. P., and Yang, B. (2012). High-efficiency TALEN-based gene editing produces disease-resistant rice. *Nat. Biotechnol.*, 30: 390–392. doi: 10.1038/nbt.2199.

[45] Shan, Q., Zhang, Y., Chen, K., Zhang, K., and Gao, C. (2015). Creation of fragrant rice by targeted knockout of the OsBADH2 gene using TALEN technology. *Plant Biotechnol. J.*, 13: 791–800. doi: 10.1111/pbi.12312.

[46] Cai, L., Cao, Y., Xu, Z., Ma, W., Zakria, M., Zou, L., Cheng, Z., & Chen, G. (2017). A transcription activator-like effector Tal7 of Xanthomonas oryzae pv. oryzicola activates rice gene Os09g29100 to suppress rice immunity. *Sci. Rep.*, 7: 5089. doi: 10.1038/s41598-017-04800-8.

[47] Li, X., Zhou, W., Ren, Y., Tian, X., Lv, T., Wang, Z., Fang, J., Chu, C., Yang, J., & Bu, Q. (2017). High-efficiency breeding of early-maturing rice cultivars via CRISPR/Cas9-mediated genome editing. *J Genet. Genomics,* 44: 175–178. doi: 10.1016/j.jgg.2017.02.001.

[48] Zheng, S., Ye, C., Lu, J., Liufu, J., Lin, L., Dong, Z., Li, J., Zhuang, C. (2021). Improving the rice photosynthetic efficiency and yield by editing OsHXK1 via CRISPR/Cas 9 system. *Int J Mol Sci.*, 22(17), 9554. https://doi.org/10.3390/ijms22179554.

[49] Li, M. R., Li, X. X., Zhou, Z. J., Wu, P. Z., Fang, M. C., Pan, X. P., ... Li, H. Q. (2016b). Reassessment of the four yield-related genes Gn1a, DEP1, GS3, and IPA1 in rice using a CRISPR/Cas9 system. *Front. Plant Sci.*, 7, 377. https://doi.org/10.3389/fpls.2016.0037.

[50] Shen, R., Wang, L., Liu, X., Wu, J., Jin, W., Zhao, X., Xie, X., Zhu, Q., Tang, H., Li, Q., Chen, L., Liu, Y. G. (2017b). Genomic structural variation-mediated allelic suppression causes hybrid male sterility in rice. *Nat. Commun.*, 8(1), 1310. https://doi.org/10.1038/s41467-017-01400-y.

[51] Li, M., Li, X., Zhou, Z., Wu, P., Fang, M., Pan, X., Lin, Q., Luo, W., Wu, G., Li, H. (2016). Reassessment of the four yield-related genes Gn1a, DEP1, GS3, and IPA1 in rice using CRISPR/Cas9 system. *Front. Plant Sci.* 7, 377.

[52] Tabassum, J., Ahmad, S., Hussain, B., Mawia, A. M., Zeb, A., Ju, L. (2021). Applications and potential of genome-editing systems in rice improvement: Current and Future perspectives. *Agronomy*, 11(7), 1359. https://doi.org/10.3390/agronomy11071359.

[53] Macovei, A., Sevilla, N. R., Cantos, C., Jonson, G. B., Slamet-Loedin, I., Čermák, T., Voytas, D. F., Choi, I.-R., Chadha-Mohanty, P. (2018). Novel alleles of rice eIF4G generated by CRISPR/Cas9-targeted mutagenesis confer resistance to Rice tungro spherical virus. *Plant Biotechnol. J.*, 16: 1918–1927.

[54] Wang, F., Wang, C., Liu, P., Lei, C., Hao, W., Gao, Y., Liu, Y. G., Zhao, K. (2016). Enhanced rice blast resistance by CRISPR/ Cas9-Targeted mutagenesis of the ERF transcription factor gene OsERF922. *PLoS ONE*, 11, e0154027.

[55] Oliva, R., Ji, C., Atienza-Grande, G., Huguet-Tapia, J. C., Perez-Quintero, A., Li, T., Eom, J.-S., Li, C., Nguyen, H., Liu, B., Auguy, F., Sciallano, C., Luu, V. T., Dossa, G. S., Cunnac, S., Schmidt, S. M., Slamet-Loedin, I. H., Vera Cruz, C., Szurek, B., … Yang, B. (2019). Broad-spectrum resistance to bacterial blight in rice using genome editing. *Nat. Biotechnol.*, 37: 1344–1350.

[56] Huang, L., Li, Q., Zhang, C., Chu, R., Gu, Z., Tan, H., Zhao, D., Fan, X., Liu, Q. (2020). Creating novel Wx alleles with fine-tuned amylose levels and improved grain quality in rice by promoter editing using CRISPR/Cas9 system. *Plant Biotechnol. J.*, 18: 2164–2166.

[57] Endo, A., Saika, H., Takemura, M., Misawa, N., Toki, S. (2019). A novel approach to carotenoid accumulation in rice callus by mimicking the cauliflower Orange mutation via genome editing. *Rice*, 12, 81. https://doi.org/10.1186/s12284-019-0345-3.

[58] Usman, B., Nawaz, G., Zhao, N., Liao, S., Qin, B., Liu, F., Liu, Y., Li, R. (2021). Programmed editing of rice (Oryza sativa l.) osspl16 gene using crispr/Cas9 improves grain yield by modulating the expression of pyruvate enzymes and cell cycle proteins. *Int. J. Mol. Sci.*, 22, 249.

[59] Zong, Y., Wang, Y., Li, C., Zhang, R., Chen, K., Ran, Y., Qiu, J. L., Wang, D., Gao, C. (2017). Precise base editing in rice, wheat and maize with a Cas9-cytidine deaminase fusion. *Nat. Biotechnol.*, 35: 438–440.

[60] Lu, Y. and Zhu, J. K. (2017). Precise editing of a target base in the rice genome using a modified CRISPR/Cas9 system. *Mol. Plant.*, 10: 523–525. doi: 10.1016/j.molp.2016.11.013.

[61] Kuang, Y., Li, S., Ren, B., Yan, F., Spetz, C., Li, X., Zhou, X., Zhou, H. (2020). Base-Editing-Mediated Artificial Evolution of OsALS1 In Planta to Develop Novel Herbicide-Tolerant Rice Germplasms. *Mol. Plant*, 13: 565–572.

[62] Liu, X., Qin, R., Li, J., Liao, S., Shan, T., Xu, R., Wu, D., Wei, P. (2020). A CRISPR-Cas9-mediated domain-specific base-editing screen enables functional assessment of ACCase variants in rice. *Plant Biotechnol. J.*, 18: 1845–1847.

[63] Hua, K., Tao, X., Liang, W., Zhang, Z., Gou, R., Zhu, J. (2020). Simplified adenine base editors improve adenine base editing efficiency in rice. *Plant Biotechnol. J.*, 18: 770–778.

[64] Xu, Y., Lin, Q., Li, X., Wang, F., Chen, Z., Wang, J., Li, W., Fan, F., Tao, Y., Jiang, Y., Wei, X., Zhang, R., Zhu, Q., Bu, Q., Yang, J., & Gao, C. (2021). Fine-tuning the amylose content of rice by precise base editing of the Wx gene. *Plant Biotechnol. J.*, 19: 11–13.

[65] Agarwal, A., Yadava, P., Kumar, K., Singh, I., Kaul, T., Pattanayak, A., Agrawal, P. K. (2018). Insights into maize genome editing via CRISPR/Cas9. *Physiol Mol Biol Plants*, 24(2): 175-183. https://doi.org/10.1007/s12298-017-0502-3.

[66] Zhang, Q., Zhang, Y., Lu, M.-H., Chai, Y.-P., Jiang, Y.-Y., Zhou, Y., Wang, X.-C., Chen, Q.-J. (2019). A novel ternary vector system united with morphogenic genes enhances CRISPR/Cas delivery in maize. *Plant Physiol.*, 181(4): 1441-1448. https://doi.org/10.1104/pp.19.00767.

[67] Zhu, J., Song, N., Sun, S., Yang, W., Zhao, H., Song, W., Lai, J. (2016). Efficiency and inheritance of targeted mutagenesis in maize using CRISPR-Cas9. *J Genet Genomics.*, 43(1): 25-36. https://doi.org/10.1016/j.jgg.2015.10.006.

[68] Feng, C., Su, H., Bai, H., Wang, B., Liu, Y., Guo, X., Liu, C., Zhang, J., Yuan, J., Birchler, J. A., Han, F. (2018). High-efficiency genome editing using a dmc1 promoter-controlled CRISPR/Cas9 system in maize. *Plant Biotechnol J.*, 16(11): 1848-1857. https://doi.org/10.1111/pbi.12920.

[69] Tiwari, M., Trivedi, P. K., Pandey, A. (2020). Emerging tools and paradigm shift of gene editing in cereals, fruits, and horticultural crops for enhancing nutritional value and food security. *Food and Energy Secur.*, 10(1), e258. https://doi.org/10.1002/fes3.258.

[70] Dong, L., Qi, X., Zhu, J., Liu, C., Zhang, X., Cheng, B., Mao, L., Xie, C. (2019). Supersweet and waxy: meeting the diverse demands for specialty maize by genome editing. *Plant Biotechnol J.*, 17(10): 1853-1855. https://doi.org/10.1111/pbi.13144.

[71] Liu, H. J., Jian, L., Xu, J., Zhang, Q., Zhang, M., Jin, M., Peng, Y., Yan, J., Han, B., Liu, J., Gao, F., Liu, X., Huang, L., Wei, W., Ding, Y., Yang, X., Li, Z., Zhang, M., Sun, J., … Yan, J. (2020). High-throughput CRISPR/Cas9 mutagenesis streamlines trai gene identification in maize. *Plant Cell*, 32(5): 1397-1413. https://doi.org/10.1105/tpc.19.00934.

[72] Zhang, J., Zhang, X., Chen, R., Yang, L., Fan, K., Liu, Y., Wang, G., Ren, Z., & Liu, Y. (2020b). Generation of transgene-free Semidwarf maize plants by gene editing of gibberellin-oxidase20-3 using CRISPR/Cas9. *Front. Plant Sci.*, 11: 1–9.

[73] Ku, H. K., Ha, S. H. (2020). Improving nutritional and functional quality by genome editing of crops: status and perspectives. *Front. Plant Sci.* https://doi.org/10.3389/fpls.2020.577313.

[74] Shukla, V. K., Doyon, Y., Miller, J. C., DeKelver, R. C., Moehle, E. A., Worden, S. E., Mitchell, J. C., Arnold, N. L., Gopalan, S., Meng, X., Choi, V. M., Rock, J. M., Wu, Y.-Y., Katibah, G. E., Zhifang, G., McCaskill, D., Simpson, M. A., Blakeslee, B., Greenwalt, S. A., … Urnov, F. D. (2009). Precise genome modification in the crop species Zea mays using zinc-finger nucleases. *Nature,* 459: 437–441. doi: 10.1038/nature07992.

[75] Sato, K. (2020). History and future perspectives of barley genomics. *DNA Res.*, 27(4): dsaa023. https://doi.org/10.1093/dnares/dsaa023.

[76] Han, Y., Broughton, S., Liu, L., Zhang, X.-Q., Zeng, J., He, X., Li, C. (2021). Highly efficient and genotype-independent barley editing based on anther culture. *Plant Commun.*, 2(2), 100082. https://doi.org/10.1016/j.xplc.2020.100082.

[77] Schreiber, M., Mascher, M., Wright, J., Padmarasu, S., Himmelbach, A., Heavens, D., Milne, L., Clavijo, B., Stein, N., Waugh, R. (2020). A genome assembly of the barley 'transformation reference' cultivar golden promise. *G3 (Bethesda)*, 10(6): 1823-1827. https://doi.org/10.1534/g3.119.401010.

[78] Gerasimova, S. V., Korotkova, A., Hertig, C., Hiekel, S., Hoffie, R., Budhagatapalli, N., Otto, I., Hensel, G., Shumny, V. K., Kochetov, A. V., Kumlehn, J., Khlestkina, E. K. (2019). Targeted genome modification in protoplasts of a highly regenerable Siberian barley cultivar using RNA-guided Cas9 endonuclease. *Vavilov Journal of Genetics and Breeding*, 22(8): 1033-1039. https://doi.org/10.18699/VJ18.447.

[79] Hisano, H. and Sato, K. (2016). Genomic regions responsible for amendability to Agrobacterium-mediated transformation in barley. *Sci. Rep.*, 6, 37505. https://doi.org/10.1038/srep37505.

[80] Li, Y., Liu, D., Zong, Y., Jiang, L., Xi, X., Cao, D., Shen, Y., Zhang, H., Liu, B. (2020). New D hordein alleles were created in barley using CRISPR/Cas9 genome editing. *Cereal Res. Commun.*, 48: 131-138. https://doi.org/10.1007/s42976-020-00023-2.

[81] Kumar, N., Galli, M., Ordon, J., Stuttmann, J., Kogel, K.-H., Imani, J. (2018). Further analysis of barley MORC1 using a highly efficient RNA-guided Cas9 gene-editing system. *Plant Biotechnol. J.*, 16(11): 1892-1903. https://doi.org/10.1111/pbi.12924.

[82] Low, Y. C., Lawton, M. A., Di, R. (2020). Validation of barley 2OGO gene as a functional orthologue of Arabidopsis DMR6 gene in *Fusarium* head blight susceptibility. *Sci. Rep.*, 10: 9935. https://doi.org/10.1038/s41598-020-67006-5.

[83] Kis, A., Hamar, E., Tholt, G., Bán, R., Havelda, Z. (2019). Creating highly efficient resistance against wheat dwarf virus in barley by employing CRISPR/Cas9 system. *Plant Biotechnol. J.,* 17(6). https://doi.org/10.1111/pbi.13077.

[84] Galli, M., Martiny, E., Imani, J., Kumar, N., Koch, A., Steinbrenner, J., Kogel, K.-H. (2021). CRISPR/*Sp*Cas9-mediated double knockout of barley Microrchidia MORC1 and MORC6a reveals their strong involvement in plant immunity, transcriptional gene silencing, and plant growth. *Plant Biotechnol. J.* https://doi.org/10.1111/pbi.13697.

[85] Mason, A. S., Wendel, J. F. (2020). Homoeologous exchanges, segmental allopolyploidy, and polyploid genome evolution. *Front. Gemet*. https://doi.org/10.3389/fgene.2020.01014.

[86] Panchy, N., Lehti-Shiu, M., Shium S.-H. (2016). Evolution of gene duplication in plants. *Plant Physiol.*, 171(4): 2294-2316. https://doi.org/10.1104/pp.16.00523.

[87] Lino, C. A., Harper, J. C., Carney, J. P., Timlin, J. A. (2018). Delivering CRISPR: a review of the challenges and approaches. *Drug Deliv*., 25(1): 1234-1257. https://doi.org/10.1080/10717544.2018.1474964.

[88] Arora, L. and Narula, A. (2017). Gene editing and crop improvement using CRISPR-Cas9 system. *Front. Plant Sci*. https://doi.org.10.3389/fpls.2017.01932.

[89] Altpeter, F., Springer, N. M., Bartley, L. E., Blechl, A., Brutnell, T. P., Citovsky, V., Conrad, L., Gelvin, S. B., Jackson, D., Kausch, A. P., Lemaux, P. G., Medford, J. I., Orozo-Cardenas, M., Tricoli, D., VanEck, J., Voytas, D. F., Walbot, V., Wang, K., Zhang, Z. J., & Stewart, C. N. (2016). Advancing crop transformation in the era of genome editing. *Plant Cell*., 28(7): 1510-1520. https://doi.org/10.1105/tpc.16.00196.

[90] Mitra, D., Guerra Sierra, B. E., Khoshru, B., De Los Santos Villalobos, S., Belz, C., Chaudhary, P., Shahri, F. N., Djebaili, R., Olatunbosun Adeyemi, N., El-Ballat, E. M., El-Esaw, M. A., Moradi, S., Mondal, R., Senapari, A., Panneerselvam, P, Mohapatra, P. K. D. (2021). Impacts of arbuscular mycorrhizal fungi on rice growth, development, and stress management with a particular emphasis on strigolactone effects on root development. *Commun Soil Sci. Plant Anal*., 52(14): 1591-1621. https://doi.org/10.1080/00103624.2021.1892728.

Chapter 13

Genome Editing in Horticultural Crops

Aditi Thakur[1,*]
Suman Sanju[2]
Reena Kumari[1]
Karuna Dhiman[1]
and Sneh Sharma[1]
[1] College of Horticulture and Forestry, Neri, Hamirpur, (H. P),
Dr. Yashwant Singh Parmar University of Horticulture and Forestry, Nauni, Solan (H. P), India
[2] Department of Botany, Abhilashi University, Mandi (H. P), India

Abstract

The biggest challenge in front of the globe today is changing climatic conditions and providing food security to the increasing human population. Horticultural crops comprise a wide range of crops, including flowers, fruits, vegetables, nuts, medicinal plants, and aromatic plants that are responsible to fulfil the fuel, food, nutritional, medicinal, and aesthetic needs of the human race. However, these horticultural crops are attenuated by several biotic and abiotic factors such as diseases, pests, and harsh climatic conditions. These challenges can be overcome through the development of high yielding and resistant varieties to the abiotic/biotic stresses. Conventional breeding approaches have been applied for crop improvement, but these are resource as well as time-intensive. Genome editing technology, such as Transcription Activator-like Effector Nucleases (TALENs) and Zinc Finger Nucleases (ZFNs) comes as a rescue to overcome all these challenges, but these techniques are complicated, time-consuming and not cost-effective. Recently, CRISPR/Cas9 is the most adapted and promising gene editing tool through which many horticultural crop species have been modified to meet desirable traits with enhanced target specificity and low off-target effects. In this chapter, we summarise the principles of CRISPR/Cas9 technology, its applicability to the horticultural crops, and its underlying challenges in crop improvement.

Keywords: CRISPR/Cas9, genome editing, biosafety regulations, biotic stress, abiotic stress, horticulture crops

---

* Corresponding Author's Email: thakuraditi574@gmail.com.

In: Genome Editing for Crop Improvement
Editors: Rahul Kumar, Ashutosh Sharma, Yogesh Kumar et al.
ISBN: 979-8-88697-740-0

Introduction

One of the greatest challenges for the scientific community is the sustainable production of horticultural crops resistant to biotic as well as abiotic stress due to climate change. By plant breeding approaches, breeders introgressed the good traits through back/cross breeding to develop high yielding and superior varieties. But the wide use of these superior varieties has led to a remarkable decrease in genetic diversity. Loss in genetic diversity has resulted to increase in sensitivity to biotic/abiotic stress. Moreover, plant breeding approaches are laborious and time-consuming because of the large-scale screening of germplasm. Crop improvement through physical as well as chemical means i.e., treatment of gamma radiations or EMS (ethylmethane sulfonate) leads to the random, multiple and undesirable changes. Transgenic approaches (genetically modified crops) have been widely used for the improvement of horticultural crops, but the encouragement of genetically modified crops is not relatively easy due to the communal concern and the exchange of the genetic material among the non-target organisms. RNAi (RNA interference) is a well-known technique for silencing genes. Gene regulation in RNAi is controlled through endogenous microRNAs. Displacement of these endogenous miRNAs from exogenous miRNAs can trigger off-target phenotype as well as hypomorphic mutations [1].

On the other hand, genome editing technology consists of accurately and precisely recognizing and locating a specific sequence within a genome and then altering the targeted sequences for various uses. Genome editing technology needs a specific restriction nuclease that can recognize and cut specific sequences. Till now, four major families of nucleases used for genome editing are known, namely: meganucleases, ZFN (zinc finger nucleases), TALENs (transcription activator-like effector-based nucleases) and the CRISPR/Cas (clustered regularly interspaced short palindromic repeats associated nucleases). Meganucleases were the first nucleases used for genome editing and discovered in the late 1980s, followed by ZFNs and TALENs for genome editing [2, 3, 4]. These three genome-editing technologies comprise of meganucleases, ZFNs, and TALENs have been studied in detail and extensively applied as a gene editing tool in agricultural crops, i.e., wheat, rice, and maize. Although, these genome editing tools have limitations for the wide applications in plants due to the complex construction process and cost.

In comparison to TALEN and ZFNs, the CRISPR/Cas9 genome editing tool has gained recognition in the scientific community. CRISPR/Cas9 technology is low-cost, faster, and more specific or precise because it permits the editing of multiple and different sites across the genome at the same time. This technology can be applied to edit the plants that were ignored previously. CRISPR/ Cas9 technology was adapted from an immune system naturally present in bacteria to fight viruses [5]. In the viral attack, the bacteria capture the DNA segments and make use them to produce the segments as CRISPR arrays. These CRISPR arrays undergo transcription and form CRISPR-RNA (crRNA) fragments that are base-paired with trans-activating crRNA (tracrRNA). These two RNA structures will direct Cas9 nucleases to the target DNA [6]. The CAS9 nuclease consists of two domains: one is a RuvC-like domain, and the other is a HNH domain. Each domain of Cas9 cuts one DNA strand. Natural dual tracrRNA: crRNA structure is designed into a single-guided RNA chimera (sgRNA), targeted to specific genomic loci for practical or specific purposes. Once the DNA is cut at the specific site, nonhomologous end-joining (NHEJ) and homologous recombination (HR) pathways repair the cut DNA and make possible the development of the novel plant varieties resistance to biotic as well as abiotic stress. Nonhomologous end-joining (NHEJ) is error–prone, and the blunt ends are repaired with the help of DNA ligase IV. The gene knockout and/or frameshift mutations can occur during repair due to insertion or deletion of the base pairs in NHEJ [7]. However, homologous recombination (HDR) pathways repair the cut sites either through the endogenous and natural phenomenon of using sister chromatids as repair templates or exogenously introduced repair template DNA that can be single-stranded or double-stranded DNA, to facilitate specific gene editing [8, 9, 10].

The CRISPR/CAS system is classified into two classes and five types. Class I consists of three types of systems: type I, type III, and type IV; and class II consist of type II and type V. The systems of class I are more common in archaea compared to bacteria [11]. Type I and III use greater than one Cas protein and have complex effector module while the type IV utilizes rudimentary Cas-protein. The single Cas protein is utilized by type II and type V systems with comparatively simpler effector modules. Cas9, the signature protein for type II, was first discovered as a novel, large protein [12]. Major differences between CRISPR classes are based on the composition of crRNP complexes. Other than the CRIPSR system developed from the bacterium *Streptococcus pyogenes* (*SpCas9*) which is most extensively used, the smaller Cas9 isolated from *Staphylococcus aureus* (*SaCas9*) and a Cas9 derived from *Streptococcus thermophilus* (*StCas9*), were discovered to work competently in plants [13]. Besides Cas9 enzyme, type V CRISPR/Cas12a (Cpf1) enzyme from *Prevotella* and *Francisella* has recently been discovered [14]. Cpf1 enzyme has some advantages over Cas9; it produces sticky or cohesive ends unlike Cas9. Due to the generation of sticky or cohesive ends, the efficiency of the insertion of the gene of interest increases. Since the first use of CRISPR/Cas9 technology for plant gene editing in 2013, the scientific community has concentrated on its application in increasing crop yield, quality, and biotic as well as abiotic stress resistance. To date, CRISPR/Cas9-mediated genome editing has been reported in 41 food crop species, 15 industrial crops, 6 oil crops, 8 ornamental crops, 1 fibre crop, and feed crop [15].

Mechanism of CRISPR/Cas9

The defence response against the invaders in the microbial adaptive immune structure mediates in three stages: immunization, expression and interference. The first phase of immunization involves the incorporation of small sequences of ~21-48 bp of the invading DNA called as spacer into the CRISPR locus. The protospacer adjacent motif (PAM) follows the spacer in the desired gene. The spacers serve as acquired genetic memory of previous infections for the host. The effective DNA-binding and cleavage are completely dependent upon the recognition of the accompanying PAM sequence in the host. Second phase is the expression, which includes the transcription of the CRISPR sequence along with the spacers and Cas genes resulting in the formation of single stranded RNA termed CRISPR RNA. The CRISPR RNA comprises copies of invading DNA sequences. In the third step of interference, the CRISPR RNA identifies the invading DNA, and the CRISPR associated Cas9 proteins cleave and eliminate the targeted DNA by creating DSBs at ~3 bp upstream of the PAM [6]. CRISPR/Cas9 technique was first of all utilized in mammalian cells since its discovery [6]. The CRISPR/CAS9 in plants has also been used for various genome editing processes recently, like gene knockout, gene knockdown, gene knocking, and also gene expression activation. Also, multiple genes editing through this technique has also enabled the pathway level research in plants. The stages involved in harnessing the CRISPR/Cas9 within plants (Figure 1) include:

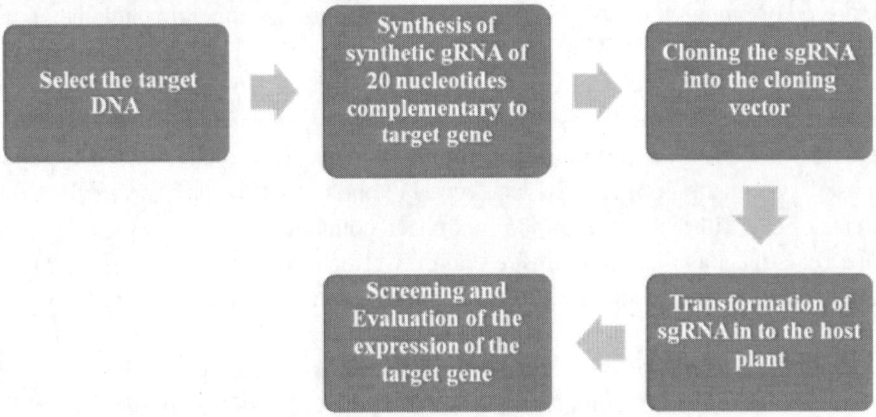

Figure 1. Steps involved in the CRISPR/CAS9 genome editing technique.

1. *Selection of target DNA*: For the gene editing using CRISPR/Cas9, firstly the reference genome of the crop to be amended is located. The whole genome sequences are available in the databases for some horticultural crops like potato, watermelon, apple, etc. The target gene sequence can be identified as per the research objective and retrieved from these databases. It is then sequenced and aligned with the reference genome.
2. *Synthesis of a 20-nucleotide-long synthetic gRNA complementary to the target DNA*: The synthesis of gRNA is done using various software for example, CRISPR-P [16], CRISPR-PLANT [17], CRISPR direct [18], Chop-Chop [19], and Benchling [20].
3. *Cloning the sgRNA into appropriate binary vector*: To make the gRNA and CRISPR/Cas9 cassette, the appropriate vector is searched from the plasmid repositories such as Addgene [165]. The vectors such as pHSN501, pHSE401, and pHSN401 are used for transformation in watermelon, and the vector pTC217 for tomato [21, 22]. The constructs are either prepared through ligation-dependent or ligation-independent methods, and sequencing is followed by proper alignment of the constructs [17, 23, 24]. For instance, the software Snapgene and Benchling [20, 166] can be used to stimulate the vector construction.
4. *Transformation into host species*: The transformation of the gRNA: CRISPR/Cas9 cassette in an appropriate vector into the host plant is initiated through various methods, e.g., biolistics [25, 26], *Agrobacterium* mediated transformation [23, 27, 28], and carbon nanotube nanoparticles [29, 30]. RNA or DNA viruses have also been exploited to transfer gRNA in to the host species [31, 32].
5. *Screening and validation of gene–transformed plants*: The putative transformants (T0) are screened for proper transfer of the gRNA into the host plant and for mutations and edits through various techniques, i.e., polymerase chain reaction (PCR), restriction digestion, and Sanger sequencing. The T7 endonuclease I (T7EI) assay can also be used for the screening of the transformants, and the software such as Snapgene and

TIDE are used for analysis [33, 34, 35, 36, 37]. The T_1, T_2, and further generations of the CRISPR/Cas9-edited plants are grown and are validated through PCR and whole genome sequencing for presence or absence of transgenic elements, and for other research objectives [38, 39, 40]. The model plant *Arabidopsis*, tobacco, and rice along with a few other crop species, have been transformed and analysed using CRISPR/Cas9 technology [27].

CRISPR/Cas9 for the Improvement of Horticultural Plants : Vegetables and Fruits

So far 20 crop species have been improved for various traits like crop yield, biotic and abiotic factors by adopting CRISPR technology [41]. CRISPR/Cas9 methodology has been harnessed to amplify the crop disease resistance against the various biotic factors in addition to the numerous abiotic stresses like drought stress, salinity stress etc. Tomato is the most extensively used crop for CRISPR/Cas9 technique where tomato has been used for 18 different genes independently. In 2014, the first gene modification in tomato was performed using CRISPR/Cas9 for the development of needle leaf mutants by knocking out *Argonaute7* [42]. Several developmental genes have been targeted in tomato using CRISPR/Cas9 for their functional validation like SHORTROOT (SHR) for root development [43]. BLADE-ON-PETIOLE (SlBOP) for inflorescence development [44] and SELF PRUNING 5G (SP5G) and SELF PRUNING (SP) for plant development [45] (Table1).

In addition, [33] explored the function of the arabinosyl-transferase gene and found it as a novel component in the CLAVATA signaling pathway, which is crucial to regulate the meristem size in tomato using CRISPR/Cas9 technology. With the massive demand for tomato fruits in the processing industry, two different groups developed parthenocarpic tomato fruits simultaneously. Klap et al. [46] developed mutant plants through knockout of the Slagamous-like 6 (*SlAGL6*) gene that were capable of producing parthenocarpic tomato fruits under heat stress conditions that would otherwise severely obstruct fertilization-dependent fruit set. Mutagenesis did not show any other effect that made SlAGL6 a desirable gene for facultative parthenocarpy. The genes of different biosynthetic pathways in tomato were mutated using CRISPR/CAS9, such as the gene Anthocyanin 1 (ANT1) of the anthocyanin biosynthetic pathway [22]. Similarly, the genes involved in the carotenoid biosynthesis pathway, such as Phytochrome interacting factor (SlPIF4), Phytoene desaturase (*SlPDS*) [47], and Phytoene synthase (*PSY1*) [48] were mutated through the CRISPR/Cas9 technique.

The shelf life determines the life span of a fruit and, hence, is a very important trait in controlling the post-harvest damage of the crop. Therefore, the role of enzymes pectate lyase, polygalacturonase 2a, and β-galactanase in cell wall modification during fruit ripening is decoded using CRISPR/CAS9 [49]. Zsogon et al. [50] edited six loci of the wild landrace of tomato *S. Pimpinelli folium* to develop a new variety with improved yield and productivity by CRISPR/CAS9 technique. The improved tomato varieties by deciphering the role of the *ENO* gene in terms of locule number and size have also been developed through CRISPR/Cas9 [51].

Table 1. Application of the CRISPR-based genome editing approach in horticultural plants for crop improvement and abiotic stress tolerance

Crop plant	Trait	Target gene	Reference
Tomato	Anthocyanin biosynthesis	Anthocyanin 1 (*ANT1*)	[22]
	Phytochrome interacting factor (*SlPIF4*)	Phytoene desaturase (*SlPDS*)	[47]
	Carotenoid biosynthesis	Phytoene synthase (*PSY1*)	[48]
	Leaf development	ARGONAUTE7 (*SlAGO7*)	[42]
	Root development	SHORTROOT (*SHR*)	[43]
	Inflorescence	BLADE-ON-PETIOLE (*SlBOP*)	[44]
	Plant development	SELF PRUNING 5G (*SP5G*) and SELF PRUNING	[45]
	Parthenocarpic fruits under heat stress	Slagamous-like 6 (*SlAGL6*)	[46]
	Seedless tomatoes	*SlARF7* (auxin response factor 7)	[46]
	Seedless tomatoes	*SlIAA9* (indole-3-acetic acid inducible 9)	[46]
	Increased production of GABA	γ-aminobutyric acid (*GABA*)	[52]
	Cell wall modification during fruit ripening	pectate lyase, polygalacturonase 2a, and β-galactanase	[49]
	Number and size of locules	*ENO* gene	[51]
	Drought stress (Protecting cell membrane from oxidative damage)	Mitogen-activated protein kinases 3	[75]
	Heat stress tolerance	*BZR1*	[73]
	Decrease in chilling stress tolerance	*SlCBF1*	[74]
	Decrease in drought stress tolerance	*SlMAPK3*	[75]
	Controls meristem size	Arabinosyl-transferase gene	[84]
Potato	Starch quality	*Granule bound starch synthase* (*GBSS*)	[53]
		ACETOLACTATE SYNTHASE1 (*StALS1*)	[54]
	Identification of gene function (Reverse genetics)	*StMYB44*	[55]
	Reduced enzymatic browning	Polyphenol Oxidase Gene, *StPPO* gene family	[56]
	Overcome self-incompatibility	S-RNase	[85, 86]
	α-solanine-free hairy roots	*St16DOX* gene.	[87]
Sweet potato	Enhance Fusarium wilt resistance	*IbBBX24* gene	[88]
Citrus	Carotenoid biosynthesis pathway	*Phytoene desaturase* (*CsPDS*)	[61]
Watermelon		*Phytoene desaturase*	[62]
	Validate function of vacuolar sugar transporter gene	*Sugar transporter VST1*	[89]
	Obtain gynoecious genotypes	*ClWIP1 gene*	[90]
	Functional characterization of a gene in fruit flesh sugar accumulation		[91]
	Resistance against herbicide	*ALS*	[92]
	Herbicide resistance	*ClALS*gene	[92]
Grapes		Phytoene desaturase gene	[63]
Lotus japonicas ornamental plant	Symbiotic nitrogen fixation	genes such as symbiosis receptor-like kinase and leghemoglobin loci *(LjLb1, LjLb2, and LjLb3)*	[93]
Chrysanthemum morifolium		the yellowish-green fluorescent protein (*CpYGFP*)	[77]
Salvia miltiorrhiza	Tanshinone biosynthesis	diterpene synthase gene (*SmCPS1*)	[78]
Cassava	Drought stress (Maintaining osmotic balance)	*MeKUPs*	[94]
	Drought stress (Tissue development)	*MeMAPKK*	[95]
Sugarcane	Catalyzing phospholipids response	*ScNsLTP*	[96]
Papaya	Drought stress	*CpDreb2*	[97]
Cotton	Drought stress (controls cell growth and development)	*GhPIN1-3* *GhPIN2*	[98]

Crop plant	Trait	Target gene	Reference
	Drought stress	*GhRDL1*	[99]
Cucumber	Generation of gynoecious plant	*WIP1*	[100]
Groundcherry	Introduction of traits associated with morphology, flower production, and fruit size	*SP, SP5G, CLV1*	[101]
Kiwifruit	Generation of a compact plant with rapid terminal flower and fruit development	*CEN4, CEN*	[66]
Banana	Short height	*RAS-PDS* genes	[71]
	Short height	*MaPDS* gene	[69]
	Gibberellin biosynthesis	*MaGA20ox2* gene	[70]
	β-carotene-enriched banana	lycopene epsilon-cyclase gene	[71]
Oilseed rape	Increased seed oil content	*BnSFAR4* and *BnSFAR5*	[60]

Table 2. Application of the CRISPR-based genome editing approach in horticultural plants for crop improvement and biotic stress tolerance

Pathogen	Name of Plant	Target Gene	Name of Pathogen	Results	Reference
Virus	*Arabidopsis*	Virus RNA genome disruption	Virus RNA genome Turnip mosaic virus	Indels in virus RNA	[108]
		Gene disruption *EIF4E*	Turnip mosaic virus	Resistance to virus	[113]
	Cucumber	*EIF4E*	Cucumber vein yellowing virus (pomovirus), Zucchini yellow mosaic virus, and Papaya ring spot mosaic virus-W (potyviruses)	Resistance to virus	[112]
	Cassava	*AC2* and *AC3*	African cassava mosaic virus	Indels in virus DNA but no virus resistance	[109]
		nCBP-1, nCBP-2	Cassava brown streak disease	Suppressed disease	[114]
	Banana	Virus DNA disruption	Endogenous banana streak virus	70% of plants remain asymptomatic	[106]
	Rice	Biomimicking *Eif4g*	Rice tungro spherical virus	Resistance to virus	[115]
		Virus RNA genome	Southern rice black streaked dwarf virus, Tobacco mosaic virus	Reduction in virus levels and disease symptoms	[107]
	Tomato	Virus DNA *Rep, IR,* and *Cp*	Virus DNA Rep, IRandCp	Resistance to virus	[103]
	N. benthamiana	Virus RNA genome	Turnip mosaic virus	Indels in virus RNA	[37]
		Virus DNA *Rep, IR,* and *Cp*	Beet curly top virus, Merremia mosaic virus, Tomato yellow leaf curl virus	Indels in virus DNA	[31]
		Virus DNA and satellite sequences	Cotton leaf curl Kokhran virus, Tomato yellow leaf curl Sardinian virus, Tomato yellow leaf curl virus, Merremia mosaic virus, BCTV-Logan, BCTV-Worland	Indels in virus DNA	[102]
		Virus DNA Rep A/Rep and LIR	Bean yellow dwarf virus	Indels in virus DNA, resistance to virus	[104]
		Multiplex editing at Rep and IR	Cotton leaf curl Multan virus	Significantly low virus accumulation and decreased disease symptoms	[140]

Table 2. (Continued)

Pathogen	Name of Plant	Target Gene	Name of Pathogen	Results	Reference
		Multiplex editing at virus DNA Rep, IR, and Cp	Chili leaf curl virus	Significantly low virus accumulation and decreased disease symptoms	[139]
Fungus	Tomato	Multiplex gRNA at *Pmr4*	Powdery mildew caused by *Oidium neolycopersici*	Significant reduction in mildew symptoms	[135]
		SlMapk3	Botrytis cinerea	Increased resistance to *B. cinerea*	[132]
		Solyc08g075770	Fusarium wilt	Tolerance to fusarium wilt	[130]
		SlMlo1	Powdery mildew Resistance	Resistance to powdery mildew [88]	[38]
	Banana	*RGA2,Ced9*	Fusarium wilt caused by *Fusarium oxysporum* f. sp. *cubense* tropical race 4 (TR4)	Significant reduction in disease.	[136]
	Grape	*VvWRKY52*	*B. cinerea*	Increased resistance to *B. cinerea*	[131]
		Mlo-7	Powdery mildew	Indels in the target	[137]
	Rice	Gene disruption Single and multiplex gRNA at *OsERF922*	Rice blast caused by *Magnaportheoryzae*	Significantly decreased blast lesions	[76]
		OsMPK5	Fungal (*Magnaporthe grisea*) and bacterial (*Burkholderia glumae*) pathogens	Indels in the target; resistance not confirmed	[138]
Bacteria	Apple	Gene disruption *DIPM-1, DIPM-2, and DIPM-4*	Fire blight disease (caused by *Erwinia amylovora*)	Indels in the target	[137]
	Tomato	*Jaz2*	Bacterial speck disease caused by *Pseudomonas syringae* pv. tomato DC 3000	Resistance to bacterial speck disease	[119]
		Dmr6	*Pseudomonas syringae, Phytophthora capsici, and Xanthomonas* spp.	Resistance to *P. syringae, P. capsici,* and *Xanthomonas*spp	[120]
	Grapefruit	*CsLOB1*	Citrus canker	Significantly reduced canker symptoms	[122]
	Wanjincheng orange	*CsLOB1*	Citrus canker	Disease severity decreased by 83.2–98.3%	[121]
	Rice	*OsSWEET11, OsSWEET13, and OsSWEET14*	Bacterial blight	Increased resistance to bacterial blight	[118]
Oomycete	Papaya	*PpalEPIC8*	*Phytophthora palmivora*	Increased resistance against *P. palmivora*	[133]
	Theobroma cacao	*TcNPR3*	*Phytophthora tropicalis*	Increased resistance against *P. tropicalis*	[134]

Very recently, in 2017, CRISPR/Cas9 was reported to be applied for actual improvement of vegetable crops. The metabolic modification of a pathway by targeting multiple genes has also been established recently in vegetables. In tomato, the five key genes of γ-aminobutyric acid (GABA) shunt (the four-carbon, non-protein amino acid) were targeted simultaneously through multiplexed CRISPR/CAS9 technique resulting in the enhanced accumulation of

GABA, both in the leaves and fruits of the tomato plant [52]. The quality of vegetables has also been improved through the CRISPR/CAS9 technique, like the quality of starch in potato. The mutation carried out in the gene *granule-bound starch synthase* (*GBSS*) through CRSIPR/Cas9 led to the development of a waxy genotype in hexaploid potato similar to the amylopectin-containing starch in maize [53]. One of the genome-modified lines confirmed the knock-out of all four alleles of the gene GBSS after the starch characterization confirmed the production of only amylopectin and amylase was completely lacking. The multi-allelic mutagenesis through CRISPR/CAS9 is established through this experiment in polyploidy crops. Similarly, the gene *ACETOLACTATE SYNTHASE1* (*StALS1*) has also been mutated through this multi-allelic mutagenesis in potato [54].

CRISPR also served as the tool for reverse genetics in potato for deciphering the function of *StMYB44* gene [55]. It has been reported that reduction in the enzymatic browning and thereby, improved nutritional quality of tubers of potato was possible because of the mutation in the member of the *StPPO* gene family [56]. Developmental genes in other vegetables like *Brassica oleracea* and *Lactuca sativa* [58] have been modified through CRISPR/Cas9. CRISPR/Cas9 -mediated mutants of oilseed rape containing low phytic acid and improved protein content in oil were developed [59]. Phytic acid is antinutritive and hinders mineral absorption in the human gut. In oilseed rape, by mutating gene *BnSFAR4* and *BnSFAR5* using CRISPR/Cas9, the oil biosynthesis gene expression has been altered, and plants with increased seed oil content have been generated [60]. This research confirmed that the gene editing tools widen a novel perspective of improving quality, and yield in major oilseed crops.

Fruits are a rich source of nutrition, but there is limited improvement of the fruits owing to their perennial nature and the breeding programmes are time-consuming. Hence, the application of CRISPR/CAS9 in genome editing of fruits is still in its infancy with a few reports so far [61] mutated the gene *Phytoene desaturase* (*CsPDS*) of carotenoid biosynthesis pathway through CRISPR/CAS9 in citrus. Similarly, *phytoene desaturase* gene has been modified using CRISPR/CAS9 in both the crops watermelon and grapes [62, 63]. In the grape crop, a specific database of the target sites for CRSIPR/Cas9-based genome editing called "Grape-CRISPR" has been developed and published [64]. These databases help in the identification of target sites in the selected gene. The early induction of flowering has been successfully attempted by reducing the juvenile period of the apple, and pear crops by editing of the *TFL1* gene through CRISPR/Cas9 [65].

Kiwifruit (*Actinidia chinensis*) the sweet and unique flavoured edible berry, is commonly cultivated in China, New Zealand, Italy, and Iran. Through the CRISPR/CAS9 tool in kiwifruit, a revolutionary achievement in gene editing was accomplished. The CEN-like genes *AcCEN4* and *AcCEN* were mutated, and the habit of the kiwifruit plant was transformed from a climbing, and woody perennial with axillary inflorescence into a compact plant with a terminal flower. The woody perennial plant took many years to develop fruits, while the compact plant started fruiting within a year [66]. Similarly, bananas are grown in more than 130 countries around the globe, making it an economically significant tropical fruit. The banana crop faces the challenge of weak lodging because the high yielding varieties of banana are longer than 2 m. Hence, the short-height varieties are favourable for crop maintenance and machine harvesting. Gene editing is the most feasible alternative in banana for genetic manipulation, since banana genome is triploid and leads to plant sterility [67]. Also, the genetic improvement through plant breeding methods in banana is tough owing to its triploid genome. Thereby, the CRISPR/Cas9 technique was used successfully in banana cv. Rasthali to mutate the two *RAS-PDS* genes [68] and

MaPDS gene in banana cv. Cavendish [69]. Shao et al. [70] edited the gibberellin biosynthesis gene *MaGA20ox2* successfully in the "Gros Michel" cultivar using CRISPR/Cas9 resulting in a semi dwarf mutant cultivar of banana. Recently, CRISPR/Cas9 tool was used to modify the gene lycopene epsilon-cyclase to develop the β-carotene-enriched banana in the Cavendish cultivar [71].

Abiotic Stress Tolerance

Abiotic stresses such as cold, heat, drought, flooding, and water stress affect crop plants adversely, and to enhance the crop production and yield, the assistance of new techniques is required. CRISPR/Cas9 is the most promising technology to increase the crop yield as well as production by regulating their ability to deal with various abiotic challenges [72]. It has been found that Brassinazole resistant 1 (*BZR1*) is involved in thermotolerance in tomato plants by regulating the action of Feronia (*FER*) genes [73]. Cbf1 mutants developed by CRISPR/Cas9 technique by mutating C-repeat binding factor 1 (*CBF1*) helps the plants to tolerate the cold stress [74]. Also, *MAPK3* mutants protect the tomato plant against drought stress by preventing the oxidative damage of the cell membranes [75] (Table1).

CRISPR/Cas9 in Other Horticultural Crops

The horticultural crops other than fruits and vegetables, like medicinal and ornamental plants, have also been subjected to genome editing through CRISPR/CAS9. The first such attempt of CRISPR/Cas9 mediated genome editing in ornamental plants was done in the model legume plant *Lotus japonicus,* where the symbiotic nitrogen fixation-associated genes, i.e., symbiosis receptor-like kinase and leghemoglobin loci (LjLb1, LjLb2, and LjLb3), were targeted [93]. Consequently, single- and multiple-gene mutant plants were attained. In the *Chrysanthemum morifolium* (chrysanthemum), the hexaploid ornamental plant, two sgRNAs mutated the *yellowish-green fluorescent protein* (*CpYGFP*) gene at different locations [77]. In *Salvia miltiorrhiza*, the gene diterpene synthase (*SmCPS1*) has been knocked out, which is concerned with tanshinone biosynthesis, thereby plants lacking tanshinones were developed. Although other phenolic metabolites remained unaffected [78]. In soybean, the characterization of a locus encoding GmPRR3b gene that regulates the vigorous growth and early flowering in the soybean plants using CRISPR/Cas9 facilitated the development of a variety suitable for the high-altitude areas [79]. However, Wang et al. [80] demonstrated that the gene *GmPRR37* is responsible for the flowering time and photoperiodism in soybean, and gene modification by CRISPR/Cas9 enabled the soybean varieties to adapt in the hill region. The soybean plants with increased branch number and node number per plant were also developed by mutations in four genes of *SPL* family [81]. Similarly, by mutating the gene *GmFT2a* using CRISPR/Cas9 soybean plants with delayed photoperiodic responses and flowering time were developed [82, 83].

Similarly, genome editing will enable the study of biosynthetic pathways of medicinally important compounds in medicinal plants, and their use at the commercial level can also be scaled up.

Improving Biotic Tolerance via CRISPR/Cas9 Tools

Resistance against Plant Viruses: Targeting Virus Genomes

To develop virus resistant crop varieties, the CRISPR/Cas9 approach has been extensively used. Research has focused on two main approaches to developing virus resistance varieties: 1) targeting the genome of virus directly, and 2) targeting the susceptibility plant genes, which are important for the progression of viral disease. The earliest work to develop virus resistance crops has focussed on the genome targeting of virus inside the plant cells. CRISPR/Cas9 was successfully testified in *Cotton leaf curl kokhran virus*, *Beet curly top virus* and *Tomato yellow leaf curl virus* having ssDNA (single-stranded DNA) [31, 102, 103]. Three regions were targeted within the viral genome i.e., the protein associated with replication process, capsid formation, and the intergenic zone. As a result of the baiting of these regions, indels are generated in the virus's genome, resulting in a low titer viral load and disease symptoms. *Beet severe curly top virus* and *Bean yellow dwarf virus* were also targeted, with low titer values and reduced disease symptoms achieved [104, 105].

In comparison to transgenics, the CRISPR approach targeted for virus resistance showed a high success rate. Resistant banana varieties to the *Banana streak virus* developed through the CRISPR mechanism showed 75% of the banana asymptomatic in comparison to the non-edited banana varieties [106]. The genome of an RNA virus was specifically targeted using the CRISPR strategy and Cas9 RNA binding variants. *Arabidopsis* and *Nicotiana benthamiana* plants developed resistance to CMV (cucumber mosaic virus) and TMV (*Tobacco mosaic virus*) via the expression of FnCas9. These plants demonstrated the remarkable reduction in viral load as well as in disease progression [36]. Rice resistance to the *Southern rice black-streaked dwarf virus* were also demonstrated though similar strategy [107]. *Arabidopsis* and *N. Benthamiana* [108] were also shown to express RNA targeting LwaCas13a to *Turnip mosaic virus*. A study done by [108] demonstrated the great efficacy and potential of using the Cas13 systems to bring about the development of plant immunity against RNA viruses. Cas13d (Cas Rx), ortholog of Cas13 is the most efficient to target the different RNA viruses i.e., *Potato virus X* (PVX), *Tobacco mosaic virus* (TMV) and *Turnip mosaic virus*. The efficiency and efficacy of Cas13 to provide resistance against RNA viruses could be applied against the plant infecting viroids.

Another study demonstrated the problem associated with virus escape from the plants edited with CRISPR [102]. Furthermore, virus escape in transgenic cassava lines that were ineffective in demonstrating resistance to the *African cassava mosaic virus* was reported [109]. Further studies demonstrated that the percentage in the range of 33-48% the CRISPR edited virus genome carry a single mutation in the conserved region responsible for the virus escape as well as resistance to CRISPR/Cas9 cleavage [109]. The introduction of DSBs into virus DNA is a strategic mechanism for circumventing the limitation of escape mutants. Double-strand breaks in the viral DNA makes viruses less prone to the non-homologous end-joining (NHEJ), which in turn leads to the development of escape mutants [110]. A recent study showed the targeting at multiple sgrNA sites of a geminivirus, *Chilli leaf curl virus* that leads to significant decrease in viral load and pathogenicity. More crop research and field trials will be required to validate this data. As a result, a better substitute should be sought to address the issue of virus escape as well as the evolution of CRISPR resistance [110, 111].

Resistance against Plant Viruses: Targeting Plants' S Genes

S genes in the plants overcome the issues related to the direct targeting of the genomes of viruses. Targeting of the S genes can be carried out through the CRISPR approach with transgene-free protocols. S genes that are widely targeted through CRISPR mechanism for viral immunity are the *elf4E* genes. The *elF4E* encodes cap binding proteins that play an important role in the infection of several RNA potyviruses. CRISPR-mediated targeting of the *eIF4E* gene conferred successful resistance against various viruses, including TYMV (Turnip yellow mosaic virus), ZYMV (*Zucchini yellow mosaic virus),* CVYV (*Cucumber vein yellowing virus*), and PRSMV (*Papaya ring spot mosaic virus*-W) in *Arabidopsis* and cucumber [112, 113]. Further studies demonstrated that targeting *eLF4E* isoforms provides broad spectrum immunity against viruses in economically significant crop species. In cassava, the isoforms of *eLF4E* i.e., nCPB-1 (novel cap binding protein-1) and nCPB-2 (novel cap binding protein-2) was targeted that reduced the disease progression of *Cassava brown streak virus* [114]. Through biomimicking of alleles of *eLF4G*, resistance against rice TSV (*Tungro spherical virus*) was successfully demonstrated in rice [115]. To summarize, the use of S genes in multiple CRISPR systems to attain resistance against viruses has been effectively demonstrated in a number of crops.

Bacteria

CRISPR-mediated bacterial resistance was successfully demonstrated through *OsSWEET* gene/s. The *OsSWEET* gene plays an indispensible role for the progression of bacterial blight disease caused by *Xanthomonas Oryzae*. The rice SWEET clade III family comprises the OsSWEET11, OsSWEET13, and OsSWEET14 that is responsible to encode sugar transporters. Induction of the *OsSWEET* gene(s) through transcription activator-like effectors (TALEs, the bacterial proteins used to develop TALENS) is responsible for release of sugar into apoplastic region. This sugar released into the apoplast is responsible to supply nutrients to the pathogen. Genome editing of *OsSWEET* through TALENS has already been successful against the bacterial blight in rice [116, 117]. Gene editing in the promoter of *OsSWEET* gene through CRISPR-Cas9 have contributed towards extensive resistance to bacterial blight disease [118]. *S1JAZ2* gene edited through CRISPR provides resistance to bacterial speck disease due to *Pseudomonas syringae* in tomato [119]. *SIDmr6-1* edited gene through CRISPR has been used for the wide range resistance against many pathogens viz., *P. syringae*, *Phytophthora capsici,* and *Xanthomonas* spp. [120]. Citrus canker disease caused by the *Xanthomonas citri* subsp. Citri, is the most destructive disease that leads to the great losses worldwide. CRISPR mediated editing of citrus canker-susceptibility gene i.e., LATERAL ORGAN BOUNDARIES 1 (*CsLOB1)* considerably reduced symptoms in two species, grapefruit (Citrus × paradisi) [121], and Wanjincheng orange (*Citrus sinensis* Osbeck) [122].

Fungi

To attain the resistance against fungus, CRISPR technology has been used to disrupt and bait the S genes. Gene *Mlo* (Barley Mildew resistance locus O) encodes a membrane protein, which is responsible for the penetration of the fungus into epidermal cells of host [123]. Mutation in *Mlo* gene activates immune system of plants against infection caused due to powdery mildew [124, 125, 126]. In tomato, mutation in *Mlo* through CRISPR-Cas9 also provides resistance to powdery mildew [38], and wheat [127, 128]. *Edr1* (Enhanced disease resistance 1) S gene of powdery mildew encodes a Raf protein (mitogen-activated protein) has also been targeted through CRISPR-Cas9, resulting in a remarkable decrease in disease progression in powdery mildew in wheat [129]. sgRNAs (single or multiple) generated *OsERF922* mutant lines showed resistance against rice blast disease. Knock-out and complementation of *Solyc08g075770*, through CRISPR showed its role in *Fusarium* wilt tolerance [130]. Role of *VvWRKY52* (transcription factor gene) that is responsible for the biotic stress in the grape (*Vitis vinifera*) is evaluated through CRISPR approach [131]. Disruption of the mitogen-activated protein kinase 3 (*SlMAPK3*) through CRISPR strategy stimulate resistance to *B. cinerea* [132].

Oomycete

Besides the major pathogens (viral, bacterial, and fungal), the CRISPR approach has also been used to tackle oomycete infection. *Phytophthora palmivora* is one of the notorious pathogens of papaya. A CRISPR-Cas9 approach has been used to develop mutants for a functional *PpalEPIC8* (cysteine protease inhibitor) in papaya that provide resistance against *P. palmivora* [133]. Resistance against *Phytophthora tropicalis*, an oomycete that causes black pod disease in *Theobroma cacao*, has also been demonstrated through CRISPR [134].

CRISPR/Cas9 Tools

Designing of sgRNA for editing of the target gene in CRISPR/Cas9 is an important aspect. For efficient and precise functioning of SgRNA, many online software and tools have been developed. An online platform called CRISPR Design has been developed at the Broad Institute by Zhang and his co-workers for the evaluation of off targets as well as designing of sgRNA. This platform has two modes, i.e., Single sequence mode and Batch mode. Single sequence mode can predict sgRNAs of 23-500 nucleotides in length, whereas batch mode can design many sgRNAs. Off-target as well as mismatches can also be evaluated by using this platform while designing sgRNAs. Besides this, other tools i.e., E-CRISPR [141], CRISPR-P [16], Cas-Designer [142], CasOT [143], Cas-OFFinder [144], and SSFinder [145] are also available through which the designing of sgRNA becomes more accessible. There are various methods for the construction of expression vectors in CRISPR/Cas9. Some scientists or researchers have constructed the expression vector through the combination of Cas9 and gRNA, while some have constructed Cas9 vectors and gRNA separately, and then induce the modification of target gene with consecutive transformation [146, 147].

Different delivery methods in plant genetic transformation are an important factor for efficient genome editing. Various delivery methods employed in plant transformation include the *Agrobacterium* approach, biolistic or bombardment methods, floral dip, and PEG mediated protoplast are generally. The proficiency of *Agrobacterium* mediated genetic transformation was enhanced in the absence of chemical selection i.e., kanamycin. In the absence of kanamycin, plant regeneration increases, this in turn enhances the mutation in callus as well as in the multiplication of shoots. SAM (shoot apical meristems) was used in biolistic or bombardment methods of plant genetic engineering, through which germ cells develop during floral organogenesis.

For genome editing through CRISPR/Cas9 in biolistic methods, gold particles are coated with plasmids carrying elements of CRISPR/Cas9. These particles are then bombarded with SAM exposed embryos to induce mutations. The easiest and generally used method for plant transformation in *Arabidopsis* is the floral-dip method. In the floral dip method, an unopened flower is immersed in an *Agrobacterium* suspension containing Silwet-77, a strong surfactant. Due to the presence of surfactant, *Agrobacterium* cells penetrate into the flower, and the T-DNA is transferred into the plant cell. As a result of fertilization, the T-DNA will transfer to the embryos that will ultimately develop into transgenic plants.

Regulations: Genome-Edited Crops

As the genome editing market is expanding up to billion dollars industry, therefore for the regulation of genome editing technology a clear guideline should be framed. DNA-based genome editing is divided into three types, i.e., SDN1, SDN2, and SDN3. Changes or editing in one or few base pairs comprises of SDN1 [148]. SDN-2 involves changes up to 20 bp [149], while the changes of long DNA fragments are characterized as SDN-3 [149] (draft document on GE India). Reports submitted by NAAS 20020 demonstrated that the editing / mutation in SDN-1 and SDN-2 are identical to mutational breeding. So, these should be made available to the farmers in the timeliest manner possible.

For public consultation, Department of Biotechnology (DBT) under the Ministry of Science and Technology, Government of India, developed the draft guidelines in January 2020: "Draft document on Genome Edited Organisms: Regulatory Framework and Guidelines for Risk Assessment". Genome-edited crops (GE) and genetically modified crops (GM) are distinctly different from each other. In genome-edited crops, there is specific editing of endogenous genes, while on the other side; GM crops contain foreign DNA from related or different organisms, which is responsible to give distinct character to the crop. Draft document on genetic engineering advised that the risk assessment would be in accordance with the intricacies associated with changes in base pair edits in SDN-1, SDN-2, and SDN-3 respectively, except that a single base pair is responsible for herbicide or weedicide tolerance. Traits related to herbicide or weediness will need additional biosafety considerations.

In accordance with the draft guidelines 2020, data on the comprehensive biosafety consideration with aspects to biology, delivery method, molecular basis of edits, group to which edits belong, molecular characterisation, integration pattern of donor DNA or cassette, study of off targets, phenotype, and biosafety among others, have to be submitted to the committee. Previously, the genetically modified organisms (GMOs) are considered hazardous under the rules for the manufacture, use/import/export and storage of hazardous microorganisms/ genetically engineered organisms or cells, 1989, in the EPA (Environment Protection Act, 1986). The final instructions or guidelines related to the use of the CRISPR-Cas9 approach for genome editing in crops are still awaited at the time of framing this book chapter. Therefore, all new biotechnological interventions, including all genome editing as well as the CRISPR-

Cas9 gene editing technologies, are still governed under the existing biosafety regulatory framework [150].

CRISPR/Cas9: Biosafety Issues

Social concerns regarding the biosafety issues continue to exist for genome editing of plants. Generally, the concerns are related to the selection of target gene, designing of Guide RNA, effects of off targets and the methods for delivery. Due to the off-target mutations, the risk of undesired changes in genetic material is of great concern [151, 152]. Through the delivery of *in vitro* assembled CRISPR/Cas9 ribonucleoproteins, the risk associated with off-target mutations can be avoided [34, 137, 151, 153]. This technique has previously been applied to many crops, but due to the low stability, high requirements at the technical level, practical utility is not possible [137, 154, 155]. In-depth and considerable research has also been done to reduce the effects of off-target, improving the design of the RNA guide, delivery of ribonucleoproteins, protein engineering, using gRNAs/Cas9 through a lot of chemical or environmental inducers, or using synthetic genetic circuits that manipulate the function of CRISPR according to predetermined logic [151, 153]. Base editing is also being altered to enhance specificity through the use of deaminase activity or deaminase effectors outside the Cas9 binding [156].

Concerns regarding the specificity of Cas9 and the low number of sites that can be targeted through PAM are also there [157]. Through protein engineering and modifications in guide RNA design and Cas9, such as Fokl fusions, the use of truncated RNAs and the paired nicking method, improvements can be done to enhance its fidelity and specificity [158, 159, 160, 161]. Novel proteins and Cas9 variants like cpfi can also be used [152, 162]. Due to the lack of information about the principles and methodology in genome editing, a social conscience arises among the public. There should be clear cut demarcation between genome edited and genome modified plants [163, 164]. Environmental, health and social concerns in the use of genetically modified crops can be prevented through the use of genetically edited crops. As cited above, if there is the removal of transgenes through the segregation, the crop should not be considered as transgenic. Spreading awareness among masses regarding the methodology and principles of CRISPR/ Cas9 can prevent the misbeliefs on the use of this technology to edit the genome of horticulture crops.

Future Perspectives and Conclusion

Plants are the key source of food, fibre, medicines, feed, and fuel. Research has been devoted to improving the quality and quantity of crops through conventional breeding strategies as well as biotechnological interventions. Traditional breeding approaches although contributed for the introduction of new varieties of crops resistance to abiotic as well as biotic stress, but due to time-consuming, less precision, high level of heterozygosity's has made the release of new varieties highly resource-demanding. On the other hand, transgenics technology has overcome the conventional breeding strategy, but due to the regulatory issues and societal concern regarding health and environment, this technology faces many obstacles in the release of GMO crops.

In a short span of time, CRISPR/Cas9 technology has become one of the most promising gene editing tools to improve the quality and quantity of crops. The CRISPR/Cas9 technology hopes for a very encouraging future in making designer plants. Gene of interest/ candidate gene isolated from a wild type species can directly incorporate at the correct location, which in turn provides direction to breeders to make designer plants that could withstand the abiotic as well as biotic stress. As the crops developed through this technology have no transgenes, they are more acceptable to customers in comparison to transgenics. The majority of plant species still face the challenge of developing a more efficient, precise, and robust CRISPR/Cas9 system, including a CRISPR/Cas9 regent delivery system, with less off-target effects. Recent multiple omics, including next generation sequencing, small RNA-seq transcriptomics, metabolomics along with the bioinformatics tools will help to find more gene targets for crop improvement based on CRISPR/Cas9. But due to the different regulatory policies in different countries regarding gene-edited crops, there are different viewpoints. Scientific concordance and uniform ordinance/policies across countries might add to the efficient use of the CRISPR/Cas9/gene-editing technology beyond the research.

References

[1] Khan, A. A., Betel, D., Miller, M. L., Sander, C., Leslie, C. S., and Marks, D. S (2009). Transfection of small RNAs globally perturbs gene regulation by endogenous microRNAs. *Nature Biotechnol.*, 27(6) : 549–555.

[2] Gaj, T., Gersbach, C. A., Barbas, C. F 3rd. (2013). ZFN, TALEN, and CRISPR/Cas-based methods for genome engineering. *Trends Biotechnol.*, 31(7) :397-405.

[3] Silva, G., Poirot, L., Galetto, R., Smith, J., Montoya, G., Duchateau, P., Pâques, F. (2011). Meganucleases and other tools for targeted genome engineering : perspectives and challenges for gene therapy. *Curr Gene Ther.*, 11(1) :11-27.

[4] Demirci, Y., Zhang, B., Unver, T. (2018). CRISPR/Cas9: An RNA-guided highly precise synthetic tool for plant genome editing. *J Cell Physiol.*, 233(3):1844-1859.

[5] Weidenheft, B., Sternberg, S. H. and Doudna, J. A. (2012). RNA guided genetic silencing system in bacteria and archaea. *Nature*, 482 :331.

[6] Jinek, M., Chylinski, K., Fonfara, I., Hauer, M., Doudna, J. A. and Charpentier, E. (2012). A programmable dual-RNA–guided DNA endonuclease in adaptive bacterial immunity. *Science,* 337: 816–821.

[7] Ran, F. A., Hsu, P. D., Wright, J., Agarwala, V., Scott, D. A. and Zhang, F. (2013). *Nat. Protoc.,* 8: 2281–2308.

[8] Voytas, D. F. (2013). Plant Genome Engineering with Sequence-Specific Nucleases. *Annu. Rev. Plant. Biol.*, 64: 327–350.

[9] Knoll, A., Fauser, F. and Puchta, H. (2014). DNA recombination in somatic plant cells: Mechanisms and evolutionary consequences. *Chromosome Res.*, 22: 191–201.

[10] Hsu, P. D., Lander, E. S. and Zhang, F. (2014). Development and applications of CRISPR-Cas9 for genome engineering. *Cell,* 157(6):1262-1278.

[11] Makarova P. D., Lander E. S. and Zhang F. (2014). Development and applications of CRISPR-Cas9 for genome engineering. *Cell,* 157:1262–1278.

[12] Bolotin, A, Quinquis, B., Sorokin, A., Ehrlich, S. D. (2005). Clustered regularly interspaced short palindrome repeats (CRISPRs) have spacers of extrachromosomal origin. *Microbiology*, 151:2551-2561.

[13] Magadán, A. H, Dupuis, M-È., Villion, M. and Moineau, S. (2012) Cleavage of Phage DNA by the *Streptococcus thermophilus* CRISPR3-Cas System. *PLoS ONE*, 7(7): e40913.

[14] Zetsche, B., Gootenberg, J. S., Abudayyeh, O. O., Slaymaker, I. M., Makarova, K. S., Essletzbichler, P., Volz, S. E., Joung, J., van der Oost, J., Regev, A., Koonin, E. V. and Zhang, F. (2015). Cpf1 is a single RNA-guided endonuclease of a class 2 CRISPR-Cas system. *Cell,* 163: 759–771.

[15] Liu, Q., Yang, F., Zhang, J., Liu, H., Rahman, S., Islam, S., Ma, W. and She, M. (2021). Application of CRISPR/Cas9 in Crop Quality Improvement. *Int. J. Mol. Sci.* 22(8):4206.

[16] Lei, Y., Lu, L., Liu, H. Y., Li, S., Xing, F., Chen, L. L. (2014). CRISPR-P: a web tool for synthetic single-guide RNA design of CRISPR-system in plants. *Mol Plant,* 7:1494–6.

[17] Xie, K., Minkenberg, B. and Yang, Y. (2015). *Proc. Natl. Acad. Sci. USA*, 112: 3570–3575.

[18] Naito, Y., Hino, K., Bono, H. and Ui-Tei, K. (2015). CRISPR direct: Software for designing CRISPR/Cas guide RNA with reduced off-target sites. *Bioinformatics,* 31: 1120–1123.

[19] Labun, K., Montague, T. G., Gagnon, J. A., Thyme, S. B. and Valen, E. (2016). CHOPCHOP v2: A web tool for the next generation of CRISPR genome engineering. *Nucleic Acids Res.,* 44: 272–276.

[20] Davies, K. (2020). From the Bench to Benchling. *GEN Edge* 2, 1: 303-309.

[21] Xing, H. L., Dong, L., Wang, Z. P., Zhang, H. Y., Han, C. Y., Liu, B., Wang, X. C. and Chen, Q. J. (2014). A *CRISPR*/Cas9 toolkit for multiplex genome editing in plants. *BMC Plant. Biol.,* 14: 327.

[22] Cermák, T., Baltes, N. J., Cegan, R., Zhang, Y. and Voytas, D. F. (2015). High-frequency, precise modification of the tomato genome. *Genome Biol.,* 16: 232.

[23] Ma, X., Zhang, Q., Zhu, Q., Liu, W., Chen, Y., Qiu, R., Wang, B., Yang, Z., Li, H., Lin, Y., Xie, Y., Shen, R., Chen, S., Wang, Z., Chen, Y., Guo, J., Chen, L., Zhao, X., Dong, Z. and Liu, Y. G. (2015). A Robust CRISPR/Cas9 System for Convenient, High-Efficiency Multiplex Genome Editing in Monocot and Dicot Plants. *Mol. Plant.,* 8(8):1274-84.

[24] Khan, A. A., El-Sayed, A., Akbar, A., Mangravita-Novo, A., Bibi, S., Afzal, Z., Norman, D. J. and Ali, G. S. (2017). A highly efficient ligation-independent cloning system for CRISPR/Cas9 based genome editing in plants. *PlantMethods,* 13: 86.

[25] Sun, Y., Zhang, X., Wu, C., He, Y., Ma, Y., Hou, H., Guo, X., Du, W., Zhao, Y. and Xia, L. (2016). Engineering Herbicide-Resistant Rice Plants through CRISPR/Cas9-Mediated Homologous Recombination of Acetolactate Synthase. *Mol. Plant,* 9: 628– 631.

[26] Hamada, H., Liu, Y., Nagira, Y., Miki, R., Taoka, N. and Imai, R. (2018). Biolistic-delivery-based transient CRISPR/Cas9 expression enables in planta genome editing in wheat. *Sci. Rep.*, 8: 14422.

[27] Jiang, W., Zhou, H., Bi, H., Fromm, M., Yang, B., and Weeks, D. P. (2013). Demonstration of CRISPR/Cas9/sgRNA-mediated targeted gene modification in *Arabidopsis,* tobacco, sorghum and rice. *Nucleic Acids Res.*, 41:e188.

[28] Char, S. N., Neelakandan, A. K., Nahampun, H., Frame, B., Main, K., M, Spalding M. H., Becraft, P. W., Meyers, B. C., Walbot, V., Wang, K., Yang, B. (2017). An *Agrobacterium*-delivered CRISPR/Cas9 system for high-frequency targeted mutagenesis in maize. *Plant Biotechnol J.*, 15(2): 257-268.

[29] Demirer, G. S., Zhang, H., Goh, N. S., González-Grandío, E. and Landry, M. P. (2019). Carbon nanotube-mediated DNA delivery without transgene integration in intact plants. *Nat. Protoc.*, 14: 2954–2971.

[30] Chandrasekaran, R., Rajiv, P. and Abd-Elsalam, K. A. (2020). Carbon nanotubes: Plant gene delivery and genome editing. In *Carbon Nanomaterials for Agri-Food and Environmental Applications*; Elsevier BV: Amsterdam, The Netherlands, 279–296.

[31] Ali, Z., Abulfaraj, A., Idris, A., Ali, S., Tashkandi, M. and Mahfouz, M. M. (2015). CRISPR/Cas9-mediated viral interference in plants. *Genome Biol.,* 16: 238.

[32] Yin, K., Han, T., Liu, G., Chen, T., Wang, Y., Yu, A. Y. L. and Liu, Y. (2015). A geminivirus-based guide RNA delivery system for CRISPR/Cas9 mediated plant genome editing. *Sci. Rep.*, 5: 14926.

[33] Xu, R. F., Li, H., Qin, R. Y., Li, J., Qiu, C. H., Yang, Y. C., Ma, H., Li, L., Wei, P. C. and Yang, J. B. (2015). Generation of inheritable and "transgene clean" targeted genome-modified rice in later generations using the CRISPR/Cas9 system. *Sci. Rep.*, 5: 11491.

[34] Zhang, Y., Liang, Z., Zong, Y., Wang, Y., Liu, J., Chen, K., Qiu, J. L. and Gao, C. (2016). Efficient and transgene-free genome editing in wheat through transient expression of CRISPR/Cas9 DNA or RNA. *Nat. Commun.,* 7: 12617.

[35] Hua, Y., Wang, C., Huang, J. and Wang, K. (2017). A simple and efficient method for CRISPR/Cas9-induced mutant screening. *J. Genet. Genom.,* 44: 207–213.

[36] Zhang, T., Zheng, Q., Yi, X., An H., Zhao, Y., Ma, S., Zhou G. (2018). Establishing RNA virus resistance in plants by harnessing CRISPR immune system. *Plant Biotechnol J.* 16(8):1415-1423.

[37] Brinkman, E. K., Chen, T., Amendola, M. and Van Steensel, B. (2014). Easy quantitative assessment of genome editing by sequence trace decomposition. *Nucleic Acids Res.*, 42 (22): e168.

[38] Nekrasov, V., Wang, C., Win, J., Lanz, C., Weigel, D. and Kamoun, S. (2017). Rapid generation of a transgene-free powdery mildew resistant tomato by genome deletion. *Sci. Rep.,* 7: 482.

[39] Feng, C., Su, H., Bai, H., Wang, R., Liu, Y., Guo, X., Liu, C., Zhang, J., Yuan, J., Birchler, J. A. and Han, F. (2018). High-efficiency genome editing using a dmc1 promoter-controlled CRISPR/Cas9 system in maize. *Plant Biotechnol J.* 16(11):1848-1857.

[40] Tang, X., Liu, G., Zhou, J., Ren, Q., You, Q., Tian, L., Xin, X., Zhong, Z., Liu, B., Zheng, X., Zhang, D., Malzahn, A., Gong, Z., Qi, Y., Zhang, T., Zhang, Y. (2018). A large-scale whole-genome sequencing analysis reveals highly specific genome editing by both Cas9 and Cpf1 (Cas12a) nucleases in rice. *Genome Biol.* 4:19(1):84.

[41] Ricroch, A., Clairand, P., Harwood W. (2017). Use of CRISPR systems in plant genome editing: toward new opportunities in agriculture. *Emerg Top Life Sci.* 1(2):169-182.

[42] Brooks, C., Nekrasov, V., Lippman, Z. B. and Van Eck, J. (2014). Efficient gene editing in tomato in the first generation using the clustered regularly interspaced short palindromic repeats/CRISPR-associated9 system. *Plant Physiol.*, 166: 1292–1297.

[43] Ron, M., Kajala, K., Pauluzzi, G., Wang, D., Reynoso, M. A., Zumstein, K., Garcha, J., Winte, S., Masson, H., Inagaki, S., Federici, F., Sinha, N., Deal, R. B., Bailey-Serres, J. and Brady, S. M. (2014.) Hairy root transformation using *Agrobacterium* rhizogenes as a tool for exploring cell type-specific gene expression and function using tomato as a model. *Plant Physiol*., 166(2):455-69.

[44] Xu, C., Park, S. J., Van Eck, J., and Lippman, Z. B. (2016). Control of inflorescence architecture in tomato by BTB/POZ transcriptional regulators. *Genes Dev.*, 30: 2048–2061.

[45] Soyk, S., Müller, N. A., Park, S. J., Schmalenbach, I., Jiang, K, Hayama R, Zhang L, Van Eck J, Jiménez-Gómez JM, Lippman ZB. (2017). Variation in the flowering gene SELF PRUNING 5G promotes day-neutrality and early yield in tomato. *Nat Genet.*, 49(1):162-168.

[46] Klap, C., Yeshayahou, E., Bolger, A. M., Arazi, T., Gupta, S. K., Shabtai, S., Usadel, B., Salts, Y. Barg, R. (2017). Tomato facultative parthenocarpy results from SlAGAMOUS-LIKE 6 loss of function. *Plant Biotechnol J.*, 15(5):634-647.

[47] Pan, C., Ye, L., Qin, L., Liu, X., He, Y., Wang, J., Chen, L. and Lu, G. (2016) CRISPR/Cas9-mediated efficient and heritable targeted mutagenesis in tomato plants in the first and later generations. *Sci Rep.*, 6:24765.

[48] Hayut, S. F., Bessudo, C. M. and Levy, A. A. (2017). Targeted recombination between homologous chromosomes for precise breeding in tomato. *Nat. Commun.*, 8:15605.

[49] Wang, D., Samsulrizal, N. H., Yan, C., Allcock, N. S., Craigon, J., Blanco-Ulate, B., Ortega-Salazar, I., Marcus, S. E., Bagheri, H. M., Perez Fons, L., Fraser, P. D., Foster, T., Fray, R., Knox, J. P., and Seymour, G. B. (2019). Characterization of CRISPR mutants targeting genes modulating pectin degradation in ripening tomato. *Plant Physiol*., 179: 544–557.

[50] Zsogon, A., Cermak, T., Naves, E. R., Notini, M. M., Edel, K. H., Weinl, S., Freschi, L., Voytas, D. F., Kudla, J. and Peres, L. E. P. (2018). De novo domestication of wild tomato using genome editing. *Nature Biotechnol.,* https://doi.org/10.1038/nature24049.

[51] Yuste-Lisbona, F. J., Fernandez-Lozano, A., Pineda, B., Bretones, S., Ortiz-Atienza, A., Garcia-Sogo, B., Muller, N. A., Angosto, T., Capel, J., Moreno, V., Jimenez-Gomez, J. M. and Lozano, R. (2020). ENO regulates tomato fruit size through the floral meristem development network. *Proc. Natl. Acad. Sci. USA*, 117: 8187–8195.

[52] Li, R., Li, R., Li, X., Fu, D., Zhu, B., Tian, H., Luo, Y. and Zhu H. (2018) Multiplexed CRISPR/Cas9-mediated metabolic engineering of γ-aminobutyric acid levels in *Solanum lycopersicum. Plant Biotechnol J.,* 16(2):415-427.

[53] Andersson, M., Turessoarabidopsisn, H., Nicolia, A., Fält, A. S., Samuelsson, M. (2017). Efficient targeted multiallelic mutagenesis in tetraploid potato (*Solanum tuberosum*) by transient CRISPR-Cas9 expression in protoplasts. *Plant Cell Rep.*, 36: 117–128.

[54] Butler, N. M., Baltes, N. J., Voytas, D. F. and Douches, D. S. (2016). Geminivirusmediated genome editing in potato (*Solanum tuberosum L.*) using sequence specific nucleases. *Front. Plant Sci.,* 7:1045.

[55] Zhou, X., Zha, M., Huang, J., Li, L., Imran, M and Zhang, C. (2017). StMYB44 negatively regulates phosphate transport by suppressing expression of PHOSPHATE1 in potato. *J. Exp. Bot.,* 68: 1265–1281.

[56] González, M. N., Massa, G. A., Andersson, M., Turesson, H., Olsson, N., Fält A-S, Storani, L., Décima Oneto, C. A., Hofvander, P. and Feingold, S. E. (2020). Reduced Enzymatic Browning in Potato Tubers by Specific Editing of a Polyphenol Oxidase Gene via Ribonucleoprotein Complexes Delivery of the CRISPR/Cas9 System. *Front. Plant Sci.*, 10:1649.

[57] Lawrenson, T., Shorinola, O., Stacey, N., Li, C., Ostergaard, L., Patron, N., Uauy, C. and Harwood, W. (2015). Induction of targeted, heritable mutations in barley and *Brassica oleracea* using RNA-guided Cas9 nuclease. *Genome Biol*., 16:258.

[58] Woo, J. W., Kim, J., Kwon, S. I, Corvalán, C., Cho, S. W., Kim, H., Kim, S. G., Kim, S. T., Choe, S. and Kim, J. S. (2015). DNA-free genome editing in plants with preassembled CRISPR-Cas9 ribonucleoproteins. *Nat Biotechnol.,* 33(11):1162-4.

[59] Sashidhar, N., Harloff, H.J., Potgieter, L. and Jung, C. (2020). Gene editing of three BnITPK genes in tetraploid oilseed rape leads to significant reduction of phytic acid in seeds. *Plant Biotechnol J.* 18(11):2241-2250.

[60] Karunarathna, N. L., Wang, H., Harloff, H. J., Jiang, L. and Jung, C. (2020). Elevating seed oil content in a polyploid crop by induced mutations in SEED FATTY ACID REDUCER genes. *Plant Biotechnol J.* 18(11):2251-2266.

[61] Jia, H. and Wang, N. (2014). Targeted genome editing of sweet orange using Cas9/sgRNA. *PLOS ONE,* 9: e93806.

[62] Tian, S., Jiang, L., Gao, Q., Zhang, J., Zong, M., Zhang, H., Ren, Y., Guo, S., Gong, G., Liu, F. And Xu Y. (2017). Efficient CRISPR/Cas9-based gene knockout in watermelon. *Plant Cell Rep.* 36(3):399-406.

[63] Nakajima, I., Ban, Y., Azuma, A., Onoue, N., Moriguchi, T., Yamamoto, T., Toki, S. and Endo M. (2017). CRISPR/Cas9-mediated targeted mutagenesis in grape. *PLoS One,* 18;12(5):e0177966.

[64] Wang, Y., Liu, X., Ren, C., Zhong, G., Yang, L., Li, S. and Liang, Z. (2016). Identification of genomic sites for CRISPR/Cas9-based genome editing in the *Vitis vinifera* genome. *BMC Plant Biol.,* 16: 96.

[65] Charrier, A., Vergne, E., Dousset, N., Richer, A., Petiteau, A. and Chevreau, E. (2019). Efficient Targeted Mutagenesis in Apple and First Time Edition of Pear Using the CRISPR-Cas9 System. *Front Plant Sci.*, 10:40.

[66] Varkonyi-Gasic, E., Wang, T., Voogd, C., Jeon, S., Drummond, R. S. M., Gleave, A. P. and Allan, A. C. (2019). Mutagenesis of kiwifruit CENTRORADIALIS-like genes transforms a climbing woody perennial with long juvenility and axillary flowering into a compact plant with rapid terminal flowering. *Plant Biotechnol. J.*, 17:869–880.

[67] Dash, P. K. and Rai, R. (2016). Translating the "Banana Genome" to delineate stress resistance, dwarfing, parthenocarpy and mechanisms of fruit ripening. *Front Plant Sci.,* 7: 1543.

[68] Kaur, N., Alok, A., Shivani, Kaur, N., Pandey, P., Awasthi, P and Tiwari, S. (2018). CRISPR/Cas9-mediated efficient editing in phytoene desaturase (PDS) demonstrates precise manipulation in banana cv. Rasthali genome. *Funct. Integr. Genomics.,* 18(1):89-99.

[69] Naim, F., Dugdale, B., Kleidon, J., Brinin, A., Shand, K., Waterhouse, P., and Dale, J. (2018). Gene editing the phytoene desaturase alleles of Cavendish banana using CRISPR/Cas9. *Transgenic Res.,* 27: 451–460.

[70] Shao, X., Wu, S., Dou, T., Zhu, H., Hu, C., Huo, H., He, W., Deng, G., Sheng, O., Bi, F., Gao, H., Dong, T., Li, C., Yang, Q., and Yi, G. (2020). Using CRISPR/Cas9 genome editing system to create MaGA20ox2 gene-modified semi-dwarf banana. *Plant Biotechnol. J.,* 18: 17–19.

[71] Kaur, N., Alok, A., Shivani., Kumar, P., Kaur, N., Awasthi, P., Chaturvedi, S., Pandey, P., Pandey, A., Pandey, A. K. and Tiwari, S. (2020). CRISPR/Cas9 directed editing of lycopene epsilon-cyclase modulates metabolic flux for β-carotene biosynthesis in banana fruit. *Metab Eng.*, 59:76-86.

[72] Haque, E., Taniguchi, H., Hassan, M. M., Bhowmik, P., Karim, M. R., Śmiech, M., Zhao, K., Rahman, M. And Islam T. (2018). Application of CRISPR/Cas9 Genome Editing Technology for the Improvement of Crops Cultivated in Tropical Climates: Recent Progress, Prospects, and Challenges. *Front. Plant Sci.,* 9:617.

[73] Yin, Y., Qin, K., Song, X., Zhang, Q., Zhou, Y., Xia, X. and Yu J. (2018). BZR1 Transcription Factor Regulates Heat Stress Tolerance Through FERONIA Receptor-Like Kinase-Mediated Reactive Oxygen Species Signaling in Tomato. *Plant Cell Physiol.,* 59(11):2239-2254.

[74] Li, R., Zhang, L., Wang, L., Chen, L., Zhao, R., Sheng, J. and Shen, L. (2018). Reduction of Tomato-Plant Chilling Tolerance by CRISPR-Cas9-Mediated SlCBF1 Mutagenesis. *J. Agric. Food Chem.*, 66(34):9042-9051.

[75] Wang, L., Chen, L., Li, R., Zhao, R., Yang, M., Sheng, J. and Shen, L. (2017). Reduced Drought Tolerance by CRISPR/Cas9-Mediated SlMAPK3 Mutagenesis in Tomato Plants. *J. Agric. Food Chem.,* 65(39):8674-8682.

[76] Wang, F., Wang, C., Liu, P., Lei, C., Hao, W., Gao, Y., Liu, Y. and Zhao, K. (2016). Enhanced Rice Blast Resistance by CRISPR/Cas9-Targeted Mutagenesis of the ERF Transcription Factor Gene *OsERF922. PLoS ONE*, 11(4): e0154027.

[77] Kishi-Kaboshi, M., Aida, R. and Sasaki, K. (2017). Generation of gene-edited *Chrysanthemum morifolium* using multicopy transgenes as targets and markers. *PlantCell Physiol.*, 58: 216–226.

[78] Li, B., Cui, G., Shen, G., Zhan, Z., Huang, L., Chen, J. and Qi, X. (2017). Targeted mutagenesis in the medicinal plant Salvia miltiorrhiza. *Sci Rep.,* 7:43320.

[79] Li, C., Li, Y. H., Li, Y., Lu, H., Hong, H., Tian, Y., Li, H., Zhao, T., Zhou, X., Liu, J., Zhou, X., Jackson, S. A., Liu, B. and Qiu, L. J. A. (2020). Domestication-Associated Gene GmPRR3b Regulates the Circadian Clock and Flowering Time in Soybean. Mol. Plant., 13(5):745-759.

[80] Wang, L., Sun, S., Wu, T., Liu, L., Sun, X., Cai, Y. and Han, T. (2020). Natural variation and CRISPR/Cas9-mediated mutation in GmPRR37 affect photoperiodic flowering and contribute to regional adaptation of soybean. *Plant Biotechnol. J.,* https://doi.org/10.1111/pbi.13346.

[81] Bao, A., Chen, H., Chen, L., Chen, S., Hao, Q., Guo, W., Qiu, D., Shan, Z., Yang, Z., Yuan, S., Zhang, C., Zhang, X., Liu, B., Kong, F., Li, X., Zhou, X., Tran, L. P. and Cao, D. (2019). CRISPR/Cas9-mediated targeted mutagenesis of GmSPL9 genes alters plant architecture in soybean. *BMC Plant Biol.*, 19: 131.

[82] Cai, Y., Chen, L., Liu, X., Guo, C., Sun, S., Wu, C., Jiang, B., Han, T. and Hou, W. (2018). CRISPR/Cas9-mediated targeted mutagenesis of GmFT2a delays flowering time in soya bean. *Plant Biotechnol. J.,* 16:176–185.

[83] Liu, W., Jiang, B., Ma, L., Zhang, S., Zhai, H., Xu, X., Hou, W., Xia, Z., Wu, C., Sun, S., Wu, T., Chen, L. and Han, T. (2018). Functional diversification of Flowering Locus T homologs in soybean: GmFT1a and GmFT2a/5a have opposite roles in controlling flowering and maturation. *New Phytol.*, 217(3):1335-1345.

[84] Xu, C., Liberatore, K. L., MacAlister, C. A., Huang, Z., Chu, Y. H., Jiang, K., Brooks, C., Ogawa-Ohnishi, M., Xiong, G., Pauly, M., Van Eck J., Matsubayashi, Y., van der Knaap E. and Lippman, Z. B. (2015). A cascade of arabinosyltransferases controls shoot meristem size in tomato. *Nat Genet., 47*(7):784-92.

[85] Ye, M., Peng, Z., Tang, D., Yang, Z., Li, D., Xu, Y., Zhang, C. and Huang, S. (2018). Generation of self-compatible diploid potato by knockout of S-RNase. *Nat. Plants,* 4: 651–654.

[86] Enciso-Rodriguez, F., Manrique-Carpintero, N. C., Nadakuduti, S. S., Buell, C. R., Zarka, D. and Douches, D. (2019). Overcoming Self-Incompatibility in Diploid Potato Using CRISPR-Cas9. *Front. Plant. Sci.,* 10:376.

[87] Nakayasu, M., Akiyama, R., Lee, H. J., Osakabe, K., Osakabe, Y., Watanabe, B., Sugimoto, Y., Umemoto, N., Saito, K., Muranaka, T. and Mizutani M. (2018). Generation of α-solanine-free hairy roots of potato by CRISPR/Cas9 mediated genome editing of the St16DOX gene. *Plant Physiol. Biochem.*, 131:70-77.

[88] Zhang, H., Zhang, Q., Zhai, H., Gao, S., Yang, L., Wang, Z., Xu, Y., Huo, J., Ren, Z., Zhao, N., Wang, X., Li, J., Liu Q. and He, S. (2020). IbBBX24 Promotes the Jasmonic Acid Pathway and Enhances Fusarium Wilt Resistance in Sweet Potato. *Plant Cell*, 32(4):1102-1123.

[89] Ren, Y., Sun, H., Zong, M., Guo, S., Ren, Z., Zhao, J., Li, M., Zhang, J., Tian, S., Wang, J., Yu, Y., Gong, G., Zhang, H., He, H., Li, L., Zhang, X., Liu, F., Fei, Z. and Xu, Y. (2020). Localization shift of a sugar transporter contributes to phloem unloading in sweet watermelons. *New Phytol.*, 227(6):1858-1871.

[90] Zhang, J., Guo, S., Ji, G., Zhao, H., Sun, H., Ren, Y., Tian, S., Li, M., Gong, G., Zhang, H. and Xu Y. A unique chromosome translocation disrupting ClWIP1 leads to gynoecy in watermelon. *Plant J.* 101(2):265-277.

[91] Guo, S., Zhao, S., Sun, H., Wang, X., Wu S, Lin T, Ren Y, Gao L, Deng Y, Zhang J, Lu X, Zhang H, Shang J, Gong G, Wen C, He N, Tian S, Li M, Liu J, Wang Y, Zhu Y, Jarret R, Levi A, Zhang X, Huang S, Fei Z, Liu W, Xu Y. (2020). Resequencing of 414 cultivated and wild watermelon accessions identifies selection for fruit quality traits. *Nat Genet.,* 51(11):1616-1623.

[92] Tian, S., Jiang, L., Cui, X., Zhang, J., Guo, S., Li, M., Zhang, H., Ren, Y., Gong, G., Zong, M., Liu, F., Chen, Q. and Xu, Y. (2018). Engineering herbicide-resistant watermelon variety through CRISPR/Cas9-mediated base-editing. *Plant Cell Rep.*, 37(9):1353-1356.

[93] Wang, L., Wang, L., Tan, Q., Fan, Q., Zhu, H., Hong, Z., Zhang, Z. and Duanmu, D. (2016). Efficient Inactivation of Symbiotic Nitrogen Fixation Related Genes in Lotus japonicus Using CRISPR-Cas9. *Front Plant Sci.,* 7:1333.

[94] Ou, W., Mao, X., Huang, C., Tie, W., Yan, Y., Ding, Z., Wu, C., Xia, Z., Wang, W., Zhou, S., Li, K. and Hu, W. (2018). Genome-Wide Identification and Expression Analysis of the KUP Family under Abiotic Stress in Cassava (*Manihot esculenta* Crantz). *Front Physiol.,* 9:17.

[95] Ye, J., Yang, H., Shi, H., Wei, Y., Tie, W., Ding, Z., Yan, Y., Luo, Y., Xia, Z., Wang, W., Peng, M., Li, K., Zhang, H. and Hu, W. (2017). The MAPKKK gene family in cassava: Genome-wide identification and expression analysis against drought stress. *Sci. Rep.,* 7(1):14939.

[96] Yun, C., Ma, J., Zhang, X., Yang, Y., Zhou, D., Yu, Q., Que, Y., Xu, L. and Guo, J. (2017). A novel non-specific lipid transfer protein gene from sugarcane (NsLTPs), obviously responded to abiotic stresses and signaling molecules of SA and MeJA. *Sugar Tech.,* 19:17-25.

[97] Arroyo-Herrera, A., Figueroa-Yanez, L., Castano, E. and Santamaria, J. M. (2016). A novel Dreb2-type gene from *Carica papaya* confers tolerance under abiotic stress. *Plant. Cell Tissue Org. Cult.*, 125(1):119–133.

[98] He, P., Zhao, P., Wang, L., Zhang, Y., Wang, X., Xiao, H., Yu, J. and Xiao, G. (2017). The PIN gene family in cotton (Gossypium hirsutum): genome-wide identification and gene expression analyses during root development and abiotic stress responses. *BMC Genom.*, 18(1):507.

[99] Dass, A., Zainul, M., Abdin, Reddy, V. S. and Leelavathi, S. (2017). Isolation and characterization of the dehydration stress inducible GhRDL1 promoter from the cultivated upland cotton (*Gossypium hirsutum*). *J. Plant Biochem. Biotechnol.*, 26(1):113–119.

[100] Hu, B., Li, D., Liu, X., Qi, J., Gao, D., Zhao, S., Huang, S., Sun, J., Yang, L. (2017). Engineering Non-transgenic Gynoecious Cucumber Using an Improved Transformation Protocol and Optimized CRISPR/Cas9 System. *Mol Plant.*, 10(12):1575-1578.

[101] Lemmon, Z. H., Reem. N. T., Dalrymple. J., Soyk, S. (2018). Rapid improvement of domestication traits in an orphan crop by genome editing. *Nat. Plants,* 4: 766–770.

[102] Ali, Z., Ali, S., Tashkandi, M., Zaidi, S. S. and Mahfouz, M. M. (2016). CRISPR/Cas9-Mediated Immunity to Geminiviruses: Differential Interference and Evasion. *Sci Rep.,* 6:26912.

[103] Tashkandi, M, Ali, Z., Aljedaani, F., Shami, A. and Mahfouz, M. M. (2018). Engineering resistance against Tomato yellow leaf curl virus via the CRISPR/Cas9 system in tomato. *Plant Signal Behav.*, 13(10):e1525996.

[104] Baltes, N. J., Hummel, A. W., Konecna, E., Cegan, R., Bruns, A. N., Bisaro, D. M. and Voytas, D. F. (2015). Conferring resistance to geminiviruses with the CRISPR-Cas prokaryotic immune system. *Nat Plants,* 1(10):15145.

[105] Ji, X., Zhang, H., Zhang, Y., Wang, Y. and Gao, C. (2015). Establishing a CRISPR-Cas-like immune system conferring DNA virus resistance in plants. *Nat Plants*, 1:15144.

[106] Tripathi, J. N., Ntui, V. O., Ron, M., Muiruri, S. K., Britt, A. and Tripathi, L. (2019). CRISPR/Cas9 editing of endogenous *banana streak virus* in the B genome of *Musa* spp. overcomes a major challenge in banana breeding. *Commun. Biol.*, 2: 46 https://doi.org/10.1038/s42003-019-0288-7.

[107] Zhang, T., Zhao, Y., Ye, J., Cao, X., Xu, C., Chen, B., An, H., Jiao, Y., Zhang, F., Yang, X. and Zhou, G. (2019). Establishing CRISPR/Cas13a immune system conferring RNA virus resistance in both dicot and monocot plants. *Plant Biotechnol J.* 17(7) :1185-1187.

[108] Aman, R., Mahas, A., Butt, H., Aljedaani, F. and Mahfouz, M. (2018). Engineering RNA Virus Interference via the CRISPR/Cas13 Machinery in Arabidopsis. *Viruses,* 10(12):732.

[109] Mehta, D., Stürchler, A., Anjanappa, R. B., Zaidi, S. S., Hirsch-Hoffmann, M., Gruissem, W. and Vanderschuren, H. (2019). Linking CRISPR-Cas9 interference in cassava to the evolution of editing-resistant geminiviruses. *Genome Biol.* 20(1): 80.

[110] Rybicki, E. P. (2019). CRISPR-Cas9 strikes out in cassava. *Nat. Biotechnol.*, 37(7): 727–8.

[111] Zaidi, S. S., Tashkandi, M., Mansoor, S. and Mahfouz, M. M. (2016). Engineering Plant Immunity: Using CRISPR/Cas9 to Generate Virus Resistance. *Front Plant Sci.*, 7:1673.

[112] Chandrasekaran, J., Brumin, M., Wolf, D., Leibman, D., Klap, C., Pearlsman, M., Sherman A., Arazi, T. and Gal-On A. (2016.) Development of broad virus resistance in non-transgenic cucumber using CRISPR/Cas9 technology. *Mol. Plant Pathol.*, 17(7):1140-53.

[113] Pyott, D. E., Sheehan, E., Molnar, A. (2016). Engineering of CRISPR/Cas9-mediated potyvirus resistance in transgene-free *Arabidopsis* plants. *Mol Plant Pathol.,* 17(8) :1276–88.

[114] Gomez, M. A., Lin, Z. D., Moll, T., Chauhan, R. D., Hayden, L., Renninger, K., Beyene, G., Taylor, N. J., Carrington, J. C., Staskawicz, B. J. and Bart, R. S. (2019). Simultaneous CRISPR/Cas9-mediated editing of cassava eIF4E isoforms nCBP-1 and nCBP-2 reduces cassava brown streak disease symptom severity and incidence. *Plant Biotechnol J.*, 17(2) :421-434.

[115] Macovei, A., Sevilla, N. R., Cantos, C., Jonson, G. B., Slamet-Loedin, I., Čermák, T., Voytas, D. F., Choi, I. R. and Chadha-Mohanty, P. (2018). Novel alleles of rice eIF4G generated by CRISPR/Cas9-targeted mutagenesis confer resistance to Rice tungro spherical virus. *Plant Biotechnol. J.* 16(11):1918-1927.

[116] Blanvillain-Baufumé, S., Reschke, M., Solé, M., Auguy, F., Doucoure, H., Szurek, B., Meynard, D., Portefaix, M., Cunnac, S., Guiderdoni, E., Boch, J., Koebnik, R. (2017). Targeted promoter editing for rice resistance to Xanthomonas oryzae pv. oryzae reveals differential activities for SWEET14-inducing TAL effectors. *Plant Biotechnol. J.* 15(3): 306-317.

[117] Li, T., Liu, B., Spalding, M. H., Weeks, D. P. and Yang, B. (2012). High-efficiency TALEN-based gene editing produces disease-resistant rice. *Nat Biotechnol.*, 30(5):390-2.

[118] Oliva, R., Ji, C., Atienza-Grande, G., Huguet-Tapia, J. C., Perez-Quintero, A., Li, T., Eom, J. S., Li, C., Nguyen, H., Liu, B., Auguy, F., Sciallano, C., Luu, V. T., Dossa, G. S., Cunnac, S., Schmidt, S. M., Slamet-Loedin, I. H., Vera Cruz, C., Szurek, B., Frommer, W. B., White, F. F. and Yang B. Broad-spectrum resistance to bacterial blight in rice using genome editing. *Nat. Biotechnol.* 37(11):1344-1350.

[119] Ortigosa, A., Gimenez-Ibanez, S., Leonhardt, N., Solano, R. (2019). Design of a bacterial speck resistant tomato by CRISPR/Cas9-mediated editing of SlJAZ2. *Plant.*

[120] de Toledo Thomazella, D. P., Brail, Q., Dahlbeck, D. and Staskawicz, B. (2016). CRISPR-Cas9 mediated mutagenesis of a DMR6 ortholog in tomato confers broad spectrum disease resistance. BioRxiv: doi: https://doi.org/10.1101/064824.

[121] Peng, A., Chen, S., Lei, T., Xu, L., He, Y., Wu, L., Yao, L. and Zou, X. (2017). Engineering canker-resistant plants through CRISPR/Cas9-targeted editing of the susceptibility gene CsLOB1 promoter in citrus. *Plant Biotechnol. J.,* 15(12):1509-1519.

[122] Jia, H., Zhang, Y., Orbović, V., Xu, J., White, F. F., Jones, J. B. and Wang, N. (2017). Genome editing of the disease susceptibility gene CsLOB1 in citrus confers resistance to citrus canker. *Plant Biotechnol. J.* 15(7): 817-823.

[123] Acevedo-Garcia, J., Kusch, S. and Panstruga, R. (2014). Magical mystery tour: *MLO*proteins in plant immunity and beyond. *New Phytol.*, 204(2): 273–81.

[124] Freisleben, R., Lein, A. (1942). Über die Auffindung einer mehltauresistenten Mutante nach Röntgenbestrahlung einer anfälligen reinen Linie von Sommergerste [*On the detection of a powdery*

mildew resistant mutant after X-ray irradiation of a susceptible pure line of spring barley]. *Naturwissenschaften*, 30(40), 608-608.

[125] Jørgensen, I. H. (1992). Discovery, characterization and exploitation of *Mlo* powdery mildew resistance in barley. *Euphytica,* 63(1–2):141–52.

[126] van Schie, C. C. and Takken, F. L. (2014). Susceptibility genes 101: how to be a good host. *Annu Rev Phytopathol.*, 52:551–81.

[127] Shan, Q., Wang, Y., Li, J., Zhang, Y., Chen, K., Liang, Z., Zhang, K., Liu, J., Xi, J. J., Qiu, J. L. and Gao, C. (2013). Targeted genome modification of crop plants using a CRISPR-Cas system. *Nat Biotechnol.* 31(8):686-8.

[128] Li, S., Lin, D., Zhang, Y., Deng, M., Chen, Y., Lv, B., Li, B., Lei, Y., Wang, Y., Zhao, L., Liang, Y., Liu, J., Chen, K., Liu, Z., Xiao, J., Qiu, J. L. and Gao, C. (2022). Genome-edited powdery mildew resistance in wheat without growth penalties. *Nature*, 602(7897):455-460.

[129] Zhang, Y., Bai, Y., Wu, G., Zou, S., Chen, Y., Gao, C. and Tang D. (2017). Simultaneous modification of three homoeologs of TaEDR1 by genome editing enhances powdery mildew resistance in wheat. *Plant J.*, 91(4) :714-724.

[130] Prihatna, C., Barbetti, M. J. and Barker, S. J. (2018). A novel tomato fusarium wilt tolerance gene. *Front Microbiol.,* 9 :1226.

[131] Wang, X., Tu, M., Wang, D., Liu, J., Li, Y., Li, Z., Wang, Y. and Wang, X. (2017). CRISPR/Cas9-mediated efficient targeted mutagenesis in grape in the first generation. *Plant Biotechnol J.* 16(4): 844-855.

[132] Zhang, S., Wang, L., Zhao, R., Yu, W., Li, R., Li, Y., Sheng, J. and Shen, L. (2018). Knockout of SlMAPK3 Reduced Disease Resistance to Botrytis cinerea in Tomato Plants. *J. Agric. Food Chem.* 66(34) :8949-8956.

[133] Gumtow, R., Wu, D., Uchida, J. and Tian, M. A. (2018). Phytophthora palmivora Extracellular Cystatin-Like Protease Inhibitor Targets Papain to Contribute to Virulence on Papaya. *Mol Plant Microbe Interact.,* 31(3) :363-373.

[134] Fister, A. S., Landherr, L., Maximova, S. N. and Guiltinan, M. J (2018). Transient expression of CRISPR/Cas9 machinery targeting TcNPR3 enhances defense response in Theobroma cacao (Original Research). *Front Plant Sci.,* 9:268.

[135] Santillán Martínez, M. I., Bracuto, V., Koseoglou, E., Appiano, M., Jacobsen, E., Visser, RGF., Wolters, A. A. and Bai, Y. (2020). CRISPR/Cas9-targeted mutagenesis of the tomato susceptibility gene PMR4 for resistance against powdery mildew. *BMC Plant Biol.,* 20(1) :284.

[136] Dale, J., James, A., Paul, J. Y., Khanna, H., Smith, M., Peraza-Echeverria, S., Garcia-Bastidas, F., Kema, G., Waterhouse, P., Mengersen, K. and Harding, R. (2017). Transgenic Cavendish bananas with resistance to Fusarium wilt tropical race 4. *Nat. Commun., 8*(1) :1496.

[137] Malnoy, M., Viola, R., Jung, M. H., Koo, O. J., Kim, S., Kim, J. S, Velasco, R. and Nagamangala, K. C. (2016). DNA-free genetically edited grapevine and apple protoplast using CRISPR/Cas9 ribonucleoproteins. *Front Plant Sci.* 7:1904.

[138] Xie, K. and Yang, Y. (2013). RNA-guided genome editing in plants using a CRISPR–Cas system. *Mol Plant*, 6(6):1975–83.

[139] Roy, A., Zhai, Y., Ortiz, J., Neff, M., Mandal, B., Mukherjee, S. K. and Pappu, H. R. (2019). Multiplexed editing of a begomovirus genome restricts escape mutant formation and disease development. *PLoS One,* (10): e0223765.

[140] Yin, K., Han, T., Xie, K., Zhao, J. and Liu, Y. (2019). Engineer complete resistance to Cotton Leaf Curl Multan virus by the CRISPR/Cas9 system in *Nicotiana benthamiana*. *Phytopathol Res.,* 1:9.

[141] Heigwer, F., Kerr, G. and Boutros, M. (2014). E-CRISPR: fast CRISPR target site identification. *Nat Methods*, 11:122–3.

[142] Park. J., Bae. S. and Kim, J. (2015). Cas-Designer: a web-based tool for choice of CRISPR-Cas9 target sites. *Bioinformatics*, 31:4014–6.

[143] Xiao, A., Cheng, Z., Kong, L., Zhu, Z., Lin, S., Gao, G. and Zhang, B. (2014). CasOT: a genome-wide Cas9/gRNA off-target searching tool. *Bioinformatics,* 30(8):1180-1182.

[144] Bae, S., Park, J. and Kim, J. (2014). Cas-OFFinder: a fast and versatile algorithm that searches for potential off-target sites of Cas9 RNA-guided endonucleases. *Bioinformatics*, 30:1473–5.

[145] Upadhyay, S. K. and Sharma, S. (2014). SSFinder: high throughput CRISPR-Cas target sites prediction tool. *BioMed Res Int.*, 2014:742482.

[146] Mikami, M., Toki, S. and Endo, M. (2015). Parameters affecting frequency of CRISPR/Cas9 mediated targeted mutagenesis in rice. *Plant Cell Rep.*, 34:1807–15.

[147] Hyun, Y., Kim, J., Cho, S. W., Choi, Y. and Kim, J. S. (2015). Coupland G. Site -directed mutagenesis in *Arabidopsis thaliana* using dividing tissue-targeted RGEN of the CRISPR/Cas system to generate heritable null alleles. *Planta,* 241:271–84.

[148] Mishra, R., Joshi, R. K. and Zhao, K. (2019). Base editing in crops: current advances, limitations and future implications. *Plant Biotechnol. J.*, 18: 20–31.

[149] Gao. (2018). The future of CRISPR technologies in agriculture. *Nat. Rev. Mol. Cell Biol.*, 19:275–276.

[150] Friedrichs, S., Takasu, Y., Kearns, P., Dagallier, B., Oshima, R., Schofield, J., et al. (2019). Meeting report of the OECD conference on "genome editing: applications in agriculture—implications for health, environment and regulation." *Transgenic Res.,* 28:419–463.

[151] Liang, Z., Chen, K., Zhang, Y., Liu, J., Yin, K., Qiu, J. L. and Gao C. Genome editing of bread wheat using biolistic delivery of CRISPR/Cas9 *in vitro* transcripts or ribonucleoproteins. *Nat Protoc.*, 13(3):413-430.

[152] Pineda, M., Lear, A., Collins, J. P. and Kiani, S. (2019). Safe CRISPR: challenges and possible solutions. *Trends Biotechnol.,* 37:389–401.

[153] Svitashev, S., Schwartz, C., Lenderts, B., Young, J. K., and Mark Cigan, A. (2016). Genome editing in maize directed by CRISPR-Cas9 ribonucleoprotein complexes. *Nat. Commun.,* 7:13274. doi: 10.1038/ncomms13274.

[154] Subburaj, S., Chung, S. J., Lee, C., Ryu, S. M., Kim, D. H., Kim, J. S., Bae, S. and Lee, G. J. (2016.) Site-directed mutagenesis in Petunia × hybrida protoplast system using direct delivery of purified recombinant Cas9 ribonucleoproteins. *Plant Cell Rep.* 35(7):1535-44.

[155] Murovec, J., Gucek, K., Bohanec, B., Avbelj, M. and Jerala, R. (2018). DNA-free genome editing of Brassica oleracea and B. rapa Protoplasts using CRISPRCas9 Ribonucleoprotein complexes. *Front. Plant Sci.,* 9:1594.

[156] Shimatani, Z., Kashojiya, S., Takayama, M., Terada, R., Arazoe, T., Ishii, H., Teramura, H., Yamamoto, T., Komatsu, H., Miura, K., Ezura, H., Nishida, K., Ariizumi, T. and Kondo, A. (2017). Targeted base editing in rice and tomato using a CRISPR-Cas9 cytidine deaminase fusion. *Nat Biotechnol.*, 35(5):441- 443.

[157] Spencer, J. M. and Zhang, X. (2017). Deep mutational scanning of S. pyogenes Cas9 reveals important functional domains. *Sci. Rep.,* 7:16836.

[158] Kleinstiver, B. P., Prew, M. S., Tsai, S. Q., Topkar, V. V., Nguyen, N. T., Zheng, Z., Gonzales, A. P., Li, Z., Peterson, R. T., Yeh, J. R., Aryee, M. J., Joung, J. K. (2015). Engineered CRISPR-Cas9 nucleases with altered PAM specificities. *Nature,* 523(7561):481-5.

[159] Kleinstiver, B. P., Pattanayak, V., Prew, M. S., Tsai, S. Q., Nguyen, N. T., Zheng, Z. and Joung, J. K. (2016.) High-fidelity CRISPR-Cas9 nucleases with no detectable genome-wide off-target effects. Nature, 529(7587) : 490-5.

[160] Leenay, R. T. and Beisel, C. L. (2017). Deciphering, communicating, and engineering the CRISPR PAM. *J. Mol. Biol.,* 429 :177–191.

[161] Wyvekens, N., Topkar, V. V., Khayter, C., Joung, J. K. and Tsai, S. Q. (2015). Dimeric CRISPR RNA-Guided FokI-dCas9 nucleases directed by truncated gRNAs for highly specific genome editing. *Hum. Gene Ther.*, 26 :425–431.

[162] Nakade, S., Yamamoto, T. and Sakuma, T. (2017). Cas9, Cpf1 and C2c1/2/3- What's next? Bioengineered, 8: 265–273.

[163] Garcia Ruiz, M. T., Knapp, A. N. and Garcia-Ruiz, H. (2018). Profile of genetically modified plants authorized in Mexico. *GM Crops Food*, 9: 152–168.

[164] Eckerstorfer, M. F., Engelhard, M., Heissenberger, A., Simon, S. and Teichmann, H. (2019). Plants developed by new genetic modification techniques—comparison of existing regulatory frameworks in the EU and non-EU countries. *Front. In Bioeng. Biotechnol.*, 7:26.

[165] Addgene CRISPR Plasmids: Plants. Available online: http://www.addgene.org/crispr/plant/.

[166] Snapgene Tutorial Videos. Available online: https://www.snapgene.com/support/tutorial-videos/ Arora L. and A. Narula (2017). Gene editing and crop improvement using CRISPR-CAS9 system. *Front. Plant Sci,.* https://doi.org/10.3389/fpls.2017.01932.

Chapter 14

Genome Editing in Vegetables and Fruit Crops: Present Status and Future Prospects

Baljeet Singh[1,2]
Urvashi Mittal[1]
Sarvjeet Kukreja[3]
Vijay Kumar[1]
Harmeet Kaur[1]
Vinay Bhardwaj[2]
Sundaresha Sidappa[2]
and Umesh Goutam[1,4]
[1] Department of Biotechnology, School of Bioengineering and Biosciences, Lovely Professional University, Punjab, India
[2] ICAR-Central Potato Research Institute, Shimla
[3] Department of Agronomy, School of Agriculture, Lovely Professional University, Punjab
[4] Department of Molecular Biology and Genetic Engineering
School of Bioengineering and Biosciences, Lovely Professional University, Punjab, India

Abstract

Horticultural crops include a large range of plants that include vegetables, flowers and fruit crops that provide essential nutrients and valuable products to humans. These crops also play a crucial role in global nutrition security. Ever since domestication, endless efforts have been made to ameliorate the qualitative and quantitative traits of horticultural crops because of which, better varieties of these crops are available these days. However, in most horticultural crops, the yield has reached a plateau and it is difficult to further enhance the crop yield through traditional approaches. Moreover, with the change in time in some horticultural crops, improvements in qualitative traits have become a priority.

These traits can be improved by traditional breeding by identifying the useful traits in the crop germplasm and through a series of backcrossing, these can be introgressed into cultivated varieties, which is a laborious and time taking process. However, with the discovery of genome-editing technologies, and their potential applications in horticultural crops the above targets can be achieved in very less time. Recently the Clustered Regularly Interspaced Short Palindromic Repeats (CRISPR) technique has been employed in several

In: Genome Editing for Crop Improvement
Editors: Rahul Kumar, Ashutosh Sharma, Yogesh Kumar et al.
ISBN: 979-8-88697-740-0

crops for different purposes because it is a simple, rapid and highly efficient approach for crop improvement. This chapter summarizes the recent progress and present status of applicability of CRISPR/Cas9 for horticultural crop improvement.

Keywords: CRISPR/Cas, genome editing, vegetables, fruits, horticultural crops, fruit quality

Introduction

It is projected that the global population will increase to 9.7 billion by 2050 (UN-DESA, 2019). The demand for food is rising with the ever-increasing population and to meet the global food demand better high-yielding crop varieties are today's need. Horticultural crops are rich in cellulose, vitamins and minerals. However, due to the changing climate sustainable production of these horticultural crops is challenging. In addition to abiotic stresses such as heat, drought and salinity, new races of pests and diseases are also mushroomed up under the climate change scenario (Velásquez et al. 2018). The development of biotic and abiotic stress-resistant varieties of fruit and vegetable crops is one way to fight the battle against food scarcity and hidden hunger. Traditional plant breeding was extensively used to introduce important genetic variations for crop improvement. However, crop germplasms have a finite pool of desirable genes. Moreover, the development of better crop varieties through traditional breeding is time-consuming and laborious. The development of transgenic crops could be a solution to the problems associated with conventional breeding. However, due to a number of legal and ethical issues, it is very difficult to release a transgenic crop variety commercially (Bhatta and Malla, 2020). Under all these scenarios, genome-editing approaches especially clustered regularly interspaced short palindromic repeats (CRISPR) have emerged as the most promising tool to develop the crops for the future.

With the emergence of next-generation sequencing (NGS) technologies and advancements in bioinformatics tools, nowadays an ample amount of genetic information is available for different phenotypic traits. The availability of crop reference genomes also catalyzed crop improvement programs. At present, gene modification, gene knockout and gene editing technologies have become the most useful assets for crop improvement (Kumar et al. 2021). Genome-wide analysis studies (GWAS) take advantage of NGS technologies to decipher the genomic regions controlling the complex phenotypic traits (Uffelmann et al. 2021; Soto-Cerda et al. 2022). Gene modifications and gene knockout techniques are being used for validating the gene functionalities (Singh et al. 2019a; Rehman et al. 2022). The gene-editing techniques allow the researchers a precise editing of target genes either to knock them out or to alter their function. Among all editing gene-editing techniques, CRISPR-Cas9 is considered to be the simplest and most effective gene-editing technique. Recognition of suitable Cas proteins, design and selection of guide RNAs (gRNAs), and selection of regulatory elements to express gRNAs and Cas proteins are all basic steps in the CRISPR construct design process (Lu et al. 2020). In recent years, this approach has been used in many crops to improve different traits such as crop yield (Rasheed et al. 2021; Liu et al. 2021), nutrients (Achary et al. 2021; Shoeb et al. 2021), and biotic (Bashar et al. 2021; Zhou et al. 2022) as well as abiotic stress tolerance (Razzaq et al. 2022; Park et al. 2022). This technique has proved its potential for crop improvement and with time it is also evolving and being widely used for crop improvement.

Mechanism of CRISPR/Cas System

After repeated breeding and selection for agronomic traits, most of the present crop varieties have less genetic diversity. For major crops, it is challenging to enhance their agronomic performance via traditional plant breeding. Moreover, it is a laborious and time-consuming task. In several cases, the crop germplasm does not have the genes to further improve that particular trait such as for nutritional concentrations and biotic stress through conventional breeding. Alterations in the targeted genes can be done through RNA interference (RNAi), virus-induced gene silencing (VIGS), Mutations, transfer DNA (T-DNA) insertions and engineered endonucleases (Singh et al. 2019a, b). These techniques have been used rigorously to improve the crops. However, these approaches have some limitations such as low accuracy and low efficiency to modify the target gene in the desired manner. Then with the advent of genome editing approaches, both the accuracy as well as the efficiency of gene editing were improved significantly.

Until 2013, the gene-editing methods were mainly revolving around zinc-finger nucleases (ZFNs), transcription activator-like effector nucleases (TALENs), etc. These technologies using zinc finger nuclease (ZFNs) and transcription activator-like effector nuclease (TALEN) were labor-intensive and expensive as they were dependent on the interaction of protein-nucleic acid and require a custom-designed protein for all gene locus of interest. CRISPR based systems use the interactions between RNA-DNA to guide DNA targeting and cleavage, making it a simple, efficient, and low-cost genetic editing technique, whereas ZFNs and TALENs, employ protein–DNA interactions to establish target specificity. CRISPR technology has already established itself as the most widely used genome-editing tool, with applications in a wide range of plant species (Talakayala et al. 2022; Ahmad et al. 2022).

For basic research and crop improvement, structured editing of the genome has been performed in several dicots and monocot plants utilizing several CRISPR/Cas systems, and the use of CRISPR/Cas technology in plants has risen substantially in recent years (Zhu et al. 2020; Huang et al. 2021; Karlson et al. 2021). It allows tissue-specific gene editing therefore one can focus on the commercially important plant parts (Singha et al. 2022). Yoshizumi Ishino discovered CRISPR in *Escherichia coli* in 1987. They were studying the alkaline phosphatase conversion gene when they accidentally cloned an unusual group of repetitive sequences intermingled with the spacer sequences (Ishino et al. 2018). CRISPR systems of three types I, II, and III have been found in a variety of archaeal and bacterial hosts, and each system is made up of a cluster of CRISPR-associated Cas genes, direct repeats, and noncoding RNAs. Clustered regularly interspaced short palindromic repeat (CRISPR) and its accompanying protein (Cas-9) which is used in many practical areas.

The two fundamental components of the CRISPR/Cas-9 system are guide RNA (gRNA) and CRISPR-associated (Cas-9) proteins. The discovery of the novel nuclease Cas9, a class II bacterial adaptive immune system from *Streptococcus pyogenes*, led to the concept of CRISPR/Cas9-mediated genome editing (Hang et al. 2007). Cas9 nuclease, crRNA that encodes guide RNA, and required tracrRNA (trans-activating crRNA that stimulates the processing of crRNA into distinct units) make up the Type II CRISPR System (Barranhou et al. 2007). Mojica and colleagues observed that bacteriophages, plasmids, and viruses all have comparable spacer sequences and that viruses cannot infect bacteria with homologous spacer sequences, hinting that these sequences are involved in prokaryotes adaptive immune system.

When a virus infects prokaryotes, the spacer sequence in CRISPR arrays is translated into short (crRNA) CRISPR RNA, which instructs (Cas) CRISPR-associated sequences to cleave corresponding DNA or RNA viral sequences via the CRISPR- Cas system. The crRNA and tracrRNA can also be joined to form a chimeric single-guide RNA (sgRNA) (Jinek et al. 2012) (Figure 1).

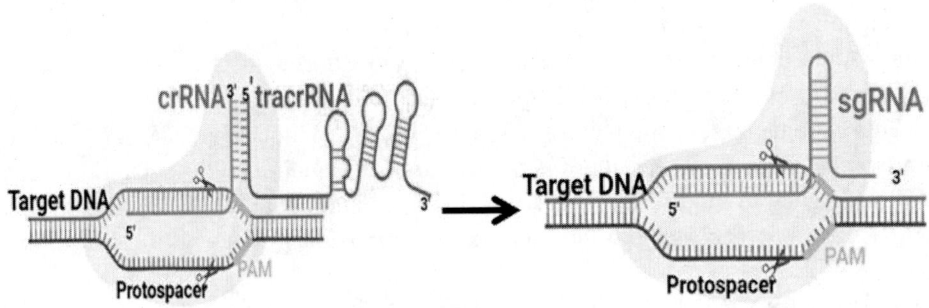

Figure 1. A powerful complex of CRISPR-Cas system.

For cleavage, the presence of a conserved protospacer adjacent motif (PAM) downstream of target DNA with the normal sequence of 5'-NGG-3' and a rare sequence of NAG is essential. The specificity is provided by the 'seed sequence', which is 12 bases upstream of the protospacer adjacent motif (PAM) and fits between target DNA and RNA (Gasiunas et al. 2012; Hsu et al. 2013; Zhang et al. 2021). Jinek et al. (2012) described the molecular mechanism by which the CRISPR array and Cas protein functioned, and the ability of CRISPR Cas was acknowledged as a result. Nowadays, CRISPR-Cas based gene editing has been a routine activity in plant molecular laboratories. This revolutionary technique has proved its potential in several crops for numerous traits. This chapter is focused on CRISPR-based gene editing in horticultural crops because of its simple structure, high performance, design versatility, and relevance to a wide range of species, CRISPR-Cas has emerged as the most effective tool for genetic engineering.

The Basic Methodology Used for CRISPR/Cas System-Based Gene Editing in Plants

In plants, selecting a novel target in the genome that is positioned upstream 5'-NGG-3' of the PAM sequence is required for efficient genome editing. The sgRNA must be homologous to the target sequence (approximately 20bp) and hold a 'scaffold' sequence for Cas9 binding (Jiang et al. 2013). If it is similar to sgRNA, then a complex form containing sgRNA and Cas9 which cleaves off the target by ligating itself to the PAM sequence in the genome. Recruitment of Cas9 is through sgRNA to a certain locus and for successful genome editing it is important to avoid any mismatching/alterations, so the best designs must be generated (Doudna, 2014; Erpen et al. 2019). After that, the sgRNA and Cas9 transcription cassettes are inserted into the plant's genome, most typically by Agrobacterium-mediated transformations (Zhou et al. 2014; Fan et al. 2015; Erpen et al. 2019; Kaur et al. 2021). Sequencing and restriction assays are performed to analyze the guise of Cas9 and sgRNA, so to determine the nicks in dsDNA and also to detect the mutations in the transformed transgenic plant lines which were mended by the homology-directed repair and non-homologous end-joining (Soda and Verma, 2018). Transgenic organisms are known to have a broad spectrum of edits and mutations. Few plant species are found to be unaltered as they do not have an expression of Cas9 and sgRNA, as well as inefficient target cleavage (Xingliang and Yaoguan, 2015; Erpen et al. 2019).

Importance of Genome Editing in Horticultural Crops

Fruit and vegetables provide several essential nutrients to the human diet. In general, these are rich in minerals and vitamins. A large portion of the global population relies on staple cereal crops, which often do not have these minerals and vitamins in sufficient amounts. Therefore, about two billion people across the globe are suffering from micronutrient deficiencies. Increased intake of these horticultural crops can reduce it. Moreover, these crops have great importance in the food industry. Numerous food products are prepared from them and also from their by-products (Ahmad and Khan 2019; Durazzo and Lucarini 2022). Moreover, people eating habits are changing, nowadays people prefer a healthy and balanced diet.

The demand for horticultural crops has up and continues to surge with the ever-increasing population. Therefore, the crop yield has to keep up with the increasing demand to reduce food scarcity. Crop development seeks to boost production, biotic and abiotic stress resistance, as well as quality and nutritional value. Over several decades, innovative agricultural technology has considerably enhanced crop productivity. Consumers are more concerned about crop quality since it provides many nutrients such as proteins, fiber, vitamins, minerals, and bioactive substances, all of which are directly linked to human health (Ramesh et al. 2020). In addition, scientists and breeders have steadily switched their attention away from boosting output and toward enhancing quality. Traditional crossover breeding, chemical and radiation-mediated mutation breeding, molecular marker-assisted breeding, and genetic engineering breeding have all proven effective in improving various crop properties. Traditional mutagenesis-based breeding procedures, on the other hand, are time-consuming and labor-intensive, especially for polyploid crop development (Chaudhary et al. 2019).

CRISPR-Cas-Based Gene Editing

The most effective, efficient, and precise technique of genome editing tool in all live cells is clustered regularly interspaced short palindromic repeat (CRISPR) and its accompanying protein (Cas9) which is used in many practical areas. The two fundamental constituents of the CRISPR/Cas9 system are guide RNA (gRNA) and CRISPR-associated (Cas9) proteins. The discovery of the novel nuclease Cas9, a class II bacterial adaptive immune system from *S. pyogenes*, has guided the concept of CRISPR/Cas9-mediated genome editing (Hang et al. 2007).

For dealing with the dangers of genetic invaders such as viruses, certain prokaryotic species, notably the archaea, have developed a bevy of weaponry, including restriction endonucleases and the Cas9 system. A cluster of short DNA repeats separated by a "spacer" and a succession of adjacent genes producing Cas proteins is found in around one-third of the bacterial population and about 90% of known archaea (Rath et al. 2015). These spacers serve as a kind of memory for bacteria, allowing them to recognize viruses and plasmids. However, nothing is known about the evolutionary mechanism by which bacteria acquire this knowledge. According to certain research, the archaea *Sulfolobus tokodaii* allocates about 1% of its genome to spacers for CRISPR/Cas immunity (Gophna et al. 2012; Ledford, 2017). Cas9-gRNA detects nuclease-specific PAM sequence (5′-NGG-3′) in the genome in general. Despite the fact that the Cas9-gRNA complex detects other PAM sequences including 5′-NAG-3′ and 5′-NGA-3′, these PAMs caused a lot of off-target mutagenesis in the plant (Zhang et al. 2014; Nishimasu et al. 2014).

Despite its extensive usage and demonstrated efficacy in genome editing across a wide range of species, SpCas9 has certain limitations. For starters, it frequently detects DNA sequences that have a high degree of sequence similarity to the target location, leading to the editing of the undesired gene. Second, PAM requirement restricts the desired DNA which SpCas9 may alter. Finally, because of its relatively large size, transporting SpCas9 inside the cells through a virus-derived vehicle is problematic as the cargo capacity of the vector is limited. To address these constraints, various natural and artificial SpCas9 variants that detect alternate PAMs have been created. Staphylococcus aureus Cas9 (SaCa9) is a noteworthy variation among them. It detects 5′-NNGRRT and has a shorter coding sequence than SpCas9, making it appropriate for use with virus-based vectors (Ran et al. 2015; Steinert et al. 2015).

Cas9-NG, xCas9, and iSpyMacCas9 (Endo et al. 2019) are just a few of the modified SpCas9 variants that have been used in plant genome editing. SpRY, a SpCas9 variation capable of targeting virtually all PAM sites (NRN>NYN) (Walton et al. 2020; Kleinstivr et al. 2021) and used in plant genome editing, is a remarkable designed variant of SpCas9. The SpCas9 has also been created in a high-fidelity version with little off-target activity. The use of paired Cas9 nickase can also help to limit off-target effects. The Cas9 variants that identify non-canonical PAM (e.g., SpCas9-NRRH) have also been used recently. Cas9's targeting capability has been expanded thanks to the introduction of modified Cas9 proteins with less stringent PAM requirements, yet there are several possible problems. According to a recent study, SpCas9-NG can modify gRNA existing in T-DNA, resulting in novel gRNA variants and increasing the chance of off-target editing. However, by modifying the gRNA scaffold sequences, this might be partially avoided (Hu et al. 2018; Hua et al. 2019; Qin et al. 2020). The basic methodology for CRISPR-based gene editing is described in Figure 1.

The most commonly utilized Cas, after the Cas9 is Cas12, with Cas12a (previously Cpf1) being employed in a wide range of plant species. It can edit AT-rich genomic areas because it identifies AT-rich PAMs. It generates lurch DSBs, making it perfect for HDR research. Cas12a needs only one short CRISPR RNA (crRNA, 42 nt), making it cost-friendly to produce as well as employ for multiplex editing. Furthermore, RNase activity of Cas12a has the ability to split a CRISPR array into separate crRNAs, allowing for multiplex genome editing. Cas12a may constantly cleave DNA until the edits preclude crRNA binding since its cleavage site is remote from the PAM sequence, which results in increased editing efficiency and huge deletions. Cas12b is another member Cas12 family that has been used in plant genome editing. (Tang et al. 2017).

Base editors are another modification under CRISPR genome editing technology in which desired alterations in bases without any double-strand breaks and DNA donor templates are possible. Base editors are designed by combining a base modification enzyme, such as deaminase, with a catalytically dormant Cas9 (dCas9) or a partly active Cas9, known as Cas9 nickase (nCas9), that can only cut one strand of DNA. dCas9 and nCas9 may be utilized in a base editor, but newer base editors prefer to employ nCas9 due to its superior performance (Kommor et al. 2016; Gaudelli et al. 2017). Cytosine base editors (CBEs) and adenine base editors (ABEs) are two current base editing programmes that can convert C/G to T/A and vise-versa. A DNA base editor capable of converting C/G to G/C DNA bases was recently discovered. C to G base editors is the name for this new type of base editor (CGBEs). CGBE is yet to be tested in plants, although ABEs and CBEs have been used successfully in a variety of plants. In recent years, significant progress has been achieved, and many variants of basic editors have been built to increase efficacy and specificity. Base editor 1 (BE1), base editor 2 (BE2), base editor 3 (BE3), and base editor 4 (BE4) are the different versions of CBE (BE4) (Molla et al. 2020; Zhao et al. 2021).

Prime editing is a flexible precise genome editing strategy that does not require DSBs or DNA donors and is able to perform all 12 potential base conversions and can produce tiny indels. A designed reverse transcriptase enzyme, a Cas9 nickase, and a prime editing guide RNA are all included in a conventional prime editor (pegRNA). Unlike base editors, which employ a Cas9D10A nickase to improve editing efficiency, prime editors use a Cas9H840 nickase. Cas9 nickase, rather than dCas9, appears to boost the DNA repair mechanism and enhance the result of base and primer edits (Anzalone et al. 2019; Mishra et al. 2020). Prime editors have been produced in several versions, including primer editor 1 (PE1), prime editor 2 (PE2), and prime editor 3 (PE3). Prime editing has been proven in human cells as well as plants like tomato (*Solanum lycopersicum*) (Lu et al. 2021) and potato (*Solanum tuberosum*) (Veillet et al. 2019).

Recognition, cleavage, and repair are the three phases of the CRISPR/Cas-9 genome editing system. A complementary base pair in the designed sgRNA identifies the target sequence in the gene of interest. While the Cas-9 nuclease creates double-stranded breaks at a 3 base pairs upstream position from a protospacer neighbouring motif, the double-stranded break is repaired by either non-homologous end joining or homology-directed repair processes. It has a wide range of applications in the majority of sectors. It might aid in the development of novel grains with higher nutritional value in agriculture.

CRISPR/Cas-Based Gene Editing in Vegetable Crops

Gene-editing technology has lately gained popularity in molecular plant breeding. The CRISPR/Cas9 technology, which is capable of producing DSBs, is mostly employed to develop knockout mutants. In the recent decade, a number of gene editing studies have been performed by employing the CRISPR technology in vegetable crops for different traits (Table 1). Base editors have recently expanded the breadth of genome editing by permitting accurate and efficient nucleotide replacements. Recently, CRISPR-Cas9 components have been utilized to silence the amylose-producing *StGBSSI* gene in the two cultivars of the tetraploid potato (*Solanum tuberosum* L.). They were able to develop tetra-allelic mutants with defective amylose production, demonstrating the *StGBSSI* protein's loss of function. By using a cytidine base editor (CBE), they produced DNA editions in the KTGGL-encoding locus which resulted in discrete variation in the sequence of amino acid and a loss-of-function allele.

Base editing is a success in the case of tetraploid potato which opens up new paths for genome engineering in this species (Veillet et al. 2019). *IbGBSSI* (encoding granule-bound starch synthase I) and *IbSBEII* (encoding starch branching enzyme II), two starch biosynthetic pathway genes, were used as a target in the Xushu22 (starch cultivar). Taizhong6, a carotenoid-rich cultivar, was used to test the efficiency of CRISPR-Cas9 technology in the root crop sweet potato (*Ipomea batatas*) was transformed using a binary vector in which the Arabidopsis *AtUBQ* promoter drives the Cas9 gene and the Arabidopsis *AtU6* promoter controls the guide RNA. A total of 35 Taizhong6 and 72 Xushu22 transgenic lines were created and their mutations were examined. In both cultivars, multiallelic mutations had mutation effectiveness of 62–92 percent.

Table 1. The utilisation of CRISPR/Cas9 technology in enhancement of different crops

Crop	Target gene	CRISPR/Cas	Associated trait	References
Potato	Mutation in *S-RNase*	CRISPR/Cas9	Self-compatible lines were generated	Qin et al. 2018
Potato	*StGBSS*	CBE	Low amylose content	Veillet et al. 2019
Sweet potato	*IbGBSSI*	Cas9	Low amylose content	Wang et al. 2019
Potato	*StSBE1, StSBE2*	Cas9	High amylose content	Tuncel et al. 2019
Sweet potato	*IbGBSSI, IbSBEII*	Cas9	High amylose content	Wang et al. 2019
Ground cherry	Disruption of *SP, SP5G and CLV1*	CRISPR/Cas9	Plants were shorter, have more flowers and fruits	Lemmon et al. 2018
Watermelon	*WIP1*	CRISPR/Cas9	Parthenocarpic phenotype	Zhang et al. 2020
Banana	*LCYε*	Cas9	High β-carotene content	Kaur et al. 2020

Nucleotide substitutions caused the majority of the mutations, which mostly resulted in alterations in amino acids. Additionally, both *IbGBSSI* and *IbSBEII* included short nucleotide insertions or deletions. In addition, one *IbSBEII* transgenic line had a 2658 bp deletion. In comparison to the wild-type control, total starch contents in *IbGBSSI* and *IbSBEII*-knockout transgenic lines were not substantially different. The *IbGBSSI*-knockout lowered the amylose percentage in the allopolyploid sweet potato, but the *IbSBEII*-knockout boosted it. These findings showed that CRISPR/Cas9 technology is a useful tool for improving sweet potato starch quality and breeding polyploid root crop breeding (Wang et al. 2019). To test whether tetraploid potatoes produce tuber starches with variations, mutagenesis of starch-branching enzymes (SBEs) was achieved by CRISPR/Cas9. Agrobacterium-mediated or PEG-mediated delivery of machinery having the Cas9 and sgRNAs targeting SBE1, SBE2, or both genes were used to introduce the Cas9 and sgRNAs targeting SBE1, SBE2, or both genes into protoplasts. When DNA was delivered into protoplasts, mutants were detected with no Cas9 gene, which indicates that foreign DNA was not present. Selected mutants with defective starch granules showed tuber SBE1 and/or SBE2 protein decreases that were largely consistent with genotype analyses. As with low-SBE potato tubers, a significant drop in both SBE isoforms resulted in an excessive starch phenotype.

In comparison to wild-type starch, HPLC-SEC and 1 H NMR demonstrated a decrease in short amylopectin chains, an increase in long chains, and a substantial drop in branching frequency. SBE2 protein-deficient mutants displayed nearly normal amylopectin chain-length distributions and minor branching frequency decreases (Tuncel et al. 2019). Genome editing has a lot of potential for enhancing agricultural output, and there is a lot of interest in breeding orphan crops, which are typically plagued with undesired traits that are similar to wild cousins.

They used CRISPR- Cas9 to mutate orthologues of tomato domestication and improved the genes that control the architecture of the plant, production of flowers, and size of the fruit in the orphan Solanaceae crop 'groundcherry' (*Physalis pruinosa*) (Lemmon et al. 2018). Zhang et al. (2019) employed CRISPR/Cas9 to modify the *WIP1* gene in watermelon, resulting in gynoecious watermelon lines. Because of its parthenocarpic origin and triploid DNA, banana improvement has been difficult. The advancement of CRISPR/Cas9 technology allows for the introduction of desired features in bananas. Plants with the exon region of lycopene epsilon-cyclase (*LCY*) inactive collected up to 6-fold more carotene (Kaur et al. 2020).

CRISPR/Cas-Based Gene Editing in Fruit Crops

In recent years, CRISPR/Cas has been employed in improving crop yield by making changes in fruit size, fruit color and improving the quality of minerals and bioactive substances in order to increase the nutritional content of the crop (Figure 2). Crop yield refers to the quantity of agricultural output produced per square meter of land. The size of the fruit, architecture of the plant and fruit set all have an impact on production per plant in fruit crops. By increasing the seed compartments, the fruit size can be increased which is a characteristic feature of plant domestication. In order to do the same, CRISPR was utilized to target the *CLAVATA 3 (CLV3)* gene promoter region which resulted in the larger size of the tomato having more locules (Zsogen et al; 2018, Li et al; 2018). Yuste-Lisbona et al. (2020) showed that an excessive number of floral organs (ENO) which shows a synergistic reaction with CLV3 and loss-of-function of ENO gene increases the fruit locules and size of the fruit. *MYC*2 gene knockout in tomatoes resulted in the production of smaller fruits. Multiflora (*MULT*) gene, a tomato inflorescence gene, when mutated resulted in the branching inflorescence and also increase the fruit number up to 10-fold (Zsogon et al; 2018, Leal et al; 2017).

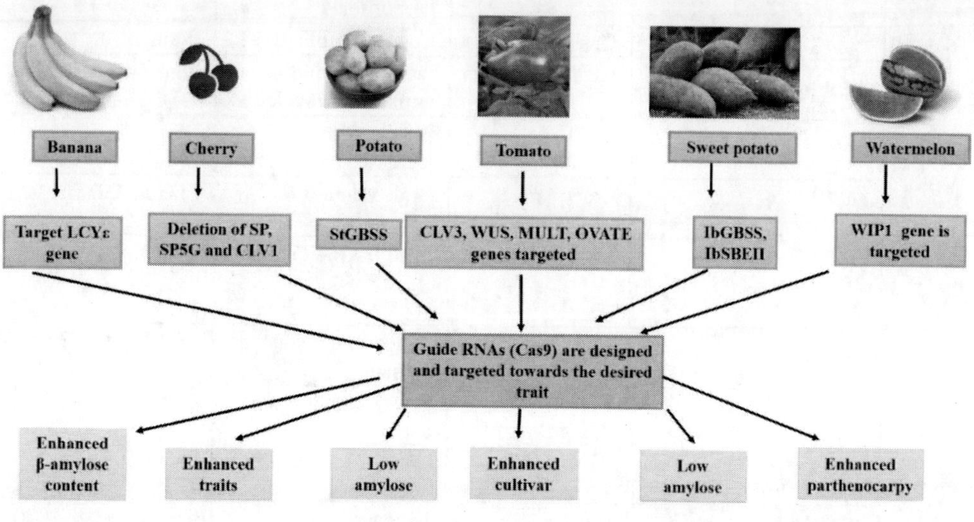

Figure 2. Some successful examples of CRISPR based gene editing for different traits in horticultural crops.

Fatty acid-derived volatile organic compounds (FA-VOCs) influence the flavour of tomato (*Solanum lycopersicum*) fruit and human preferences. In tomato fruits, short-chain FA-VOCs (C5 and C6) are among the most numerous and significant volatile chemicals. Glycerolipid cleavage produces the progenitors of these volatiles, linoleic acid (18:2) and linolenic acid (18:3). Genome-wide association research using metabolites combined with genetic mapping and functional analysis found *Sl-LIP8*, a novel class III lipase family member linked to the buildup of short-chain FA-VOCs in tomato fruit. *Sl-LIP8* has been shown to cleave 18:2 and 18:3 acyl groups from glycerolipids *in vitro*. Multiple fruit short-chain FA-VOCs were found to be considerably lower in a CRISPR/Cas9 gene-edited SILIP8 mutant, indicating that this enzyme plays an essential part in the system (Li et al. 2020). CRISPR-based gene editing studies performed on tomato crop are summarized in Table 2.

Table 2. The utilization of CRISPR/Cas9 technology in enhancement of tomatoes

Crop	Target gene	CRISPR/Cas	Associated trait	References
Tomato	Editing in *SP, SP5G, CLV3, WUS, MULT, OVATE GGP1 and Cyc 8*	CRISPR/Cas9	Attractive cultivar	Li et al. 2018; Zsogon et al. 2018
	Psy1, CrtR-b2	Cas9	Flower color (yellow)	D'Ambrosio et al. 2018
	OVATE, Fas, Fw2.2 and CLV3	Cas9	Fruit size and oval fruit shape	Zsogon et al. 2018
	fas, lc and ENO	Cas9	Fruit size	Rodríguez-Leal et al. 2017; Yuste-Lisbona et al. 2020
	slyPDS	Cas9	Increased lycopene content	Li et al. 2018
	BnFAD2	Cas9li	High oleic acid proportion	Okuzaki et al. 2018
	PL, PG2α and TBG4	Cas9	Long shelf life	Wang et al. 2019
	MYC2	CRISPR/Cas9	Fruits with high numbers of locules	Shu et al. 2020
	MULT	Cas9	Flower number increased	Zsogon et al. 2018
	GGP1	Cas9	High Vitamin-C content	Li et al. 2018
	GABA-TP1, GABA-TP2, GABA-TP3, CAT-9 and SSADH	Cas9	High GABA content	Li et al. 2018
	CycB	Cas9	Increase in lycopene content	Zsogon et al. 2018
	ANT1, ANT2, HY5	Cas9	Lower anthocyanin content	Cermak et al. 2016; Yan et al. 2020; Zhi et al. 2020; Qiu et al. 2019
	FLORAL4	Cas9	Increased phenylalanine-derived volatile content	Tikunov et al. 2020
	Lip8	Cas9	Decreased volatile content	Li et al. 2020

Despite very modest levels in food, bioactive chemicals have a significant role in the prevention of cardiovascular disease and cancer. There has been a lot of effort put into raising their levels. GABA (gamma-aminobutyric acid) is a non-protein amino acid with four carbons that is regarded as a health-promoting functional molecule. Li et al. (2018) established six target loci of five genes in the GABA metabolism pathway and put up a multiplexed CRISPR/ Cas9 system to boost GABA content in tomato. This enabled the creation of quadruple mutants with considerably higher GABA content. The carotenoid family includes lycopene. It is thought to be a bioactive component with potential use in cardiovascular disease. Lycopene β-cyclase (*CycB*) is an enzyme that can convert lycopene to β-carotene. One sgRNA was designed to target the exon region of *CycB*. The mutant plant lines generated enriched lycopene fruits, 5 times higher than the wild type.

Anthocyanins are a class of antioxidants that provide different colors in plants. Anthocyanins are considered healthy when ingested by humans. Recent research has verified the use of CRISPR/Cas9 to improve the content of numerous positive regulators of tomato anthocyanins, including Anthocyanin mutant 1 (*ANT1*) (Cermak et al. 2016), anthocyanin mutant 2 (*ANT2*) (Yan et al. 2020, Zhi et al. 2020), and Elongated hypocotyl 5 (*HY5*) (Qiu et al. 2019). The advancement of CRISPR/Cas9 technology allows for the introduction of desired features in bananas. The altered plants collected up to 6-fold more carotene when the exon region of lycopene epsilon-cyclase (*LCY*) was silenced. Aroma and flavour are also essential factors in determining the quality of fresh fruits. Aroma is affected by the amount of phenylalanine-derived volatiles (PHEVs). Tikunov et al. (2020) have discovered *FLORAL4* to be a critical regulator of PHEV build-up in tomato fruit. Fruit taste development benefits from the presence of 80 fatty-acid-derived volatile organic compounds (FAVOCs). The importance of *LIP8*, a lipase family member, in short-chain FA-VOC biogenesis has been confirmed by researchers. Short-chain FAVOCs were reduced in CRISPR/Cas9 gene-edited tomato fruits

Conclusion

CRISPR/Cas based gene editing has opened various new opportunities for crop improvement. The real importance of horticultural crops is due to their qualitative traits. The improvements in these complex traits via traditional approaches are demanding time and patience. Long plant breeding programs and numerous rounds of recurrent selection are required to gain a considerable amount of genetic gain. However, CRISPR technology allows targeted gene editing to improve complex phenotypic traits in a simple and quick way. The potential of this technique is widely proven in horticultural crops for improving yield, nutritional value, and biotic and abiotic stress tolerance. This technique will catalyze the current genetic engineering programs aimed to reduce hunger and malnutrition and will help to sustain the global food supply.

The CRISPR technology is evolving with time, in recent years several new Cas proteins have been identified. Moreover, the methodology to transfer the vectors into the plant genomes is also improving. The increasing knowledge of genes and genome sequences is setting a perfect platform to enhance the efficiency of gene editing. The use of NGS, GWAS, comparative transcriptomic studies, RNA-seq, and modern bioinformatics tools are providing potential candidate genes controlling different phenotypic traits for editing. Nowadays, several candidate genes for controlling desirable traits are known and many of them are functionally validated. The positive regulators of desirable traits can be edited to develop the over-expressing lines, whereas the negative regulators should be knocked out to improve the present-day crops. A holistic approach involving different tools as mentioned above is required to take the best from this novel gene-editing technique.

References

Achary, V., and Reddy, M. K. (2021). CRISPR-Cas9 mediated mutation in GRAIN WIDTH and WEIGHT2 (GW2) locus improves aleurone layer and grain nutritional quality in rice. *Scientific Reports*, 11(1), pp. 1-13.

Ahmad, F., and Khan, S. T. (2019). Potential industrial use of compounds from by-products of fruits and vegetables. In *Health and Safety Aspects of Food Processing Technologies*, (pp. 273-307). Springer, Cham.

Ahmad, S., Shahzad, R., Jamil, S., Nisar, A., Khan, Z., Kanwal, S., Tabassum, J., Afza, N., Yousaf, M. I., Ali, Z., and Khan, A. A. (2022). CRISPR-mediated genome editing for developing climate-resilient monocot and dicot crops. In *Plant Perspectives to Global Climate Changes* (pp. 393-411). Academic Press.

Asmamaw, M., and Zawdie, B. (2021). Mechanism and applications of CRISPR/Cas-9-mediated genome editing. *Biologics: Targets & Therapy*, 15, p.353.

Bashar, K. K., and Hanif, M. A. (2021). Crop gene editing against biotic stresses via CRISPR/Cas9 tools: A review. *Archives of Phytopathology and Plant Protection*, 54(15-16), pp.1159-1181.

Ben Shlush, I., Samach, A., Melamed-Bessudo, C., Ben-Tov, D., Dahan-Meir, T., Filler-Hayut, S., and Levy, A. A. (2021). CRISPR/Cas9 induced somatic recombination at the CRTISO locus in tomato. *Genes*, 12(1), p.59.

Bhatta, B. P., and Malla, S. (2020). Improving horticultural crops via CRISPR/Cas9: current successes and prospects. *Plants*, 9(10), p.1360.

Čermák, T., Baltes, N. J., Čegan, R., Zhang, Y., and Voytas, D. F. (2015). High-frequency, precise modification of the tomato genome. *Genome Biology*, 16(1), pp.1-15.

D'Ambrosio, C., Stigliani, A. L., and Giorio, G. (2018). CRISPR/Cas9 editing of carotenoid genes in tomato. *Transgenic Research*, 27(4), pp. 367-378.

Deng, L., Wang, H., Sun, C., Li, Q., Jiang, H., Du, M., Li, C. B., and Li, C. (2018). Efficient generation of pink-fruited tomatoes using CRISPR/Cas9 system. *Journal of Genetics and Genomics*, 45(1), pp.51-54.

Durazzo, A., and Lucarini, M. (2022). Chemical Properties, Nutritional Quality, and Bioactive Components of Horticulture Food. *Horticulturae*, 8(1), p. 3.

Gophna, U., and Brodt, A. (2012). CRISPR/Cas systems in archaea: What array spacers can teach us about parasitism and gene exchange in the 3rd domain of life. *Mobile Genetic Elements*, 2(1), pp. 63-64.

Han, J., Zhou, H., Horazdovsky, B., Zhang, K., Xu, R. M., and Zhang, Z. (2007). Rtt109 acetylates histone H3 lysine 56 and functions in DNA replication. *Science*, 315(5812), pp. 653-655.

Hu, J. H., Miller, S. M., Geurts, M. H., Tang, W., Chen, L., Sun, N., Zeina, C. M., Gao, X., Rees, H. A., Lin, Z., and Liu, D. R. (2018). Evolved Cas9 variants with broad PAM compatibility and high DNA specificity. *Nature*, 556(7699), pp. 57-63.

Hua, K., Tao, X., Han, P., Wang, R., and Zhu, J. K. (2019). Genome engineering in rice using Cas9 variants that recognize NG PAM sequences. *Molecular Plant*, 12(7), pp. 1003-1014.

Karlson, C. K. S., Noor, S. N. M., Nolte, N., and Tan, B. C. (2021). *CRISPR/dCas9-based Systems: Mechanisms and Applications in Plant Sciences.*

Kaur, H., Pandey, D. K., Goutam, U., and Kumar, V. (2021). CRISPR/Cas9-mediated genome editing is revolutionizing the improvement of horticultural crops: Recent advances and future prospects. *Scientia Horticulturae*, 289, p. 110476.

Kaur, N., Alok, A., Kumar, P., Kaur, N., Awasthi, P., Chaturvedi, S., Pandey, P., Pandey, A., Pandey, A. K., and Tiwari, S. (2020). CRISPR/Cas9 directed editing of lycopene epsilon-cyclase modulates metabolic flux for β-carotene biosynthesis in banana fruit. *Metabolic Engineering*, 59, pp.76-86.

Kaur, N., Awasthi, P., and Tiwari, S. (2020). Fruit crops improvement using CRISPR/Cas9 system. In *Genome Engineering via CRISPR-Cas9 System* (pp. 131-145). Academic Press.

Klap, C., Yeshayahou, E., Bolger, A. M., Arazi, T., Gupta, S. K., Shabtai, S., Usadel, B., Salts, Y., and Barg, R. (2017). Tomato facultative parthenocarpy results from Sl AGAMOUS-LIKE 6 loss of function. *Plant Biotechnology Journal*, 15(5), pp. 634-647.

Kumar, R., Nizampatnam, N. R., Alam, M., Thakur, T. K., and Kumar, A. (2021). Genome Editing Technologies for Plant Improvement: Advances, Applications and Challenges. In *Omics Technologies for Sustainable Agriculture and Global Food Security* Volume 1 (pp. 213-240). Springer, Singapore.

Ledford, H. (2017). Five big mysteries about CRISPR's origins. *Nature News*, 541(7637), p. 280.

Lemmon, Z. H., Reem, N. T., Dalrymple, J., Soyk, S., Swartwood, K. E., Rodriguez-Leal, D., Van Eck, J., and Lippman, Z. B. (2018). Rapid improvement of domestication traits in an orphan crop by genome editing. *Nature Plants*, 4(10), pp.766-770.

Li, R., Li, R., Li, X., Fu, D., Zhu, B., Tian, H., Luo, Y., and Zhu, H. (2018). Multiplexed CRISPR/Cas9-mediated metabolic engineering of γ-aminobutyric acid levels in *Solanum lycopersicum*. *Plant Biotechnology Journal*, 16(2), pp. 415-427.

Li, T., Yang, X., Yu, Y., Si, X., Zhai, X., Zhang, H., Dong, W., Gao, C., and Xu, C. (2018). Domestication of wild tomato is accelerated by genome editing. *Nature Biotechnology*, 36(12), pp.1160-1163.

Li, X., Tieman, D., Liu, Z., Chen, K., and Klee, H. J. (2020). Identification of a lipase gene with a role in tomato fruit short-chain fatty acid-derived flavor volatiles by genome-wide association. *The Plant Journal*, 104(3), pp. 631-644.

Li, X., Wang, Y., Chen, S., Tian, H., Fu, D., Zhu, B., Luo, Y., and Zhu, H. (2018). Lycopene is enriched in tomato fruit by CRISPR/Cas9-mediated multiplex genome editing. *Frontiers in Plant Science*, 9, p. 559.

Liu, L., Gallagher, J., Arevalo, E. D., Chen, R., Skopelitis, T., Wu, Q., Bartlett, M., and Jackson, D. (2021). Enhancing grain-yield-related traits by CRISPR–Cas9 promoter editing of maize CLE genes. *Nature Plants*, 7(3), pp. 287-294.

Nishimasu, H., Ran, F. A., Hsu, P. D., Konermann, S., Shehata, S. I., Dohmae, N., Ishitani, R., Zhang, F., and Nureki, O. (2014). Crystal structure of Cas9 in complex with guide RNA and target DNA. *Cell*, 156(5), pp. 935-949.

Nonaka, S., Arai, C., Takayama, M., Matsukura, C., and Ezura, H. (2017). Efficient increase of γ-aminobutyric acid (GABA) content in tomato fruits by targeted mutagenesis. *Scientific Reports*, 7(1), pp. 1-14.

Okuzaki, A., Ogawa, T., Koizuka, C., Kaneko, K., Inaba, M., Imamura, J., and Koizuka, N. (2018). CRISPR/Cas9-mediated genome editing of the fatty acid desaturase 2 gene in Brassica napus. *Plant Physiology and Biochemistry*, 131, pp. 63-69.

Park, J. R., Kim, E. G., Jang, Y. H., Jan, R., Farooq, M., Ubaidillah, M., and Kim, K. M. (2022). Applications of CRISPR/Cas9 as New Strategies for Short Breeding to Drought Gene in Rice. *Frontiers in Plant Science*, 13.

Qin, X., Li, W., Liu, Y., Tan, M., Ganal, M., and Chetelat, R. T. (2018). A farnesyl pyrophosphate synthase gene expressed in pollen functions in S-RN ase-independent unilateral incompatibility. *The Plant Journal*, 93(3), pp. 417-430.

Qiu, Z., Wang, H., Li, D., Yu, B., Hui, Q., Yan, S., Huang, Z., Cui, X., and Cao, B. (2019). Identification of candidate HY5-dependent and-independent regulators of anthocyanin biosynthesis in tomato. *Plant and Cell Physiology*, 60(3), pp. 643-656.

Rasheed, A., Gill, R. A., Hassan, M. U., Mahmood, A., Qari, S., Zaman, Q. U., Ilyas, M., Aamer, M., Batool, M., Li, H., and Wu, Z. (2021). A critical review: recent advancements in the use of CRISPR/Cas9 technology to enhance crops and alleviate global food crises. *Current Issues in Molecular Biology*, 43(3), pp. 1950-1976.

Rath, D., Amlinger, L., Rath, A., and Lundgren, M. (2015). The CRISPR-Cas immune system: biology, mechanisms and applications. *Biochimie*, 117, pp. 119-128.

Razzaq, M. K., Akhter, M., Ahmad, R. M., Cheema, K. L., Hina, A., Karikari, B., Raza, G., Xing, G., Gai, J., and Khurshid, M. (2022). CRISPR-Cas9 based stress tolerance: New hope for abiotic stress tolerance in chickpea (Cicer arietinum). *Molecular Biology Reports*, pp. 1-9.

Rehman, F., Gong, H., Bao, Y., Zeng, S., Huang, H., and Wang, Y. (2022). CRISPR gene editing of major domestication traits accelerating breeding for Solanaceae crops improvement. *Plant Molecular Biology*, pp. 1-17.

Rodríguez-Leal, D., Lemmon, Z. H., Man, J., Bartlett, M. E., and Lippman, Z. B. (2017). Engineering quantitative trait variation for crop improvement by genome editing. *Cell*, 171(2), pp. 470-480.

Sánchez-León, S., Gil-Humanes, J., Ozuna, C. V., Giménez, M. J., Sousa, C., Voytas, D. F., and Barro, F. (2018). Low-gluten, nontransgenic wheat engineered with CRISPR/Cas9. *Plant Biotechnology Journal*, 16(4), pp. 902-910.

Shimatani, Z., Kashojiya, S., Takayama, M., Terada, R., Arazoe, T., Ishii, H., Teramura, H., Yamamoto, T., Komatsu, H., Miura, K., and Ezura, H. (2017). Targeted base editing in rice and tomato using a CRISPR-Cas9 cytidine deaminase fusion. *Nature Biotechnology*, 35(5), pp. 441-443.

Shoeb, E., Badar, U., Venkataraman, S., Hashim, G. M., and Hefferon, K. (2021). Genome Engineering and Essential Mineral Enrichment of Crops. *Genome Engineering for Crop Improvement*, pp. 210-223.

Shu, P., Li, Z., Min, D., Zhang, X., Ai, W., Li, J., Zhou, J., Li, Z., Li, F., and Li, X. (2020). CRISPR/Cas9-mediated SlMYC2 mutagenesis adverse to tomato plant growth and MeJA-induced fruit resistance to Botrytis cinerea. *Journal of Agricultural and Food Chemistry*, 68(20), pp. 5529-5538.

Singh, B., Kukreja, S., Salaria, N., Thakur, K., Gautam, S., Taunk, J., and Goutam, U. (2019b). VIGS: a flexible tool for the study of functional genomics of plants under abiotic stresses. *Journal of Crop Improvement*, 33(5), pp. 567-604.

Singh, B., Salaria, N., Thakur, K., Kukreja, S., Gautam, S., and Goutam, U. (2019a). *Functional Genomic Approaches to Improve Crop Plant Heat Stress Tolerance*. F1000Research, 8.

Singha, D. L., Das, D., Sarki, Y. N., Chowdhury, N., Sharma, M., Maharana, J., and Chikkaputtaiah, C. (2022). Harnessing tissue-specific genome editing in plants through CRISPR/Cas system: current state and future prospects. *Planta*, 255(1), pp. 1-17.

Soto-Cerda, B. J., Vasudevan, A., Laroche, A., and Ragupathy, R. (2022). Genome-wide association studies (GWAS) in cereals. In *Accelerated Breeding of Cereal Crops* (pp. 119-132). Humana, New York, NY.

Talakayala, A., Ankanagari, S., and Garladinne, M. (2022). CRISPR-Cas Genome editing system: a versatile tool for developing disease resistant crops. *Plant Stress*, p. 100056.

Tikunov, Y. M., Roohanitaziani, R., Meijer-Dekens, F., Molthoff, J., Paulo, J., Finkers, R., Capel, I., Carvajal Moreno, F., Maliepaard, C., Nijenhuis-de Vries, M., and Labrie, C. W. (2020). The genetic and functional analysis of flavor in commercial tomato: the FLORAL4 gene underlies a QTL for floral aroma volatiles in tomato fruit. *The Plant Journal*, 103(3), pp. 1189-1204.

Tuncel, A., Corbin, K. R., Ahn-Jarvis, J., Harris, S., Hawkins, E., Smedley, M. A., Harwood, W., Warren, F. J., Patron, N. J., and Smith, A. M. (2019). Cas9-mediated mutagenesis of potato starch-branching enzymes generates a range of tuber starch phenotypes. *Plant Biotechnology Journal*, 17(12), pp. 2259-2271.

Uffelmann, E., Huang, Q. Q., Munung, N. S., de Vries, J., Okada, Y., Martin, A. R., Martin, H. C., Lappalainen, T., and Posthuma, D. (2021). Genome-wide association studies. *Nature Reviews Methods Primers*. Aug 26, 1(1), 1-21.

United Nations (UN). (2019). *Probabilistic Population Projections based on the World Population Prospects 2019*; United Nations, Department of Economic and Social Affairs, Population Division: New York, NY, USA.

Veillet, F., Chauvin, L., Kermarrec, M. P., Sevestre, F., Merrer, M., Terret, Z., Szydlowski, N., Devaux, P., Gallois, J. L., and Chauvin, J. E. (2019). The Solanum tuberosum GBSSI gene: a target for assessing gene and base editing in tetraploid potato. *Plant Cell Reports*, 38(9), pp. 1065-1080.

Velásquez, A. C., Castroverde, C. D. M., and He, S. Y. (2018). Plant–pathogen warfare under changing climate conditions. *Current Biology*, 28(10), pp. R619-R634.

Vu, T. V., Sivankalyani, V., Kim, E. J., Doan, D. T. H., Tran, M. T., Kim, J., Sung, Y. W., Park, M., Kang, Y. J., and Kim, J. Y. (2020). Highly efficient homology-directed repair using CRISPR/Cpf1-geminiviral replicon in tomato. *Plant Biotechnology Journal*, 18(10), pp. 2133-2143.

Wang, D., Samsulrizal, N. H., Yan, C., Allcock, N. S., Craigon, J., Blanco-Ulate, B., Ortega-Salazar, I., Marcus, S. E., Bagheri, H. M., Perez Fons, L., and Fraser, P. D. (2019). Characterization of CRISPR mutants targeting genes modulating pectin degradation in ripening tomato. *Plant Physiology*, 179(2), pp. 544-557.

Wang, H., Wu, Y., Zhang, Y., Yang, J., Fan, W., Zhang, H., Zhao, S., Yuan, L., and Zhang, P. (2019). CRISPR/Cas9-based mutagenesis of starch biosynthetic genes in sweet potato (Ipomoea Batatas) for the improvement of starch quality. *International Journal of Molecular Sciences*, 20(19), p. 4702.

Yan, S., Chen, N., Huang, Z., Li, D., Zhi, J., Yu, B., Liu, X., Cao, B., and Qiu, Z. (2020). Anthocyanin Fruit encodes an R2R3-MYB transcription factor, SlAN2-like, activating the transcription of SlMYBATV to fine-tune anthocyanin content in tomato fruit. *New Phytologist*, 225(5), pp. 2048-2063.

Yu, Q. H., Wang, B., Li, N., Tang, Y., Yang, S., Yang, T., Xu, J., Guo, C., Yan, P., Wang, Q., and Asmutola, P. (2017). CRISPR/Cas9-induced targeted mutagenesis and gene replacement to generate long-shelf life tomato lines. *Scientific Reports*, 7(1), pp. 1-9.

Yuste-Lisbona, F. J., Fernández-Lozano, A., Pineda, B., Bretones, S., Ortíz-Atienza, A., García-Sogo, B., Müller, N. A., Angosto, T., Capel, J., Moreno, V., and Jiménez-Gómez, J. M. (2020). ENO regulates tomato fruit size through the floral meristem development network. *Proceedings of the National Academy of Sciences*, 117(14), pp. 8187-8195.

Zetsche, B., Gootenberg, J. S., Abudayyeh, O. O., Slaymaker, I. M., Makarova, K. S., Essletzbichler, P., Volz, S. E., Joung, J., Van Der Oost, J., Regev, A., and Koonin, E. V. (2015). Cpf1 is a single RNA-guided endonuclease of a class 2 CRISPR-Cas system. *Cell*, 163(3), pp. 759-771.

Zhang, D., Zhang, Z., Unver, T., and Zhang, B. (2021). CRISPR/Cas: A powerful tool for gene function study and crop improvement. *Journal of Advanced Research*, 29, pp. 207-221.

Zhang, J., Guo, S., Ji, G., Zhao, H., Sun, H., Ren, Y., Tian, S., Li, M., Gong, G., Zhang, H., and Xu, Y. (2020). A unique chromosome translocation disrupting ClWIP1 leads to gynoecy in watermelon. *The Plant Journal*, 101(2), pp. 265-277.

Zhang, Y., Ge, X., Yang, F., Zhang, L., Zheng, J., Tan, X., Jin, Z. B., Qu, J., and Gu, F. (2014). Comparison of non-canonical PAMs for CRISPR/Cas9-mediated DNA cleavage in human cells. *Scientific Reports*, 4(1), pp. 1-5.

Zhi, J., Liu, X., Li, D., Huang, Y., Yan, S., Cao, B., and Qiu, Z. (2020). CRISPR/Cas9-mediated SlAN2 mutants reveal various regulatory models of anthocyanin biosynthesis in tomato plant. *Plant Cell Reports*, 39(6), pp. 799-809.

Zhou, Y., Xu, S., Jiang, N., Zhao, X., Bai, Z., Liu, J., Yao, W., Tang, Q., Xiao, G., Lv, C., and Wang, K. (2022). Engineering of rice varieties with enhanced resistances to both blast and bacterial blight diseases via CRISPR/Cas9. *Plant Biotechnology Journal*, 20(5), pp. 876-885.

Zhu, H., Li, C., and Gao, C. (2020). Applications of CRISPR–Cas in agriculture and plant biotechnology. *Nature Reviews Molecular Cell Biology*, 21(11), pp. 661-677.

Zsögön, A., Čermák, T., Naves, E. R., Notini, M. M., Edel, K. H., Weinl, S., Freschi, L., Voytas, D. F., Kudla, J., and Peres, L. E. P. (2018). De novo domestication of wild tomato using genome editing. *Nature Biotechnology*, 36(12), pp. 1211-1216.

Chapter 15

Prospects of CRISPR/Cas9-Mediated Gene Editing in Legume Crops

Anil Kumar Singh*
Department of Agriculture, Sant Baba Bhag Singh University, Khiala, Punjab, India

Abstract

Leguminous crops are an important source of plant-based proteins, essential amino acids and health-promoting phytochemicals. These crops also contribute towards sustainable farming by promoting soil health through symbiotic nitrogen fixation and the release of organic matter in the soil. Nutritional significance and ecological importance have designated leguminous crops as the most beneficial field crop. The benefits of legumes are often curtailed due to a lack of superior varieties. Efforts are being made to overcome production hurdles through genetic gains for agronomic traits, nutritional quality and stress resilience. Recent advancements in gene-editing tools have paved a new pathway for legume crop improvement. The clustered regularly interspaced short palindromic repeat (CRISPR)/CRISPR-associated protein (Cas) gene editing system is a promising and powerful gene modification tool. The chapter focuses on recent advances in CRISPR/Cas based gene editing technology with special emphasis on legumes. The CRISPR/Cas9 system has been used successfully for precise gene editing in legumes like cowpea, chickpea and soybean. However, limitation in regeneration and *in vitro* gene transfer has created serious challenges for CRISPR/Cas9 based gene editing system. Regulatory concerns associated with CRISPR/Cas9 technology are also included in the chapter. Additionally, the chapter also discusses limitations, challenges and future perspectives of CRISPR/Cas9 based gene editing technology in legumes crop improvement.

Keywords: CRISPR, Cas9, genome editing, gRNA, ZFNs, TALENs

Introduction

Leguminous crops are an excellent source of essential amino acids and plant-based proteins for humans. Legumes are recognized as third largest family of angiosperms with more than 751

* Corresponding Author's Email: anilimtech@gmail.com.

In: Genome Editing for Crop Improvement
Editors: Rahul Kumar, Ashutosh Sharma, Yogesh Kumar et al.
ISBN: 979-8-88697-740-0

documented genera and 19500 species [1]. Dried seeds of legumes serve as a good source of food for humans and are more commonly referred to as Pulses. Grain legumes along with cereals form staple food in many regions of the world. Legumes have tremendous potential to provide a sustainable solution to global food and protein security [2].

The leguminous crops have nitrogen fixing ability, which makes them an important part of sustainable farming. Legumes develop a symbiotic relationship with nitrogen-fixing soil bacteria. Bacteria residing within the root nodules convert atmospheric nitrogen into ammonia and make it biologically available to plants. Legumes are also known to enhance soil fertility by adding several high-quality organic matters to soil [3].

In recognition of nutritional importance and their contribution to sustainability, sixty eighth session of the United Nations General Assembly declared 2016 as International Year of Pulses [2]. This initiative tried to endorse the utilization of grain legumes throughout the food system, increase awareness of their benefits, promote legume research, promote use of legumes in crop rotation, and deal with challenges in legume trade [2]. The International Year of Pulses (IYP-2016) also highlighted the potential contribution of legumes in achieving important targets under sustainable development goals such as food availability for all, decreasing malnutrition and improving the income of small farmers as well as it will support sustainable and resilient agriculture [2]. Irrespective of nutritional and ecological significance, lack of superior varieties with good yield has curtailed importance of legumes in modern day agriculture. From the larger perspective, the use of legumes in agriculture and efforts for genetic improvement of grain legumes has lagged behind several cereal crops. There is a vital need for innovations in leguminous crop breeding technology to increase agriculture productivity and hasten the efforts to develop sustainable agriculture. This is also crucial for feeding and nourishing a rapidly growing population under the realm of climate change, diminishing arable land and reducing available water resources [4]. Leguminous cultivars with enhanced resilience to adverse environment, superior quality and high yield need to be developed to support ever increasing population in a more sustainable manner [5].

Traditionally, plant breeders rely on the natural process for introducing desirable alleles by cross breeding and increasing genetic variability by recombination [6, 7]. Genetic diversity is very important in traditional trait improvement programs. The natural process of crop improvement takes several years. Moreover, thousands of years of directed evolution through breeding has fixed large parts of the genomes of major leguminous crops. This has also decreased genetic variability and reduced the chances of improving several traits [7]. Mutation breeding with physical mutagens (such as x-rays and fast neutrons) or chemical mutagens (such as sodium azide and diepoxybutane) has aided in expanding genetic variation by introducing random mutations [8]. The genetic variation introduced via traditional methods can be either autogenous or non-autogenous random variation [6]. Such random genetic variants can arise due to point mutations, deletions, translocations, and chromosome losses [6, 7]. The generation of genetic variation in legume plants by random mutation has several drawbacks such as being time-consuming, environment dependent, non-specific and laborious [7]. Moreover, random mutations generated during traditional methods are generally less effective in trait improvements due to the genetic redundancy of polyploid legumes [5]. Recent advances in genome sequencing and structural genome analysis of legumes have paved the pathway for development of fast and site-specific gene editing technology [5]. Draft genome and transcriptome assemblies of more than 35 legumes are available in scientific literature [9]. The availability of genome information has played a pivotal role in executing site-specific genome editing and subsequent trait improvement of legume crops [5].

Genome Editing

Genome editing is a method to alter the genotype and phenotype of organisms. It enables precise and targeted change in the genome with help of sequence-specific engineered nucleases and cellular DNA repair pathways [6]. The precise gene editing technologies started with the discovery of meganuclease 1-SceI, which can introduce a double-stranded break in the chromosome [10]. However, extensive application of meganuclease in gene editing was curtailed due to the low frequency of site-specific cleavage [8].

The precise gene editing technology received momentum with the introduction of zinc finger nucleases (ZFNs) and transcription activator-like effector nucleases (TALENs). ZFNs are sequence-specific nucleases consisting of a synthetic zinc finger DNA binding domain and a DNA cleavage domain [5]. The zinc finger DNA binding domain can be altered to specifically introduce cleavage at targeted double-stranded DNA. ZFNs have been used for improving traits in crops such as tobacco, wheat and maize [6]. However, there is limited availability of sites in the plant genome that can be targeted by ZFNs [6]. Wide application of ZFNs in plant genome editing has been hindered due to the complexity of construction, unpredictable off-target effects, high cost and sophisticated skill requirements [5, 6].

TALENs are sequence-specific nucleases, constructed by combining transcription activator-like effectors (TALE) to the catalytic domain of the Fok I endonuclease [6]. The DNA-binding domain of TALE monomers consists of the central repeat domain (CRD) that determines DNA binding specificity. The CRD consists of tandem repeats of 34 amino acid residues. Each amino acid residue binds to one nucleotide of the targeted nucleotide sequence. TALENs in comparison to ZFNs, offer more flexibility in designing targets. TALENs have the potential to target a greater number of sites, relative to those that can be targeted by ZFNs. TALENs have been applied for precise gene editing in crops such as barley, maize, tobacco, potato, rice, soybean, tomato and wheat [6, 11]. Haun et al. [12] used TALEN to generate soybean lines with low polyunsaturated fats content. TALEN mutated two fatty acid desaturase 2 genes (FAD2-1A and FAD2-1B) which are responsible for converting mono-unsaturated fat to poly-unsaturated fat. These mutations increased the content of mono-unsaturated fat and decreased polyunsaturated fat content. A Minnesota-based company Calyxt Inc. is selling this premium quality high-oleic soybean oil under the name of Calyno oil. Calyno oil has become the first commercialized product from a gene-edited plant [13]. Extensive application of TALENs in plant gene editing remains a challenge due to the complexity and skill requirement to construct site-specific TALENs [6].

Undoubtedly, the use of ZFNs and TALENs platforms for gene editing led to important advances in crop research but was only limited to a few crops [6]. Both ZFNs and TALENs have their own unique limitations that restrict extensive application in crop improvement programs. The most significant advancement in precise gene editing technology was achieved by the discovery of type II clustered regularly interspaced short palindromic repeat (CRISPR)/ CRISPR associated protein 9 (Cas9) systems [5]. Discovery of the CRISPR/Cas system provided plant breeders with extensive ability to control site-specific introduction of sequence variation and enable rapid improvement of crops [8].

CRISPRs were first reported downstream of the alkaline phosphatase isozyme gene (*iap*) in *Escherichia coli* [14]. However, researchers started appreciating the biological significance of CRISPRs and its potential application as an efficient gene editing tool in past three decades [6, 8]. As seen in Figure 1, there is a continuous increase in the number of publications with the term "CRISPR/Cas" and "CRISPR/Cas gene editing in plant" in the scientific literature. Thus, highlighting the importance and growing interest of researchers in the CRISPR/Cas system and CRISPR/Cas based gene editing in plants to empower present-day agriculture. Recent advances in CRISPR/Cas9 system have made genome editing a widely adopted, low-cost, easy-to-use and precise genetic manipulation tool for improving desirable traits in several crop plants [6].

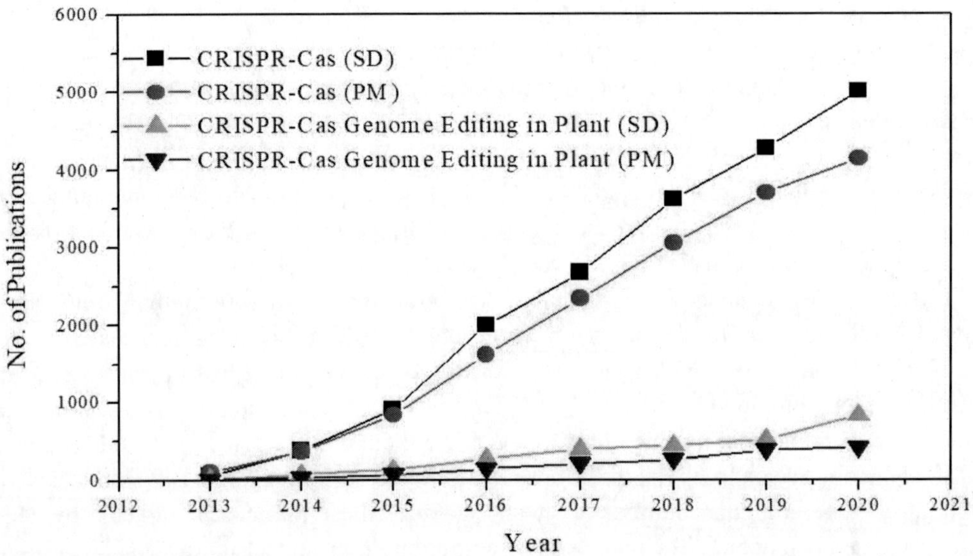

Figure 1. Increase in the number of publications related to "*CRISPR-Cas*" and "*CRISPR-Cas Genome Editing in Plant*" as seen in Science Direct (SD; www.sciencedirect.com) and PubMed ® (PM; https://pubmed.ncbi.nlm.nih.gov/).

CRISPR/Cas Technology

CRISPR/Cas9 based gene editing is the most powerful method of site-specific genetic manipulation in plants [4]. The CRISPR/Cas system is RNA mediated adaptive immune system of bacteria and archaea [6]. This system imparts defense against bacteriophages and other invasive genetic elements by cleaving the nucleic acid genome of the invader. The CRISPR/Cas systems employed for genome editing are based on RNA-guided interaction with DNA unlike earlier discussed genome editing tools where DNA-protein interaction enables targeted cleavage [15]. The original CRIPSR system employed for gene editing was developed from the bacterium *Streptococcus pyogenes* (SpCas9). To date, several Cas9 orthologs with diverse properties has been identified in different bacterial genomes [15, 16]. The CRISPR-Cas9 system is much easier to be constructed than ZFNs or TALENs as well as simple, versatile, efficient, low cost and allows targeting of multiple genes at once [6].

Typically, CRISPR/Cas system consist of two important components namely CRISPR (repeat-spacer arrays) and Cas proteins [6]. Based on *Cas* genes and the nature of the interference complex, CRISPR/Cas systems have been divided into two classes namely Class 1 and Class 2 CRISPR/Cas systems [17]. The Class 1 CRISPR/Cas system has multi-subunit crRNA-effector complexes. This class has been further divided into type I, type III and type IV CRISPR/Cas systems (Table 1). The class 2 CRISPR/Cas system utilizes interference with single effector proteins in complex with CRISPR RNAs (crRNAs). This second class has been further divided into type II, type V and type VI CRISPR/Cas systems [6, 17]. The aforementioned six types have been further divided into 33 subtypes and several variants [17].

Table 1. Classification of CRISPR/Cas9 system

Class	Type	Subtypes	Source Organism	Cas Proteins
Class I	I	I-A	*Archaeoglobus fulgidus*	Cas3, Cas8
		I-B	*Clostridium kluyveri*	Cas3, Cas8
		I-C	*Bacillus halodurans*	Cas3, Cas8
		I-D	*Cyanothece sp.*	Cas3, Cas10
		I-E	*Escherichia coli*	Cas3, Cas8
		I-F	*Yersinia pseudotuberculosis*	Cas3, Cas8
		I-U	*Geobacter sulfurreducens*	Cas3, Cas8
	III	III-A	*Staphylococcus epidermidis*	Cas10
		III-B	*Pyrococcus furiosus*	Cas10
		III-C	*Methanothermobacter thermaautotroph*	Cas10
		III-D	*Roseiflexus sp*	Cas10
	IV	IV	*Acidithiobacillus ferrooxidans*	Csf1
Class 2	II	II-A	*Streptococcus thermophilus*	Cas9
		II-B	*Legionella pneumophila*	Cas9
		II-C	*Neisseria lactamica*	Cas9
	V	V	*Francisella cf. novicida*	Cpf1
	VI	VI	*Leptotrichia shahii*	C2c2

Adapted from Markarova et al., [17].

Out of six types of CRISPR/Cas systems, more attention has been garnered by type II CRISPR-Cas system primarily due to flexibility in compositional construction. Type II CRISPR/Cas9 obtained from *Streptococcus pyogenes* was the first system to demonstrate specific DNA cleavage both under *in-vitro* conditions and in eukaryotic cells [18]. Smaller Cas9 derived from *Staphylococcus aureus* (SaCas9), *Neisseria meningitides* (NmCas9) and *Streptococcus thermophilus* (StCas9) has been demonstrated to work efficiently in plants [6, 17]. Type V CRISPR/Cas12a system has also demonstrated its efficiency in plant genome editing [6].

Type II CRISPR/Cas9 system consists of a Cas9 protein and CRISPR RNA (crRNA) molecule [4]. Cas9 protein act as an endonuclease and brings about cleavage at specific target DNA site. The crRNA molecule fuses to fixed trans-activating crRNA (tracr-RNA) to form single guide RNA (sgRNA). In the cell, Cas9 proteins bind to the sgRNA and form a binary complex. This complex scans the genome for the DNA target for executing cleavage using Watson-Crick base pairing [4]. The specificity is determined by the sgRNA. Ideally any DNA molecule with sequence complementary to the first 20 nucleotide of sgRNA can be a target, but DNA cleavage is only permitted in the presence of a specific proto-spacer adjacent motif (PAM) localized on the non-target DNA strand, directly downstream of the target DNA sequence. Cas9 from *S. pyogenes* recognizes a G-rich (NGG) proto-spacer adjacent motif (PAM) at the 3' end of the DNA targets for executing cleavage activity [6]. The Cas9 proteins have two signature nuclease domains namely HNH and RuvC. The HNH-like nuclease domain cleaves the target DNA strand complementary to the sgRNA sequence. The RuvC-like nuclease domain cleaves the non-target DNA strand [6].

The important outcome of CRISPR/Cas gene editing is the creation of double-stranded DNA breaks (DSBs) at the target site. The DSBs can be repaired either by non-homologous end joining (NHEJ) or homology-directed repair (HDR). During the repair of DSBs a variety of desirable genomic modifications can be introduced into plants [5].

The NHEJ repair pathway is a popular way to disrupt genes by creating small insertions or deletions at specific points in the target genes. This repair pathway can be employed at any phase of cell cycle and does not require a homologous repair template thus it is an efficient method for gene stack for crop improvement. NHEJ is considered highly efficient and very much suited for large-scale knockout however it lacks the desired specificity for advance-level genome editing [6].

HDR-mediated genome editing can be employed to accurately introduce specific point mutations and to insert or replace desired sequences into the targeted DNA. HDR takes place during the S- and G2-phase of the cell cycle [6]. HDR require template in form of sister chromatid or an exogenous template for carrying out repair. Generally exogenous template comprises of an exogenous DNA or single-strand DNA having a desirable sequence modification so as to incorporate change into the cleaved site [19]. Precise HDR-mediated genome modification has been successful in diverse organisms. However, HDR-mediated genome modification in plants is relatively difficult to achieve owing to the low efficiency of HDR and the limitations of delivering donor template into plant cells. Many approaches have been employed to improve HDR-mediated gene targeting in plants. Nishizawa-Yokoi et al., [20] carried out successful gene targeting in rice using CRISPR/Cas9 with help of a positive-negative selection approach, but this method is complicated and restricted to a few disease resistance genes. Increasing the amount of donor template in single cells can also increase HDR in some plants [6]. In many plant species, Geminivirus replicons based on bean yellow dwarf virus and wheat dwarf virus have been employed to increase the number of copies of the donor template and thus, increase the frequency of HDR-mediated gene insertion [21]. The efficiency of HDR in rice was also improved by the use of chimeric sgRNA molecules constructed by fusing sgRNA and repair template sequences [22].

CRISPR-Mediated Base Editing System

Base editing systems induce specific base changes without homology-directed repair or requirement for donor DNA. The system does not involve any double-stranded DNA breaks for inducing specific base changes at the target site. They are very much suitable for inducing the point mutation in the target site. Base editing systems are associated with several advantages that make it popular for gene editing in plants. They are more efficient and generate lesser undesirable products than double-stranded DNA break mediated systems. There are very less chance of large chromosomal rearrangements with base-editing systems. Base editing systems are suitable for creating nonsense mutations to avoid DSB-induced in-frame indels [6].

Komor et al. [23] developed a cytosine base-editor (CBE) system by fusing cytidine deaminase with nCas9 and uracil glycosylase inhibitor. The system was capable of replacing targeted cytosine to uracil in genomic DNA. The base-editor 3 (BE3) system created by fusing the rat cytidine deaminase APOBEC1 with nCas9 and uracil glycosylase inhibitor has been widely used for gene editing in several plant species like wheat, rice, and potato [24]. Adenine base editors (ABEs) bring about the conversion of targeted adenine to guanine in genomic DNA. ABE systems have been optimized for bringing about the conversion of targeted adenine to guanine in plants such as wheat and rice [25]. Chen et al. [6] considered base-editing systems as an important tool for advance plant genetic research.

Delivery of CRISPR/Cas Reagents to Plant Cells

CRISPR-mediated gene editing requires reagents such as DNA, RNA, and ribonucleoproteins (RNPs) to be delivered into the recipient cells. Broadly, the delivery cargo for CRISPR-mediated gene editing can be either CRISPR/Cas DNA or DNA free reagents such ribonucleoproteins (RNPs) or *in-vitro* transcripts [6].

CRISPR/Cas DNA

Delivery of gene editing reagents can be made by Agrobacterium-mediated transfer DNA (T-DNA) transformation, or particle bombardment. Agrobacterium-mediated transformation and particle bombardment are successful delivery methods for the production of edited plants [5, 6]. The DNA segment expressing Cas protein and sgRNA gets integrated into the plant genome and expresses itself to bring about genome editing. Marker-based selection aid in confirming the integration of delivered DNA segment. However, CRISPR constructs and marker genes can become integrated into the genome and cause side effects such as increased off-target changes [5, 6]. Undesirable changes have often restricted wide commercial applications of CRISPR/Cas DNA based system. To overcome aforementioned problems, researchers have obtained transgene-free derivatives through genetic segregation by selfing and crossing [6]. Genetic segregation is good method to obtain transgene-free mutants, however this method is not useful in asexually propagated crops such potato and banana [6]. Another interesting method to obtain transgene-free mutant is to use the suicide genes *CMS2* and *BARNASE*. These suicide genes kill transgene-containing pollen and embryos produced by the T_0 plants [26].

The transgene-free mutants can also be achieved by transient gene expression of CRISPR reagents. Generally, here selection steps are eliminated so that some of the regenerated plants are edited without any integration of foreign DNA into the genome [4]. Researchers have used the protoplast transfection method for obtaining transgene-free mutants by transient expression of CRISPR reagents in tetraploid potato and tobacco [27, 28].

Zhang et al., [29] used the particle bombardment method to deliver a CRISPR/Cas9 plasmid into immature wheat embryos. The resulting plants were allowed to regenerate without selection pressure. This approach significantly reduced the time required for regeneration in comparison to traditional tissue culture methods. Also, the mutation frequency was comparable to those of the conventional DNA-integration method in which selection pressure is applied during tissue culture. Interestingly, 86.8% of the T0 mutants were free from transgenes [29]. Transgene-free nucleotide substitutions have also been achieved by DNA of base editors, such as plant cytidine base editor and plant adenine base editor, delivered by the transient method [24, 25].

Transferring CRISPR-mediated gene editing reagents into legumes is a challenge for researchers. Legumes are well known for exhibiting resistance to the uptake and integration of introduced DNA. Moreover, some of the transformed legumes are recalcitrant in terms of regeneration [5]. Hence, researchers have highlighted the importance of developing a transformation protocol where the transformation vector is incorporated from the onset rather than from the development of recreation protocol. *Agrobacterium*-mediated transformation has been successfully used in several large-seeded leguminous crops like pea [30], lentil [31], chickpea [32], cowpea [33], mung bean [34] and soybean [35].

DNA-Free Genome Editing

Transient expression of CRISPR/Cas DNA helped researchers to obtain transgene-free mutants. However, this approach was unable to completely eradicate the presence of transgene [6]. There was a possibility that degraded DNA fragments may integrate into the plant genome and bring about undesirable changes. To overcome the disadvantages associated with plasmid-based expression and messenger RNA (mRNA)-based expression of Cas9/sgRNA, an efficient DNA-free genome-editing system has been developed [36, 37].

DNA-free genome-editing system employs Cas9/sgRNA ribonucleoprotein (RNP) for bringing about desirable alteration in the plant genome. RNP is a complex consisting of purified Cas protein and *in vitro* transcribed or artificially synthesized sgRNA [36]. In terms of efficiency, Cas9/sgRNA RNPs are comparable to plasmid-based expression systems. The RNP bring about cleave of the target site immediately upon delivery as they do not require the cellular transcription and translational machinery. After cutting of target site the RNP degrade quickly and thus reducing the chances of off-target effects in comparison to the DNA-based expression system [36].

Woo et al. [37] demonstrate polyethylene glycol-mediated RNP transfection of protoplasts for genome editing in crop plants like *Arabidopsis*, rice, tobacco, and lettuce. Kim et al. [38] delivered LbCpf1/crRNA and AsCpf1/crRNA RNPs into soybean protoplast by adding Cpf1 to the RNP editing toolbox. Regeneration of plants from protoplasts is challenging for most cereal and leguminous crops [6].

RNP can also be delivered by particle bombardment method [29, 39]. Zhang et al. [29] used particle bombardment method to deliver *in-vitro* transcripts of Cas9 and sgRNAs into immature wheat embryos and generated DNA-free edited wheat. Unfortunately, the editing efficiency achieved by this method was lower than the DNA expression system. Interestingly no mutations were observed at the off-target sites [29]. Svitashev et al. [39] delivered pre-assembled Cas9/gRNA RNPs into maize embryos by particle bombardment method. This method exhibited 2.4% to 9.7% editing frequency with no off-target sites [39].

CRISPR/Cas9 Technology for Improving Leguminous Crops

CRISPR/Cas9 mediated gene editing has been established in model legumes namely *Medicago truncatula* and *Lotus japonicus*. Several other leguminous crops such as soybean, cowpea and chickpea have also been successfully improved genetically through CRISPR/Cas 9 system [5].

Barrel Medic (*Medicago truncatula* Gaertn.) and Japanese Trefoil (*Lotus japonicas* Regel.)

Barrel medic and Japanese trefoil are model legume plants. Both species have been used for genetic and physiological studies of several traits associated with legume physiology. Both barrel medic and Japanese trefoil have diploidy condition, self-fertility, small genome size, short life cycle and well-established transformation method that make them an ideal choice for legume research [5, 40]. The availability of whole-genome sequences of *M. truncatula* and *L. japonicas* has further aided in molecular research [40]. Both species share several similar characteristics with leguminous crops but also differ in the type of root nodule established for nitrogen fixation. *M. truncatula* forms indeterminate root nodules very much similar to pea, chickpea, lentil and faba bean. *L. japonicus* forms determinate root nodules similar to soybean, mung bean, and cowpea [5].

CRISPR/Cas9-based gene editing system has been used efficiently to introduce targeted mutation of nodulation-associated genes in *M. truncatula* as well *L. japonicas.* CRISPR/Cas9 based gene editing system was employed by Curtin et al., [41] to establish the function of candidate nodulation genes in *M. truncatula*. The CRISPR/Cas9 was also successfully used to carry out mutational studies of five different nodule-specific PLAT domain (NPD1-5) and nitrate peptide family (NPD) genes in *M. truncatula* [42]. CRISPR/Cas9 was used to generate mutation of the Hua enhancer1 (MtHen1) gene in *M. truncatula* for studying small RNA processing [43].

CRISPR/Cas9 based gene editing system has been applied to generate several mutant plants for symbiosis receptor-like kinase (LjSYMRK), leghemoglobin (LjLb1-3), and Lotus histidine kinase 1 (LHK1)-interacting protein (LjCZF1-2) in *L. japonicas* [44, 45]. Recently, Rodas et al. [46] established SUPERMAN (MtSUP) of *M. truncatula* as a true orthologue of AtSUP of Arabidopsis by using CRISPR/Cas9 system. CRISPR/Cas9 has been used to identify the biochemical activity of protein encoded by CYP716A51-1 in *L. japonicus* [46]. The CRISPR/Cas9 system has been employed to mutate *phytoene desaturase* (PDS) gene of *M. truncatula*[47]. The CRISPR/Cas9 system with Arabidopsis UBQ10 promoter exhibited high mutation efficiency in *M. truncatula* [4]. Several recent findings provide excellent evidence to establish that CRISPR/Cas9 is a successful gene-editing system in *M. truncatula* and *L. japonicus* for studying various aspects of legume physiology [5, 46, 47 and 48]. Recently, Wang et al. [49] successfully induced haploid condition in *M. truncatula* thus paving path for haploid gene editing in legume.

Soybean (*Glycine max* Merr.)

Soybean is an important legume which also serves as source of vegetable oil. It is one of the first legume crop targeted for genetic alteration by CRISPR/Cas9 technology [50]. Soybean genome is highly duplicated and thus poses a major hurdle for conventional genetic approaches for studying gene function. Another problem with soybean is that Agrobacterium-mediated transformation efficiency is very low in soybean [5]. Some researchers have used Agrobacterium rhizogenes-mediated hairy root transformation to quickly estimate the efficacy of gRNAs before whole plant transformation [51].

Several research groups have successfully employed CRISPR/Cas9 technology for targeted gene editing in soybean. Di et al. [52] examined multiple GmU6 promoters (11 in total) for their suitability in regulating gRNA expression in soybean hairy roots. Their group reported that the GmU6-8 and GmU6-10 promoters had higher activity in enhancing editing efficiency as compared to the other nine GmU6 promoters [52].

Studies have also shown that large genomic deletions (from around 1 kb to more than 4.5 kb) is possible by using dual gRNAs to cleave two adjacent loci on the same chromosome of soybean [53, 54]. The large size deletion will not only bring about the complete knockout of targeted genes but also help in molecular research, such as understanding the function of regulatory elements or non-coding genes in soybean [50]. In one of the study, Zheng et al. [55] reported that the egg cell specific promoters can enhance gene-editing efficiency (26.8%) in stable transgenic soybean lines.

Researchers are making efforts to understand gene function and improve/manipulate agronomic traits, such as seed oil and storage proteins, flowering timing, and plant architecture [50]. Kanazashi et al. [56] used a single gRNA to edit two homologous genes *GmPPD1* and *GmPPD2* to create double mutants with severe abnormal leaf and pod phenotypes. Cai et al. [53] used CRISPR/Cas9-mediated mutagenesis to make homozygous mutants in FLOWERING LOCUS T2a (*GmFT2a*) that delayed flowering in soybean. Double mutant of *GmFT2a* created by crossing *ft2a* mutant exhibited delayed flowering under short day conditions and also produced higher number of pods and seeds per plant compared to the wild type [57]. CRISPR/Cas9-based multiplex genome editing tool has been used to mutate four *GmSPL9* genes, and created several soybean mutants containing different combinations of mutated loci [58]. These mutants exhibited altered number of node and the number of branches on the main stem. CRISPR/Cas9 system has been also used to alter seed related traits [54], isoflavone content and induce resistance to soybean mosaic virus [59] and remove the unpleasant beany flavor of soybean seed products [60]. The first soybean CRISPR library targeting more than 100 candidate genes has been constructed. A collection of mutant soybean lines was also generated using improved procedures [61]. Such efforts toward a collection of mutants of all soybean genes by the community will be pivotal for understanding soybean genetics and improving traits.

Cowpeas (*Vigna unguiculata* Walp.)

Cowpeas are a grain legume suitable for cultivation in warm and dry climatic conditions. Good protein content along with high lysine percentage makes cowpea suitable to complement cereal crop-based diets. There has been a gradual increase in global cowpeas production. As per record, globally 9.7 million tons of cowpea were produced which increased to 14.4 million tons in 2019 [5].

Cowpeas are recalcitrant for transformation, hence hindering routine practices of CRISPR/Cas9 system. Recently, Che et al., [62] improved transformation efficiencies across nine cowpea genotypes by expressing Cas9 under the soybean elongation factor (GmEF1A2) promoter and a gRNA under the VuU6 promoter. Juranic et al. [63] overcame the difficulty of testing CRISPR/Cas9 constructs in stable cowpea transformants by Agrobacterium infiltration of detached leaflets method. Using this method, the researchers were able to test a variety of constructs within 48 hours using the Arabidopsis 50S ribosomal protein (AtRPS5A) promoter (RPS5A) or the parsley ubiquitin4-2 (PcUbi4-2) promoter for Cas9 expression and either the VuU6 or AtU6 promoters for gRNA expression. A construct using the AtRPS5A and AtU6-26 promoters for VuSPO11-1 knockout showed a mutation frequency of 0.05-3.9% in the transient assay and was used to produce ten T1 plants. Two of these plants displayed 50% gene editing. Only one of these plants survived, but it displayed a sterile phenotype [62]. Recently, CRISPR/Cas9-based gene editing system was successfully used in cowpea for the disruption of symbiotic nitrogen fixation by targeting a symbiosis receptor-like kinase gene [64].

Peas (*Pisum sativum* L.)

Peas are the second most important grain legume after the common bean. It can be consumed both in dry and green form. Pea is a valuable source of dietary proteins, minerals nutrients, starch and fibers. However, the nutritional importance of pea-based food is often undermined due to some undesirable characteristic referred to as "off-flavors". Off-flavors highlight green, beany, hay-like, metallic, and astringent flavours [5]. The presence of a mixture of volatile organic compounds and non-volatile saponins are the major factors responsible for undesirable flavors. CRISPR/Cas9 based gene editing has the potential to remove off-flavors and improve the nutritional value of pea [5].

The pea genome has provided the basic information required for improving nutritional and agronomic traits. Volatile organic compounds are a collection of small organic molecules synthesized during the cultivation, transportation, and storage of pea products. Enzyme lipoxygenases (LOX) catalyze the conversion of polyunsaturated fatty acids (linoleic acid and linolenic acid) to hydroperoxide forms (HPOs) of fatty acids [65] and led to the generation of volatile organic compounds. Targeting of LOX by CRISPR/Cas9 is expected to block the release of VOC-related "off-flavors" in peas.

Lentils (*L. culinaris* Medik.)

Lentils are the excellent sources of protein, carbohydrates, dietary fiber components, minerals and vitamins. Secondary metabolites of lentils have several health benefiting compounds. Lentil is ranked sixth among the 11 primary legumes harvested as dry seeds [5].

Plant breeders are trying to develop lentil cultivars with improved nutritional quality well adapted to the environment and climate instability. Genetic diversity is crucial for developing improved lentil cultivars. To attain further breakthroughs in enhancing yield and nutritional quality, and improving the stability of cultivars, new sources of genes and alleles must be either identified or generated in lentil genetic resources. The availability of the reference genome of lentil (redberry) has provided a basic resource for its trait improvement [66]. Researchers believe that nuptial between traditional breeding techniques and CRISPR/Cas9 based gene editing can aid in improving the nutritional as well as agronomic traits of lentil [5]. Several genetic transformation methods including *Agrobacterium*-mediated genetic transformation have been attempted in lentil, but with very less transformation efficiency [67]. Success of CRISPR/Cas9-based gene editing heavily rely on the success of transformation efficiency [5].

Faba Beans (*Vicia faba* L.)

Faba beans or broad beans are one of the most important cool-season grain legumes. They have good nutritional value owing to the presence of high-protein content, dietary fibers, vitamins, minerals, and antioxidants. In recent past, the faba bean has attracted lot of global attention owing to high nutritional value, contribution in sustainable agriculture and soil conservation [5]. Faba bean has large genome of 13Gb size with more than 85% of repetitive DNA [68]. Genome sequence of faba bean is not available and this has imposed challenges for the application of CRISPR/ Cas gene editing. However, significant work has been done to generate genetic and genomic resources to facilitate molecular breeding. Faba beans share strong synteny relationships with model legume *M. truncatula* and other legumes [69]. Benefiting from this relationship, transcriptome analyses and comparative genomics approaches have been used to discover single-nucleotide polymorphisms (SNPs) for constructing of high-density consensus genetic map and predicting the candidate genes controlling various important agronomic traits of interest [69]. The lack of an efficient transformation method has also hindered the gene alteration and trait improvement process. Advancements in genetic transformation and CRISPR/Cas9 system are believed to speed up the genetic improvement of faba beans.

Mung Beans (*Vigna radiata* (L.) Wilczek)

Mung beans are a warm-season and short-duration grain legume. Seeds of mung bean are rich source of carbohydrates, proteins, dietary fiber, antioxidants and iron. As compared to other leguminous crop, mung bean has low productivity owing to non-synchronous maturity, lack of improved varieties, and susceptibility to biotic and abiotic stresses. Conventional and mutation breeding approaches have been successfully employed for improving the agronomic traits of the crop [70]. Mung bean is a diploid crop ($2n = 2x = 22$) with a relatively small genome size. However, still there are lots of genetic gains possible with molecular breeding techniques. Earlier, molecular techniques involving random amplified polymorphic DNA (RAPD), restriction fragment length polymorphism (RFLP), SNP, and SSR markers have been applied for improving yield, nutritional quality and disease resistance in mung bean [70, 71].

Kang et al., [72] sequenced cultivated *V. radiate* and a wild relative (*V. sublobata*) of mung bean. Their group reported the transcriptome sequences of 22 Vigna accessions of 18 species. There is a great scope to use CRISPR/Cas9 gene editing in mung bean breeding programs with the availability of whole-genome sequence, and the collection of 1481 mung bean entries that have been evaluated for various important agronomic traits. The success of CRISPR/Cas9 in a

Vigna system hints that gene editing can be successful employed in mung bean for improving important agronomic traits. Developing mung bean varieties that can withstand changing weather patterns would help in further expanding the mung bean cultivation area worldwide.

Chickpeas (*Cicer arietinum* L.)

Excellent nutritional value and good symbiotic nitrogen fixation ability make chickpea one of the important food legumes. According to FAOSTAT (2019), approximately 14.24 million tons of chickpeas were produced worldwide in 2019.

The lack of efficient and reproducible plant regeneration systems has curtailed efforts of developing new chickpea varieties through genomics and gene-editing approaches. Earlier Das et al. [73], has reported the regeneration of the transgenic chickpea plants expressing a chimeric gene-encoding insecticidal crystal protein Cry1Aabc of *Bacillus thuringiensis.* The transgenic chickpea plants were resistant to pod borer (*Heliothis armigera*). Recently, Bhowmik et al. [74], reported transformation system to obtain stable transgenic lines in chickpea. The system was successfully employed to generate transgenic lines expressing GUS (*uidA*), stress tolerance (*AtBAG4* and *TlBAG*), and Fe biofortification (*OsNAS2* and *CaNAS2*) genes [74].

Currently, researchers have great opportunity to use gene-editing technologies to characterize the functions of genes and improve agricultural traits with the availability of reference genome in chickpea, transcriptome sequences and transformation methods. The International Crops Research Institute for the Semi-Arid Tropics (ICRISAT) is making efforts to employ CRISPR/Cas9-based gene editing system to increase yield, seed size, and quality of chickpea. Recently, Badhan et al. [75] reported the use of CRISPR/Cas9-based gene editing of chickpea protoplasts to create knockouts of 4-coumarate ligase (*4CL*) and Reveille 7 (*RVE7*) genes. Some of the recent studies have increased the feasibility of CRISPR/Cas9-based gene editing of chickpea and thus laying a technical foundation for future trait discovery and improvement.

Concerns Associated with Gene Editing in Legumes

Undoubtedly, CRISPR/Cas9-based gene editing is a promising and powerful technology for improving traits in leguminous crops. Since its advent in 2013, there have been hundreds of reports exhibiting the usefulness of CRISPR/Cas9-based gene editing in crop improvement [4]. However, the achievements of CRISPR/Cas9-based gene editing technology have been overshadowed by controversy surrounding the technology. The main concern associated with the technology is that whether CRISPR/Cas9-based gene-edited leguminous plants should be considered genetically modified organisms (GMO). Broadly, GMO are living organisms whose genetic material has been artificially manipulated using genetic engineering techniques. Here genetic material manipulation involves insertion of exogenous DNA fragments into plant so that the genetically modified plant exhibits new desirable traits such herbicide resistance, pest resistance, increase in yield or improved nutritional quality [5]. Source of inserted genetic material may be same species or non-closely related species. In later scenario, genetically modified plant is designated as Transgenic Plant since it is containing transgenes. Transgenic plants such as Bt-Cotton have contributed significantly in improving modern day agriculture. However, they have also been associated with unproven concerns over food and environmental safety. In many countries, GM crops are scrutinized through strict regulatory frameworks and exhaustive risk assessment processes to address food and environment safety concerns.

In the present scenario, the current definition of GMOs and corresponding regulatory frameworks needs to be reframed as CRISPR/Cas9-based gene editing technology is entirely different from transgenic technology. CRISPR-induced gene manipulation includes small insertion or deletion of DNA fragments rather than insertion or rearrangement of large DNA fragments. In natural condition, the insertion or deletion of small DNA fragments is very common in plants. Furthermore, such manipulation of small DNA fragments can be induced artificially by the use of physical or chemical mutagens.

Traditional GMOs have transgene as an integral part of the plant genome. In GMOs, transgenes are inherited from one generation to other so that desired traits can be passed through generations. Contrary, CRISPR system edited plants are free from the transgene. For gene editing, CRISPR constructs are expressed for a transient period without any DNA integration in the plant genome. CRISPR constructs can be stably incorporated into the plant genome and expressed for executing gene editing but after achieving a desirable product, the CRISPR constructs can be removed by genetic segregation. Alternatively, *in vitro* transcripts or ribonucleoproteins can be applied for delivering CRISPR system into the regenerative plant cells without any DNA construct.

Several countries like Australia, Argentina, Brazil and the United States advocate that genetically modified crops without any foreign genes should be considered and regulated as a conventional breed crop variety and not as GM crops. Contrary, European Union believes that all crops developed by gene editing should be considered and regulated as GM crops.

The United States Secretary of Agriculture declared that crop varieties developed through the use of new plant breeding technologies such as genome editing would be viewed as conventionally breed plants and would not require additional regulatory rules. New plant breeding technologies such as genome editing do not involve plant pests or are developed by exploitation of plant pests, and thus cannot be included under GMOs [76]. The consequence of announcement by the United States Secretary of Agriculture was that several other countries including Argentina, Australia, and Brazil announced regulatory policies that were in line with that of the United States. These countries announced that the new crop variety without any foreign genes or DNA segment would be treated as a conventional crop and would not be subjected to additional regulation.

However, the Court of Justice of the European Union ruled that crop variety developed by gene editing technologies would be considered equivalent to GMO crops and subsequently would subject to additional regulation within the EU. This ruling has noteworthy impact on gene-editing research investment in the EU. Bayer, BASF and smaller plant breeding firms relocated all of their gene-editing research investment to other countries harboring conducive rules for gene-edited crops.

Till-date, there is no globally accepted regulatory framework for CRISPR-gene edited plants. US Department of Agriculture exempted gene-edited crops from GMO regulatory policies, on the contrary, European Court of Justice has stated that gene-edited products should be considered similar to traditional GMOs. There are countries (particularly developing countries), which do not have clearly developed regulatory policies for gene-edited crops. Most of the researchers are of view that gene-edited crops may be considered as products of biological mutagenesis, very much similar to the conventional plant breeding method. Thus, should not be subjected to special regulatory policies of GMOs. Gene-edited crops must be treated in a similar manner as conventionally breed crop plants. Extensive usage of genome editing technologies is likely to bring about serious technical challenges for regulatory authorities in near future, since it is difficult to differentiate between natural and artificially induced mutants.

CRISPR tools have also been reported for off-targeted mutation especially in the field of gene therapy. There are some studies in mammalian cells and some other systems, where Cas9/sgRNA complexes have shown the ability to cut DNA sequences with an imperfect match to guide sequence [4, 76]. Contrary, very limited off-targeted genome manipulation has been observed in plants. Whole-genome sequencing was applied in *Arabidopsis,* rice, soybean and tomato to detect off-target mutations due to CRISPR based gene editing. Interestingly, a very small number of off-target mutations were identified in *Arabidopsis,* rice, soybean and tomato [6, 11].

Off-targeted mutation can be avoided by designing highly specific guide RNAs with help of software tools like CRISPR-P and CRISPR-GE. It has been observed that nucleotide sequence of PAM and PAM-proximal sgRNA are very crucial for target recognition. PAM-distal nucleotide sequence can tolerate mismatch to some extent. Some of the recent research has demonstrated that specificity of CRISPR system can be improved by applying engineered Cas9 variant such as enhanced-specificity Cas9s, high-fidelity Cas9s and xCas9 [16].

Tokyo-based Sanatech Seed Co. Ltd. is selling CRISPR-Cas9 edited tomatoes under the name of "Sicilian Rouge tomatoes". These tomatoes are genetically edited to contain high amounts of γ-aminobutyric acid (GABA). Regulators of Japan approved CRISPR–Cas9 edited tomatoes in December 2020 and paved the pathway for their market [77].

Future Scope

The potential future of agriculture relies on developing improved crop varieties with pest resistance, enhanced nutritional quality and ability to survive climate changes. Recent, advances made in CRISPR/Cas technology has provided ample prospect to develop desirable legume varieties by targeted gene editing. Direct delivery of CRISPR cassette in the recipient cell using *Agrobacterium* and viral replicon by nanoparticle has undoubtedly improved and simplified the genome editing technology in several legumes [78]. Development of an inducible Cas9 expression system can provide means to overcome the off-target effects by expressing the Cas9/sgRNA only when required.

Targeted gene editing by CRISPR/Cas9 system has the potential to serve research in functional genomics of legumes, particularly for the gene involved in roots and nodules formation [79]. However, future research must not only add desirable traits but also remove undesirable traits from grain legumes for expanding their dietary applications.

Most of the CRISPR/Cas9 based gene editing has been performed on soybean with very little work on other legume crops. Expanding CRISPR/Cas9 system for gene editing in other legumes like faba bean and mung bean will help in improving plant breeding resources. Developing an efficient protocol for legume transformation and overcoming the hurdle of whole plant regeneration can establish CRISPR/Cas9 system as an imperative part of plant breeding programs [80, 81].

Conclusion

The CRISPR/Cas based genome editing system is an innovative tool that has revolutionize agriculture sector. Gene editing system particularly, CRISPR-Cas9 has ability to create improved legume varieties with desirable traits. Designing legume crops with broader environmental tolerance, disease and pest resistance, superior nutritional quality and enhanced yield will help to meet the future food requirements of the world. However, the successful application of CRISPR-Cas9 based genome editing technology for legume crop improvement will require a proficient system for plant transformation and regeneration of whole plants. Also, the scientific community needs to create a conducive regulatory environment for increasing public acceptance of gene-edited crops. Advancements made in CRISPR/Cas9 system have revolutionized both basic and applied research in legumes. However, still lots of unexplored realms are waiting to be explored and serve research in legumes.

Acknowledgments

The author is thankful to the Chancellor and Vice-chancellor of Sant Baba Bhag Singh University (SBBSU) for their encouragement and support while preparing the chapter.

References

[1] Lewis, G. P. (2005). *Legumes of the World.* Richmond, UK: Royal Botanic Gardens, Kew.

[2] Calles, T., Xipsiti, M. and del Castello R. (2019). Legacy of the International Year of Pulses. *Environ. Earth Sci.* 78:124.

[3] Soumare, A., Diedhiou, A. G., Thuita, M., Hafidi, M., Ouhdouch, Y., Gopalakrishnan, S. and Kouisni, L. (2020). Exploiting Biological Nitrogen Fixation: A Route Towards a Sustainable Agriculture. *Plants (Basel, Switzerland),* 9(8): 1011.

[4] El-Mounadi, K., Morales-Floriano, M. L. and Garcia-Ruiz, H. (2020). Principles, Applications, and Biosafety of Plant Genome Editing Using CRISPR-Cas9. *Front. Plant Sci.* 11:56.

[5] Bhowmik, P., Konkin, D., Polowick, P., Hodgins, C. L., Subedi, M., Xiang, D., Yu, B., Patterson, N., Rajagopalan, N., Babic, V., Ro, D. K., Taran, B., Bandara, M., Smyth, S. J., Cui, Y. and Kagale S. (2021). CRISPR/Cas9 gene editing in legume crops: Opportunities and challenges. *Legume Sci.* 3(3): e96.

[6] Cheng, C. K., Wang, Y., Zhang, R., Zhang, H. and Gao C. (2019). CRISPR/Cas genome editing and precision plant breeding in agriculture. *Annu. Rev. Plant Biol*. 70:667–697.

[7] Pacher, M. and Puchta, H. (2017). From classical mutagenesis to nuclease-based breeding-directing natural DNA repair for a natural end-product. *Plant J.* 90:819–833.

[8] Mao, Y., Botella, J. R., Liu, Y. and Jian-Kang Z. (2019). Gene editing in plants: progress and challenges. *National Sci. Review* 6(3):421–437.

[9] Bauchet, G. J., Bett, K. E., Cameron, C. T., Campbell, J. D., Cannon, E. K. S.,Cannon, S. B., Carlson, J. W., Chan, A., Cleary, A., Close, T. J., Cook, D. R., Cooksey, A. M., Coyne, C., Dash, S., Dickstein,

R., Farmer, A. D., Fernandez-Baca, D., Hokin, S., Jones, E. S. and Zhao P. X. (2019). The future of legume genetic data resources: Challenges, opportunities, and priorities. *Legume Sci*, 1:e16.

[10] Rouet, P., Smih, F. and Jasin M. (1994). Introduction of double-strand breaks into the genome of mouse cells by expression of a rare-cutting endonuclease. *Mol. Cell. Biol.* 14: 8096–8106.

[11] Bao, A., Tran, L. P. and Cao D. (2020). CRISPR/Cas9-Based Gene Editing in Soybean. *Methods Mol. Biol.* 2107:349-364.

[12] Haun, W., Coffman, A., Clasen, B. M., Demorest, Z. L., Lowy, A., Ray, E., Retterath, A., Stoddard, T., Juillerat, A., Cedrone, F., Mathis, L., Voytas, D. F. and Zhang, F. (2014). Improved soybean oil quality by targeted mutagenesis of the fatty acid desaturase 2 gene family. *Plant Biotechnol J* 12: 934-940.

[13] Wilke, Carolyn. *Gene-Edited Soybean Oil Makes Restaurant Debut,* Mar 13, 2019 https://www.the-scientist.com/news-opinion/gene-edited-soybean-oil-makes-restaurant-debut-65590.

[14] Ishino, Y., Shinagawa, H., Makino, K., Amemura, M. and Nakata A. (1987). Nucleotide sequence of the iap gene, responsible for alkaline phosphatase isozyme conversion in *Escherichia coli,* and identification of the gene product. *J. Bacteriol.* 169: 5429–5433.

[15] Murugan, K., Babu, K., Sundaresan, R., Rajan, R. and Sashital D. G. (2017). The revolution continues: newly discovered systems expand the CRISPR-Cas toolkit. *Mol. Cell.* 68: 15–25.

[16] Hassan, M., Zhang, Y., Yuan, G., De K., Chen J-G., Muchero, W., Tuskan, G. A., Qi Y. and Yang X. (2021). Construct design for CRISPR/Cas-based genome editing in plants. *Trends in Plant Sci.* 26(11): 1133-1152.

[17] Makarova, K. S., Wolf, Y. I., Iranzo. J., Shmakov, S. A., Alkhnbashi, O. S., Brouns, S. J. J., Charpentier, E., Cheng, D., Haft, D. H., Horvath, P., Moineau, S., Mojica, F. J. M., Scott, D., Shah, S. A., Siksnys, V., Terns, M. P., Venclovas, Č., White, M. F., Yakunin, A. F., Yan, W., Zhang, F., Garrett, R. A., Backofen, R., van der Oost, J., Barrangou R. and Koonin E. V. (2020). Evolutionary classification of CRISPR–Cas systems: a burst of class 2 and derived variants. *Nat. Rev. Microbiol.* 18: 67–83.

[18] Jinek, M., Chylinski, K., Fonfara, I., Hauer, M., Doudna, J. A. and Charpentier E. (2012). A programmable dual-RNA-guided DNA endonuclease in adaptive bacterial immunity. *Science* 337:816–821.

[19] Salsman J. and Dellaire G. (2016). Precision genome editing in the CRISPR era. *Biochem. Cell Biol.* 95:187– 201.

[20] Nishizawa-Yokoi, A., Mikami M. and Toki S. (2020). A Universal System of CRISPR/Cas9-Mediated Gene Targeting Using All-in-One Vector in Plants. *Front. Genome Ed.* 2:604289.

[21] Gil-Humanes, J., Wang, Y., Liang, Z., Shan, Q., Ozuna, C. V., Sánchez-León, S., Baltes, N. J., Starker, C., Barro, F., Gao C. and Voytas D. F. (2017). High-efficiency gene targeting in hexaploid wheat using DNA replicons and CRISPR/Cas9. *Plant J.* 89:1251–1262.

[22] Butt, H., Eid, A., Ali, Z., Atia, M. A. M., Mokhtar, M. M., Hassan, N., Lee, C. M., Bao, G. and Mahfouz M. M. (2017). Efficient CRISPR/Cas9-mediated genome editing using a chimeric single-guide RNA molecule. *Front. Plant Sci.* 8:1441.

[23] Komor, A. C., Kim, Y. B., Packer M. S. and Liu D. R. (2016). Programmable editing of a target base in genomic DNA without double-stranded DNA cleavage. *Nature* 533:420–424.

[24] Zong, Y., Song, Q., Li, C., Jin, S., Zhang, D., Wang, Y., Qiu, J. L. and Gao C. (2018). Efficient C-to-T base editing in plants using a fusion of nCas9 and human APOBEC3A. *Nat. Biotechnol.* 36:950–953

[25] Li, C., Zong, Y., Wang, Y., Jin, S., Zhang, D, Song, Q., Zhang R. and Gao C. (2018). Expanded base editing in rice and wheat using a Cas9-adenosine deaminase fusion. *Genome Biol.* 19:59

[26] He, Y., Zhu, M., Wang, L., Wu, J., Wang, Q., Wang, R. and Zhao Y. (2018). Programmed self-elimination of the CRISPR/Cas9 construct greatly accelerates the isolation of edited and transgene-free rice plants. *Mol. Plant* 11:1210–1213.

[27] Andersson, M., Turesson, H., Nicolia, A., Falt, A. S., Samuelsson, M. and Hofvander P. (2017). Efficient targeted multiallelic mutagenesis in tetraploid potato (*Solanum tuberosum*) by transient CRISPR-Cas9 expression in protoplasts. *Plant Cell Rep.* 36:117–128.

[28] Lin, C. S., Hsu, C. T., Yang, L. H., Lee, L. Y., Fu, J. Y., Cheng, Q. W., Wu, F.H., Hsiao, H. C., Zhang, Y., Zhang, R., Chang, W. J., Yu, C. T., Wang, W., Liao, L. J., Gelvin, S. B. and Shih M. C. (2017). Application of protoplast technology to CRISPR/Cas9 mutagenesis: from single cell mutation detection to mutant plant regeneration. *Plant Biotechnol. J.* 16:1295–1310.

[29] Zhang, Y., Liang, Z., Zong, Y., Wang, Y., Liu, J., Chen, K., Qiu J. L. and Gao C. (2016). Efficient and transgene-free genome editing in wheat through transient expression of CRISPR/Cas9 DNA or RNA. *Nat. Commun.* 7:12617.

[30] Shah, Z., Iqbal, A., Khan, F. U., Khan, H. U., Durrani F. and Ahmad M. Z. (2020). Genetic manipulation of pea (*Pisum sativum* L.) with Arabidopsis's heat shock factor HsfA1d improves ROS scavenging system to confront thermal stress. *Genet. Resour. Crop Evol.* 67: 2119-2127.

[31] Das, S. K., Shethi, K. J., Hoque, M. and Sarker R. (2019). Agrobacterium-mediated Genetic Transformation of Lentil (*Lens culinaris* Medik.) with Chitinase Gene followed by *In vitro* Flower and Pod Formation. *Plant Tissue Culture and Biotechnol* 29: 99-109.

[32] Ganguly, S., Ghosh, G., Ghosh, S., Purohit, A., Chaudhuri, R. K., Das, S. and Chakraborti, D. (2020). Plumular meristem transformation system for chickpea: an efficient method to overcome recalcitrant tissue culture responses. *Plant Cell, Tissue and Organ Culture (PCTOC)* 142: 493-504.

[33] Bakshi, S., Sadhukhan, A., Mishra, S. and Sahoo L. (2011). Improved Agrobacterium-mediated transformation of cowpea via sonication and vacuum infiltration. *Plant cell reports* 30: 2281-2292.

[34] Baloda, A., Madanpotra, S. and Aiwal, P. (2017). Transformation of Mungbean Plants for Salt and Drought Tolerance by introducing a gene for an osmoprotectant glycine betaine. *J. Plant Stress Physiol.* 3:5-11.

[35] De Ronde, J., Cress, W., Krüger, G., Strasser, R. and Van Staden J. (2004). Photosynthetic response of transgenic soybean plants, containing an Arabidopsis P5CR gene, during heat and drought stress. *J. plant physiol.* 161: 1211-1224.

[36] Liang, Z., Chen, K., Zhang, Y., Liu, J., Yin, K., Qiu, J-L. and Caixia, G. (2018). Genome editing of bread wheat using biolistic delivery of CRISPR/Cas9 *in vitro* transcripts or ribonucleoproteins. *Nat. Protoc.* 13:413–430.

[37] Woo, J. W., Kim, J., Kwon, S. I., Corvalan, C., Cho, S. W., Kim, H., Kim, S-G., Kim, S-T., Choe, S. and Kim J-S. (2015). DNA-free genome editing in plants with preassembled CRISPR-Cas9 ribonucleoproteins. *Nat. Biotechnol.* 33:1162–1164.

[38] Kim, H., Kim, S. T., Ryu, J., Kang, B. C., Kim, J. S. and Kim S. G. (2017). CRISPR/Cpf1-mediated DNA-free plant genome editing. *Nat. Commun.* 8:14406.

[39] Svitashev, S., Schwartz, C., Lenderts, B., Young, J. K. and Mark, C. A. (2016). Genome editing in maize directed by CRISPR-Cas9 ribonucleoprotein complexes. *Nat. Commun.* 7:13274.

[40] Young, N. D., Debellé, F., Oldroyd, G. E. D., Geurts, R., Cannon, S. B., Udvardi, M. K., Benedito, V. A., Mayer, K. F. X., Gouzy, J., Schoof, H., Van de Peer, Y., Proost, S., Cook, D. R., Meyers, B. C., Spannagl, M., Cheung, F., De Mita, S., Krishnakumar, V., Gundlach, H., et al., (2011). The Medicago genome provides insight into the evolution of rhizobial symbioses. *Nature*, 480:520–524.

[41] Curtin, S. J., Tiffin, P., Guhlin, J., Trujillo, D. I., Burghart, L. T., Atkins, P., Baltes, N. J., Denny, R., Voytas, D. F., Stupar, R. M. and Youn, N. D. (2017). Validating genome-wide association candidates controlling quantitative variation in nodulation. *Plant Physiol* 173: 921–931.

[42] Trujillo, D. I., Silverstein, K. A. T. and Young, N. D. (2019). Nodule-specific PLAT domain proteins are expanded in the Medicago lineage and required for nodulation. *New Phytol* 222: 1538–1550.

[43] Curtin, S. J., Xiong, Y., Michno, J. M., Campbell, B. W., Stec, A. O., Čermák, T., Starker, C., Voytas, D. F., Eamens, A. L. and Stupar R. M. (2018). CRISPR/Cas9 and TALENs generate heritable mutations for genes involved in small RNA processing of *Glycine max* and *Medicago truncatula*. *Plant Biotechnol. J.* 16: 1125–1137.

[44] Cai, K., Yin, J., Chao, H., Ren, Y., Jin, L., Cao, Y., Duanmu, D. and Zhang Z. (2018). A C3HC4-type RING finger protein regulates rhizobial infection and nodule organogenesis in *Lotus japonicus*. *J. Integr. Plant Biol.* 60:878–896.

[45] Wang, L., Rubio, M. C., Xin, X., Zhang, B., Fan, Q., Wang, Q., Ning, G., Becana, M. and Duanmu, D. (2019). CRISPR/Cas9 knockout of leghemoglobin genes in *Lotus japonicus* uncovers their synergistic roles in symbiotic nitrogen fixation. *New Phytol.* 224: 818–832.

[46] Rodas, A. L., Roque, E., Hamza, R., Gomez-Mena, C., Minguet, E. G., Wen, J., Mysore, K. S., Beltran, J. P. and Canas L. A. (2021). *MtSUPERMAN* plays a key role in compound inflorescence and flower development in *Medicago truncatula*. *Plant J.* 105:816–830.

[47] Wolabu, T. W., Park, J. J., Chen, M., Cong, L., Ge, Y., Jiang, Q., Debnath, S., Li, G., Wen, J. and Wang Z. (2020). Improving the genome editing efficiency of CRISPR/Cas9 in *Arabidopsis* and *Medicago truncatula*. *Planta* 252: 15.

[48] Meng, Y., Hou, Y., Wang, H. et al. (2017). Targeted mutagenesis by CRISPR/Cas9 system in the model legume *Medicago truncatula*. *Plant Cell Rep.* 36: 371–374

[49] Wang, N., Xia, X., Jiang, T., Li, L., Zhang, P., Niu, L., Cheng, H., Wang, K. and Lin, H. (2022). In planta haploid induction by genome editing of DMP in the model legume Medicago truncatula. *Plant Biotechnol. J.* 20: 22-24

[50] Bao, A., Zhang, C., Huang, Y., Chen, H., Zhou, X. and Cao D. (2020). Genome editing technology and application in soybean improvement. *Oil Crop Sci.* 5(1):31-40.

[51] Li, C., Nguyen, V., Liu, J., Fu, W., Chen, C., Yu, K. and Cui Y. (2019). Mutagenesis of seed storage protein genes in soybean using CRISPR/Cas9. *BMC Res. Notes* 12: 176.

[52] Di, Y. H., Sun, X. J., Hu, Z., Jiang, Q. Y., Song, G. H., Zhang, B., Zhao, S. S. and Zhang H. (2019). Enhancing the CRISPR/Cas9 system based on multiple GmU6 promoters in soybean. *Biochem. Biophys. Res. Commun.* 519:819–823.

[53] Cai, Y., Chen, L., Sun, S., Wu, C., Yao, W., Jiang, B., Han, T. and Hou W. (2018). CRISPR/Cas9-Mediated Deletion of Large Genomic Fragments in Soybean. *Int. J. Mol. Sci.* 19(12):3835.

[54] Do, P. T., Nguyen, C. X., Bui, H. T., Tran, L. T., Stacey, G., Gillman, J. D., Zhang, Z. J., and Stacey, M. G. (2019). Demonstration of highly efficient dual gRNA CRISPR/Cas9 editing of the homeologous GmFAD2–1A and GmFAD2–1B genes to yield a high oleic, low linoleic and α-linolenic acid phenotype in soybean. *BMC Plant Biol.* 19: 311.

[55] Zheng, N., Li, T., Dittman, J. D., Su, J., Li, R., Gassmann, W., Peng, D., Whitham, S. A., Liu, S. and Yang, B. (2020). CRISPR/Cas9-based gene editing using egg cell-specific promoters in Arabidopsis and soybean. *Front. Plant Sci.* 11:800.

[56] Kanazashi, Y., Hirose, A., Takahashi, I., Mikami, M., Endo, M., Hirose, S., Toki, S., Kaga, A., Naito, K. and Ishimoto, M. (2018). Simultaneous sitedirected mutagenesis of duplicated loci in soybean using a single guide RNA. *Plant Cell Rep.* 37:553–563.

[57] Cai, Y., Wang, L., Chen, L., Wu, T., Liu, L., Sun, S., Wu, C., Yao, W., Jiang, B. and Yuan S. (2020). Mutagenesis of GmFT2a and GmFT5a mediated by CRISPR/Cas9 contributes for expanding the regional adaptability of soybean. *Plant Biotechnol. J.* 18: 298–309.

[58] Bao, A., Chen, H., Chen, L., Chen, S., Hao, Q., Guo, W., Qiu, D., Shan, Z., Yang, Z., Yuan, S., Zhang, C., Zhang, X., Liu, B., Kong, F., Li, X., Zhou, X., Tran, L. P. and Cao D. (2019). CRISPR/Cas9-mediated targeted mutagenesis of *GmSPL9* genes alters plant architecture in soybean. *BMC Plant Biol.* 19: 131.

[59] Zhang, P., Du, H., Wang, J., Pu, Y., Yang, C., Yan, R., Yang, H., Cheng, H. and Yu D. (2020). Multiplex CRISPR/Cas9-mediated metabolic engineering increases soya bean isoflavone content and resistance to soya bean mosaic virus. *Plant Biotechnol J.* 18(6):1384-1395.

[60] Al Amin, N., Ahmad, N., Wu, N., Pu, X., Ma, T., Du, Y., Bo, X., Wang, N., Sharif, R. and Wang, P. (2019). CRISPR-Cas9 mediated targeted disruption of *FAD2-2 microsomal omega-6 desaturase* in soybean (*Glycine max.L*). *BMC Biotechnol.* 19: 9.

[61] Bai, M., Yuan, J., Kuang, H., Gong, P., Li, S., Zhang, Z., Liu, B., Sun, J., Yang, M. and Yang, L. (2020). Generation of a multiplex mutagenesis population via pooled CRISPR-Cas9 in soya bean. *Plant Biotechnol. J.* 18:721–731.

[62] Che, P., Chang, S., Simon, M. K., Zhang, Z., Shaharyar, A., Ourada, J., O'Neill, D., Torres-Mendoza, M., Guo, Y., Marasigan, K. M., Vielle-Calzada, J.-P., Ozias-Akins, P., Albertsen, M. C. and Jones T. J. (2021). Developing a rapid and highly efficient cowpea regeneration, transformation and genome editing system using embryonic axis explants. *Plant J.* 106(3):817-830.

[63] Juranić, M., Nagahatenna, D. S. K., Salinas-Gamboa, R., Sánchez-León, M. L. N., Leong, W. H., How, T., Bazanova, N., Spriggs, A., Vielle-Calzada, J.-P. and Koltunow A. M. G. (2020). A detached leaf assay for testing transient gene expression and gene editing in cowpea (*Vigna unguiculata* [L.] Walp.). *Plant Methods* 16**:** 88.

[64] Ji, J., Zhang, C., Sun, Z., Wang, L., Duanmu, D. and Fan Q. (2019). Genome Editing in Cowpea *Vigna unguiculata* Using CRISPR-Cas9. *Int. J. Mol. Sci.* 20(10):2471.

[65] Porta, H. and Rocha-Sosa M. (2002). Plant lipoxygenases. Physiological and Molecular Features. *Plant Physiol* 130: 15–21.

[66] Kumar, J., Sen Gupta, D., Baum, M., Varshney, R. K. and Kumar S. (2021). Genomics-assisted lentil breeding: Current status and future strategies. *Legume Sci.* 3(3): e71.

[67] Sarker, R. H., Das, S. K., Shethi, K. J. and Hoque M. I. (2019). Genetic transformation. In *Lentils* (pp. 141–202). Cambridge, Massachusetts: Academic Press.

[68] Carrillo-Perdomo, E., Vidal, A., Kreplak, J., Duborjal, H., Leveugle, M., Duarte, J., Desmetz, C., Deulvot, C., Raffiot, B. and Marget P. (2020). Development of new genetic resources for faba bean (*Vicia faba* L.) breeding through the discovery of gene-based SNP markers and the construction of a high-density consensus map. *Sci. Reports* 10: 1–14.

[69] Braich, S., Sudheesh, S., Forster, J. W. and Kaur S. (2017). Characterisation of faba bean (*Vicia faba* L.) transcriptome using RNA-Seq: Sequencing, de novo assembly, annotation, and expression analysis. *Agronomy* 7:53.

[70] Chankaew, S., Somta, P., Sorajjapinun, W. and Srinives P. (2011). Quantitative trait loci mapping of Cercospora leaf spot resistance in mungbean, *Vigna radiata* (L.) Wilczek. *Mol. Breed.* 28:255–264.

[71] War, A. R., Murugesan, S., Naresh, B., Ramasamy, S. and Nair R. (2017). Mechanism of resistance in mungbean [*Vigna radiata* (L.) R. Wilczek var. radiata] to bruchids, *Callosobruchus* spp. (Coleoptera: Bruchidae). *Front. Plant Sci.* 8:1031.

[72] Kang, Y. J., Kim, S. K., Kim, M. Y., Lestari, P., Kim, K. H., Ha, B.-K., Jun, T. H., Hwang, W. J., Lee, T., Lee, J., Shim, S., Yoon, M. Y., Jang, Y. E., Han, K. S., Taeprayoon, P., Yoon, N., Somta, P., Tanya, P., Kim, K. S., et al. (2014). Genome sequence of mungbean and insights into evolution within Vigna species. *Nat. Commun.* 5:5443.

[73] Das, A., Datta, S., Thakur, S., Shukla, A., Ansari, J., Sujayanand, G., Chaturvedi, S. K., Kumar, P. and Singh N. (2017). Expression of a chimeric gene encoding insecticidal crystal protein Cry1Aabc of *Bacillus thuringiensisn* in chickpea (*Cicer arietinum* L.) confers resistance to gram pod borer (*Helicoverpa armigera* Hubner.). *Front. Plant Sci.* 8:1423.

[74] Bhowmik, S. S. D, Cheng, A. Y., Long, H., Tan, G. Z. H., Hoang, T. M. L., Karbaschi, M. R., Williams, B., Higgins, T. J. V. and Mundree S. G. (2019). Robust genetic transformation system to obtain non-chimeric transgenic chickpea. *Front. Plant Sci.* 10: 524.

[75] Badhan, S., Ball, A. S. and Mantri N.(2021). First Report of CRISPR/Cas9 Mediated DNA-Free Editing of *4CL* and *RVE7* Genes in Chickpea Protoplasts. *Int. J. Mol. Sci.* 22(1):396.

[76] Crystal, T., Morten, L., and Trine, A. K. (2021). Global Regulation of Genetically Modified Crops Amid the Gene Edited Crop Boom – A Review. *Front. Plant Sci.* 12:258.

[77] Waltz, Emily. 14 December 2021. https://www.nature.com/articles/d41587-021-00026-2.

[78] Thakur, S., Kushwah, N., Kumar, S., Singh, P., Rathore, M., Singh, N. P., and Das, A. (2021). Genome editing using CRISPR/Cas systems in legumes. In *Genome Editing in Plants* (pp. 89-107). CRC Press.

[79] Rasheed, A., Barqawi, A. A., Mahmood, A. et al. (2022) CRISPR/Cas9 is a powerful tool for precise genome editing of legume crops: a review. *Mol Biol Rep* 49:5595–5609

[80] Gayen, D., Karmakar, S. (2021). Designing, Performing, and Analyzing CRISPR-Cas9-Mediated Genome Editing Experiments in Leguminous Plants. In: Islam, M. T., Molla, K. A. (eds) *CRISPR-Cas Methods.* Springer Protocols Handbooks. Humana, New York, NY.

[81] Meng, Y., Wang, C., Yin, P., Zhu, B., Zhang, P., Niu, L. and Lin, H. (2019). Targeted mutagenesis by an optimized agrobacterium-delivered CRISPR/Cas9 system in the model legume *Medicago truncatula*. In The Model Legume Medicago truncatula, F. de Bruijn (Ed.).

Chapter 16

Genome Editing in Grain and Fodder Legumes: Current Status and Opportunities

Neha Sharma[1,*]
and Himani Sharma[2]
[1] CSK HPKV HAREC Bajaura, Kullu, India
[2] Dr. YS Parmar COH&F, Neri, Hamirpur, India

Abstract

Legumes play a major role in food and nutritional security. Their importance is well established owing to their ecological significance and healthy fodder and food choices. Breeding ameliorated cultivars has always been a challenge in Leguminaceae due to various crossability barriers genetically associated with plants in this family. Genome editing using various chemical and physical techniques has been a part of crop improvement experiments for the last three decades. Major breakthroughs have been achieved through the zinc finger nuclease (ZFN), TALEN, and CRISPR systems. Despite being a standout choice for targeted mutagenesis, the CRISPR/Cas9 system is still a poorly studied technology in legumes. Abiotic stress tolerance, biotic stress resistance, quality improvement and elimination of anti-nutritional factors can be easily achieved with the use of high throughput state of the art systems like CRISPR/Cas9 as compared to the traditional breeding approaches. This chapter critically elaborates comparison between various genome editing tools and their significant contributions to legume improvement and the reasons behind various lacunas.

Keywords: CRISPR, genome editing, legumes, precision breeding, targeted mutagenesis

Introduction

The Leguminaceae family is second only to the Gramineae in terms of its importance in agriculture. Being the third largest angiosperm family, it includes 751 genera of agroforestry, food grains, and fodder plants [1] accounts for the 27 percent of the world's primary crop

* Corresponding Author's Email: nsh9149@gmail.com.

In: Genome Editing for Crop Improvement
Editors: Rahul Kumar, Ashutosh Sharma, Yogesh Kumar et al.
ISBN: 979-8-88697-740-0

production [2]. Grain legumes are an important component of sustainable agriculture and play critical role in ensuring nutritional security [3]. Despite being ecologically significant and a healthy food and fodder choice, genetic improvement in grain legumes has been slow for decades. These crops need precise breeding programme to overcome their inherent low yielding nature and their high susceptibility to abiotic and biotic stresses, and poor edaphic conditions. Legumes have high genetic diversity. Wild species and wild relatives are substantial sources of biotic and abiotic stress tolerance and resistance genes. However, only limited genetic resources have been employed in the development of modern-day cultivars due to inherent crossability barriers associated with these genera.

Utilizing techniques like embryo rescue has not been proven to be of great success due to the low regeneration response of grain legumes to *in vitro* procedures, high genotypic specificity, and slow morphogenesis [4, 5]. The development of modern-day varieties from a limited number of parents has led to the narrowing of the genetic base of these crops, making them more susceptible to diseases, pest attacks, and the adverse effects of climate change. Many varieties of legumes have been developed through mutation breeding over the years; however, untargeted and random mutagenesis is a cumbersome process involving the screening of an immense population with a high degree of uncertainty in the results. Recalcitrant behavior of grain legumes towards *in vitro* approaches again acts as a hindrance to the doubled haploid breeding. Except in a few studies [6, 7, 8], doubled haploid breeding has not yielded any major breakthroughs.

Many advances in quality improvement [9], abiotic stress tolerance [10, 11, 12, 13, 14] and biotic stress resistance [15, 16, 17] have been achieved through genetic engineering. However, their acceptance is still debatable due to health, environmental, and legal issues. A shift from genetic engineering to gene editing is observed as the latter does not involve the integration of any foreign genes into the genetic makeup of the organism. It only edits one gene without affecting other members of the gene family. Gene editing also provides an advantage over traditional mutagenesis due to the specificity of its actions. The evolution of gene editing tools in the last three decades is discussed hereafter. This review focuses on the recent advances in genome editing with respect to legume improvement and as a precision breeding tool.

Genome Editing Tools

Since the advent of agriculture, humans have genetically modified plants to suit their requirements. The basic principle of crop improvement involves selection from the available variations. However, when the desirable variations are not available in nature, variation is created through various methods. The simplest of them all involve hybridization and mutation, whereas gene editing is part of a more complex and advanced process. Today, with the advances of science and technology we have reached a stage where we can precisely manipulate the genotypes, reducing the necessity to undergo extensive screening. It has consequently shortened the breeding cycle considerably.

Genome editing involves manipulation of genetic code, mRNA processing, and post-transcriptional processes, resulting in functional changes in the behavior of subsequent proteins (suppression, overexpression, or alteration). It uses engineered nucleases and DNA repair pathways to achieve the desired results. The first task is to identify the gene controlling the trait of interest [18], which is then modified. The plants are regenerated and screened for the targeted manipulation. There are various techniques used to manipulate the genome. With time, gene editing techniques have improved and their complexity has decreased, making them a more practical tool for precision breeding. Various methods of gene editing from the beginning to the present are discussed below.

Conventional Genome Editing Techniques

This technique includes single-stranded oligodeoxynucleotides (ssODNs) and conventional homologous recombination techniques. These are conventional methods and may not relate to emerging genome editing technologies. Homologous recombination is related to targeting genes to certain sites within the genome of the organism and involves gene intervention. The genetic information is exchanged between two identical molecules and is used to repair harmful breaks known as double-strand breaks (DSB), which are based on physiological processes. This technique is not frequently used today; however, some recent studies revealed that the RAD52 protein is important for homologous recombination mediation and is extensively used in cancer therapy with the help of the BRCA 1 and BRCA 2 repair pathways. Moreover, due to the advent of new and modified methods, this technique could not gain wide spread. Further, RNP complexes and all-in-one expression plasmids involve the concurrent delivery of two ssODNs and target, each derivative chromosome, which serves as a splint for guiding DSB ends. Thus, ssODNs promote the formation of translocation activity and have been used in combination with other tools like TALEN and CRISPR-Cas9 to achieve greater targeted gene editing in crops like *Arabidopsis* and flax [19], rather than being used in isolation. The major conventional tools used were as follows:

Chemical System for Genome Editing

The chemical system involves the use of peptide nucleic acid, an artificial molecule that is used as a DNA analogue in genome editing. It has a polypeptide backbone with nucleic acid bases attached as side chains. This method implements pseudo-complementary peptide nucleic acid (pcPNA), which specifies the cleavage sites within the chromosome or the telomeric region. Once the site is specified by pcPNA, excision takes place by a chemical mixture of cerium and EDTA, which leads to splicing [20], and the ligated DNA can be further attached with the enzyme DNA ligase within the spliced site. This technique can be carried out in high salt concentrations, which offers a great advantage. This technique has been extensively utilized in the clinical fields, but various issues like increased turnaround time and the manufacturing of site-specific pcPNA became a great hurdle.

Homing Endonuclease System

Homing endonucleases are a collection of endonucleases that are encoded as freestanding genes within introns or by self-splicing or fusion with host proteins. These are concerned with the hydrolysis of genomic DNA within the cells that synthesize it. “Homing” in homing endonuclease (HE) refers to the movement of these genes and is interpreted as lateral transmission of a genomic DNA fragment in which the site is removed by the endonucleases and results in the formation of two DNA segments [21]. These selfish genetic elements occur naturally with a size equivalent to 14 bp and leads to the splicing of larger DNA sequence [22]. The most efficient vehicles, like recombinant adeno-associated viruses (rAAVs), have been introduced, which allow them to be engineered into the cells. But the main drawback of this technology includes engineering difficulties for the nucleases’ preparation as well as vector development for entry into the cells [23]. Further, vector rAAVs also had off-target effects like less integration of DNA, a reduction in site specificity, and possible chances of host genome mutation [24]. Targeted mutagenesis has been achieved in maize through it [25], however, their DNA binding properties make them difficult to work with in research [26].

Protein-Based Nuclease System

Nucleases are the enzymes that degrade nucleic acids (DNases or RNases). Nucleases protein are incorporated for editing of DNA sequences. It includes meganucleases or molecular DNA scissors, which are large-base pair structures characterized by large reorganization sites that are sometimes found in the genome. These have the large potential to excise large pieces of DNA sequences and thus are being used as a genetic tool to modify DNA. The recognition sites are being modified to create nicks for changes in DNA sequence. Sometimes site-specific cleavage can be provided by the joining of meganucleases by proteins to create large variants like E-DreI and DmoCre [27]. The two main steps involved in this technique are the recognition of the cleavage site and the splicing out of the region by endonucleases. The main advantages of this enzyme are its low toxicity and its natural occurrence, which provide site-specific cleavage. Further, due to the advancement of new technologies far better and improved techniques have been developed, which has made them less popular.

Zinc finger nucleases (ZFNs) are generated by combinatorial approaches, and these artificial structures include restriction endonucleases that are joined with zinc finger domain binding proteins. The first targeted nucleases with widespread use were Zinc Finger Nucleases (ZFNs) [28, 29], which are basically fusions between the particularly designed cleavage domain of the FokI restriction endonuclease and the Cys2-His2-zinc-finger protein [30]. A specific half site sequence (9-18 bps DNA) along with each monomer is recognised by zinc finger nucleases (ZFNs), which function as dimers by zinc-finger DNA-binding. The FokI cleavage domain mediates the diamerization of the ZFN proteins by cleaving DNA within a 5-7 bp spacer sequence, which leads to the separation of two flanking zinc finger binding sites [31]. Three or four zinc finger domains is composed by each ZFN with each individual domain composed of 30 amino acid residues which are organized in a bba motif [32]. The DNA recognition is carried by the residues that are located within the a-helical domain and interact with 3bps of DNA, having an overlap from an adjacent domain [33]. Distinct DNA triplets by zinc finger domains are identified by using methods such as phage display [34, 35, 36]. Various selection-based methods have also been reported other than the modular assembly approach to zinc-finger construction [37], including context-dependent interactions between adjacent zinc-finger domains such as oligomerized pool engineering (OPEN) [38].

One major concern that is associated with the use of ZFNs for genome editing is the occurrence of off-target mutations [39, 31]. Thus, several approaches have been taken into consideration to enhance their specificity, like the creation of obligate heterodimeric ZFN architectures that rely on charge–charge repulsion in order to prevent unwanted homodimerization of the FokI cleavage domain [40], thus minimizing the dimerization of ZFNs at off-target sites. Further, protein engineering methods were used in order to enhance the cleavage efficiency of the FokI cleavage domain [25], and in order to increase their specificity, these were delivered into cells as proteins, which would facilitate genome editing with fewer off-target sites.

TALENs are bacterial effector proteins that resemble ZFNs in terms of mode of action and manufacturing. Unlike ZFNs, these are composed of an amino-terminal TALE DNA-binding domain, which is fused to a carboxy terminal Fok1 cleavage domain [41], which specifies its modular function and can lead to modification of any gene. The Fok1 cleavage domain is also associated with dimerization of the TALEN protein, which cuts within a 12-19 bp spacer sequence and separates each TALE binding site and enhances genome editing specificity [42]. It is based on the principle of binding restriction nucleases to a TAL effector, which is a DNA binding protein domain. The main advantage of TALENs is that it can target 1nt in one go, whereas ZFNs can address only 3nt which makes TALENs more site-specific with very few off-target effects. It has series of repeat domains with each 34 residues in length and it contacts DNA by amino acid residues at 12 and 13 positions respectively [43]. The Golden Gate assembly is the best-known approach for the construction of TALE DNA-binding domains. Moreover, various other high-throughput TALE assembly methods, like capped assembly, FLASH assembly, and ligation-independent cloning, have also been developed.

Thus, TALENs offer many more advantages for genome editing than ZFNs, as no direct evolution or selection is required to engineer TALE arrays. Further, it has reduced toxicity and also shows improved specificity, as it has increased affinity for target DNA [44]. TALENs have high repetitive structures which makes them highly efficient than ZFNs to deliver into the cells by using single adeno-associated virus (AAV) and lentivirus [45].

RNA/DNA Systems

The CRISPR-Cas9 system, which plays an important role in adaptive immunity in bacteria [46], is the most important and recent addition to genome editing systems. Two components are present in CRISPR itself, which are SPR (spacers) having varying and differing nucleotide sequences, and each one of them represents past exposure to a foreign antigen, and former is the CRI, which represents the bacterial genetic memory, and once encountered with a similar foreign antigen it can be reactivated.

Basically, in bacteria, protection is provided against DNA from invading viruses and plasmids by the type-II CRISPR system by RNA-guided DNA cleavage by Cas proteins [47]. Foreign DNA present in the form of short segments is integrated within the CRISPR locus and then transcribed into CRISPR RNA (crRNA), which is then annealed to trans-activating crRNA (tracrRNA) which further leads to sequence specific degradation of pathogenic DNA by the Cas9 protein [48]. Further target recognition by the Cas9 protein requires only a particular sequence within the crRNA and the protospacer adjacent motif (PAM), which is located upstream of the crRNA binding site [48]. Thus, in order to carry out efficient genome editing, the system is a simplified, consisting of only Cas9 nuclease and single guide RNA (gRNA), which contains the essential crRNA and tracrRNA elements. Target site recognition is the most flexible and user-friendly approach because it is mediated by the gRNA, thus eliminating the need to engineer new proteins to recognize new target sites. The Cas9 nuclease recognizes PAM motif, which is essential for DNA cleavage and is located downstream of the gRNA target site.

Moreover, structurally, DNA recognition by Cas9 forms heteroduplex gRNA, and the complementary strand of DNA is housed in a positively charged groove between two nuclease domains (RuvC and HNH) within the Cas9 protein, and PAM is recognized by an arginine-rich motif present in Cas9 [49]. So as to enable Cas9-mediated cleavage of both DNA strands, structural displacement is carried out via DNA strand displacement within the Cas9 protein, which directs the non-target DNA strand into the RuvC active site, positioning the HNH domain near target DNA [50]. In order to carry out genome editing, the Cas9 nucleases and its gRNA have to be delivered to cells using the same or different plasmids, and for this purpose several resources have been developed, such as E-CRISP, etc. [51]. Further, in order to improve its specificity, various efforts are being made, such as using paired nickases, which tend to hamper gene editing specificity by requiring the induction of two sequential and adjacent nicking events for DSB formation. Moreover, off target cleavage can also be reduced by Cas9 variants enabling conditional genome editing, such as rapamycin inducible split Cas9 architecture [52] or Cas9 variants containing strategically placed in the domains [53].

Further, transient transfection of Cas9 ribonucleoprotein complex also had shown reduced off target effects which enable DNA free gene editing in plants [54], and also by fusion of Cas9 to DNA-binding domains improvement in specificity was monitored [55]. Protein engineering can broadly enhance Cas9 specificity [56] and also alter the requirements of PAM, which creates customized variants of Cas9 for allele specific genome editing, but instead various unique specificities of PAM have been uncovered in nature [52].

The CRISPR/Cas technologies by various experimentation have divided CRISPR-Cas into 3 distinct types:

- Type 1 CRISPR/Cas system
- Type 2 CRISPR/Cas system
- Type 3 CRISPR/Cas system

The type 1 CRISPR/Cas system is concerned with the utilization of Cas5 or Cas6 for crRNA pre-processing, and for further cleavage function, Cas3, cascade, and crRNA are needed for interference. While in Type 2 CRISPR/Cas system, although Cas9 functions under crRNA guidance to target the DNA, but other three factors are involved in trimming at the 5′ end which includes RNase III, trans activating RNA (tracrRNA) and a yet-to-be-identified protein factor. Further, type 3 CRISPR/Cas system, unlike the type 1 system, uses Cas6 for crRNA processing at 3′ end trimming. This technique targets RNA with a unique complex called the type III Csm/Cmr complex.

Gene-Silencing Techniques

This method usually does not fall under genome editing techniques, but it is still capable of modifying a particular DNA sequence. The technologies included in this category are RNAi, CRISPR interference (CRISPRi) and various morpholino oligonucleotide techniques.

Advancements in Legume Breeding

The potential of genome editing technology as a tool for precision breeding is yet to be harnessed in leguminous crops as compared to cereals. Very few studies were reported where these tools have been utilized for crop improvement. ZFNs and TALENs have made no significant contribution as these systems depend upon DNA protein interactions, complex design, and assembly. Soybean genome editing is still a challenge due to its unresponsiveness to *Agrobacterium* mediated transformation and its limited gene redundancy as a diploid species evolved from a palaeotetraploid species [57]. Despite that, there are many reports of quality improvement alone in soybeans, showing that it can be explored intensively for its improvement.

Soybean is the only legume crop with transgenic cultivars (which predominantly show resistance to herbicides), and the most reports on CRISPR. Also, the gene edited product of soybean using TALEN, has also reached the markets as oil with high oleic acid and reduced saturated fat. The trans-fatty acids produced through hydrogenation pose serious health problems in soybean oil. Soybean lines with low polysaturated fats have been developed through TALENs through mutagenesis in FAD2-1D and FAD2-1B, showing increased oleic acid content and decreased linoleic acid content [58]. Modulation in oleic acid and linoleic acid content in soybean has been achieved through site directed mutagenesis of FAD2-2 microsomal omega-6 desaturase in soybean via RNA guided CRISPR-Cas 9 [59]. This technique has been efficiently utilized in increasing the shelf life (enhancement of oxidative stability) and frying stability of soybean oil [60]. High levels of polyunsaturated linoleic and linolenic acids lead to high oxidative instability. This can be managed by increasing the oleic acid content and decreasing the linoleic and linolenic acid content by creating knockout mutations within the FAD2-1A, FAD2-1B, and FAD3A genes.

CRISPR from Prevoltella and Francisella1 (CRISPR-Cpf1), similar to CRISPR Cas, has also been reported to effectively cause mutations (mainly nucleotide deletions) in the targeted site of soybean DNA [61]. *Streptococcus pyrogenes* Cas9 guided RNA (gRNA) has also been used for genome editing in soybean [62], where two genomic sites on chromosome 4 were mutagenized further, successfully creating acetolactate synthase 1 gene mutations. Target specificity in CRISPR can be easily achieved by changing single-guided RNA, which pairs with target DNA and is also low-cost as compared to ZFNs and TALENs. This system is more successful as compared to TALENs and ZFNs and facilitates the development of stable non-transgenic multiplex plant genome editing through a much simpler methodology [63]. RNAi has been used to induce resistance to various diseases, parasitic nematodes and insects in chickpea [64, 65], and abiotic stress and in pigeon peas [66].

Lotus japonicus is a wild legume widely used as a model system to study legume specific phenomena. CRISPR/Cas9 was used to modify symbiotic nitrogen fixation (SFN)-related genes in this crop via *Agrobacterium*-mediated stable or hairy root transformation [67]. A sgRNA was developed to target leghemoglobin loci (LjLb1, LjLb2, and LjLb3) and found competent to target up to triple gene lockouts, establishing its higher efficiency in causing multiple knockouts. An optimized *Agrobacterium* derived CRISPR/Cas 9 system is not available to modify *M. truncatula* [68]. Successful generation of haploids by selective mutagenesis of ZmDMP like, domains in *M. truncatula* MtDMP8 and MtDMP9 [69] is a substantial basis to conclude the effectiveness of these tools to precisely achieve the targeted results. A high rate of mutagenic efficiency has already been achieved in commercial legumes like *V. unguiculata,* targeting the SNF genes, showing the efficiency of the CRISPR-Cas system in manipulating the agronomic traits [70, 71].

Badhan et al. [72] edited two drought tolerance-related genes in chickpea *4-coumarate ligase* (4CL) and Reveille 7 (RVE7) using a DNA-free CRISPR/Cas9 approach and found that although the efficiency of editing through this approach was high, it was different in both the genes. Micro RNAs editing, regulating biological processes is successful in rice [73], however, yet to be implemented in legumes. CRISPR/CAS technology has been used to boost the immune systems of plants [74, 75, 76, 77, 78]. Earlier in *Nicotiana benthamiana*, artificial zinc nucleases were successfully used to target the pathogenic viruses using a conserved sequence motif in Geminivirus [79], but no such study has been conducted in legumes.

Table 1. List of genes edited or database development for functional genetics study and legume trait improvement using genome editing technology

Crop	Advancements/ gene or targeted location	Reference
Medicago truncatula	*MtHen1* gene, Phytoene desaturase and symbiosis receptor like kinase, Online web tool for fast identification of CRISPR/Cas9 target loci within soybean gene models, and generic DNA sequences, *Phytoene desaturase* (PDS) genes	[80, 68, 89, 82]
Lotus japonicus	Leghemoglobin 1/2/3, symbiosis receptor like kinase (LjSYMRK) and Lotus histidine kinase 1 (LHK1)- interacting protein (LjCZF1-2) regulating nodule organogenesis Representative SNF (symbiotic nitrogen fixation) related genes	[83, 67, 84]
Glycine max	Phytoene desaturase, glutamine sunthase, chalcone flavanone isomerase, glucosyltransferase, DDM1 orthologs, MET1 orthologs, dicer-like3a, double-stranded RNA-binding 2a/2b, fatty acid desaturase 2 and Squamosa promoter binder protein like GMFT2a	[59, 57, 80, 85, 86, 62, 81, 87, 88]
Cajanus cajan	CsSTOP1 characterization for Al tolerance	[90, 91]
Cicer arietinum	4-coumarate ligase (4CL) and Reveille 7 (RVE7)	[72]

Future Prospects and Conclusion

Unlike traditional mutagenesis techniques, genome editing tools like CRISPR-Cas9 target specific genes and are capable of mutating multiple loci. Controversies around crop improvement involving genetic modification have made genome editing tools a popular choice for plant breeding. These new plant breeding techniques are a faster way to develop stable, precisely modified plant genomes [92]. Although the plants developed lack the transgenic inserts and should therefore be exempt from GMO regulations, acceptance and regulation of such genetically edited crops still pose challenges and are currently under debate [92, 93]. In legumes, not much experimentation has been done till now, but the scope of genome editing in pulses is wide, keeping in mind their role in nutritional security and maintaining soil health. Like other crops' abiotic stress tolerance and biotic stress resistance, quality improvement and the elimination of non-nutritional factors can be easily achieved with the use of high-throughput state-of-the-art systems like CRISPR/Cas9 as compared to the traditional breeding approaches.

References

[1] Lewis, G.P. 2005. *Legumes of the World.* Richmond, UK: Royal Botanic Gardens, Kew.

[2] Vance, C.P., Graham, P.H. and Allan, D.L. (2000). Biological nitrogen fixation. Phosphorus: a critical future need. In FO Pedrosa, M Hungria, MG Yates, WE Newton, eds, *Nitrogen Fixation: From Molecules to Crop Productivity.* Kluwer Academic Publishers, Dordrecht, The Netherlands, 506-514.

[3] Polhill, R.M., Raven, P.H. and Stirton, C.H. (1981). Evolution and systematic of the Leguminosae. In RM Polhill, PH Raven, eds, *Advances in Legume Systematics Part 1.* Royal Botanic Gardens, Kew, UK, 1-26.

[4] Somers, D.A., Samac, D.A. and Olhoft, P.M. (2003). Recent advances in legume transformation. *Plant Physiology*, 131(3):892-0899.

[5] Pratap, A., Choudhary, A.K. and Kumar, J. (2010). *In vitro* techniques towards genetic enhancement of food legumes-A review. *Journal of Food Legumes*, 23: 169-185.

[6] Croser, J., Lulsdorf, M., Davies, P., Wilson, J., Sidhu, P., Grewal, R. and Sid-dique, K. (2005). Haploid embryogenesis from chickpea and fieldpea–progress towards a routine protocol. *Proceedings of the Australian branch of the IAPTC&B*, 71-82.

[7] Grewal, R.K., Lulsdorf, M. and Croser, J. (2009). Doubled-haploid production in chickpea (*Cicer arietinum* L.): role of stress treatments. *Plant Cell Reports* 28:1289-1299.

[8] Ochatt, S., Pech, C., Grewal, R., Conreux, C., Lulsdorf, M. and Jacas, L. (2009). Abiotic stress enhances androgenesis from isolated micro-spores of some legume species (Fabaceae). *Journal of Plant Physiology*, 166(12): 1314-1328.

[9] Thu, T.T., Dewaele, E., Claeys, M., Jacobs, M. and Angenon, G. (2007). Increasing lysine levels in pigeonpea (Cajanus cajan(L.) Millsp) seedsthrough genetic engineering. *Plant Cell, Tissue and Organ Culture*, 91(2), 135-143.

[10] Surekha, C.H., Kumari, K.N., Aruna, L.V., Suneetha, G., Arundhati, A. and Kishor, P.K. (2014). Expression of theVigna aconitifoliaP5CSF129Agene in transgenic pigeonpea enhances proline accumulation and salttolerance. *Plant Cell, Tissue and Organ Culture (PCTOC)*, 116 (1): 27-36.

[11] Do, E., Park, M., Hu, G., Caza, M., Kronstad, J.W. and Jung, W.H. (2016). The lysine biosynthetic enzyme Lys4 influences iron metabolism, mitochondrial function and virulence inCryptococcus neoformans. *Bio-chemical and Biophysical Research Communications*, 477(4): 706-711.

[12] Anbazhagan, K., Bhatnagar-Mathur, P., Vadez, V., Dumbala, S.R., Kishor, P.K. and Sharma, K.K. (2015). DREB1A overexpression in transgenicchickpea alters key traits influencing plant water budget across water regimes. *Plant Cell Reports*, 34(2), 199-210.

[13] Mishra, S., Behura, R., Awasthi, J.P., Dey, M., Sahoo, D., Bhowmik, S.S. and Sahoo, L. (2014). Ectopic overexpression of a mungbean vacuolarNa+/H+antiporter gene (VrNHX1) leads to increased salinity stresstolerance in transgenicVigna unguiculataL. walp. *Molecular Breeding*, 34: 1345-1359.

[14] Baloda, A. and Madanpotra, S. (2017). Transformation of Mungbean Plantsfor Salt and Drought Tolerance by introducing a gene for an osmo-protectant glycine betaine. *Journal of Plant Stress Physiology*, 3:5–11.

[15] Ghosh, G., Ganguly, S., Purohit, A., Chaudhuri, R.K., Das, S. and Chak-raborti, D. (2017). Transgenic pigeonpea events expressing Cry1Acand Cry2Aa exhibit resistance to *Helicoverpa armigera*. *Plant Cell Reports*, 36(7):1037-1051.

[16] Das, A., Datta, S., Thakur, S., Shukla, A., Ansari, J., Sujayanand, G.K. and Singh, N.P. (2017). Expression of a chimeric gene encoding insectici-dal crystal protein Cry1Aabc of *Bacillus thuringiensis* in Chickpea (*Cicer arietinumL.*) confers resistance to gram pod borer (*Helicoverpa armigera Hubner.*). *Frontiers in Plant Science*, 8:1423.

[17] Broglie, K.E. and Gregory, R.J. (2016). Identification of P. pachyrhizipro-tein effectors and their use in producing asian soybean rust (asr) resistant plants. U.S. Patent 9(416), 368.

[18] Garcia-Ruiz, H. (2018). Susceptibility genes to plant viruses. *Viruses*, 10:484.

[19] Sauer, N.J., Narvaez-Vasquez, J., Mozoruk, J., Miller, R.B., Warburg, Z.J., Woodward, M.J., Mihiret, Y.A., Lincoln, T.A., Segami, R.E., Sanders, S.L., Walker, K.A., Beetham, P.R., Schopke, C.R. and Gocal, G.F. (2016). Oligonucleotide-Mediated Genome Editing Provides Precision and Function to Engineered Nucleases and Antibiotics in Plants. *Plant Physiology*, 170(4):1917-28.

[20] Komiyama, M. (2014). Chemical modifications of artificial restriction DNA cutter (ARCUT) to promote its *in vivo* and *in vitro* applications. *Artiicial DNA:PNA & XNA*, 5:111-2457.

[21] Schultz, B.R. and Chamberlain, J.S. (2008). Recombinant adeno-associated virus transduction and integration. *Molecular Therapy*, 16: 1189-1199.

[22] Smith, J., Grizot, S., Arnould, S., Duclert, A., Epinat, J.C., Chames, P., Prieto, J., Redondo, P., Blanco, F.J. and Bravo, J. (2006). A combinatorial approach to create artificial homing endonucleases cleaving chosen sequences. *Nucleic Acids Research*, 34:149.

[23] Rey-Rico, A. and Cucchiarini, M. (2016) Controlled release strategies for rAAV-mediated gene delivery. *Acta Biomaterialia*, 29: 1-10.

[24] Jin, L., Lange, W., Kempmann, A., Maybeck, V., Günther, A., Gruteser, N., Baumann, A., Offenhausser, A. (2016). High-efficiency transduction and specific expression of ChR2opt for optogenetic manipulation of primary cortical neurons mediated by recombinant adeno-associated viruses. *Journal of Biotechnology*, 233: 171-180.

[25] Gao, H., Smith, J., Yang, M., Jones, S., Djukanovic, V., Nicholson, M.G., West, A., Bidney, D., Falco, S.C. and Jantz, D. (2010). Heritable targeted mutagenesis in maize using a designed endonuclease. *The Plant Journal*, 61(1):176–87.

[26] Malzahn, A., Lowder, L. and Qi, Y. (2017). Plant genome editing with TALEN and CRISPR. *Cell & Bioscience*, 7: 21.

[27] Zaslavskiy, M., Bertonati, C., Duchateau, P., Duclert, A. and Silva, G.H. (2014). Efficient design of meganucleases using a machine learning approach. *BMC Bioinformatics*, 15: 191-201.

[28] Porteus, M.H. and Carroll, D. (2005). Gene targeting using zinc finger nucleases. *Nature Biotechnology*, 23: 967-973.

[29] Urnov, F.D., Rebar, E.J., Holmes, M.C., Zhang, H.S. and Gregory, P.D. (2010). Genome editing with engineered zinc finger nucleases. *Nature Review Genetics*, 11: 636-646.

[30] Kim, Y.G., Cha, J. and Chandrasegaran, S. (1996). Hybrid restriction enzymes: Zinc finger fusions to Fok I cleavage domain. *Proceeding of the National Academy of Sciences*, 93: 1156-1160.

[31] Smith, J., Bibikova, M., Whitby, F.G., Reddy, A.R., Chandrasegaran, S. and Carroll, D. (2000). Requirements for double-strand cleavage by chimeric restriction enzymes with zinc fingerDNA-recognition domains. *Nucleic Acids Research*, 28: 3361-3369.

[32] Pavletich, N.P. and Pabo, C.O. (1991). Zinc finger-DNA recognition: Crystal structure of a Zif268-DNA complex at 2.1 A. *Science*, 252: 809-817.

[33] Wolfe, S.A., Nekludova, L. and Pabo, C.O. (2000). DNA recognition by Cys2His2 zinc finger proteins. *Annual Review of Biophysics and Biomolecular Structure*, 29:183-212.

[34] Choo, Y. and Klug, A. (1994). Toward a code for the interactions of zinc fingers with DNA: Selection of randomized fingers displayed on phage. *Proceedings of the National Academy of Sciences*, 91:11163-11167.

[35] Jamieson, A.C., Kim, S.H. and Wells, J.A. (1994). *In vitro* selection of zinc fingers with altered DNA-binding specificity. *Biochemistry*, 33: 5689-5695.

[36] Wu, H., Yang, W.P. and Barbas, C.F. (1995). Building zinc fingers by selection: Toward a therapeutic application. *Proceeding of the National Academy of Sciences*, 92: 344-348.

[37] Greisman, H.A. and Pabo, C.O. (1997). A general strategy for selecting high-affinity zinc finger proteins for diverse DNA target sites. *Science*, 275: 657-661.

[38] Maeder, M.L., Thibodeau-Beganny, S., Osiak, A., Wright, D.A., Anthony, R.M., Eichtinger, M., Jiang, T., Foley, J.E., Winfrey, R.J. and Townsend, J.A. (2008). Rapid "open-source" engineering of customized zinc-finger nucleases for highly efficient gene modification. *Molecular Cell*, 31: 294-301.

[39] Gabriel, R., Lombardo, A., Arens, A., Miller, J.C., Genovese, P., Kaeppel, C., Nowrouzi, A., Bartholomae, C.C., Wang, J. and Friedman, G. (2011). An unbiased genome-wide analysis of zinc-finger nuclease specificity. *Nature Biotechnology*, 29:816-823.

[40] Miller, J.C., Holmes, M.C., Wang, J., Guschin, D.Y., Lee, Y.L., Rupniewski, I., Beausejour, C.M., Waite, A.J., Wang, N.S. and Kim, K.A. (2007). An improved zinc-finger nuclease architecture for highly specific genome editing. *Nature Biotechnology*, 25: 778-785.

[41] Christian, M., Cermak, T., Doyle, E.L., Schmidt, C., Zhang, F., Hummel, A., Bogdanove, A.J. and Voytas, D.F. (2010). Targeting DNA double-strand breaks with TAL effector nucleases. *Genetics*, 186: 757-761.

[42] Miller, J.C., Tan, S., Qiao, G., Barlow, K.A., Wang, J., Xia, D.F., Meng, X., Paschon, D.E., Leung, E. and Hinkley, S.J. (2011). ATALE nuclease architecture for efficient genome editing. *Nature Biotechnology*, 29: 143-148.

[43] Boch, J., Scholze, H., Schornack, S., Landgraf, A., Hahn, S., Kay, S., Lahaye, T., Nickstadt, A. and Bonas, U. (2009). Breaking the code of DNA binding specificity of TAL-type III effectors. *Science*, 326:1509-1512.

[44] Mussolino, C., Alzubi, J., Fine, E.J., Morbitzer, R., Cradick, T.J., Lahaye, T., Bao, G. and Cathomen, T. (2014). TALENs facilitate targeted genome editing in human cells with high specificity and low cytotoxicity. *Nucleic Acids Research*, 42: 6762-6773

[45] Holkers, M., Maggio, I., Liu, J., Janssen, J.M., Miselli, F., Mussolino, C., Recchia, A., Cathomen, T. and Goncalves, M.A. (2013). Differential integrity of TALE nuclease genes following adenoviral and lentiviral vector gene transfer into human cells. *Nucleic Acids Research*, 41:63.

[46] Horvath, P. and Barrangou, R. (2010). CRISPR/Cas, the immune system of bacteria and archaea. *Science*, 327: 167-170.

[47] Wiedenheft, B., Sternberg, S.H. and Doudna, J.A. (2012). RNA-guided genetic silencing systems in bacteria and archaea. *Nature*, 482: 331-338.

[48] Jinek, M., Chylinski, K., Fonfara, I., Hauer, M., Doudna, J.A. and Charpentier, E. (2012). A programmable dual-RNA-guided DNA endonuclease in adaptive bacterial immunity. *Science*, 337: 816-821.

[49] Anders, C., Niewoehner, O., Duerst, A. and Jinek, M. (2014). Structural basis of PAM-dependent target DNA recognition by the Cas9 endonuclease. *Nature*, 513:569-573.

[50] Jiang, F., Taylor, D.W., Chen, J.S., Kornfeld, J.E., Zhou, K., Thompson, A.J., Nogales, E. and Doudna, J.A. (2016). Structures of a CRISPR-Cas9 R-loop complex primed for DNA cleavage. *Science*, 351: 867-871.

[51] Heigwer, F., Kerr, G. and Boutros, M. (2014). E-CRISP: Fast CRISPR target site identification. *Nature Methods*, 11: 122-123.

[52] Zetsche, B., Gootenberg, J.S., Abudayyeh, O.O., Slaymaker, I.M., Makarova, K.S., Essletzbichler, P., Volz, S.E., Joung, J., van der Oost, J. and Regev, A. (2015a). Cpf1 is a single RNA-guided endonuclease of a class 2 CRISPR-Cas system. *Cell*, 163: 759-771.

[53] Davis, K.M., Pattanayak, V., Thompson, D.B., Zuris, J.A. and Liu, D.R. (2015). Small molecule-triggered Cas9 protein with improved genome-editing specificity. *Nature Chemical Biology*, 11: 316-318.

[54] Zuris, J.A., Thompson, D.B., Shu, Y., Guilinger, J.P., Bessen, J.L., Hu, J.H., Maeder, M.L., Joung, J.K., Chen, Z.Y. and Liu, D.R. (2015). Cationic lipid-mediated delivery of proteins enables efficient protein-based genome editing *in vitro* and *in vivo*. *Nature Biotechnology*, 33: 73-80.

[55] Bolukbasi, M.F., Gupta, A., Oikemus, S., Derr, A.G., Garber, M., Brodsky, M.H., Zhu, L.J. and Wolfe, S.A. (2015). DNA-binding domain fusions enhance the targeting range and precision of Cas9. *Nature Methods*, 12:1150-1156.

[56] Kleinstiver, B.P., Pattanayak, V., Prew, M.S., Tsai, S.Q., Nguyen, N.T., Zheng, Z. and Joung, J.K.(2016). High-fidelity CRISPR-Cas9 nucleases with no detectable genome-wide off-target effects. *Nature*, 529: 490-495.

[57] Bao, A., Zhang, C., Huang, Y., Chen, H., Zhou, Z. and Dong, C. (2020). Genome editing technology and application in soybean improvement. *Oil Crop Science*, 5: 31-40.

[58] Zhongsen, Li., Zhan-Bin, L., Aiqiu, X., Bryan, P., Moon., Jessica, P., Koellhoffer., Lingxia Huang, R., Timothy, W., Elizabeth, C., Carl, F. and Mark, Cigan. (2015). Cas9-Guide RNA Directed Genome Editing in Soybean. *Plant Physiology*, 169(2): 960-970.

[59] Amin, al N., Ahmad, N., Wu, N. et al. (2019). CRISPR-Cas9 mediated targeted disruption of FAD2–2 microsomal omega-6 desaturase in soybean (Glycine max.L). *BMC Biotechnology*, 19 (9). https://doi.org/10.1186/s12896-019-0501-2.

[60] Demorest, Z.L., Coffman, A., Baltes, N.J., Stoddard, T.J., Clasen, B.M., Luo, S., Retterath, A., Yabandith, A., Gamo, M.E. and Bissen, J. (2016). Direct stacking of sequence-specific nuclease-induced mutations to produce high oleic and low linolenic soybean oil. *BMC Plant Biology*, 16(1):225.

[61] Kim, H., Kim, S.T., Ryu, J., Kang, B.C, Kim, J.S. and Kim, S.G. (2017). CRISPR/Cpf1-mediated DNA free plant genome editing. *Nature Communications*. 8:14406.

[62] Li, Z., Liu, Z.B., Xing, A., Moon, B.P., Koellhoffer, J.P., Huang, L., et al. (2015). Cas9-guide RNA directed genome editing in soybean. *Plant Physiology*. 169: 960–970. doi:10.1104/pp/15.00783.

[63] Khatodia, S., Bhatotia, K., Passricha, N., Khurana, S.M.P., Tuteja, N. (2016). The CRISPR/Cas Genome-Editing Tool: Application in Improvement of Crops. *Frontiers in Plant Science*, 7: 506.

[64] Nahid, N., Amin, I., Briddon, R.W. and Mansoor, S. (2011). RNA interference-based resistance against a legume mastrevirus. *Virology Journal*, 8:499.

[65] Khatodıa, S. (2017). Helicoverpa Resistant Chickpea Plants: From Bt Toxins to Plant-Mediated RNAi . *Ekin Journal of Crop Breeding and Genetics*, 3 (1): 52-60.

[66] Meng, D., Dong, B., Niu, L., Song, Z., Wang, L., Amin, R., Cao, H., Li, H., Yang, Q. and Fu, Y. (2021). The pigeonpea CcCIPK14-CcCBL1 pair positively moducles drought tolerance by enhancing flavonoid biosynthesis. *The plant Journal*, 1278-1297.

[67] Wang, L., Tan, Q., Fan, Q., Zhu, Hong Z. and Zhang, Z. (2016). Efficient inactivation of Symbiotic Nitrogen Fixation related genes in Lotus japonicas using CRISPR-Cas 9. *Frontiers in Plant Science*, 7: 1333.

[68] Meng, Y., Hou, Y., Wang, H., Ji, R., Liu, B., Wen, J., Niu, L. and Lin, H. (2017). Targeted mutagenesis by CRISPR/Cas9 system in the model legume *Medicago truncatula*. *Plant Cell Reports*. 36(2):371-374.

[69] Wang, N., Xia, Z., Jiang, T., Li, L., Zhang, P., Niu, L., Cheng, H., Wang, K. and Lin, H. (2021). *In planta* haploid induction by genome editing of DMA in the model legume *Medicago truncatula*. *Plant Biotechnology Journal*. 20(1):22-24.

[70] Ji. J., Zhang, C., Sun, Z., Wang, L., Duanmu, D. and Fan, Q. (2019). Genome Editing in Cowpea *Vigna unguiculata* Using CRISPR-Cas9. *International Journal of Molecular Sciences*, 20(10):2471.

[71] Uranić, M., Nagahatenna, D., Salinas-Gamboa, R., Hand, M.L., Sánchez-León, N., Leong, W.H., How, T., Bazanova, N., Spriggs, A., Vielle-Calzada, J.P., & Koltunow, A. (2020). A detached leaf assay for testing transient gene expression and gene editing in cowpea (*Vigna unguiculata* [L.] Walp.). *Plant methods*. 16: 88. https://doi.org/10.1186/s13007-020-00630-4.

[72] Badhan, S., Ball, A.S., Mantri, N. (2021). First Report of CRISPR/Cas9 Mediated DNA-Free Editing of 4CL and RVE7 Genes in Chickpea Protoplasts. *International Journal of Molecular Sciences*, 22(1):396.

[73] Mangrauthia, S.K., Maliha, A. and Prathi, N.B. (2017). MicroRNAs: potential target for genome editing in plants for traits improvement. *Indian Journal of Plant Physiology*, 22: 530-548.

[74] Ali, Z., Abulfaraj, A., Idris, A., Ali, S., Tashkandi, M. and Mahfouz, M.M. (2015). CRISPR/Cas9-mediated viral interference in plants. *Genome Biology*, 16:238.

[75] Ji, X., Zhang, H., Zhang, Y., Wang, Y., Gao, C. (2015) Establishing a CRISPR-Cas-like immune system conferring DNA virus resistance in plants. *Nature Plants*, 1:15144.

[76] Baltes, N.J., Hummel, A.W., Konecna, E., Cegan, R., Bruns, A.N. and Bisaro, D.M. (2015). Conferring resistance to geminiviruses with the CRISPR-Cas prokaryotic immune system. *Nature Plants*, 1:15-145.

[77] Garcia, Chaparro A., Kamoun, S. & Nekrasov, V. (2015). Boosting plant immunity with CRISPR/Cas. *Genome Biology*, 16: 254. https://doi.org/10.1186/s13059-015-0829-4.

[78] Leonetti, P., Paolo, A.G., Hanafy, M.S. and Pantaleo, V. (2018). Viruses and Phytoparasitic Nematodes of *Cicer arietinum* L.: Biotechnological Approaches in Interaction Studies and for Sustainable Control. *Frontiers in Plant Science*, 9: 319.

[79] Chen, W., Qian, Y., Wu, X., Sun, Y., Wu, X. and Cheng, X. (2014). Inhibiting replication of begomoviruses using artificial zinc finger nucleases that target viral-conserved nucleotide motif. *Virus Genes*, 48:494-501.

[80] Curtin, S.J., Xiong, Y., Michno, J.M., Campbell, B.W., Stec, A.O., Cermak, T., Starker, C., Voytas, D.F., Eamens, A.L., Stupar, R.M. (2018). CRISPR/Cas9 and TALENs generate heritable mutations for genes involved in small RNA processing of Glycine max and *Medicago truncatula*. *Plant Biotechnology Journal*, 6:1125-1137.

[81] Michno, J.M., Virdi, K., Stec, A.O., Liu, J., Wang, X., Xiong, Y., et al. (2020). Integration, abundance, and transmission of mutations and transgenes in aseries of CRISPR/Cas9 soybean lines. *BMC Biotechnology*. 20:10. doi:10.1186/s12896-020-00604-3.

[82] Wolabu, T.W., Park, J.J., Chen, M. et al. (2020). Improving the genome editing efficiency of CRISPR/Cas9 in *Arabidopsis* and *Medicago truncatula*. *Planta* 252, 15. https://doi.org/10.1007/s004 25-020-03415-0.

[83] Yin, J., Guan, X., Zhang, H. et al. (2019). An MAP kinase interacts with LHK1 and regulates nodule organogenesis in Lotus japonicus. *Science China Life Sciences*. 62, 1203–1217. https://doi.org/10.100 7/s11427-018-9444-9.

[84] Wang, L., Maria, Carmen R., Xian, X., Baoli, Z., Qiuling, F., Qiang, W., Guogui, N., Manuel, B. and Deqiang, D. (2019). CRISPR/Cas9 knockout of leghemoglobin genes in Lotus japonicus uncovers their synergistic roles in symbiotic nitrogen fixation. *New Phytologist*. 224(2): 818-832. https://doi.org/ 10.1111/nph.16077.

[85] Du, H., Zeng, X., Zhao, M., Cui, X., Wang, Q., Yang, H., Cheng, H. and Yu, D. (2016). Efficient targeted mutagenesis in soybean by TALENs and CRISPR/Cas9. *Journal of Biotechnology*, 217:90-7.

[86] Jacobs, T.B., LaFayette, P.R., Schmitz, R.J., and Parrot, W.A. (2015). Targeted genome modifications in soybean with CRISPR/Cas9. *BMC Biotechnology*. 15:16. Doi: 10.1186/s12896-015-0131-2.

[87] Sun, X., Hu, Z., Chen, R., Jiang, Q., Song, G., Zhang, H., et al. (2015). Targetedmutagenesis in soybean using the CRISPR-Cas9 system. *Science Reports* 5:10342. https://doi.org/10.1038/srep10342.

[88] Cai, Y., Chen, L., Liu, X., Guo, C., Sun, S., Wu, C., Jiang, B., Han, T. and Hou, W. (2018). CRISPR/Cas9-mediated targeted mutagenesis of GmFT2a delays flowering time in soya bean. *Plant Biotechnology Journal*, 1:176-185.

[89] Michno, J.M., Wang, X., Liu, J., Curtin, S.J., Kono, T.J., & Stupar, R.M. (2015). CRISPR/Cas mutagenesis of soybean and *Medicago truncatula* using a new web-tool and a modified Cas9 enzyme. *GM Crops & Food*. 6(4), 243–252. https://doi.org/10.1080/21645698.2015.1106063.

[90] Daspute, A.A., Kobayashi, Y., Panda, S.K., Fakrudin, B., Kobayashi, Y., Tokizawa, M., Iuchi. S., Choudhary, A.K., Yamamoto, Y. and Koyama, H. (2018). Characterization of CcSTOP1; a C2H2-type transcription factor regulates Al tolerance gene in pigeon pea. *Planta*, 247:201-214.

[91] Che, P., Chang, S., Simon, Marissa K., Zhang, Z., et al. (2021). Developing a rapid and highly efficient cowpea regeneration, transformation and genome editing system using embryonic axis explants. *The Plant Journal.* https://doi.org/10.1111/tpj.15202.

[92] Araki, M. and Ishii, T. (2015). Towards social acceptance of plant breeding by genome editing. *Trends in Plant Science*, 20:145-149.

[93] Kanchiswamy, C.N., Sargent, D.J., Velasco, R., Maffei, M.E. and Malnoy, M. (2015). Looking forward to genetically edited fruit crops. *Trends in Biotechnology*, 33:62-64.

Index

A

B

C

D

E

N

O

P

Q

R

S

T

U

V

W

About the Editors

Dr. Rahul Kumar (M.Sc., Ph.D.)
Associate Professor
Faculty of Agricultural Sciences,
DAV University, Jalandhar, Punjab, India

Dr. Rahul Kumar completed his Ph.D. from CSIR-IHBT, Palampur. Later, he worked as RA in CSKHPKV, Palampur and availed a DBT-RAship in NRCPB, IARI campus in New Delhi. His research interests include plant genetics and genomics, plant biotechnology and genome mapping. Dr. Kumar has published 22 research articles in journals of international repute and seven book chapters. He was awarded the Start-up Young Scientist Project (2016-2019) by SERB-DST, GoI. He is currently investigating two EMR research projects funded by CSIR-HRDG, New Delhi, and one in-house research project funded by DAV University.

Dr. Ashutosh Sharma (M.Sc., M.Phil., Ph.D., PGDAEM)
Associate Professor
Faculty of Agricultural Sciences,
DAV University, Jalandhar, Punjab, India

Dr. Ashutosh Sharma completed his M.Sc. and M.Phil. in Botany from Guru Nanak Dev University, his Ph.D in Biotechnology from Guru Nanak Dev University, and his PGDAEM from Manage, Hyderabad. His area of research includes plant biotechnology, plant stress biology and plant microbe interaction. Dr. Sharma has published more than 26 research papers in journals of international repute and around 30 book chapters. He also handled two major research projects as Co-PI, funded by CSIR-HRDG, New Delhi and DST-SERB, Govt. of India, and received one research project funded by DAV University as PI.

Dr. Yogesh Kumar (M.Sc., Ph.D.)
Assistant Professor
Department of Biotechnology,
DAV University, Jalandhar, Punjab, India

Dr. Yogesh Kumar completed his M.Sc. in Botany from GNDU Amritsar and his Ph.D. in Biotechnology from CSIR-IHBT, Palampur, India. Later he worked as an assistant professor in Amity University, Noida, India. His research interests include virology and plant biotechnology. Dr. Kumar has published more than 20 research articles in journals of international repute and 3 chapters in edited books. He has been granted major research projects (as Principal Investigator) by CSIR-HRDG, and DST, Govt. of India.

Dr. Indu Sharma (M.Sc., Ph.D.)
Associate Professor
Department of Life Sciences, University Institute of Sciences,
Sant Baba Bhag Singh University,
Khiala, Jalandhar, Punjab, India

Dr. Indu Sharma completed her M.Sc. in Botany from Guru Nanak Dev University and her Ph.D. in Botany (with DST-INSPIRE Fellowship) from Guru Nanak Dev University, India. Her area of research includes plant physiology and plant stress biology. Dr. Sharma has published more than 10 research papers in journals of international repute and more than 25 chapters in edited books. She also handled one research project as Principal Investigator, funded by Sant Baba Bhag Singh University, Khiala, Jalandhar.